A Handbook on IRRIGATION AND DRAINAGE

A Handbook on IRRIGATION AND DRAINAGE

BALRAM PANIGRAHI
Professor and Head
Department of Soil and Water Conservation Engineering
Orissa University of Agriculture and Technology
Bhubaneswar, Odisha

nipa®
NEW INDIA PUBLISHING AGENCY
New Delhi – 110 034

nipa®

101, Vikas Surya Plaza, CU Block, LSC Market
Pitam Pura, New Delhi 110 034, India
Phone: + 91 (11)27 34 17 17 Fax: + 91(11) 27 34 16 16
Email: info@nipabooks.com
Web: www.nipabooks.com

Feedback at feedbacks@nipabooks.com

ISBN: 978-93-81450-88-8

Composed and Designed by NIPA

Visit us at: ouat.ac.in
Tel : (0674)2397700 (O), 2561606 (R)
FAX : 0674-2397780
Email: vcouat@gmail.com
ORISSA UNIVERSITY OF AGRICULTURE & TECHNOLOGY
BHUBANESWAR-751003, ODISHA

Prof. M. Kar
VICE CHANCELLOR

Foreword

The ever increasing population is ushering tremendous pressure on water resources resulting in a rapid declination of per cepita water availability day by day. The water resources in the country are also not being efficiently utilized resulting a large mismatch between the irrigation potential created and that actually utilized. There is fast depletion of underground water on one side, while there is huge wastage of costly irrigation water due to faulty management practices on the other side. In this context, a thorough research is required with emphasis on optimum utilization of water resources of the country. It is worthy to mention that both irrigation and drainage play crucial role in judicious utilization of the scarce water resources and enhancing production and productivity of different crops.

The book on "A Handbook on Irrigation and Drainage" written by the author contains useful information on both aspects of irrigation and drainage. The content of this book has a very broad coverage especially in different aspects of irrigation. The book contains a chapter on drainage also which broadly describes design of both surface and sub-surface drainage. As groundwater forms an important part of irrigation, the book contains two chapters for exploring the groundwater through tubewells and open wells. The book will be useful to the students at both undergraduate and post graduate level, especially in the field of Agriculture and Civil Engineering. The book is an invaluable resource for academicians, practicing engineers and persons working in the department of agriculture, soil and water conservation, irrigation and command area development. The information compiled in the book would be of great help to the teachers, researchers, extension workers, students as well as farming communities in the country

I complement the commendable endeavour of the author and wish the publication a great success.

NKar 7/4/2013

M. Kar
Vice Chancellor
OUAT, Bhubaneswar, Odisha

Preface

This is a text book for Agriculture and Agricultural and Civil Engineers and will be very much helpful for the beginning students in irrigation. It is designed to guide students from a basic knowledge of soil, water and plant, hydrologic and hydraulics to the state-of-the-art of irrigation system design and management. The goal of the author is to present the principles and concepts of farm irrigation in a simple manner to maximize the students learning, understanding and motivation. The method and order of presentation have been carefully developed and classroom tested to make this book a useful and effective teaching tool. The book will not only be a helping tool to the students and teachers in Agricultural and Civil Engineering but also to all the practicing irrigation engineers, agriculturists, soil conservationists and agricultural extension workers who deal directly or indirectly with water management and other associated farm development works. The book is written covering syllabus of irrigation engineering which is taught in different State Agricultural Universities as well as in the department of Civil Engineering of different Engineering colleges. However, the design of complex hydraulic structures including dams and reservoirs are beyond the scope of this book.

The book contains 95 solved problems, 638 short and long type questions, 95 tables, 188 figures and more than 180 references which will be immensely helpful to the students and design engineers. Several field experimental results have also been incorporated in the book at appropriate sections to make the book interesting for the readers.

The book contains 16 chapters and each chapter contains several sections and sub-sections. The title of the various chapters and the contents of each chapter are as follows.

Chapter 1: Irrigation and Water Resources Development in India: This chapter deals with necessities, advantages and disadvantages of irrigation on crop production and productivity and on environment. It also gives an overview of large scale investment incurred on irrigation sector and the mismatch between the irrigation potential created and that utilised in different places in the country in different plan periods. Irrigation development in pre-plan, plan and post-plan periods are briefly mentioned in this chapter. Scope and various techniques of augmentation of water resources are also discussed in this chapter.

Chapter 2: Sources of Irrigation Water: Chapter two deals with hydrologic cycle and different components of hydrologic cycle including rainfall, its measurement and computation of average depth of rainfall by various methods. Runoff is a major part of hydrologic cycle which has been discussed here. Factors affecting runoff, estimation of runoff by different methods, estimation of peak runoff rate by rational and S.C.S. Curve Number method are also presented in this chapter.

Chapter 3:Storage of Irrigation Water: Dams and reservoirs form major storage for irrigation water. Selections of dam site with different surveys are required for construction of the dam and reservoir. All these aspects have been covered in this chapter. Further, the chapter includes area-capacity determination of the reservoir, different storage zones of the reservoir, design capacity of the reservoir including mass curve method and procedures for estimation of seepage and evaporation loss of water in the reservoir including the means to reduce these losses. Finally sedimentation of reservoir including factors affecting sedimentation and various preventive and curative measures required for reducing reservoir sedimentation are all presented in this chapter.

Chapter 4: Discharge Measurement: Chapter 4 discusses various methods of discharge measurement that includes volume, velocity-area, slope-area, gauge-discharge relationship, weirs including rectangular, trapezoidal and triangular weirs, spillway and siphon spillway, proportional weir, orifices and mouthpieces, meter gate, water meters, parshall flume, cutthroat flume, chemical dilution and radioisotope methods. Measurement of velocity by float method and current meter are presented in this chapter. Discussions on discharge site selection, types of gauges are also presented in chapter 4.

Chapter 5: Water Requirement of Crops: This chapter deals with many important terms like duty, delta, relationship between duty and delta, factors affecting duty and measures to improve duty of water. Various methods to measure crop water requirement like lysimeter method, field experimental plots, soil moisture depletion studies and water balance method have been discussed in this chapter. Estimation methods including aerodynamic method, combination method, and empirical methods like Thornwaite, modified Blaney-Criddle and Pan evaporation method are also presented in Chapter 5. Finally, estimation of actual crop evapotranspiration from potential evapotranspiration using values of crop coefficient, Kc are discussed.

Chapter 6: Basic Soil Water Relationship: Chapter 6 contains different terms related to soil-water like density, unit weight, water content, porosity, void ratio, degree of saturation and their interrelation. Physical properties and classification of soil are also mentioned in this chapter. Determination of soil moisture content by various methods like oven dry, sand bath, alcohol, neutron probe, resistance block, tensiometer, calcium carbide, pycnometer, radiation and torsion balance method are presented in this section. Other important topics like soil moisture suction, soil moisture potential curve, measurement of soil water potential, field capacity, wilting point, readily available soil moisture, flow through soils in horizontal and vertical direction in unsaturated condition and soil water v/s yield of crops relationship are briefly discussed in this chapter.

Chapter 7: Irrigation Scheduling: Irrigation scheduling and methods of irrigation (Chapter 8) are the two major chapters of this book. Chapter 7 discusses when to irrigate a crop and how much to irrigate the crop whereas Chapter 8 discusses how to irrigate i.e. methods of irrigation of the crop. Different approaches like plant indicators, physiological growth stages, soil moisture regime concept, water balance technique and climatological approach based on IW:CPE method are well discussed in this section. Water management practice of rice including water saving irrigation technique presented in this chapter will help the rice growers to adopt the right technique of irrigation that will save a lot of costly irrigation water and at the same time will enhance its water productivity.

Chapter 8: Methods of Irrigation: Methods of irrigation is very important in irrigation scheduling. The right crop is to be irrigated by the right method of irrigation. Different types of irrigations like uncontrolled flooding, controlled flooding including free flooding, contour laterals, border strip method, check basin, ring method, furrow irrigation, micro irrigation like sprinkler and drip methods are all discussed in this chapter. Hydraulics of border, check basin and furrow irrigation and design of border, check basin and furrow irrigation are presented in this section. Adaptability of sprinkler and drip irrigation, their components, layout and design are all discussed in this chapter. Moreover, the chapter includes evaluation methods of both sprinkler and drip irrigation that will help in estimating the efficiency of the system in the fields and adjustment of the system if required any in the practical field conditions.

Chapter 9: Conveyance of Irrigation Water: Chapter 9 discusses various types of canal systems and canal alignments, distribution system for canal irrigation, inundation and perennial irrigation, bandhara irrigation with advantages and disadvantages, various types of canal sections etc. Computation of discharge capacity of the channels along with Manning's, Chezy's and Kutter's formulae is described. Derivation of conditions for the most economical channel section is also a part in this chapter. Design conditions for unlined and lined channels under both alluvial and non-alluvial cases, design of regime channel including Kennedy and Lacey's theory are also discussed in Chapter 9. Garret's and Lacey's diagram are presented for design of channel sections. Moreover underground conveyance system, pipeline systems along with materials of pipes, testing of irrigation pipes, structures for underground pipeline, design of underground pipeline irrigation systems, are presented in this chapter. Finally, the various structures in conveyance systems, including drop and chute spillway, and cross-drainage works are described in chapter 9.

Chapter 10: Water Lifting Device: Various types of water lifting devices both indigenous and pumps are included in this chapter. Both manually operated and animal operated water lifts with low, medium and high head types including sedua, don, swing basket, Archimedean screw, water/pedal wheels, counter poise bucket, tenda, chain pump, rope and bucket lift, circular mot, sakia and Persian wheel are discussed in this section. In addition, water lifts powered by other sources like, wind, solar, hydraulic ram are also presented in this chapter. Different pumps like reciprocating, centrifugal, rotary type, vertical turbine, submersible, jet type, air lift, propeller and mixed flow pump and low lift hand pumps are discussed and included in Chapter 10. Various classifications of centrifugal pump, principles and operation of centrifugal pumps, installation of pumps and operation, maintenance and common trouble shooting of centrifugal pumps are also presented in this chapter. Further, the chapter includes procedures for estimation of power requirements and characteristics of pumps, criteria and procedure for selection of pumps and techniques to compute the irrigation pumping costs.

Chapter 11: Losses of Water in Canal and Control: The two major losses of water in canal and reservoirs are seepage and evaporation which have been discussed in this chapter. Measurement of seepage loss by ponding, inflow-outflow and seepage meter are presented.

Various methods to reduce seepage losses in canals are described. Discussions on various methods of seepage control including different sealants like clay, cement concrete, pre-cast, cement mortar, lime concrete, soil cement, asphalt, shotcrete, boulder, brick, bentonite and polythene lining are set forth. Economics of lining and its justification are also discussed in this section.

Chapter 12: Salt Problem in Irrigated Agriculture: Descriptions on water quality problems, causes of formation of saline and alkaline soils, factors affecting soil salinity and alkalinity, effects of salinity on crop, soil and environment, tolerance of crops, vegetables, fruits and forest species to salinity etc. are some of the principal components in this chapter. Methods to reclaim salt affected soils including soil amendments, leaching, cultivation practices etc, are also presented in this section. Major and minor dissolved constituents in water and their functions are also discussed. Different water quality standard for irrigation and water quality guidelines of irrigation water and trace elements present in water also form parts of this chapter. Finally, the chapter includes the ways how to use the poor quality irrigation water in crop fields that includes reclamation by amendments of soil, selection of salt tolerant crops, using optimum dose of fertilisers and manures, proper irrigation and soil management practice and proper seed placement technique.

Chapter 13: Waterlogging and Drainage: The chapter contains causes and effects of waterlogging and various measures; both preventive and curative to control it. Land drainage including planning, layout and classifications are presented in this section. Both surface and sub-surface drainage; their various classifications and design procedures are discussed. Open ditches; their design discharge, capacity calculation, design criteria, construction and maintenance are also dealt in this section. Methods of sub-surface drainage; especially tile drains, their layout and design procedures based on Hooghoudt's equation and installation techniques in fields are also set forth.

Chapter 14: Groundwater and Wells: Like canal water groundwater forms a major part in irrigation. Geologic formations for groundwater supply, different types of aquifers including confined, unconfined and semi confined are dealt in this section. Well irrigation, their advantages, disadvantages, various types of wells, site selection for installation of wells and yield tests are presented in this chapter. The chapter also includes tubewell irrigation, site selection for installation of tubewells, various classifications of tubewells like cavity type and strainer type tubewells. Different types of strainers used in strainer type tubewells are discussed. Design of tubewells including gravel packing and pumping arrangements are also presented in chapter 14. The chapter further, contains various investigations for groundwater development, procedures for construction of tubewells, and methods of tubewell development. Moreover, life of tubewell and causes of failures of tubewells are discussed in this chapter.

Chapter 15: Well Hydraulics: Chapter 15 contains descriptions of groundwater movement, groundwater flow directions, general flow equations under both transient and steady state saturated flow conditions and procedures for construction of flow nets. Well hydraulics including steady radial flow to a well in unconfined and confined aquifers and unconfined aquifer with uniform recharge are dealt in this section. Discharge of wells and combined drawdown under interference case are also discussed and presented in Chapter 15. Moreover, the chapter includes discussions on well loss, specific capacity and efficiency of wells.

Chapter 16: Water Productivity: Various methods to enhance the water productivity that includes irrigation management practice under deficit supply, provision of supplemental irrigation, farming system approach, correct methods of irrigation, land drainage, on-farm development, land levelling and grading etc. are presented in Chapter 16. Water productivity in rice is discussed in this section. Results of water productivity of some field investigations are presented. Study of performance of farm irrigation systems, various types of irrigation efficiencies including conveyance, storage, distribution, application and water use efficiencies are also discussed and presented in this chapter.

Balram Panigrahi

Acknowledgements

The author is indebted to Dr. M. Kar, Hon'ble Vice-Chancellor, Orissa University of Agriculture and Technology, Bhubaneswar for his kind blessings and encouragement to write this book.

The moral supports and encouragements provided by the Dean, College of Agricultural Engineering and Technology, Dean PGF-Cum-DRI, Dean of Research, Dean of Extension, Director Planning, Monitoring and Evaluation and Chief Librarian of Central Library, Orissa University of Agriculture and Technology, Bhubaneswar for writing the book is duly acknowledged.

The author is grateful to Dr. S.N.Panda, Professor, Department of Agricultural and Food Engineering and Coordinator School of Water Resources, Indian Institute of Technology, Kharagpur, Professor B.C. Mal, Hon'ble Vice-Chancellor, Chhattisgarh Swami Vivekanand Technical University, Chhattisgarh, Professor D.P. Ray, Dr. K. Pradhan and Dr. I.C. Mahapatra, Ex- Hon'ble Vice-Chancellors, Orissa University of Agriculture and Technology, Bhubaneswar for their moral supports and helps to write this book.

The inspirations and encouragements of the Director, Directorate of Water Management, Bhubaneswar and all his scientists especially, Dr. A. Mishra, Dr. S.K. Jena and Dr. S. Mohanty to accomplish this hard job of writing the book is unforgettable and is acknowledged.

It will be injustice if the author does not pay his heartfelt obligations to all the faculty members of the College of Agricultural Engineering and Technology, Orissa University of Agriculture and Technology, Bhubaneswar especially to Dr. J.C. Paul, Dr. B.P. Behera, Dr. N Sahoo, Dr. A.P. Sahu and Dr. J.N. Mishra without whose helps this book would have not been published.

A word of praise and forbearance is also due to my wife, Smita and daughters Kajal and Komal who have helped a lot and constantly encouraged to finish writing this book. The author is also woeful to his parents for their blessings to undertake this writing of the book.

Special thanks are due to the **New India Publishing Agency, New Delhi,** for kindly accepting my proposal to publish this book and for their tireless efforts in bringing out the book in the present shape in a short time.

The author gratefully acknowledges the helps drawn from various publications and text books on the subject. Suggestions and constructive criticisms for the improvement of the book are most welcome and will be duly acknowledged.

Balram Panigrahi

Conversion Table

Unit	Multiply by	To obtain
Acre	4.05×10^{3}	Square meter, m^2
Acre	0.405	Hectare, ha ($10^4 m^2$)
Acre	4.05×10^{-3}	Square kilometer, km^2
Angstrom	0.1	Nanometer, nm (10^{-9} m)
Atmosphere	0.101	Mega pascal, M pa (10^6 Pa)
Bar	10^{-1}	Mega pascal, M pa (10^6 Pa)
British thermal unit	1.05×10^{2}	Joule, J
Calorie	4.19	Joule, J
Calorie per square centimeter / minute	698	Watt per square meter, W m^{-2}
Centimeter per day (elongation rate)	0.116	Micrometer per second, μ ms^{-1} (10^{-6} ms^{-1})
Cubic feet	0.028	Cubic meter, m^3
Cubic inch	1.64×10^{-5}	Cubic meter, m^3
Cubic inch	16.4	Cubic centimeter, cm^3 (10^{-6} m^3)
Curie	3.7×10^{10}	Becquerel, Bq
Degrees (angle)	1.75×10^{-2}	Radian , rad
Dyne	10^{-5}	Newton, N
Erg	10^{-7}	Joule, J
Foot	0.305	Metre,m
Foot-pound	1.36	Joule, J
Bar	10^{3}	Kilo pascal, K pa
Gallon (U.S)	3.78	Litre, L (10^{-3} m^3)
U.S. gallon per acre	9.35	Liere per hectare, 1it ha^{-1}
Gram per cubic centimeter	1.00	Mega gram per cubic metre, Mg m^{-3}
Inch	25.4	Millimeter, mm (10^{-3} m)
Micromole (H_2O) per square centimeter second (transpiration)	180	Milligram (H_2O) per square metre per second, mg m^{-2} s^{-1} (10^{-3} g m^{-2} s^{-1})
Unit	Multiply by	To obtain
Micron	1.00	Micro metre, μm (10^{-6} m)
Mile	1.61	Kilometer, km (10^3 m)
Miles per hour	0.477	Metre per second, m s^{-1}
Milli mhos per centimeter	1.00	Deci siemens per metre, ds m^{-1}
Ounce	28.4	Gram, g (10^{-3} kg)
Ounce (FLUID)	2.96 x 10-2	Litre, (10^{-3} m^3)
Pint (fluid)	0.473	Litre, (10^{-3} m^3)

(*Contd.*)

Unit	Multiply by	To obtain
Pound	454	Gram, g (10^{-3} kg)
Pound per acre	1.12	Kilogram per hectare, kg ha^{-1}
Pound per acre	1.12 x 10-3	Mega gram per hectare, Mg ha^{-1}
Pound per cubic feet	16.02	Kilogram per cubic metre, kg m^{-3}
Pound per cubic inch	2.77 x 104	Kilogram per cubic metre, kg m^{-3}
Pound per square feet	47.9	Pascal, pa
Pound per square inch	6.9 x 103	Pascal, pa
Quart (liquid)	0.946	Litre, (10^{-3} m^3)
Quintal (metric)	102	Kilogram, kg
Square centimeter per gram	0.1	Square metre per kilogram, m^2 kg^{-1}
Square feet	9.29 x 10-2	Square metre, m^2
Square inch	645	Square millimeter, mm^2 (10^{-6} m^2)
Square mile	2.59	Square kilometer, km^2
Square millimeter per gram	10^{-3}	Square metre per kilogram, m^2 kg^{-1}
Temperature (^{0}F-32)	0.556	Temperature, ^{0}C
Temperature (^{0}C+273)	1	Temperature, K
Ton (metric)	10^3	Kilogram, kg
Ton (2,000 lb)	907	Kilogram, kg
Ton (2,000 lb) per acre	2.24	Mega gram per hectare, Mg ha^{-1}

Conversion Factors for Different Units of Length	
1 metre (m)	3.208 feet (ft)
1 metre (m)	39.37 inches (in)
1 centimetre (cm)	0.3937 inch
1 kilometre (km)	1,000 metres (m)
1 kilometre (km)	0.621 mile (statute mile, mi) = 0.5396 nautical mile (n mi)
1 foot	0.305 metre
1 inch (in)	2.54 centimetre
1 mile (mi)	5,280 feet
1 mile (mi)	1609 kilometre
1 micrometre (µm.)	10^{-6} m
1 angstrom (^{0}A)	10^{-10} m
1 metre	1.09361 yard
1 kilometre	1093.61 yard
1 inch	0.02778 yard
1 foot	0.33333 yard
1 mile	1760 yard
Conversion Factors for Different Units of Area	
1 square metre	10.764 square feet
1 square centimetre	0.155 square inch
1 square kilometre	100 hectares
1 square kilometer	0.3861 square mile
1 hectare	10,000 square meres
1 hectare	107,640 square feet
1 hectare	2.741 acres
1 square foot	0.0929 square metre
1 square inch	6.452 square centimeter

(Contd.)

1 acre	43,560 square feet
1acre	0.4047 hectare
1square mile	640 acres
1 square mile	258.99 hectares
1 square mile	2.59 square kilometres
1 square metre	1.19599 square yard
1 are	119.599 square yard
1 hectare	11960 square yard
1 acre	4840 square yard
1 hectare	100 are
1 square metre	0.01 are
1 square kilometre	10000 are

Conversion Factors for Different Units of Volume

1 cubic metre	35.314 cubic feet
1 cubic metre	1.308 cubic yards
1 cubic metre	1,000 litres
1 cubic metre	106 cubic centimeters (cm^3)
1 litre	0.035 cubic feet
1 litre	0.2642 U.S. gallon (gal.)
1 litre	0.2201 imperial gallon
1 litre	10^{-3} cubic metre (m^3)
1 cubic centimetre	0.061 cubic inch
1 cubic foot	0.0283 cubic metre
1 cubic foot	23.32 litres
1 cubic foot	7.4 U.S. gallons
1 cubic foot	6.23 imperial gallons
1 cubic inch	16.39 cubic centimeters
1 cubic yard	0.7645 cubic metre
1 U.S. gallon	3.7854 litres
1 U.S .gallon	0.833 imperial gallon
1 imperial gallon	1.201 U.S. gallon
1 imperial gallon	4.5436 litres
1 acre foot	43,560 cubic feet
1 acre foot	1233.5 cubic metres
1 acre inch	3,630 cubic feet
1 acre inch	102.8 cubic metres

Conversion Factors for Different Units of Weight

1 gram (gm)	0.0353 ounce (oz)
1 gram (gm)	0.0022 pound (lb)
1 kg	2.205 pounds (lb)
1 quintal	100 kg
1 metric ton	2,205 pounds (lb)
1 metric ton	1,000 kg
1 metric ton	10 quintals
1 metric ton	1.1 ton
1 kilogram (kg)	1,000 gm
1 ounce	28 gm
1 pound	0.45 kg
1 ton	0.9 metric ton (m.t)

(Contd.)

Conversion Factors for Different Units of Pressure

1 bar	10^6 dynes / cm^2
1 bar	1,023 cm of water column
1 atmosphere	1,032 cm of water column
1 atmosphere	76 cm Hg
1 atmosphere	1.013 bar
1 inch Hg	0.0334 atmosphere
1 inch H2O	2.49 millibar (mb)
1 mbar (mb)	0.75mm Hg
1 lb/in2 (psi)	51.72 mm Hg.
1 atm	14.70 lb/in^2 (psi)
1 atm	29.92 in Hg
1 millibar (mb)	0.0145 lb/in^2 (psi)
1 millibar (mb)	1,000 dynes/cm^2
1 millibar (mb)	1000 newton/m^2 (N/m^2)
1 millibar (mb)	100 Pascal (Pa)

Conversion Factors for Different Units of Velocity

1 m/sec	3.60 km/hr
1 m/sec	2.24 mi/hr
1m/sec	1.94 knots
1 knots	1 n mi/hr
1 knots	1.15 mi/hr
1 knots	0.515m/sec

Conversion Factors for Different Units of Energy

1 calorie (cal)	4.187 joules (J)
1 calorie	4.187 x 10^7 ergs
1 calorie	1.16 x 10^{-6} killowatt hour (Kwh)
1 calorie	3.97 x 10^{-3} British thermal units (Btu)
1 calorie	1 langely-centimetre^2 (ly-cm^2)

Conversion Factors for Power

1 watt (W)	1 joule / sec
1 watt (W)	0.239 cal/sec
1 watt (W)	0.0569 Btu/min
1 watt (W)	0.00134 electrical horse power
1 watt (W)	2.39 x 10-5 ly-m^2 / sec

Conversion Factors for Radiation to Equivalent Depth Evaporation

1 cal/cm^2	1/59 mm
1 cal/cm^2 /min^{-1}	1 mm/hr
1 cal/cm^2/ min (24 hr)	24 mm / day
1 mw /cm^2	1/70 mm/hr
1 mw/cm^2 (24hr)	0.344 mm/day
1 joule / cm^2 / min (24 hr)	5.73 mm/day

Conversion Factor for Energy Flux Density

1 cal cm^{-2} min^{-1}	1 langely min^{-1} = 69.8 milli watt cm^{-2}

Contents

1

Irrigation and Water Resources Development in India

1.1 Introduction

Like animals, plants are living beings and so require water and air for their survival. In addition, they need water for preparation of food by the process of photosynthesis. Majority of this water requirement of the crops is met from different forms of precipitation. However, when the water requirement is not fully met from precipitation, then growth of the crops is affected. This results in reduction of yield of the crops. The reduction in yield depends on the amount of scarcity of water requirement of the crops. It also depends on the stage of crop growth and also depends on the types of the crops. Generally all the crops are sensitive to water stress during reproductive stage. Water requirement of crops depend on many factors. It depends on the type of crops, irrigation schedules practiced, soil texture, position of groundwater table below effective root zone of the crop and management practice followed in crop cultivation. The pulses, oilseeds, vegetables etc. which are also otherwise termed as dry land crops need less water compared to rice which is a water loving crops and perennial crops like sugarcane, banana etc.

Water requirement of crops is partly met by the upward flux of groundwater. The capillary rise of shallow water table in the form of upward flux helps in meeting the water requirement of crops to some extent. The upward flux depends on the type of the soil, effective root zone depth of the crop, depth of water table below effective root zone depth of the crop and potential gradient of soil moisture suction. If the soil is heavy like clay, effective root zone depth of the crop is more (deep rooted crop) and position of ground water table is near to ground level (shallow water table), then the upward flux is more.

The residual soil moisture helps in supporting water requirement of crops also. The heavy soil like clay has more moisture retention capacity than the light soils like sandy soil. The residual soil moisture held in clay soil is released slowly and supports more soil moisture to the crops for a longer period than the sandy soil which has less soil moisture retention capacity. After meeting the water requirement of crops from all sources including different forms of precipitation, ground water contribution by capillary rise and residual soil moisture, if there is any shortage of water, then this shortage should be met from other sources which are called as irrigation. It is to be noted that irrigation is not the natural phenomenon of supply of water; rather it is the artificial means of supply of water to the crops. Irrigation may, therefore, be defined as the science of artificial application of water to the crop, in accordance with the crop requirement for better nourishment of the crops.

1.2 Necessity of Irrigation

India is an agrarian country where more than 70 percent population depends on agriculture. It has a vast diversity of climate, topography and vegetation. The agriculture is almost entirely dependent on amount and distribution of rain; mostly by the south-west monsoon. Depending on the nature of monsoonal rain, the cropping season is mostly divided into three seasons i.e. *kharif* (rainy season), *rabi* (winter season) and summer. There is a considerable seasonal and temporal variation of rainfall in the country. The distribution of rainfall is highly erratic. Sometimes, there is high intensity of rainfall occurring for a longer period at some places creating serious flood. The flood causes waterlogging and subsequently inundates the crops for a number of days that damages the crops to a great extent. At other times, there is dry spell for a number of days causing drought that hinders the crop growth and hence productivity of the crops. It is to be noted that both floods and drought are the extreme hydrologic events that affect the crop. Out of these two extreme events, crops need artificial supply of water in the form of irrigation during drought.

As mentioned above, agriculture in India mostly depends on the vagaries of monsoon. The erratic behaviour of the monsoon along with its uncertain occurrence of onset and withdrawal in conjunction with in-season critical dry spells has necessitated the provision of supplemental irrigation for agricultural operation. Studies have indicated that there is 80% probability of exceedance that there will be a critical dry spell of 12 days duration coinciding with critical growth stage of rice in every *kharif* season in eastern India, which is detrimental for rice (Panigrahi, 2001). Under such circumstances, farmers have to arrange and apply irrigation from any source otherwise yield will be affected too much. Irrigation development has enabled crops to be grown all year round.

The rainfall is highly uneven in distribution and thus there is a high necessity of assured supply of irrigation water to raise crops successfully. It is no doubt that the country has achieved tremendous growth in food grain production raising from 51 M tons during the early fifties to present level of 245 M tons. However, feeding ever increasing population in the coming years remains a challenging task for the country. The country will have to feed about 1.64 billion people by 2050 AD, which requires an increase of about 5 M tons of additional food grains every year. As per the recent National Policy on Agriculture, the country requires to achieve a growth rate of 4.0 per cent in food grains production from the existing rate of 2.0 per cent. The irrigated agriculture contributes about 60% of the total food grain production and supports 60% of the total population in the country.

At present about 95% of the area under coarse cereals, 91% under pulses, 80% under oil seeds, 65% under cottons, and 53% under rice are rainfed and rest are irrigated (Bhatia and Singh, 2001). These figures show that very less amount of cultivated areas in the country have been brought under irrigation. Since, in the coming years, the per capital share of land holdings for agricultural production is going to decrease, and there is a necessity to produce more crops to feed all the populations, there is a great concern to increase the irrigation potential of the country. Further, many new high yielding and hybrid varieties of seeds of different crops have come up to the market for boosting the agricultural production. These new varieties give optimum result only under assured irrigation.

1.3 Advantages of Irrigation

Introduction of irrigation has acted as a prime mover to the agricultural development programme by way of increasing and stabilizing crop production. It has increased the cropping intensity and hence the total crop production and productivity of agricultural lands. It is needless to mention that the productivity of the agricultural lands has been enhanced by two to three times more than that of the rainfed lands. In the coming years, the success of the second green revolution would become feasible by increasing the irrigation potential and decreasing the gap between the irrigation potential created and that actually used. There is hardly any doubt in emphasizing the importance and advantages of irrigation during the periods of acute food shortages and growing population of the country. Even then, it is worthwhile to mention some of the salient advantages of irrigation that are summarized as below:

i) **Increase in Food Production**: Irrigation helps to add water to soil to supply the moisture essential for plant growth. It cools soil and atmosphere, thereby making more favourable environment for plant growth. It thus helps in increasing the total crop production and productivity, and hence, to attain self-sufficiency in food.

ii) **Optimum Benefits**: Irrigation helps in obtaining the optimum utilization of water. Optimum utilization of water means to achieve maximum crop yield with a given amount of water. Any amount of water either less or more than the optimum amount would result in getting less crop yield than the optimum value.

iii) **Elimination of Mixed Cropping**: Mixed cropping means sowing two or more than two crops together in the same field. Though mixed cropping is beneficial in rainfed farming systems where weather induced conditions are not favourable for growing a single crop, it has many disadvantages also. Mixed cropping creates hindrance since different crops need different types of field and intercultural operations, different types of manuring and watering. However, if assured irrigation supply is provided, a single superior crop can be sown in a field with no problem in carrying out farming operation, watering and manuring.

iv) **Domestic Water Supply**: Irrigation helps in augmenting the water supply in towns where water is available with great difficulty. It also provides water for other uses like swimming, recreational uses, bathing, cattle drinking etc.

v) **Communication:** Embankments of irrigation channels are used as pathways for facilitating communication to the villages.

vi) **Inland Navigation**: Sometimes larger irrigation canals are used for navigation purposes.

vii) **Industrial Uses**: Irrigation water is used to meet the water demand of industries, thus facilitating the industrial growth.

viii) **Generation of Hydro-electric Power**: Cheaper power generation can be obtained on projects, primarily designed for irrigation alone. Canal falls can be used for power generation.

ix) **Crop Insurance**: Irrigation helps to provide crop insurance against short duration droughts.

x) **Dilution of Salts**: Irrigation helps in washing out or diluting salts in the soil which is called as leaching. Thus, it helps in reclamation of saline soils to some extent.

1.4 Disadvantages of Irrigation

Besides many advantages, irrigation causes many disadvantages as outlined below:

i) **Outbreak of Diseases:** Irrigation causes swampy and marshy climatic conditions which in turns acts as breeding places of mosquitoes and flies and cause outbreak of diseases like malaria.

ii) **Pollution Problems:** Sometimes, farmers maintain deep ponding in paddy fields, which not only causes more losses of precious irrigation water but also decreases the yield of crop. It decreases the soil health, increases the iron toxicity in paddy field and thus decreases the yield. Besides, deep ponding in paddy field, decreases the nitrogen and other fertilizer efficiency in field, and causes groundwater pollution/contamination by increasing the nitrogen level in groundwater. Contaminated groundwater causes many diseases like anemia, cancer etc. Moreover, deep ponding in paddy field increases methane emission and thus creates environmental pollution. In irrigated areas, farmers use more agro chemicals, pesticides etc which washes away from the field, and enhances groundwater pollution and soil pollution, Washing away of these chemicals from the fields ultimately reaches the rivers and seas and affects marine lives.

iii) **Waterlogging and Salinity:** Over irrigation causes rise of groundwater table in canal commands which causes twin problems of waterlogging and salinity and reduces crop yields.

iv) **Increasing Flooding in High Water Table Areas:** The continuous rise in water table in canal commands reduces the capacity of under ground storage of water in the aquifers. There is a reduction of ground water recharge in a waterlogged tract due to non-availability of storage space in the aquifers beneath the tract. The reduced amount of natural recharge adds to rainfall excess, thereby enhancing the occurrence of floods.

v) **Sea Water Intrusion in Coastal Areas:** Over exploitation of ground water near the sea coast creates imbalance between the sea water and contiguous ground water level: consequently the sea water intrudes the coastal aquifers and deteriorates the ground water quality. It also increases the salinity level of soil and water in the field and thereby affects the crop yields.

vi) **Declining Groundwater Table:** Massive groundwater exploitation in many areas like Punjab, Haryana, Gujarat and some parts of Tamil Nadu, due to heavy pumpage of groundwater has caused ground water mining. Ground water mining causes decline of water level and cause many wells to go dry in summer. Sometimes unplanned ground water pumpage causes landslide also.

1.5 Effect of Irrigation on Crop Yield

The expansion of irrigation has not only directly enabled yield increases of all the crops including rice, it has also enabled the farmers to use more advanced technology, field inputs including high yielding and hybrid seeds, more use of chemical fertilizers and pesticides, adoption of farm mechanization etc. About 60% of total agricultural production comes from irrigated agriculture. Furthermore, the spread of irrigation has also enabled crop diversification. Crop yields have increased substantially and irrigated crops produce much more than rainfed crops. Consequently, the food production has quadrupled during the last 50 years resulting in the annual per capita availability of food grain of 200 kg in the country. Fig. 1.1 (Food and Agriculture Organisation, 1999) represents the influence of irrigation on yield of various crops.

1.6 Irrigation and Water Resources Development in Early Period

Irrigation is an age-old art. Historically, civilization has followed the development of irrigation. All ancient civilizations have grown up on the banks of rivers and all great epics written. In the Vedas, the earliest sacred books of Aryans mentions have been made about the wells, canals

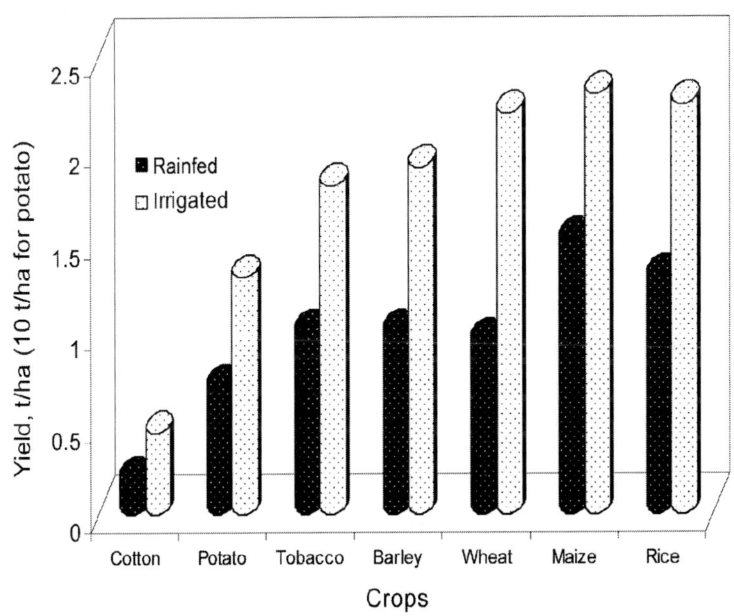

Fig.1.1: Influence of irrigation on crop yields

and dams. In "Rigved" four types of water sources are mentioned i.e. the water that comes from the sky or rainwater, those which flow in the rivers and streams, those which are obtained by digging and those which ooze out of springs. In "Yajurved" there are mentions about the canals and dams. "Atharvaved" gives description of digging canals from the rivers. River is mentioned as cow and canal as calf. The ancient historical work mentions an account of gigantic feat of king Bhagiratha and his engineers of diverting the course of water of the sacred Ganga from the altitude of the Himalayas towards the present Indo-Gangetic plain, the granary of India.

Writings at 300 B.C. indicate that all the main lands in India were under irrigation and the country was very prosperous because of the double harvests which the people were able to reap each year. Besides the various ancient books which confirm the above fact, there are ruins of various ancient irrigation works, a few of which still exist. The most famous old irrigation works which is functioning even today is the "Grand Anicut" which was built by Chola rulers in the first century A.D. on the river Cauvery (Garg, 1987).

In China, where reclamation was begun more than 4000 years ago, the success of early kings was measured by their wisdom and progress in water-control activities. King Yu of Hsia dynasty (2200 B.C.) was elected king by the people as a reward for his outstanding work in water control. The famous Tu-Kiang dam still a successful dam of today, was built by Mr. Li and his son in the China dynasty (200 B.C.) and provides irrigation to about one-half million acres of rice lands. The water ladder, a widely used pumping device in China and neighbouring countries, is believed to have been invented about the same time. The Grand canal, 1125 m long was built by the Sui empire during 589 to 618 A.D (Hansen *et al.,* 1979).

Hammurabi, the king of Babylon, developed laws indicating that the people had to depend upon irrigation for existence. One of the laws of Hammurabi states that if a man neglects to strengthen his bank of the canal and waters carry away the meadow, the man in whose bank the breach is opened, shall render back the corn which he has caused to be lost. An ancient Assyrian Queen, supposed to have lived before 2000 B.C. is credited with directing her government to divert the water of the Nile to irrigate the desert lands of Egypt. Irrigation canals supposed to have been built under this Queen of Assyria are still delivering water. Thus, there are records and evidence of continuous irrigations for thousands of years in the valleys of the Nile and for comparatively long periods likewise in Syria, Persia, Java, and Italy (Hansen *et al.,* 1979).

Egypt claims to have had the world's oldest dam built about 5000 years ago to store water for irrigation and domestic uses. The Spaniards on their first entrance into Mexico and Peru found elaborate provisions for storing and conveying water supplies which had been used for many generations. Extensive irrigation works also existed at that time in the south western United States. The early Spanish missionaries sprayed knowledge of irrigation from their Mediterranean homes (Tripathy, 2008). Thus, there are thousands of records on origin and development of centuries of irrigation in many other countries across the gloves The pressure of survival and the need for additional food productions have necessitated a rapid expansion of irrigation throughout the world.

1.7 Irrigation and Water Resources Development in Pre-Plan Period

During the British regime, considerable attention was paid towards utilizing water for irrigation, so as to overcome the frequent famines in India. Lot of works was done on various rivers such as Cauvery, Godavari, Ganga, Krishna etc. and construction of various projects and blue prints of many small and large irrigation projects were prepared. The first Irrigation Commission, (1901-1903) reviewed the performances of the completed projects and progresses made on the-then on-going projects. Table 1.1 gives description of irrigation projects constructed in different state of India in pre-plan periods before 1950. Irrigation from wells received less attention than that from canals and tanks. An area of 4.73 M ha was being irrigated in India through wells by the year 1925-26 which got increased to 5.75 M ha in 1940. Following the recommendations of the Commission after the Bengal famine to dig more wells and tanks throughout the country, several new wells and tanks were dug in Uttar Pradesh, Bihar, Punjab, Bengal and Gujarat (Lenka, 2001).

Table 1.1: Irrigation projects constructed in India in pre-plan periods

State	Name of project	Name of river on which constructed	Period	Area irrigated (M ha)
Andhra Pradesh	Godavari Delta Project	Godavari	1890	0.35
Madras Presidency	Krishna Delta Project	Krishna	1898	0.60
Bihar	Sone Canal	Sone	1869	0.29
Punjab	West Yamuna Canal	Yamuna	1856	0.52
	Sirhind Canal	Sutlej	1887	0.95
	Upper Baridoab Canal	Beas	1879	0.40
Rajasthan	Ganga Canal	Sutlej	1928	0.22
Uttar Pradesh	Upper Ganga Canal	Ganges	1854	0.80
	Lower Ganga Canal	Ganges	1878	0.60
	Sarada Canal	Sarada	1926	0.80
	Agra Canal	Yamuna	1873	0.18

During the repression of 1930 and second World War, little development took place in the field of irrigation. The net irrigated area in undivided India was about 28.2 M ha (1944-47) which was the largest in any country of the world. Out of 28.2 M ha, canal irrigation contributed to the highest share (54%), followed by wells (23%), tanks (12%), and other sources contributed to 11%. Though, correct records regarding irrigation are not available, on the basis of the records from undivided India, data on area irrigated (cumulative) in India alone during different pre-plan periods are presented in Table 1.2 (Lenka, 2001).

Table 1.2: Cumulative irrigation development (M ha) during different pre-plan periods

Time period	Sources of irrigation				Area irrigated (% of cultivated area)
	Canal	Well	Others	Total	
1911-15	4.0	4.4	6.1	14.5	17.9
1915-20	4.1	4.7	7.0	15.7	19.0
1920-25	4.4	4.7	6.9	16.0	17.4
1925-30	4.6	4.8	6.8	16.2	17.1
1930-35	5.0	4.8	7.3	17.1	17.6
1935-40	5.6	5.2	8.2	18.0	18.6
1940-45	6.0	5.4	7.6	19.0	19.2
1945-50	6.6	6.3	9.7	19.4	19.1

1.8 Irrigation and Water Resources Development in Plan Periods

The work which was started by the Britishers got tremendous amount of momentum with our country's independence. Top priority on irrigation was given during the first plan period (1951-1956) and especially due to the large scale efforts of late Pt. Jawahar Lal Nehru, the then Prime Minister of India, numerous multi purpose projects at huge costs were planned and implemented. The second Irrigation Commission was appointed during 1969-1972 during the regime of the-then Prime Minister Smt. Indira Gandhi. Importance on full scale development of country's vast water resources was given (Garg, 1987). Several multipurpose dam projects were constructed of which Hirakud, Bhakra Nangal, Damodar Valley Corporation (D.V.C.) were important. Table 1.3 gives names of some important irrigation projects constructed during plan periods in India. It is to note that in addition to the projects listed in Table 1.3, many more projects have been constructed/under construction in different stats in India.

Table 1.3: Some important irrigation projects constructed during plan periods in India

State	Project	River on which constructed	Area to irrigate (M ha)
Andhra Pradesh	Nagarjuna Sagar	Krishna	0.81
	Panchampad	Godavari	0.28
Bihar	Gandak	Gandak	1.31
	Kosi	Kosi	1.06
	Sone (high level)	Sone	0.12
Gujarat	Kakrapada	Tapi	0.23
	Ukai	Tapi	0.16
	Mahi (South)	Mahi	0.19
	Ekadana		0.18
	Narmada	Narmada	0.40
	Sabarmati	Sabarmati	0.12
	Panam	Panam	—
	Karjan	Karjan	—
Karnataka	Bhadra	Bhadra	0.28

(Contd.)

State	Project	River on which constructed	Area to irrigate (M ha)
	Upper Krishna	Krishna	0.61
	Ghataprabha	Ghataprabha	0.12
	Malaprabha	Malaprabha	0.12
	Tungabhadra	Tungabhadra	0.41
	Cauvery	Cauvery	-
Madhya Pradesh	Tawa	Tawa	0.32
	Chambal	Chambal	—
	Gandhi Sagar	Chambal	0.45
	Rana Pratap Sagar	Chambal	0.12
	Kota Barrage	Chambal	0.56
	Mahanadi Reservoir Project	Mahanadi	0.41
Maharashtra	Bhima	Bhima	0.12
	Jayakawadi	Godavari	0.14
	Kukadi	Kukadi	0.25
	Krishna	Krishna	
Odisha	Hirakud	Mahanadi	0.74
	Mahanadi Delta I and II	Mahanadi	0.68
	Salandi	Salandi	0.35
	Anandapur Barrage	Baitarani	0.16
Punjab	Bhakra Nangal	Sutlej	0.88
	Beas	Beas	0.42
	Their	Ravi	
Rajasthan	Rajasthan Canal (Indira Gandhi Canal)	Sutlej	1.06
Tamil Nadu	Parambikulum Aliyar	Integrated project for eight rivers	-
Uttar Pradesh	Sarda	Ghagra	
	Sahayak	Barrage Sarada, Gomati, Sai	
	Ramganga	Ramganga	0.69
	Left Bank Ghagra canal	Ghagra	0.40
	Tehri Dam	Bhagirathi	0.08
Kerala	Kalada	Kalada	0.11
West Bengal	Farakka	GangaBhagirathi	
	Mayurakshi	Mayurakshi	0.25
	Kangasabati	Kangasabati	0.35
	Durgapur Barrage	Damodar	0.42
	Damodar Valley Project	Damodar	0.42

The Ministry of Water Resources, Government of India (1993) has divided the country into 20 river basins (Fig. 1.2) out of which 12 are major river basins and eight are medium/small river basins. The 12 major river basins are:

a) Indus
b) Mahanadi
c) Pennar
d) Brahmani-Baitarani
e) Godavari
f) Krishna

g) Ganga-Brahmaputra-Meghna
h) Cauvery
i) Sabarmati
j) Mahi
k) Narmada
l) Tapi

The eight medium/small river basins are:

a) Subarnarekha-combining small rivers between Subarnarekha and Baitarani
b) East flowing rivers between Mahanadi and Pennar

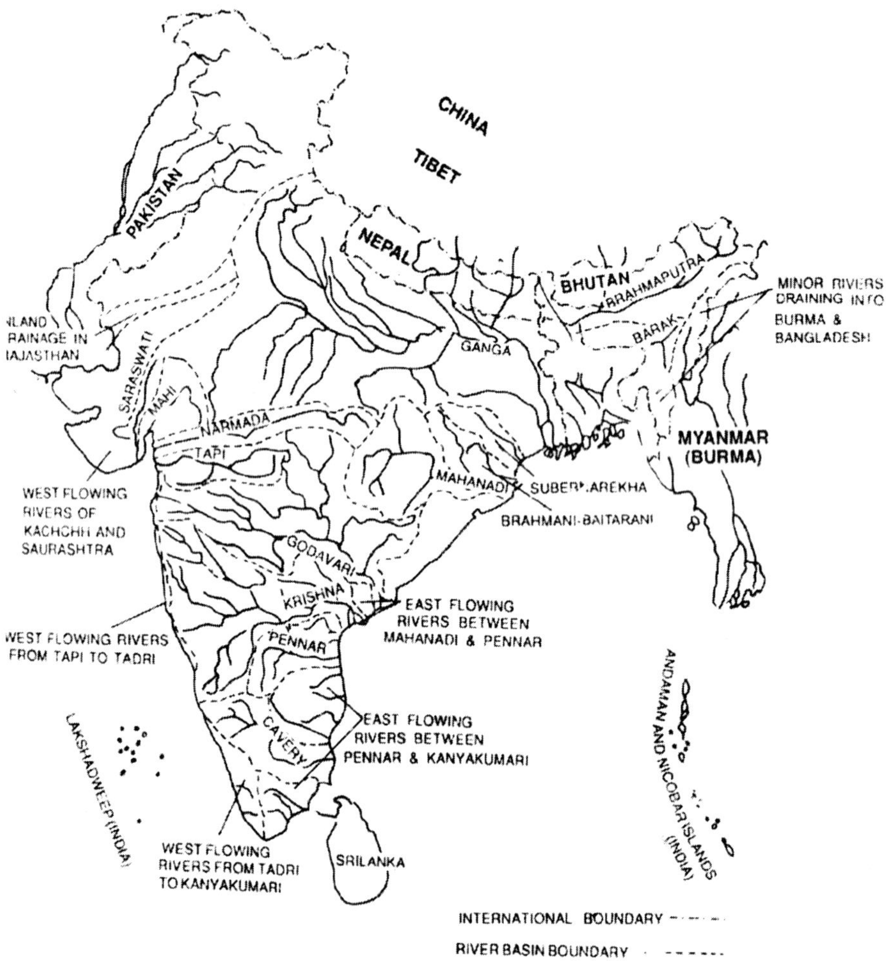

Fig. 1.2: River basins of India

c) East flowing rivers between Pennar and Kanyakumari
d) West flowing rivers of Kutch and Saurashtra including Luni
e) West flowing rivers from Tapi to Dadri
f) Inland drainage area in Rajasthan desert
g) West flowing rivers from Dadri to Kanyakumari
h) Minor rivers draining into Myanmar and Bangladesh.

The Central water Commission is maintaining 500 gauges and discharge observation stations in almost all inter-stats-river systems in the country. On the basis of data available, the water resources potential of 20 river basins of India is presented in Table 1.4. Fig. 1.3 represents the basin wise distribution of estimated utilizable water resources of the country.

Construction of major, medium and minor irrigation projects has enabled to create large irrigation potential in the country. The annual gross area under irrigation upto1951 (pre-plan period) from different sources was about 22.6 M ha. It rose to 54 M ha by 1983-84 which means that about 30% of the cropped area came under irrigation in the year 1983-84. It further rose to 81 M ha in the year 2002. An idea of large scale efforts made by the Indian Government after independence can be obtained from the fact that large scale investments have been spent on irrigation sector during successive plan periods. Beginning with an expenditure of Rs. 366 crore. during first plan, it rose to Rs. 10,786 crores during sixth plan. During seventh plan, an outlay of Rs. 17861 crores has been made on irrigation projects. Out of Rs. 17861 crore, major and medium projects shared Rs. 11556 crore and minor projects shared Rs. 6305 crore. The tenth Five year Plan had an outlay of Rs. 95743.42 crores for the irrigation sector. Expenditures incurred by Government of India on irrigation sector during successive years is presented in Fig. 1.4. From

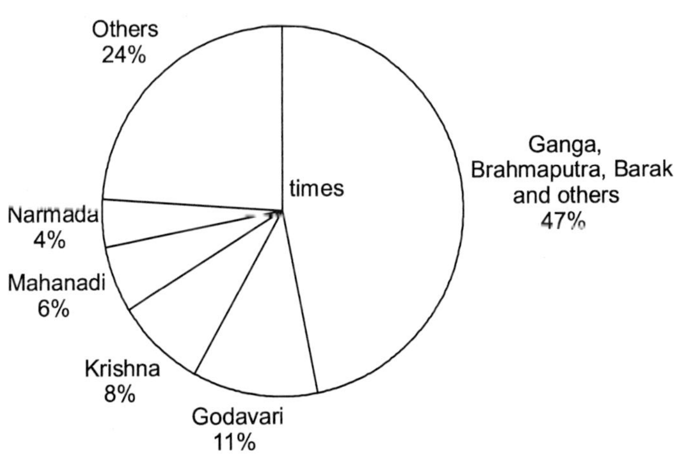

Fig. 1.3: Basin wise distribution of estimated utilisable water resources of the country

Table 1.4: Water resources of river basins in India

Sl. No.	River basin	Catchment area (km^2)	Water resources potential (Mm^3)		Groundwater potential (Mm^3)
			Average	Dependable	
1.	Indus	321289*	7305	25543	
2.	Ganga-Brahmaputra Meghna				
	Ganga	861452*	525023	436322	171725
	Brahmaputra	194413*	537240	441736	27857
	Barak and others	41723	48357	-	1795
3.	Godavari	312812	110540	80545	46762
4.	Krishna	258948	78124	69411	26646
5.	Cauvery	81155	21358	19375	13598
6.	Subarnerekha	29196	12368	9855	2185
7.	Brahmani-Baitarani	51822	28477	20051	5879
8.	Mahanadi	141589	66879	53786	21293
9.	Pennar	55213	6316	4393	5047
10.	Mahi	34842	11020	5713	4875
11.	Sabarmati	21674	3809	3146	3033
12.	Narmada	98796	45639	30829	11890
13.	Tapi	65145	14879	8860	8173
14	West flowing rivers (Tapi to Dadri)	44940	87411	65663	9479
15.	West flowing rivers (Dadri to Kanykumari)	56177	113532	85285	8810
16.	East flowing rivers (Mahanadi - Pennar)	86643	22520	18768	22788
17.	East flowing rivers (Pennar – Kanykumari)	100139	16458	13930	20907
18.	West flowing rivers (Kutch-Saurastra-Luni)	321851	15098	—	13948
19.	Inland drainage area in Rajasthan district	—	—	—	—
20.	Minor rivers draining into Myanmar and Bangladesh	36302	31000	—	—

* Area in Indian Territories

Source: Central Ground Water Board Report-1993

this figure, it can be seen that the share in total plan expenditure has decreased from 23% in first Plan to 6.3% in the tenth Plan. The cost of creating per hectare of major and medium irrigation potential has increased from Rs. 1200 during First Plan to Rs.15018 during Seventh Plan. In minor irrigation projects, per hectare cost averaged Rs. 2452 till Fourth Plan. It then increased to Rs. 4574 and Rs. 7331 during Sixth and Seventh Plans, respectively. Thus, the cost per hectare under major and medium irrigation projects during Seventh Plan has been 4.8 times to that of minor irrigation projects (Singh, 2001).

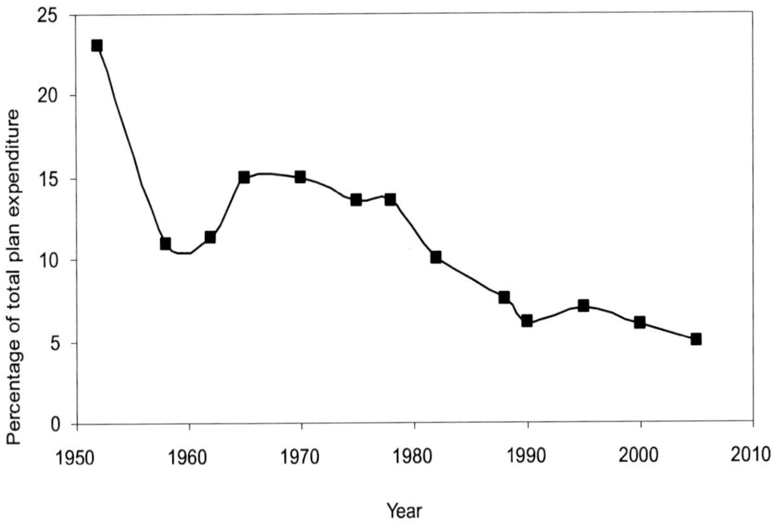

Fig. 1.4: Expenditure incurred on irrigation projects during various years

1.9 Irrigation Potential Created and Actually Used

The ultimate irrigation potential created in India from different major, medium and minor schemes is 140 M ha. The ultimate irrigation potential varies from state to state. It varies from 29.64 M ha (highest) for Uttar Pradesh to 0.070 M ha (lowest) for both Sikkim and Mizoram. It is observed that there is a large mismatch between the irrigation potential created and that utilized in all the states of India. Upto the end of sixth Plan, the gap between the potential created and that utilized from major, medium and minor irrigation projects was about 7 M ha which rose to 19.70 M ha by the end of 10th Plan. The physical achievement on potential created and potential actually utilized till the end of 10th Plan is 109.10 and 90.42 M ha, respectively With an average irrigation intensity of 140%, actual net irrigated area upto the end of 10th Plan is likely to be 64.8 M ha which is only 45% of net sown area. Considering the 10th Plan period alone, it is revealed that the target of irrigation potential was 15.15 M ha out of which only 10.36 M ha was utilized (4.79 M ha short of the potential created). Up to the end of 10th Plan, the country has achieved the irrigation potential of 78% of the ultimate value. Out of 28 states of India 12 states have already achieved the irrigation potential more than 78% of the ultimate irrigation potential value and 8 states between 50 to 78% of the ultimate irrigation potential value. The remaining eight states which have realized less than 50 percent of the ultimate irrigation potential are Assam (37.62%), Jammu and Kashmir (43.23%), Madhya Pradesh (24.88%), Manipur (45.11%), Meghalaya (37.88%), Mizoram (26.22%), Odisha (44.28%) and Sikkim (49.5%).

The country has utilized 82.88% of the irrigation potential that has been created and 64.63% of ultimate irrigation potential till the end of 10th Plan. 19 states in the country have utilized about 80% of the potential created whereas the remaining 9 states have used less than 80% of the created irrigation potential till end of 10th Plan. Table 1.5 gives cumulative irrigation potential.

Table 1.5: Plan wise cumulative irrigation potential created and utilized

Plan	Potential created, M ha	Potential utilized, M ha
Up to 1951 (Pre-Plan)	22.6	22.6
First Plan (1951-56)	26.26	25.04
Second Plan (1956-61)	29.08	27.8
Third Plan (1961-66)	33.57	32.17
Annual Plan (1966-69)	37.1	35.75
Fourth Plan (1969-74)	44.2	41.89
Fifth Plan (1974-78)	52.02	48.46
Annual Plan (1978-80)	56.61	52.64
Sixth Plan (1980-85)	65.22	58.82
Seventh Plan (1985-90)	76.44	68.59
Annual Plan (1990-92)	81.09	72.85
Eighth Plan (1992-97)	86.26	77.21
Ninth Plan (1997-2002)	93.95	81.00
Tenth Plan (2002-2007)	108.9	

Source: 11th Five Year Plan, Planning Commission Report, Government of India, 2007

created and that utilized in different periods in India. From this table it is revealed that there is gradual increase in the irrigation potential created and that utilized in different years. Fig. 1.5 represents the present distribution of land resources including the irrigation potential created and that used in India. From this figure it is observed that about 27.4 % of the total geographical area and 47.4% of the total cultivable cropped area of India are presently under irrigation. Table 1.6 gives the state-wise physical achievements of irrigation potential created and utilized from different irrigation projects of India.

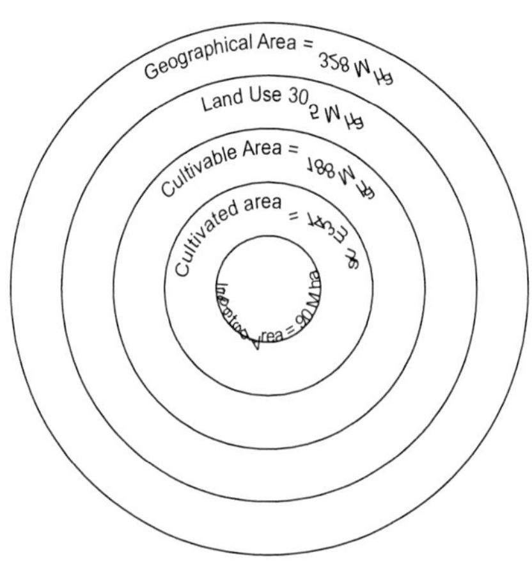

Fig. 1.5: Land resources including irrigated area of India

Table 1.6: Physical achievements of irrigation potential created and utilised (X10^3 ha) in India

Sl. No.	State/Union Territories	Ultimate irrigation potential	Achievements upto IX plan		Target for X plan	
			Potential created	Potential utilised	Potential created	Potential utilised
1.	Andhra Pradesh	11260.00	6322.68	5832.79	935.28	637.22
2.	Arunachal Pradesh	168.00	99.52	77.40	24.00	18.60
3.	Assam	2870.00	847.54	668.48	232.20	168.35
4.	Bihar	10887.00	7396.44	5474.29	1213.02	611.68
5.	Chhatisgarh	1717.93	1410.20	1083.60	360.00	242.25
6.	Goa	116.00	40.31	36.33	31.20	16.66
7.	Gujrat	6163.00	3429.14	3176.97	1970.00	1282.60
8.	Haryana	4512.00	3730.44	3428.09	161.50	111.35
9.	Himachal Pradesh	353.00	174.35	145.81	18.00	13.20
10.	Jharkhand	2460.00	943.34	701.54	315.00	240.75
11.	Jammu & Kashmir	1358.00	562.19	535.55	25.00	16.25
12.	Karnataka	5974.00	3706.52	3386.56	1221.18	826.93
13.	Kerala	2679.00	1249.51	1162.63	140.00	98.50
14.	Madhya Pradesh	16214.07	3643.03	3025.11	390.30	227.20
15.	Maharastra	8952.00	6181.60	4704.96	1433.43	956.08
16.	Manipur	604.00	231.49	173.34	42.60	29.86
17.	Meghalaya	168.00	50.97	47.31	12.50	10.00
18.	Mizoram	70.00	16.69	14.08	1.66	1.33
19.	Nagaland	85.00	76.56	65.63	9.43	7.54
20.	Odisha	8803.00	3300.68	3131.72	597.44	408.20
21.	Punjab	5967.00	5970.04	5853.81	160.30	104.20
22.	Rajsthan	5128.00	4929.25	4675.67	463.80	308.97
23.	Sikkim	70.00	29.67	23.61	5.00	4.00
24.	Tamil Nadu	5532.00	3672.69	3668.81	18.40	13.32
25.	Tripura	281.00	114.55	100.59	32.40	25.92
26.	Uttar Pradesh	29635.00	29509.49	23613.62	4617.56	3543.93
27.	Uttaranchal	864.00	781.28	586.21	18.00	13.53
28.	West Bengal	6918.00	5475.81	4625.24	700.00	455.00
	Union Territories	144.00	50.20	39.40	4.40	3.50
	All India Total	**139893.00**	**93946.38**	**80058.10**	**15154.60**	**10360.90**

Source: Annual Document 2003-04, Planning Commission of Govt. of India

1.10 Per Capita Water Availability

Per capita water availability in India is low compared to many other countries. The per capita water availability will further decrease in the years to come. Table 1.7 represents the per capita annual water availability in some selected countries till 2025 AD. From this table, it is marked that in 2025 AD, the per capita annual water availability in the country would be 1400 m^3 which would be less than China and our neighbouring countries like Pakistan and Nepal. The per capita water availability in the country is reducing progressively due to increasing population. The par capita water availability in 1955 was 5300 m^3 which has come down to 1800 m^3 in 2007 as against the world average and Asian average of 7420 and 3240 m^3, respectively. With

Table 1.7: Per capita annual water availability in some selected countries

Country	Per capita annual water availability (x 10^3 m^3) Year		
	1955	2000	2025
China	3.0	2.2	1.9
India	3.1	1.9	1.4
Pakistan	5.6	2.7	1.6
UK	1.3	1.2	1.2
USA	11.3	8.9	7.6
Bangladesh	15.8	9.4	6.8
Nepal	16.4	8.8	5.5

the projected population growth, urbanization and industrialization, the average will dwindle to about 1200 m^3 by 2050 AD (Fig. 1.6).

1.11 Scopes of Water Resources Development

As the population of India continues to increase, the demand for food and fiber for the people will also increase. The country will have to feed about 1.3 billion people by 2020 AD and 1.64 billion people by 2050 A.D. which will require about 350 and 450 M tons of food grains by 2020 and 2050 A.D., respectively. Considering the present level of annual food grains production, an additional food grain of 5 M tons/year has to be produced in the country (Sahu, 2008). The average size of farm holdings is gradually decreasing, and there is no scope for

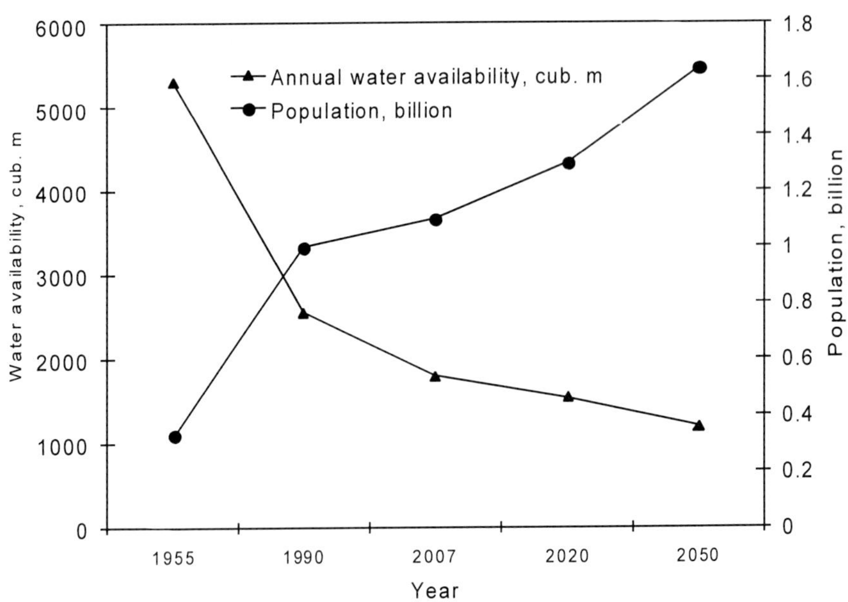

Fig. 1.6: Population and water availability trends

adding any more area to crop cultivation since demand for other sectors especially for dwelling purposes are increasing rapidly. On the other hand, the water resources of the country is finite and the share of water for irrigation in agriculture is depleting at faster rate because of demand in other sectors like industrialization, community development, recreational uses etc.

Considering an average annual rainfall of 1250 mm over 328 M ha area, about 400 M ha m of water is received by the country. About 200 M ha m of water enters the soil which constitutes soil moisture and ground water recharge. A considerable part of remaining 200 M ha m of water is flowing in the rivers inundating adjoining areas before entering seas and oceans. Annual water resources in the various river basins in the country is estimated to be about 187 M ha m. In addition, there is a dynamic rechargeable ground water resources whose potential has been estimated as 43.2 M ha m. This means that the total water available in the country is about 230 M ha m and per capita availability is about 2100 m^3/year. When the population will reach to 1.64 billion in 2050, according to U.N. agencies, per capita availability will decrease to 1200 m^3/year. The country will be water stressed but not a water scarce country by 2050 according to the World Bank and other agencies (Sivanappan, 2002).

There are vast variations in the average per capita availability of water in various parts of the country. This varies from about 18147 m^3 in Brahmaputra, 3133 m^3 in Mahanadi about 666 m^3 in Cauvery and as low as 364 m^3 in some east flowing rivers of Tamil Nadu considering both surface and ground water. Thus, there is an urgent need to inter link the rivers and to allow the water flow from excess to scarce areas. Of the 187 M ha m of water resources available in different river basins in the country, about 70 M ha m of surface water are available for use. Besides about 43.5 M ha m of ground water are also available for use. This creates an irrigation potential of 113.5 M ha . The national productive envisages an additional potential of 25 M ha through surface storage and inter basin transfer of river water. Thus, the ultimate irrigation potential comes to about 140 M ha in the country. To enable 1.64 billion people to get square meal, the current irrigation potential of 109 M ha has to be upgraded to 140 M ha by 2050 A.D. (Sivanappan, 2002).

There is one more thing to be noted. The water use efficiency in Indian agriculture is about 40-45% which is very less compared to many countries in the world i.e. for Israel it is 80%. The water use efficiency of the irrigated agriculture has to be increased to about 60% so that the irrigated land could be increased without putting more pressure on creating new water resources. FAO report indicates that our land and water resources are capable of sustaining a population of 2.6 billion people provided they are properly managed (Sharma and Sharma, 2002).

In order to increase the irrigation potential, it is imperative to increase the water storage capacity of all the reservoirs. At present the river storage capacity of all the rivers is about 20 M ha m. Considering the storage capacity of the projects under construction and consideration, the total usable water stored in various reservoirs will come to about 38-40 M ha m. To cater the needs of agriculture, domestic, industrial and energy sectors, storage capacity of all the reservoirs should be increased to 60-65 M ha m. Comprehensive control and management of rainwater, ground water, surface water and soil moisture would be necessary for maximum utilization of water resources and increasing the irrigation potential. There is also considerable scope for further development through economic utilization of available water resources by proper water resources planning and management and augmentation of the supplies by tapping new sources of irrigation.

1.12 Augmentation of Water Resources

Water resources of the country is finite, but the demand of water in various sectors including irrigation for agricultural crops and industrial uses are expanding at higher rate. Fig. 1.7 shows the estimated annual water requirement in various sectors in the country up to the year 2050 A.D. (Central Water Commission, 2000). Data of Fig. 1.7 reveals that multi dimensional use of water

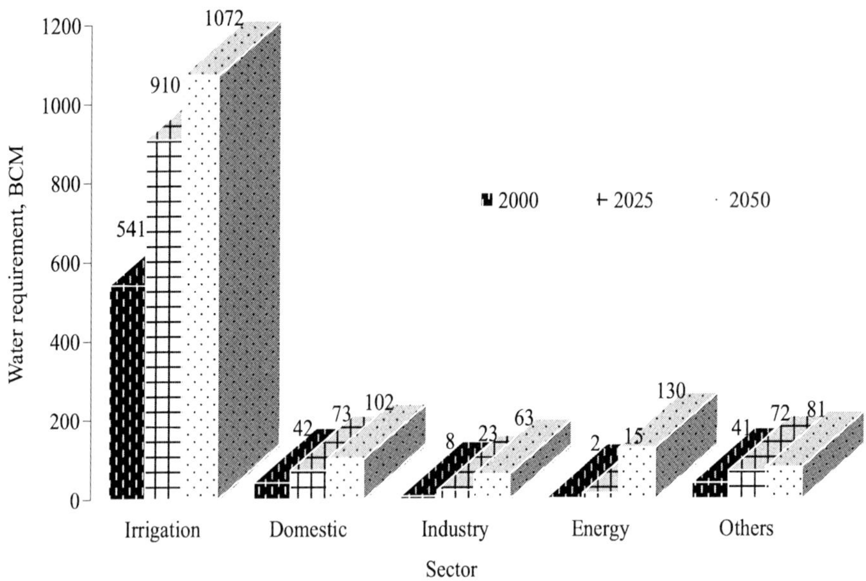

Fig. 1.7: Annual water requirement in various sectors in India

water is expanding at a faster rate especially in the field of industrial and energy sector. Hence, it is imperative to augment the existing water resources and search for new water resources. Some of the methods for augmentation of water resources are (Sharma, 1987):

- **Conservative Water Use:** Conservative use of water can save a lot of costly irrigation water which will help in irrigating more areas. Using appropriate irrigation scheduling, proper methods of irrigation in different crops, decreasing seepage and percolation in crop fields, using water saving irrigation techniques in crop fields, increasing canal conveyance efficiencies by low cost lining, using proper dry land technologies in rainfed areas are some of the means to conserve water and augment the water resources.
- **Equitable Water Distribution:** Equitable distribution of water in canal commands is necessary to increase the overall irrigation efficiencies. It helps farmers to get assured and adequate supply and prevent them to practice over-irrigation. For better equitable

distribution of irrigation water, it is essential to use volumetric supply of water and charge water rates based on volumetric basis.

- **Evaporation Reduction:** Evaporation accounts for considerable loss of water particularly in hot, arid and semi-arid regions. Reduction of evaporation from the soil surface either bare or cropped field, can be reduced through different types of mulching. Two to three irrigations can be reduced in vegetables and fruit orchards by mulching especially in tail end of canal areas. Similarly, the reduction of evaporation in large storage reservoirs can be achieved by covering the water surface with dyes, thin mono-molecular films, hexadeconal or cetyl alcohol etc. A reduction of wind speed by wind breaks can also control lake/reservoir evaporation. Biological shading in small on-farm reservoirs helps to reduce considerable evaporation losses.

- **Snow Melting:** Augmentation of water resources by melting of glaciers in the Himalayan region to increase the dry weather flow in rivers is under study in India. Melting of snow can be accomplished by heating, using chemicals, salts, compounds like calcium chloride etc. The heat used for induced melting of ice may be solar energy absorbed by a coating of materials spread over the snow. The material sued for the purpose may be soot, finely powered coal and industrial slag.

- **Artificial Recharge:** In-situ rainfall as well as flow of water over the soil surface recharges the soil surface in natural ways. However, there is scope for artificial recharge of aquifers and there by augmenting the water resources in underground reservoirs. These underground reservoirs serve as a dependable source of water so as to supplement the water available from canals. Artificial recharge of ground water reservoirs is used for:

i) Storage of excess surface water in the ground water reservoirs and there by increase the ground water mound.

ii) Improvement of the quality of underground water through mixing of local ground water with recharge water.

iii) Increase the water pressure through the raised water table and to reduce sea water intrusion.

iv) Decrease the drainage congestion in low lying areas and deltaic regions

v) Prevention of land subsidence due to increased water table

vi) Reduction in pumping costs since water can be pumped at lower lifts and there by decreasing pumping costs.

Different methods of artificial recharge are shown in the flow chart (Fig. 1.8) as shown below.

- **Flood Water Use:** High intensity of rainfall for longer periods causes floods in several parts in the country causing great economic losses. However, some amount of flood water can be stored in under ground reservoirs through artificial recharge techniques. Creation of major and medium reservoirs can store considerable flood water and reduce flood damage in downstream areas. Storage of flood water in local depressions, on-farm reservoirs and tanks can be advocated where feasible, and should be used as irrigation in *rabi* crops.

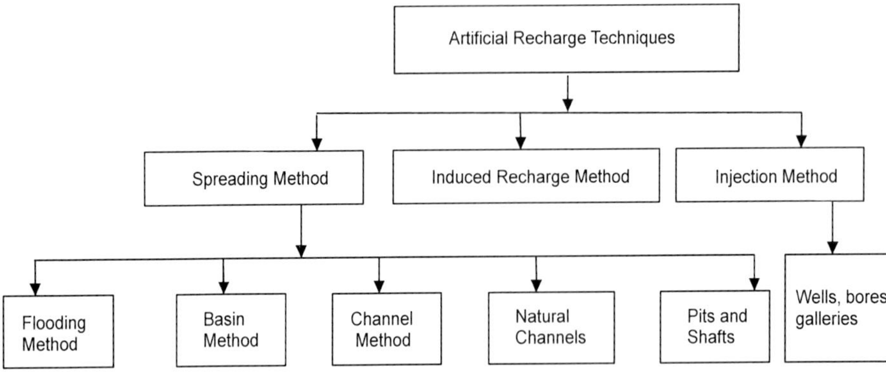

Fig. 1.8: Flow chart showing different methods of artificial recharge techniques

- **Weather Modification:** Weather modification helps in producing rain from clouds and to form new clouds by increasing evaporation from water bodies. Seeding of clouds is a new technique to produce rain. Seeding involves intensification of the condensation of cloud drops that subsequently collide and coalesce to form larger drops or change in the dynamic properties of the clouds themselves. Clouds can be made unstable by the addition of dry ice, silver iodide, common salt or other chemical agents.

- **Command Area Development:** Command area development aims to accelerate the process of utilization of irrigation potential already created and to increase the water productivity so that optimum yield is obtained per unit of water, per unit of land and per unit of time. Construction of water courses and field channels and irrigating the crops through them, adoption of rotational methods of irrigation, volumetric supply of water are some of the pre-requisites for command area development. Multi-disciplinary approach in this context requires adequate land reforms, including consolidation of land holdings, land drainage, soil survey, adjustment of new cropping patterns suitable to soil, water and climatic conditions, land leveling and grading etc.

- **Grid Water System:** Inter-basin transfer of water resources from areas where surplus water is available to water scarce areas is a promising means to augment the water resources. Israel has developed a piped water grid throughout the country and as a result it has utilized about 90% of the water resources.

- **Desalination:** India has a long coast line of 6400 km. It has vital scope for developing desalination of sea water for irrigation and public water supply. Desalination of saline water for use in agriculture is costly. However, the advances in nuclear reactor technology have opened vistas for large scale desalination of sea water.

- **Harnessing Rainwater:** India receives considerable amount of rainwater in rainy season. About 90% of this water flows past the soil and reaches sea/ocean. Before reaching the sea/ocean, it causes a lot of soil and water erosion and nutrient losses from the cropped field. Efforts should be made for harnessing this abundant resources in cost effective farm reservoirs/ponds and recyling the harvested water for supplemental irrigation to crops especially non-paddy crops.

- **Waste Water Reclamation and Utilisation:** In many parts of the country, people are facing scarcity of water both for drinking and irrigation. At the same time huge quantities of waste water from domestic and industrial effluents are discharged into the rivers every day which creates water pollution. About 1660 Ml of effluent water is discharged into the Ganges every day. Similarly, huge quanties of sewage water is dumped with water bodies spoiling the land and water resources. The waste water should be reclaimed and then reused for crop production. Israel is present using 30% of treated waste water in agriculture.

1.13 Water Resources Planning and Management

Water is the most essential input for irrigation. Presently about 80% of the total water resources of the country is used for irrigation. However, due to increased population, industrialization and urbanization, it is apprehended that the share will decrease to 70% or so by 2050 A.D. On one hand it is hard pressured to boost the agricultural production whereas on the other hand, the percentage of water for use in agriculture is bound to decrease. In this juncture, there is no other way then adopting the scientific water management practice including water saving irrigation techniques in irrigated agriculture. There are many methods for irrigation system planning and management. Some salient points to be borne in mind are:

- Irrigating rice fields through field channels instead of field to field to irrigation
- Irrigating the rice by intermittent irrigation method instead of maintaining standing water in field always
- Encouraging crop diversification instead of rice-rice cultivation
- Using accurate irrigation scheduling including proper methods of irrigation
- Adopting water saving irrigation techniques in the crops
- Proper land smoothing and grading using appropriate farm machineries
- Using high tech irrigations like sprinkler and drip especially in fruits, orchards and vegetables which has high application efficiency
- Reusing the waste water after reclamation
- Harnessing the abundant rainwater and reusing it as supplemental irrigation to crops
- Multiuse of harvested water
- Reduction of seepage and percolation in crop fields
- Using cost effective mulching in crops fields to check soil evaporation and reduce the frequency of irrigation in crop fields

- Decreasing the conveyance losses of canals by lining the canal systems
- Volumetric supply of water in canal commands
- Charging water rates on volumetric basis

1.14 Economics of Irrigation

Economics is important in evaluating irrigation practices, for irrigation is largely for the purpose of increasing profit. Higher profits resulting from more efficient production will ultimately result in lower prices for consumers, and lower prices result in more consumption of food and fiber, greater availability of food and fiber results in higher standards of living of people. These factors should be kept in mind before designing any multi-purpose irrigation projects. If the capacity of a multi-purpose reservoir is increased, it can irrigate more area, can generate more power. Hence, as the capacity increase, benefits tend to increase, but at the same time costs also increase. In all such cases, a thorough analysis of benefit cost considerations must be made before deciding about the size of a project.

Question Banks

Q1. Define irrigation. How it is necessary for agriculture?

Q2. What are the advantages and disadvantages of irrigation?

Q3. Irrigation is both bone and course to mankind. Justify.

Q4. How does irrigation helps in elimination of mixed cropping?

Q5. How does irrigated agriculture differ from rainfed agriculture? How does irrigation affect production and productivity of crops?

Q6. Describe the early development of irrigation in India and other countries of the world.

Q7. Irrigation has developed tremendously in plan periods in India. How?

Q8. Explain briefly how has the irrigation potential been developed and utilized in India.

Q9. With diagram, describe the land and water resources of the country.

Q10. There is enough scope for water resources development for irrigation in the country. How can these scopes be achieved?

Q11. Describe different methods for augmenting water resource for use in irrigation.

Q12. What do you mean by artificial recharge? What are its benefits?

Q13. Describe various methods to increase artificial recharge.

Q14. Write different methods used for water resources planning and management to increase the system productivity.

Q15. Why it is important to study the economics of irrigation?

References

Bhatia, P.C. and Singh, H.P. 2001. Rainfed Agriculture: Research and Development Perspective. Yojana, January, 45 (1): 57-62.

Food and Agriculture Organisation. 1999. Irrigation in Asia in figures. Water Report No. 18,, Rome, pp. 91-106

Garg, S.K. 1987. Irrigation Engineering and Hydraulic Structures. Khanna Publishers, Delhi.

Government of India .1993. Annual Document. The Ministry of Water Resources, New Delhi.

Government of India. 2004. Annual Document 2003-04. Planning Commission of Government of India, New Delhi.

Government of India. 2007. 11th Five Year Plan, Planning Commission Report, Government of India, New Delhi.

Hansen, V.E., Israelsen, O.W. and Stringham, G.E. 1979. Irrigation Principles and Practices. John Wiley and Sons Inc., New York.

Lenka, D. 2001. Irrigation and Drainage. Kalyani Publishers, New Delhi.

Panigrahi, B. 2001. Water Balance Simulation for Optimum Design of On-Farm Reservoir in Rainfed Farming System. Ph.D. Thesis, Indian Institute of Technology, Kharagpur.

Sahu, G.C. 2008. Interlinking of Rivers for National Integration. Technical Annual of Institution of Enginers (India), Orissa State Centre: 40-49.

Sharma, R.K., 1987. A Text book of Hydrology and Water Resources. Dhanpat Rai and Sons, Delhi.

Sharma, R.K. and Sharma, T.K. 2002. Irrigation Engineering including Hydrology, S. Chand and Company Ltd. New Delhi.

Singh, S.R. 2001. Management of Land and Water Resources for Increasing Agricultural Productivity. Theme Paper Presented in the Annual Convention of Agricultural Engineers (Indian Society of Agricultural Engineers), Bhubaneswar.

Sivanappan, R.K. 2002. Expansion of Storage Capacity, Critical. Survey of Indian Agriculture, The Hindu.

Tripathy, L.K. 2008. Sustainable Water Management. Technical Annual of Institution of Enginers (India), Orissa State Centre: 55-60.

2

Sources of Irrigation Water

2.1 Introduction

Rain and snow are the two chief sources of water of the earth. In-situ conservation of rainwater helps the plants to meet some of their consumptive uses directly. Not all the amount of rainwater that falls at a point is conserved by the soil reservoir to support the plant growth. Some part of it flows over the soil surface as surface runoff and some seeps down as infiltration and finally adds to the under ground reservoir. Therefore, rain or snow, which is not used directly by the plants, becomes a potential source of either surface or ground water for irrigation. In areas with extreme low temperature, snow contributes major source of water supply wherever, in humid and sub humid regions, rain is the principal source. Precipitation, rain or snow, is the first stage of the continual cycle of water in its round from atmosphere to the earth and return by evaporation from water bodies, snow surfaces and land surfaces. The continual cycle of distribution of water is governed by the process called as *hydrologic cycle*. Thus, precipitation or atmospheric water is the first form of water above the earth's surface, the second form being the surface water on earth and the third form is the subsurface water or underground water. The study of occurrence, distribution and circulation of precipitation through the unending hydrologic cycle forms the basic part of study of *hydrology*. As mentioned above, the part of precipitation that is not used by the plants in-situ flows over the soil surface as runoff and contributes to either surface or ground water supply.

Engineering hydrology is the science that deals with the estimation of runoff and its transmission from one place to the other, besides dealing with the properties of water in nature. For, an engineer, surface and groundwater are the two most important components to deal with. But due to lack of data of stream flow, it is usually necessary to study precipitation, evaporation and percolation to indirectly determine the availability of surface and groundwater. For planning and management of water resources systems, an engineer has to estimate (i) water resources potential of a basin, (ii) dependable yield for irrigation and hydel power generation and industrial water demand (iii) probability of occurrence of floods with their magnitude, (iv) sediment yield and transportation (v) hydrometeorology (vi) new sources of water for irrigation (vii) stream flow forecasting by use of precipitation data (viii) water control, navigation, control of soil erosion and control of pollution of water bodies and (ix) operation and maintenance of river valley projects. For estimation of the above mentioned parameters, it is important to understand the *hydrologic cycle* properly (Garg, 1987).

2.2 Hydrologic Cycle

Earth's water sources are mainly rivers, oceans, lakes and underground sources. All these sources get water by precipitation including rain. The rain depends on evaporation from water bodies which subsequently condenses forming clouds and hence rain. The hydrologic cycle (sometimes also called as water cycle) consists of a number of processes as shown in Fig. 2.1. The hydrologic cycle useful from engineering point of view is represented by Fig. 2.2 (Eagleson, 1970). The essential components of hydrologic cycle are:

Precipitation: Precipitation refers to all forms of water derived from atmospheric vapour and deposited on the earth's surface. It occurs due to condensation of water vapors that have evaporated from the water bodies like ocean, sea, rivers, lakes, ponds etc. The different forms of precipitation are rain, snow, hail, mist, dew, sleet and fog. Very often the term precipitation is used indiscriminately to mean rainfall.

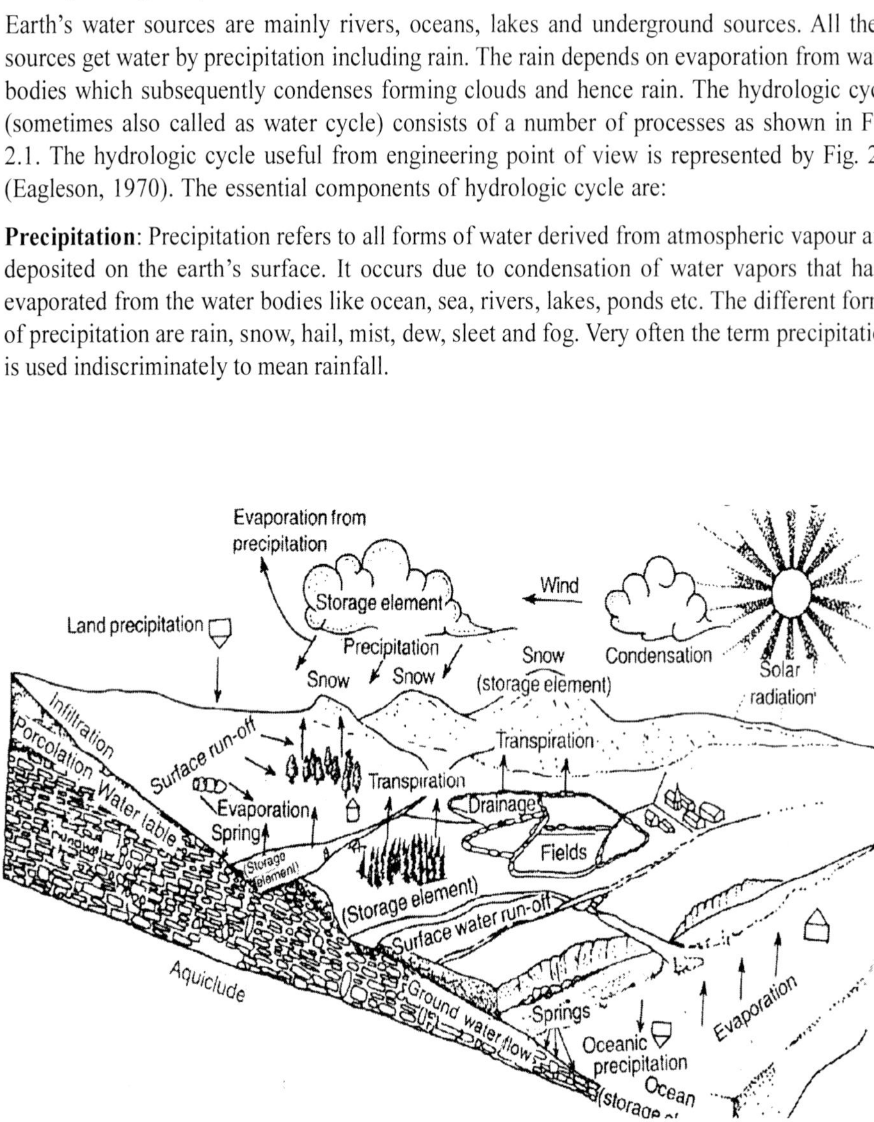

Fig. 2.1: Schematic representation of hydrologic cycle

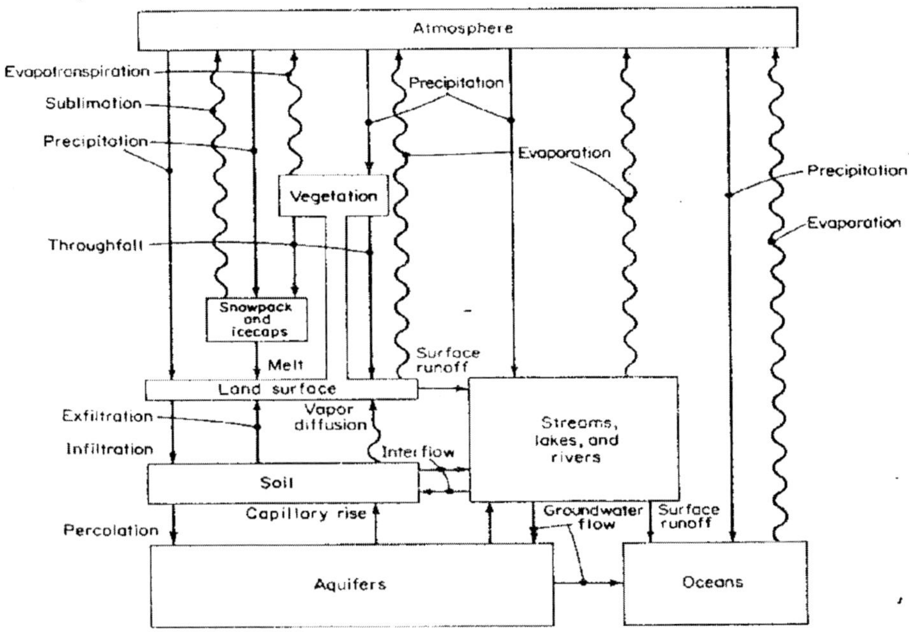

Fig. 2.2: Hydrologic cycle from engineering point of view

Evaporation: The conversion of water in the liquid form to vapour is termed as evaporation. Evaporation occurs due to rise in temperature such that increasing kinetic energy of the water molecules at liquid-air interface overcomes the cohesive force of water molecules that hold them together and as a result the water molecules gets evaporated. Evaporation occurs from both water surfaces as well as from the land surfaces. Moreover, the part of precipitation that is intercepted by the plants' canopy and gets intercepted by other sources such that they do not reach the land surfaces gets evaporated and returned back to the atmosphere. About 80% of precipitation in India is assumed to be contributed by evaporation from seas and oceans.

Transpiration: Transpiration is the process by which water in the soil is transferred to the atmosphere by plants as water vapour. Plants require water for their growth. This water along with minerals is absorbed by the roots from soil through diffusion process. Out of the amount of water absorbed by the plants, a substantial amount is returned to the atmosphere by transpiration whereas a little amount of about 1% is retained by the plants for meeting metabolic growth.

Infiltration: When precipitation falls on land surfaces, not all the portion flows over the land. The part of the precipitation that does not flow over the land but moves down the land surface is known as infiltration. Infiltration occurs due to the porosity of the soil mass.

Percolation: Water that infiltrates into the soil is retained by it as soil moisture content. Some of this soil moisture is evaporated while the other is used by plants for transpiration. When the soil moisture content is more than the field capacity, then the soil moisture in excess of field capacity moves down as percolation. The percolated water adds to the ground water reservoir.

Runoff: The portion of the precipitation that makes its way towards streams, lakes, rivers or any other water bodies as surface or subsurface flow is known as runoff. Surface runoff refers to the flow of water over the land surface whereas the subsurface runoff refers to the flow below the land surface. Very often, the term runoff when used alone refers to the surface runoff only. Surface runoff is sometimes referred as overland flow. The infiltrated water that enters into the land surface flows as subsurface runoff. A part of the subsurface runoff flows laterally as interflow and subsequently joins the stream. The other part of the infiltrated water moves vertically down and adds to ground water (Murty, 1998).

The hydrologic cycle can be represented by the mathematical equation as:

$$P = R + I + E + T + S \tag{2.1}$$

where, P = precipitation, R = surface runoff, I = interception loss, E = evaporation from both land and water surfaces, T = transpiration loss by plants and S = storage on or below the surface of the earth.

Precipitation is the most important component of hydrologic cycle. It is vital to have a thorough knowledge of precipitation in estimating runoff, planning erosion control measures, planning and management of water resources systems and for many more other allied activities. Data of other components of hydrologic cycle like runoff, infiltration, percolation, evaporation and transpiration are essential for planning irrigation and drainage systems.

2.2.1 Precipitation

It is the general term for all forms of moisture emanating from clouds and then falling to the earth's surface. It is thus the total supply of water derived from the atmosphere in the forms of rain, snow, hail, mist, dew, sleet and fog etc. The liquid form of precipitation is called as rainfall whereas the frozen forms of precipitations are snow, hail and sleet. Generally it is expressed as the depth water on a horizontal surface over any specified time interval. If the specified time interval is day, then the amount of precipitation is called as daily precipitation. Similarly, the precipitation being contributed over time interval of a month, season and year. is termed as monthly, seasonal or annual precipitation.

2.2.2 Occurrence and Causes of Precipitation

There are three essential requirements that cause precipitation to occur. In the first requirement, it is necessary for the air to get sufficiently cooled by some mechanisms so that condensation occurs and growth of the droplet occurs. In the second requirement, condensation nuclei are necessary which are present in the atmosphere in adequate quantities. In the third requirement, large scale cooling is needed for significant amounts of precipitation to occur. Because of evaporation, the water vapours from the water bodies are lifted up. At the higher altitudes because of lower temperature, the water vapours get cooled which in turn saturates and sufficiently cools the surrounding air causing condensation. Condensation results in the formation of small droplets and hence cloud formation. These droplets will not in themselves form into rain drops and may remain indefinitely as droplets in a condition of colloidal suspension. When the minute droplets of moisture collect on particles in the air called as hygroscopic nuclei, rain drops or other forms of precipitation occurs. The very common hygroscopic nuclei are salt particles from ocean and products of combustion in atmosphere (Eagleson, 1970). The type of precipitation to occur depends on the temperature of the air in which precipitation is formed and the temperature of the lower air layers through which it falls. Precipitation may be convective, orographic and cyclonic types. Cyclonic precipitation may be divided into frontal and non-frontal types. The frontal precipitation can further be divided into warm-front and cold-front precipitation. At times, precipitation may cause by turbulent ascent of air masses caused due to increased friction of earth surface after it travels over ocean.

2.2.3 Rainfall and its Variation

The total liquid product of precipitation or condensation from the atmosphere which is measured in a raingauge is called as rainfall. A day receiving 2.5 mm or more rainfall is called as a rainy day. The amount of the rain that reaches the earth is usually expressed in millimeters of depth of water over an area. The daily rainfall pertains to the rainfall from 0830 hours of previous day to 0830 hours IST of date. The average annual rainfall in India is about 120 cm. The average annual rainfall is computed by adding the annual rainfall of several years which is then divided with the total number of years. Since, the annual rainfall is different from year to year, the average annual rainfall will vary depending on the numbers of years whose annual rainfalls are taken into account for assessing average rainfall. For example, the average annual rainfall of 50 years may be different than the average annual rainfall of 60 years. Even for the same number of years, the average annual rainfall will be different for different periods of time. Average annual rainfall of 50 years from 1950 to 1999 may be different than the average annual rainfall of 50 years from 1958 to 2007. The term normal rainfall refers to the average annual rainfall of 30 years. The annual rainfall suffers from wide departure from the normal from year to year. Practically all parts of Assam, Meghalaya and the adjoining areas, the Western Ghats and the adjoining coastal strips and parts of the Himalayas receive heavy rainfall of more than 2000 mm. Cherrapunji of India records the highest rainfall in the world (about 1142 cm annual rainfall). On the other hand, Rajasthan, Kutch and the high peaks of Ladakh receives low annual rainfall ranging from 10 to 50 cm (Sharma, 1987).

Rainfall in India is seasonal in character. Most of the rainfall (about 75 to 85% of total annual rainfall) is contributed by South-West monsoon and the contribution is only from four months (June to September). The rest amount of annual rainfall is contributed by North-East monsoon (November and December). Some parts of the country receive hot weather rainfall between March and May due to thunderstorm activity called North-Westers. About 8 percent of total geographical area of India receives more than 250 cm annual rainfall, 50 percent area receives 100 cm or more, 35 percent area receives less than 75 cm and 12 percent areas receives annual rainfall less than 40 cm. The distribution of rainfall in India is very erratic. It has both spatial and temporal variation. The coefficient of variation of annual rainfall over large part of the country is about 30 percent. In some of the interior districts, the coefficient of variation may be as high as 100 percent. The coefficient of variation of rainfall is higher during the individual months compared to the annual values (Sharma, 1987). The variation of rainfall during different parts of the year is due to the difference in seasonal temperature between land and sea.

2.2.4 Measurement of Rainfall

For crop planning, design of different soil and water conservation structures, irrigation and drainage planning and/ or any other engineering studies, analysis of rainfall is required which necessitates to measure it. The instruments used to measure rainfall are called as rain gauges. Rain gauges are divided into two types i.e. (i) recording type gauges and (ii) non-recording type gauges.

2.2.4.1 Non-recording Rain Gauge

It is also termed as non-automatic rain gauge. It is used to measure the total rainfall during a period. Observations are taken at the end of 24 hours period or at lesser intervals during rain. It cannot give the rainfall intensity during different time intervals during a day. The Symons rain gauge is a common type non-recording rain gauge (Fig. 2.3). The rain gauge consists of a cylindrical vessel of 127 mm diameter with a suitable base and gun metal rim. It essentially has a collector to intercept the rainfall to be measured and a receiver consisting of a base and a bottle to collect and store the rainfall until measured. A funnel of exactly 127 mm diameter is inserted into the cylindrical vessel at the top and is used as the collector. The shank of the funnel is put into the receiving bottle having diameter 76 to 100 mm. The bottle has sufficient capacity to hold extremes of rainfall likely to occur in 24 hours at a location. The usual time to measure rainfall is 8.30 A.M. The rainfall collected in the receiving bottle is measured by a graduated measuring cylinder which can read to the nearest 0.1 mm. Recently Indian Meteorological Department has introduced fiberglass reinforced polyester rain gauge which is an improvement over Symon's gauge. In remote areas, storage gauge is used to measure total seasonal precipitation. It consists of a vertical 60 cm diameter steel pipe of sufficient length to place its 20 cm catch ring above maximum accumulated snow. To convert the snow which falls into the gauge to liquid state, an antifreeze solution can be placed in the receiver.

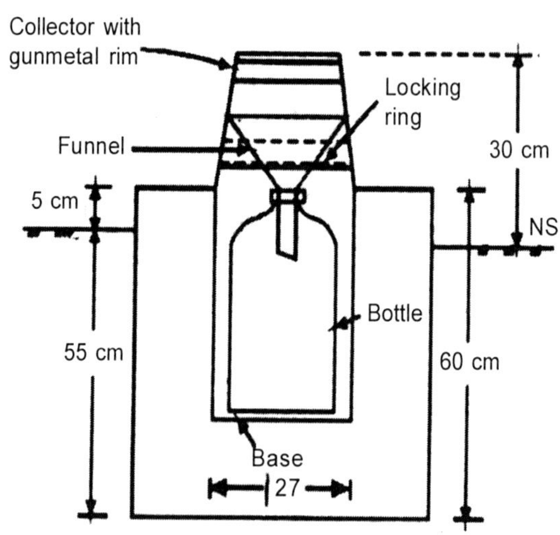

Fig. 2.3: Standard raingauge

2.2.4.2 Recording Rain Gauge

Recording rain gauge also called as automatic rain gauge is an improvement over the non-recording rain gauge. Since this rain gauge represents cumulative rainfall, it is also termed as integrated rain gauge or continuous rain gauge. It is especially suitable to inaccessible areas where it is used to give accumulated rainfall over an appreciable long period. In this rain gauge, the total rainfall received by a container over certain time period is recorded by a pen over a clock driven chart attached to the rain gauge. The graph of the cumulative rainfall is called as mass curve of the rainfall. The mass curve represents the total amount of rainfall received since the start of recoding at the station, duration of rainfall and intensity of rainfall. Recording rain gauges are of following four types i.e. (i) weighing bucket type (ii) tipping bucket type (iii) float recording type and (iv) automatic radio reporting rain gauge.

2.2.5 Average Depth of Rainfall

As discussed, there is large variation of rainfall over a large area. The average depth of rainfall over the whole area is computed by recording the rainfall at different stations covering the area under consideration and employing any one of the following three methods i.e. (i) Arithmetic mean method, (ii) Thiessen method, and (iii) Isohyetal method.

2.2.5.1 Arithmetic Mean Method

In this method, the average rainfall is computed by adding the rainfall of individual stations and then dividing the addition with the number of stations.

2.2.5.2 Thiessen Method

This is also called as weighted mean method. In this method, location of rain gauges is plotted on a map and the rain gauge locations are joined by straight lines. Perpendicular bisectors are drawn on each of the straight lines forming polygons around each rain gauge (Fig. 2.4). Measuring the area of each polygon and with recorded rainfall of the rain gauge located inside each polygon, the average rainfall depth can be computed. The formula used for computation of average rainfall is:

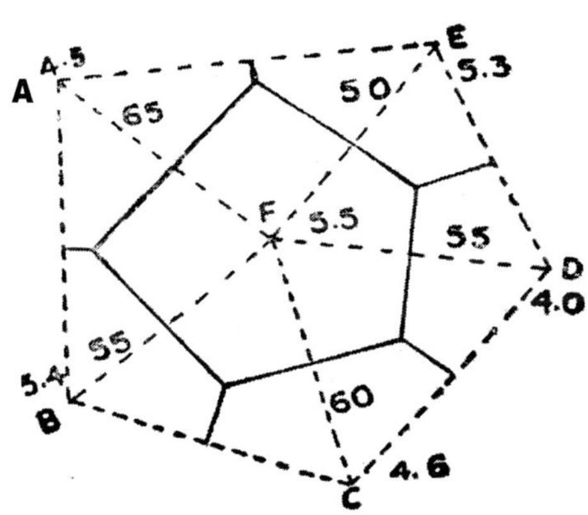

Fig. 2.4: Thiessen polygon method

$$P = \frac{\sum_{i=1}^{n} P_i A_i}{\sum_{i=1}^{n} A_i} \qquad (2.2)$$

where P = rainfall recorded by station, i and A = area of the polygon enclosing the *station* i and n refers to the number of stations of the whole basin whose average depth of rainfall is to be computed.

2.2.5.3 Isohyetal Method

Isohyets are the lines joining places of equal rainfall. In this method, location of rain gauges with recorded amount of rainfalls are plotted on a map/graph paper and then isohyets are joined by interpolation (Fig. 2.5). The areas between the successive isohyets are measured by a planimeter.

The same formula as used in Theissen method (Eqn. 2.1) is used to compute the average rainfall However, in this method, A_1, A_2 ... represents the areas between two successive isohyets and P_1, P_2... represent the average rainfall of the areas between two successive isohyets. In both Isohyetal and Theissen method, data of rainfall of stations located slightly beyond the coverage region can be used to compute the average rainfall.

Sample Calculation 1: Estimate the average depth of rainfall occurring in a basin of area 245 ha having 10 rain gauges. The rainfall recorded by each rain gauge and the polygon area enclosing each rain gauge are given as follows:

Station	A	B	C	D	E	F	G	H	I	J
Rainfall, cm	50	45	40	23	35	20	12	5	33	20
Polygon area, ha	10	13	18	15	26	25	20	23	40	55

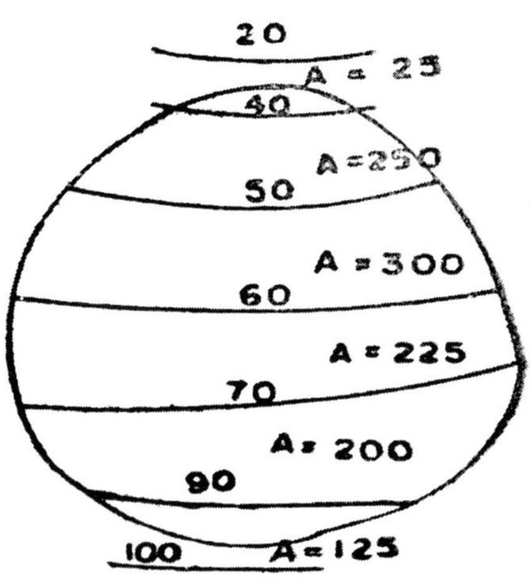

Fig. 2.5: Isohyetal method

Solution: Average depth of rainfall over a catchment by Thiessen polygon method is given as

$$P = \frac{\sum_{i=1}^{n} P_i A_i}{\sum_{i=1}^{n} A_i}$$

$$= \frac{50 \times 10 + 45 \times 13 + 40 \times 18 + 23 \times 15 + 35 \times 26 + 20 \times 25 + 12 \times 20 + 5 \times 23 + 33 \times 40 + 20 \times 55}{10 + 13 + 18 + 15 + 26 + 25 + 20 + 23 + 40 + 55}$$

$= 25.86$ cm

Sample Calculation 2: Estimate the average rainfall over an area with the following data:

Rainfall between isohyets, cm	Area, ha	Rainfall between isohyets, cm	Area, ha
125 - 130	20	140 – 145	22
130 - 135	25	145 – 150	35
135 - 140	28	150 - 155	40

Solution: Average rainfall by isohyetal method is:

Average rainfall between isohyets, *P* (cm)	Area, *A* (Ha)	Product (*P.A*)
(125 + 130)/2 = 127.5	20	2550
(130 + 135) / 2 = 132.5	25	3312.5
(135 + 140) / 2 = 137.5	28	3850
(140 + 145) / 2 = 142.5	22	3135
(145 + 150) / 2 = 147.5	35	5162.5
(150 + 155) / 2 = 152.5	40	6100
Total	170	24110

= 24110 / 170 = 141.82 cm

2.3 Runoff

Runoff is defined as that portion of precipitation which is not absorbed by the deep soil strata but finds its way into the stream after meeting the demands of evapotranspiration, interception, infiltration, surface storage, and surface as well as channel detention. It includes surface runoff received into the channels after precipitation occurs, delayed runoff that enters the stream after passing through the land surface, and other delayed runoff from the temporarily detained in pits, depressions, swamps or detained as snowfall. It is thus the flow collected from the drainage basin and appearing at the outlet of the basin. The runoff process is demonstrated in Fig. 2.6 as flow chart.

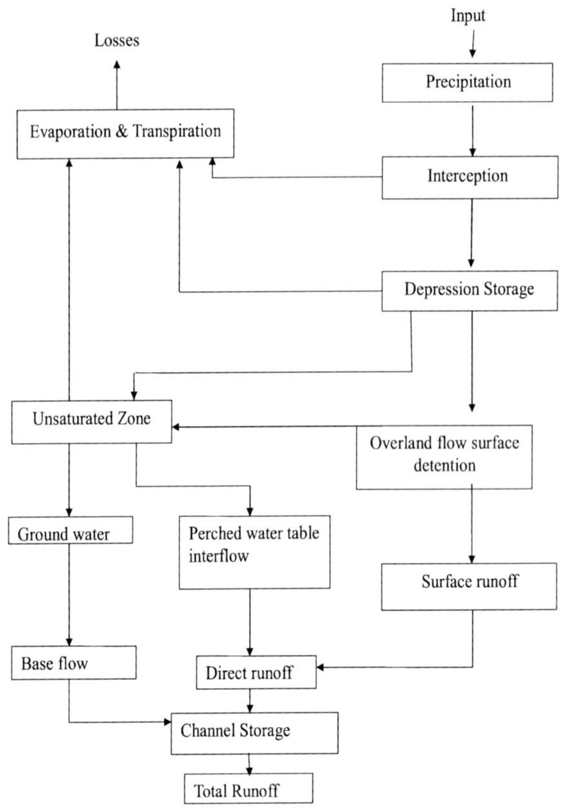

Fig. 2.6: Flow chart for computation of runoff

2.3.1 Factors Affecting Runoff

Rate and volume of runoff from an area are influenced mainly by two factors i.e. (i) climatic factors and (ii) physiographic factors. The various climatic factors that affect the runoff process are:

(i) Nature of precipitation
(ii) Interception
(iii) Evaporation and
(iv) Transpiration

Nature of precipitation includes:

(i) Rainfall intensity
(ii) Frequency of intense rainfall
(iii) Dration of rainfall
(iv) Distribution of rainfall and

(v) Direction of storm movement

The physiographic factors consist of both the watershed characteristics and channel characteristics. The watershed characteristics include

(i) Size

(ii) Shape

(iii) Slope

(iv) Orientation

(v) Land use

(vi) Infiltration characteristics

(vii) Soil texture and

(viii) Topography

Channel characteristics relate mostly to hydraulic properties of the channel that governs the water flow. These factors include

(i) Size

(ii) Slope

(iii) Cross section and

(iv) Roughness coefficient of channel

Some of the factors mentioned as above are interrelated to each other which makes it difficult to quantify the affect of each individual factor on runoff. However, some dominant factors that affect the runoff of an area are discussed below.

2.3.1.1 Nature of Precipitation

Surface runoff is produced when the rainfall intensity exceeds the infiltration capacity of the soil. Rainfall intensity influences both the rate and volume of runoff. Even though two storms have same total precipitation, an intense storm of short duration that exceeds the infiltration capacity of a soil at a faster rate produces greater volume of runoff than does a lesser intense storm of longer duration. The intensity-frequency relationship of a storm affects the runoff process in a way that an intense storm occurring frequently produces higher volume of runoff. Similarly a storm of given intensity produces higher amount of runoff if it occurs for a longer duration. In the initial stage of occurrence of a storm, the infiltration capacity may decrease with time and thus a storm of short duration may produce little runoff whereas a storm of longer duration but having same intensity will result in producing larger runoff. Distribution of rainfall also affects the rate and volume of runoff. When there is a higher intense storm occurring uniformly over the whole catchment, the entire watershed contributes and hence, the rate and volume of runoff becomes more. Further if there is an intense storm on one portion of the watershed, then it may produce more runoff than a moderate storm occurring over the entire watershed. If the topography, soil and other conditions in the watershed are uniform, then for any given total volume of rain falling on the watershed, the more non-uniform the distribution is the greater is the peak runoff. The minimum peak is produced by a uniform distribution.

Direction of storm movement also affects the peak rate of runoff. A storm moving in the direction of a stream produces a higher peak in a shorter period of runoff than a storm moving upstream. Interception, evaporation and transpiration depend on the meteorological parameters like temperature, humidity wind velocity, and solar radiation and thus influence the runoff. More is the magnitude of these meteorological parameters, more will be losses due to interception, evaporation and transpiration and thus less will be the amount of runoff.

2.3.1.2 Physiographic Factors Affecting Runoff

Regarding the size of the catchment, all the factors including depth and intensity of rainfall remaining constant, two catchments irrespective of their sizes will produce the same amount of total runoff expressed as centimeter depth. However, for a catchment with larger area, it takes more time for the runoff to pass through the outlet hence the peak flow expressed as depth will be smaller. Runoff is influenced by the shape of the catchment. Two catchments having same area, fern-leaf shaped catchment gives lesser runoff because the tributaries of different lengths meet the main stream at regular intervals thereby the discharge are likely to be distributed over a large period of time On the contrary, fan shaped compact catchments tend to produce higher peak runoff rate than long and narrow catchments as in the former case, all parts of the catchment contributes to runoff at the outlet simultaneously and thus, the time of concentration to peak runoff is less.

In addition to the above mentioned two types of catchments, the catchments may be classified depending on the shape as tree like, rectangular type, radial type, parallel type or trellis type. The slope of the catchment has pronounced affect on infiltration, surface runoff, soil moisture and ground water contribution to stream flow. It affects the period of over land flow. A watershed with flat slope having inadequate drainage systems results in getting more infiltration and hence less runoff than the watershed with steep slope and having well defined drainage systems. Orientation i.e. general direction of the slope of a catchment, with respect to prevailing winds, storm movement, and sun's position has considerable affect on the runoff. Soil texture and its permeability influences runoff greatly.

Shallow pervious soil and coarse textured soil like sandy soil tend to have more infiltration and thereby reducing runoff. Runoff is more for hard compacted subsoil since infiltration is less. Fault zones, fissures and cracks result in enhancing infiltration of in-*situ* rainfall and hence reducing runoff. However, the infiltrated water may contribute to the stream as base flow which is a delayed flow. Land treatment measures like contour ploughing, strip cropping, bench and terrace cultivation and other soil and water conservation measures increase the water holding capacity of the soil and thus reduces runoff and soil loss in a watershed. Vegetated watersheds, due to, surface retention, and storage produces less runoff than bare watersheds. Storage characteristics like lakes, reservoirs, check dams, pits etc. act as flood modulator and reduce the runoff of a watershed. A channel having rough surface with higher values of roughness coefficient will retard the velocity of flow and thereby reduces the surface runoff contributed to a stream. Topographical factors like altitude and drainage density also affect the runoff. Higher catchments usually have steep slopes and produce greater total runoff. At higher altitudes, though the evaporation arte of the atmosphere is less, the rainfall, under certain condition, is more and hence total runoff may be more. Finally, a watershed with higher drainage density will produce more runoff than a watershed with less drainage density.

2.3.2 Runoff Estimation

Runoff can be estimated from two sources i.e. due to rainfall and due to snow melt. Various methods used to estimate runoff due to rainfall (Sharma and Sharma, 2002) are:

(i) Runoff coefficient

(ii) Runoff percentage

(iii) Strange's tables and curves

(iv) Empirical formula

(v) Watershed routing technique

(vi) Synthetic unit hydrograph method

(vii) Infiltration method

2.3.2.1 Runoff Coefficient

Runoff is a function of rainfall. More is the amount of rainfall, more is the runoff. Thus, the runoff (R) and rainfall (P) can be correlated by a coefficient called as runoff coefficient or runoff factor (K). Runoff coefficient is defined as the ratio of volume of stream flow above the base flow to the volume of rainfall. The rainfall-runoff relation by this approach is considered as linear as given below:

$$R = K P \tag{2.3}$$

where R = runoff, cm and P = rainfall, cm.

Runoff coefficient (K) depends on all those factors that affect the runoff. Values of K are presented in Table 2.1.

Table 2.1: Values of runoff coefficient (K) for various soil and vegetation condition

Type of soil/vegetation condition	Values of K	Type of soil/vegetation condition	Values of K
Plateau lightly covered	0.70	Loam largely cultivated and suburbs with garden, lawns	0.30
Clayey soils stiff and clayey soil lightly covered	0.55	Sandy soils, light growth	0.20
Loam lightly cultivated or covered	0.40	Parks, lawn, meadows, gardens	0.05-0.20

2.3.2.2 Runoff Percentage

(i) **Binnie's Percentage:** Sir Alexander Binnie has suggested the following percentages of runoff to rainfall based on the data of two rivers of Madhya Pradesh.

Annual rainfall (cm)	Runoff (%)	Annual rainfall (cm)	Runoff (%)
50	15	90	35
60	21	100	38
70	25	110	40
80	29		

(ii) **Barlow's Percentage:** T.G. Barlow's runoff percentage, based on studies of catchments mostly under 130 km^2 in Uttar Pradesh are given below.

Class	Description of catchment	Runoff (%)
A	Flat, cultivated and black cotton soils	10
B	Flat, partly cultivated various soils	15
C	Average	20
D	Hills and plains with little cultivation	35
E	Very hilly and steep, with hardly any cultivation	45

The above values of runoff percentages are for average type of monsoon. They are modified by the application of following coefficients according to the nature of season.

Nature of season	Class of catchment				
	A	B	C	D	E
Light rain	0.70	0.80	0.80	0.80	0.80
Average or varying rainfall with no continuous downpour	1.0	1.0	1.0	1.0	1.0
Continuous downpour	1.50	1.50	1.60	1.70	1.80

2.3.2.3 Strange's Tables and Curves

Strange, based on the data of Maharastra, has provided following tables and curves between rainfall and runoff. Strange's table (Table 2.2) and curves (Fig. 2.7) give runoff for daily

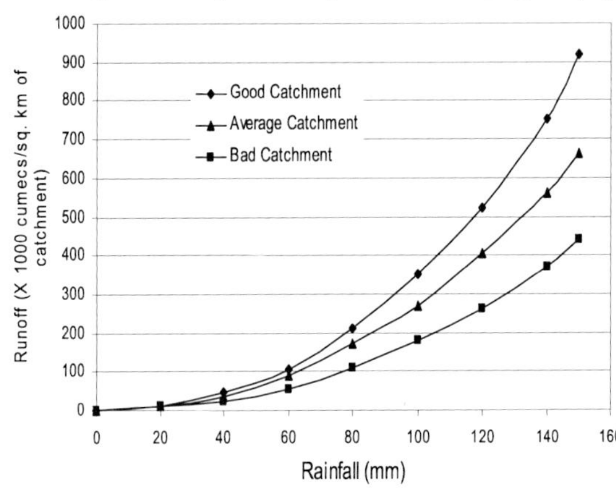

Fig. 2.7: Strange's curves to compute runoff

rainfall for good, average and bad catchments and surface conditions being dry, damp and wet prior to rain.

Table 2.2: Daily runoff and percentage of runoff according to Strange for average catchment

Daily rainfall (mm)	Original stage of ground surface					
	Dry		Damp		Wet	
	Percentage	Runoff (mm)	Percentage	Runoff (mm)	Percentage	Runoff (mm)
5	—	—	4	0.20	7	0.35
10	1	0.10	5	0.50	10	1.0
20	2	0.40	9	1.80	15	3.0
25	3	0.75	11	2.75	18	4.5
30	4	1.20	13	3.90	20	6.0
40	7	2.80	18	7.20	28	11.20
50	10	5.00	22	11.0	34	17.0
60	14	8.46	28	16.8	41	24.6
75	20	15.0	37	27.75	55	41.25
100	30	30.0	50	50.0	70	70.0

N.B. Strange has suggested to add or deduct 25 per cent of runoff values for good or bad catchments.

2.3.2.4 Empirical Formulae

There are large numbers of empirical formulae developed for different regions and countries which are used for estimation of runoff. Most of these formulae relate rainfall with runoff. There are some formulae which relate runoff with other parameters in addition to rainfall like temperature and duration of rainfall. The empirical formulae have limitations that they are suitable for the particular regions for which they are developed and cannot be used confidently for other regions. Some of the important empirical formulae are mentioned below (Sharma and Sharma, 2002).

(i) Lacey's Formula:

$$R = \frac{P}{1 + \frac{304.8}{P}\left(\frac{F}{S}\right)} \tag{2.4}$$

where, R = runoff, P = rainfall, F = monsoon duration factor varying between 0.5 – 1.5 and S = catchment factor depending upon the slope; values of which vary from 0.25 for flat cultivated areas to 3.45 for very hilly areas.

(ii) Khosla's Formula:

$$R = (P - T/3.74) \tag{2.5}$$

where, T = temperature, ^{0}C, P = rainfall, cm and R = runoff, cm.

(iii) Inglis and Desouza's Formula:

$R = 0.85\ P - 30.5$ for ghat areas of western India (2.6)

$$R = \frac{(P-17.8)P}{254} \text{ for Deccan plateau.}$$

where, R = runoff , cm and P =rainfall, cm..

2.3.3 Methods for Estimation of Runoff Rate

Following methods are used for estimation of runoff rate from a catchment.

(i) Rational Method

(ii) Crook's Method

(iii) Curve Number Method

Out of the above mentioned three methods, the rational and US curve number method are mostly used for computation of runoff rate and hence are discussed in this book.

2.3.3.1 Rational Method

Rational method is widely used in computation of the peak runoff rate in small watersheds. In this method, the peak rate of runoff is given by the equation as:

$$Q = \frac{CIA}{360}$$

where, Q = peak runoff rate, m^3/sec; I = rainfall intensity, mm/hr; C = runoff coefficient, and A = catchment area generating runoff, hectares.

Values of runoff coefficient (C) for different slopes and land use conditions as determined from the field experiments are given in Table 2.3.

Table 2.3: Values of runoff coefficient, *C* used in rational method

Soil Types	Land Use		
	Cultivation	Pasture	Forest
Sandy or gravelly soil having high infiltration rate	0.29	0.15	0.10
Loamy soil and soils with no clay pans near the surface having average infiltration rate	0.40	0.35	0.30
Heavy clay soil or soils with clay pan near the surface or shallow soil above impervious rock having below average infiltration rate	0.50	0.45	0.40

For catchment with different values of C, the weighted values of C as estimated below are considered for the whole catchment.

$$C = \frac{\sum_{i=1}^{n} C_i A_i}{A} \qquad (2.7)$$

where, C = runoff coefficient of the whole catchment, A = area of the whole catchment, C_i = runoff coefficient for catchment i and A_i = area of catchment i for which runoff coefficient is $C_{i,}$ and i = index for catchment ranging from number 1 to n.

The value of the intensity of rainfall to be used in Eq. 2.6 should be calculated for the period equal to the time of concentration of the catchment. While using the rainfall intensity in the rational method, the rainfall records at different places are analysed and the 1 hour rainfall intensity are calculated. For small water storage structures generally 10 years return periods is considered and hence, the rainfall intensity at 10 years frequency is considered for computation of runoff in rational formula.

The time of concentration of the catchment is defined as the time required for water to flow from the remotest point of the catchment to the outlet of the catchment. When the duration of the rainfall equals to the time of concentration, all parts of the catchment will take part in contributing to the runoff and hence the discharge would be maximum.

Kirpich (1940) has given the following empirical formula to compute the time of concentration.

$$T_c = 0.02\, L^{0.77}\, S^{-0.385} \qquad (2.8)$$

where, T_c = time of concentration, min; L = length of the channel reach, m and S = average slope of the channel reach, m/m.

For small catchment with area less than 5 sq. km having less developed network of channels and where overland flow dominates the time of concentration is given by the Eq. 2.9 (Haan *et al.*, 1982) as:

$$T_c = 0.02\, L^{0.77}\, S^{-0.385} + \left(\frac{2Lon}{\sqrt{So}}\right)^{0.467} \qquad (2.9)$$

where, L_o = length of the overland flow, m; S_o = slope along the flow path, m/m and n = average roughness coefficient of the channels contributing overland flow.

The second term of Eq. 2.9 accounts for the overland flow. The value of this term has a maximum value regardless of watershed size. The values of n suggested by Haan *et al.* 1982 is given in Table 2.4.

Table 2.4. Values of roughness coefficient n used in overland flow

Nature of the surface	Values of n
Smooth, impervious surface	0.02
Smooth, bare packed surface	0.10
Cultivated row crops of moderately rough bare soil	0.20
Pasture or average grass	0.40
Deciduous forest areas	0.60
Timber land with deep forest litter or dense grass	0.80

Time of concentration can also be computed as:

$$T_p = T_c^{0.5} + 0.6\, T_c \qquad (2.10)$$

where, T_p = time from beginning of rise of hydrograph to the peak.

Limitations of Rational Method

The limitations of rational method as proposed by (Singh, 1964) are:

(i) Rational method assumes that the rainfall intensity over the whole cathcment is uniform during the duration of the storm which is rarely satisfied.

(i) Before the occurrence of runoff, the losses due to depression storage and infiltration must be fulfilled this is not taken care by the rational method.

(ii) When there is flow in channels, the retardance affect caused due to surface detention and channel storage is not taken care in rational method.

Sample Calculation 3. A watershed of area 50 hectares having 20 hectares under cultivation, 10 hecatres under pasture and 20 hectares under forest. The cultivated area has sandy soil, whereas the pasture and forest area have heavy clay soil with very low infiltration rate. The watershed has a fall of 10 m over the longest reach of length 1 km; the reach being connecting the remotest point to the outlet point of the watershed. Etsimate the peak runoff rate if the 1 hour rainfall intensity for 10 years frequency of the watershed is 100 mm/hr. Assume that the ratio of 1 hour rainfall intensity to rainfall intensity of any other duration is 1.2,

Solution: Using Table 2.4, the values of C for cultivated, pasture and forest areas are calculated as 0.29, 0.45 and 0.40, respectively. The weighted value of C for the whole watershed is calculated as;

$$C = \frac{20 \times 0.29 + 10 \times 0.45 + 20 \times 0.40}{50} = 0.366$$

For L = 1 km = 1000 m, slope S = 10/1000 = 0.01

Using Eq. 2.8, time of concentration, T_c = 24 min.

Given the I hour rainfall intensity of 10 year frequency is 100 mm/hr and the ratio of 1 hour rainfall intensity to that of any other duration is 1.2, the raimnfall intensity for 24 minutes of rainfall is 100 x 1.2 = 120 mm/hr.

Hence peak runoff rate = $$Q = \frac{CIA}{360} = \frac{0.366 \times 120 \times 50}{360} = 6.1\ m^3/\text{sec}.$$

2.3.3.2 Curve Number Method

This method is also called as Hydrologic Soil Cover Complex Number method or US Soil Conservation Service (SCS) method. It is based on the recharge capacity of the watershed. The recharge capacity is determined by the antecedent moisture condition (*AMC*) and by the physical characteristics of the watershed. Let I_a be the initial quantity of interception, depression storage and infiltration that must be satisfied by any rainfall before runoff can occur. It is assumed that the ratio of the direct runoff Q and the rainfall minus initial losses $(P - I_a)$ and the storage capacity Sr are related by:

I_a is assumed to be certain percentage of *Sr*. Assuming it to be 0.2 *Sr*, the runoff is given as

$$\frac{Q}{P-I_a}=\frac{P-Q-I_a}{Sr} \tag{2.11}$$

$$Q=\frac{(P-0.2Sr)^2}{(P+0.8Sr)}=P>0.2Sr \tag{2.12}$$

$$Q=0 \quad \text{if } P \leq 0.2Sr \tag{2.13}$$

where, Q = surface runoff, mm; P = rainfall, mm and *Sr* = potential maximum retention, mm.

The parameter *Sr* (value in mm) is related to the curve number (CN) as:

$$Sr=254\left(\frac{100}{CN}-1\right) \tag{2.14}$$

The curve number for antecedent moisture condition (*CN* for *AMC II* or $CN = CN_2$) for different land use conditions and hydrologic soil groups are given by USDA (1972) (Table 2.5). Fig.2.8 helps in the computation of runoff from rainfall for different values of curve number (*CN*) for *AMC II*. A heterogeneous basin may be divided into sub areas with different values of *CN* and in such a case, a weighted value of *CN* for *AMC* II is calculated for the whole watershed. The corresponding values of *CN* for dry (CN for *AMC I* = CN_I) and wet (CN for *AMC III* = CN_3) conditions are given by Sharpley and Williams (1990) as:

$$CN_1=CN_2-\frac{20(100-CN_2)}{100-CN_2+\exp\{2.533-0.0636(100-CN_2)\}} \tag{2.15}$$

$$CN_3=CN_2+10\exp\{0.00673(100-CN_2)\} \tag{2.16}$$

and Sr given by them is:

$$Sr=Sr_1\left[1-\frac{FFC}{FFC+\exp(W_1-W_2.FFC)}\right] \tag{2.17}$$

where, Sr_1 = value of Sr associated with CN_1 and FFC = availability of soil water expressed as fraction of field capacity given as:

$$FFC=\frac{SMC_i-WP}{FC-WP} \tag{2.18}$$

where, SMC_i = soil moisture content on day i, FC and *WP* = soil moisture content at field capacity and wilting point, respectively.

Table 2.5: Curve numbers for hydrologic soil cover complexes

(For watershed condition 11, I[a]=0.2S) Land Use or cover	Treatment or Practice	Hydrologic Condition	Hydrologic Soil Group A	B	C	D
Faiiow- Row	Straight row		77	86	91	94
Crops	Straight row	Poor	72	81	88	91
	Straight row	Good	67	78	85	89
	Contoured	Poor	70	79	84	88
	Contoured	Good	65	75	82	86
Contour & terraced		Poor	66	74	80	82
Contour & terraced		Good	62	71	78	81
Small Grain	Straight Row	Poor	65	76	84	88
	Straight	Good	63	75	83	87
	Contour	Poor	63	74	82	85
	Contour	Good	63	75	83	87
	Contoure & terraced	Poor	61	72	79	82
	Contour & terraced	Good	59	70	78	81
Seeded Legumes	Straight row	Poor	66	77	85	89
or Rotation	Straight row	Good	58	72	81	85
Meadow	Contour	Poor	64	75	83	85
	Contour	Good	55	69	78	83
	Contour & terraced	Poor	63	73	80	83
	Contour & terraced	Good	66	67	76	80
Pasture or range		Poor	68	79	86	89
		Fair	49	69	79	84
		Good	39	61	74	80
	Contour	Poor	47	67	81	88
	Contour	Fair	25	59	75	83
	Contour	Good	06	35	70	79
Meadow (permanent)		Good	30	58	71	78
Woodlands		Poor	45	66	77	83
(farm woodlots)		Fair	36	60	73	79
Farmsteads			59	74	82	86
Roads(dirt)			72	82	87	89
Roads(Hard surface)			74	84	90	92

W_1 and W_2 are given as:

$$W_1 = \ln \frac{Sr_3}{Sr_1 - Sr_3} + W_2 \tag{2.19}$$

$$W_2 = \ln \left(\frac{0.5\, Sr_2 \left(Sr_1 - Sr_3 \right)}{Sr_3 \left(Sr_1 - Sr_2 \right)} \right)^2 \tag{2.20}$$

where Sr_2 and Sr_3 are the values of Sr when FFC = 0.5 and 1, respectively. Sr_1, Sr_2 and Sr_3 are the retention parameters corresponding to CN_1, CN_2 and CN_3, respectively. Once the parameters W_1, W_2 and FFC are estimated, value of Sr can be obtained from equation 2.17. The estimates of peak runoff rate from total runoff computed by Eq. 2.21 as suggested by US Soil Conservation Service (1964) is:

$$Q_{peak} = \frac{0.0208AQ}{T_p} \tag{2.21}$$

where, Q_{peak} = peak rate of runoff, m³/sec; A = area, ha; T_p = time to peak, hr; T_c = time of concentration, hr as discussed earlier in Eqs. 2.8 and 2.9 and Q = runoff, cm.

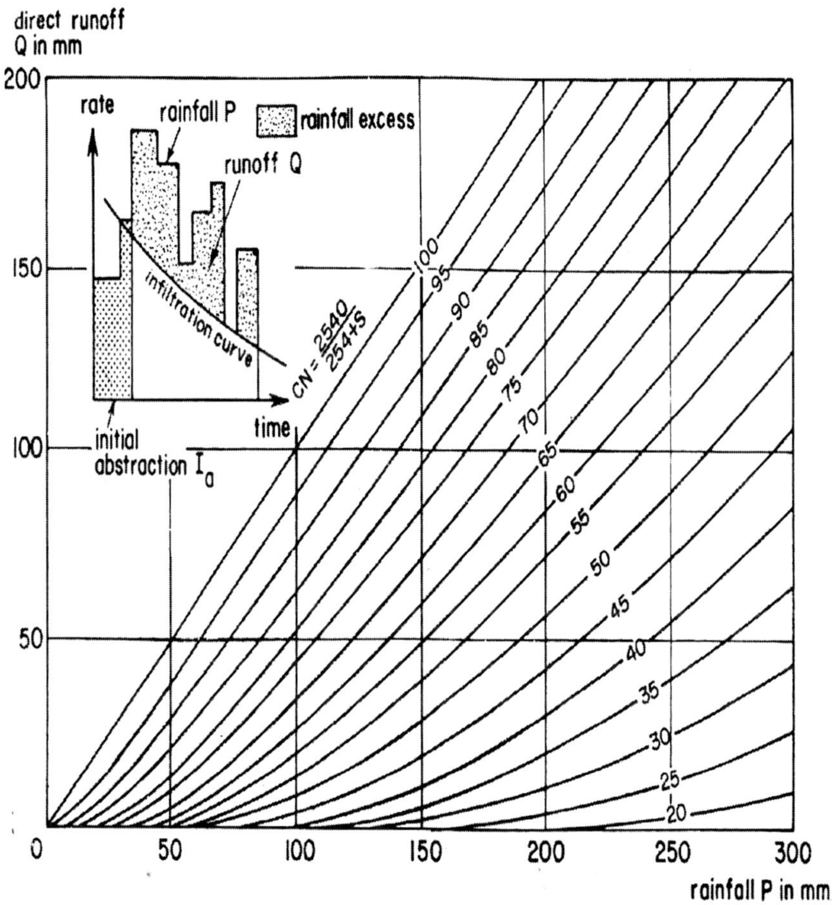

Fig. 2.8: Curve number to convert rainfall to runoff

T_p is given as:

$$T_p = \sqrt{T_c} + 0.6T_c \qquad (2.22)$$

Antecedent moisture condition (AMC) is an index for watershed wetness. The three levels of AMC are:

AMC I: Lowest runoff potential with dry condition of watershed.

AMC II: Average condition.

AMC III: Highest runoff potential with wet condition of watershed. The AMC class is determined using 5 days antecedent rainfall in dormant or crop growing season as given below (Table 2.6).

Table 2.6: Values of 5 days antecedent rainfall in deciding AMC class

AMC	5 days antecedent rainfall, cm	
	Dormant season	Growing season
I	< 1.25	< 3.5
II	1.25 - 2.75	3.5 – 5.25
III	> 2.75	> 5.25

The hydrologic soil groups are divided into 4 classes depending on runoff potential as:

Hydrologic Soil Group A (Low runoff potential): Soil having high infiltration rate and consisting chiefly of deep well to excessively drained sands or gravels.

Hydrologic Soil group B (Moderately low runoff potential): Soil having moderate infiltration rate and consisting chiefly of moderately deep to deep, moderately well to well drained soils having moderate coarse texture.

Hydrologic Soil group C (Moderately high runoff potential): Soil having low infiltration rate and consisting chiefly of moderately deep to deep, moderately well to well drained soils having moderate coarse texture.

Hydrologic Soil group D (High runoff potential): Soil having very low infiltration rate and consisting chiefly of clay soils with high swelling potential and soil with clay layer at or near the surface.

Hydrologic soil groups as mentioned above is given by Gupta *et al.* (1970) and is given in Table 2.7.

Table 2.7: Hydrological grouping of different soil types

Sl. No.	Soil type	Tentative hydrological groups
1.	Alluvial soil non calcareous to moderately calcareous	B
2.	Alluvial soil highly calcareous	B
3.	Alluvial soil developed on coast as alluvium	A
4.	Alluvial soil developed on deltaic alluvium	B
5.	Alluvial soil reverine affected by salinity and alkalinity	C
6.	Pedocal sicrozem of alluvial origin	C
7.	Pedocal brown soil of alluvial origin	C
8.	Grey and brown (desert soil)	D
9.	Desert soil	A
10.	Deep black soil	D
11.	Medium black soil	D
12.	Shallow black soil	D
13.	Black soil affected by salinity and alkalinity	D
14.	Black soil undifferentiated	D
15.	Mixed red and black soil	C
16.	Ferruginous red soil	B
17.	Ferruginous red gravely soil	B
18.	Red and yellow soil	C
19.	Laterite	B
20.	Laterite and black soil	D
21.	Brown soil under deciduous forest	B
22.	Forest soil	B
23.	Podsolic soil	B
24.	Foothill soil	C
25.	Mountain and hill soil	B
26.	Mountain meadow soil	B
27.	Peat	B

Sample Calculation 4. A watershed of area 50 hectares consist of 20 hectare with row crops cultivated in good terrace land and 30 hectares of good pasture land. The soil of the watershed belongs to hydrologic soil group C. Values of field capacity and wilting point soil moisture contents are 20 and 10%, respectively and soil moisture content of the soil on a particular day is 16%. Find the runoff amount on that day when a rainfall amount of 100 mm occurs in the watershed.

Solution: Using the data of land use, treatment conditions, hydrologic soil group and hydrologic condition, values of CN (AMC II) for 20 and 30 hectares lands are found from Table 2.5 as 78 and 74, respectively. The weighted CN value is given as

$CN_2 = (78 \times 20 + 74 \times 30)/50 = 75.6$

Using Eq. 2.15, $CN_1 = 57.68$ and Eq. 2.16, $CN_3 = 87.40$

Value of Sr_1 for $CN_1 = 57.68$ using Eq. 2.14 (replacing CN as CN_1) = 186.36. Similarly using Eq. 2.14 and replacing CN as CN_3, value of Sr_3 is obtained as 36.62.

Value of FFC (Eq. 2.18) = (16-10)/(20-10) = 0.6

Using Eq. 2.20, W_2 is obtained as 0.944 and using Eq. 2.19, W_1 is obtained as -0.461.

Using Eq. 2.17, value of Sr is obtained as 69.63 mm

Using Eq. 2.21, Q is obtained as

$$Q = \frac{(100 - 0.2 \times 69.63)^2}{100 + 0.8 \times 69.63} = 47.6\ mm.$$

Question Banks

Q1. What is hydrologic cycle? With a labeled diagram, describe the different parameters of the hydrologic cycle.

Q2. What is engineering hydrology? What forms the basic part of study of engineering hydrology.

Q3. How is the hydrologic cycle helpful for the study of hydrology?

Q4. Define the followings:

(i) Precipitation (ii) Evaporation (iii) Transpiration (iv) Runoff (v) Infiltration (vi) Percolation

Q5. What are the differences between precipitation and rainfall?

Q6. What are the different forms of precipitation? Which form of precipitation is important and why?

Q7. What is the difference between precipitation and rainfall?

Q8. What is interflow? How does it help as a source to irrigation water?

Q9. Write a mathematical equation for hydrologic cycle and explain the various terms of it.

Q10. Which component of the hydrologic cycle is very important in study of water resources and hydrology? How is it helpful in the fields of hydrolygy and water resources engineering?

Q11. Describe the mechanism for the occurrence of precipitation.

Q12. What are the different types of precipitation?

Q13. What is average annual rainfall? Describe the variation of average annual rainfall across the country.

Q14. Rainfall in India is seasonal. Justify the statement.

Q15. Write short notes on (i) Recording type rain gauges and (ii) Non-recording type rain gauges.

Q16. What do you mean by mass curve of rainfall? How is it helpful in analysis of rainfall records?

Q17. What are the different types of recording rain gauges? Write short notes on each of them with labeled diagram.

Q18. What are the points to be considered while selecting sites for setting rain gauges?

Q19. Describe the different methods to compute the average rainfall of a place.

Q20. The annual rainfall of a gauging site is measured for 9 years and the data are presented in the tabular from as given below. The rainfall data for one year is missing. Find the missing data if the average annual rainfall of the place is 1250 mm.

Year	1999	2000	2001	2002	2003	2004	2005	2006	2007
Annual rainfall, mm	1200	1340	1155	1205	1430	1250	—	1320	1190

Q21. Data of average depth of rainfall occurring in a basin of area 250 ha having 10 rain gauges.are measured and given in table as below. The polygon area enclosing each rain gauge is also given in the table. However, rainfall, data of one gauge is missing. Find the missing data if the average depth of rainfall estimated by Thiessen method is 26 cm.

Station	A	B	C	D	E	F	G	H	I	J
Rainfall, cm	50	45	40	23	30	—	12	15	30	20
Polygon area, ha	13	15	18	15	26	25	20	23	40	55

Q22. Estimate the average rainfall of a site by isohyetal method with the help of the following recorded data.

Rainfall between isohyets, cm	120- 130	130-140	140-150	150-160	160-170	170 -180	180-190	190-200
Area enclosed between the isohyets, ha	22	24	30	23	25	33	37	18

Q23. How does the snow contribute as a source to irrigation water?

Q24. What do you mean by runoff? What are the factors on which the runoff of catchments depends on?

Q25. How is the nature of precipitation affect the runoff of a catchment?

Q26. Explain the different physiographic factors that influence the runoff of a catchment?

Q27. Which one of the following will produce more runoff and explain why?

(i) Fan shaped catchment and (ii) Fern shaped catchment.

Q28. What do you mean by rainfall-runoff modeling? Explain the watershed routing method to estimate the runoff from a given rainfall mount.

Q29. Briefly explain the different methods used to compute the runoff produced from a catchment.

Q30. Write notes on the following empirical formulae to estimate the runoff from a catchment.

(i)Lacey's formula (ii) Khosla's formula (iii) Inglis and Desouza's formula.

Q31. Describe the rational method to estimate the runoff rate of a catchment.

Q32. What are the limitations to use the rational method for computation of runoff rate of a catchment?

Q33. A watershed has an area of 100 ha under pasture, 120 ha under cultivation and 50 ha under the forest. The cultivated area has loamy soil whereas the pasture and forest lands have heavy clay soil with very low infiltration rate. The average longitudinal slope of the watershed with the longest reach contributing runoff is 0.1 percent. The one hour rainfall intensity for 10 years frequency interval of the watershed is 60 mm/hr. Estimate the runoff rate of the watershed. Assume the ratio of one hour rainfall intensity to rainfall intensity of any other duration as 1.3.

Q34. Explain how the runoff rate of a catchment is estimated by the curve number method.

Q35. Write about the four hydrologic soil groups. What are their characteristic features?

Q36. Which one of the following will produce more runoff rate and explain why?

(i) Hydrologic soil group A and (ii) Hydrologic soil group B.

Q36. A watershed has an area of 100 ha out of which 60 ha is under the row crop with good terrace and the rest is under good pasture land. Soil of the watershed belongs to the hydrologic soil group B. Values of field capacity and wilting point soil moisture contents are 22 and 10%, respectively and soil moisture content of the soil on a particular day is 18%. Find the runoff amount on that day when a rainfall amount of 120 mm occurs in the watershed.

Q37. What do you mean by the term AMC? What are the factors on which the AMC depends on?

Q38. What do you mean by the term CN? Write in short the factors on which the CN depends on?

Q39. What is lag time and how is it determined?

References

Beven, K.J. 2001. Rainfall-Runoff Modelling: *The Primer*. Wiley.

Eagleson, , P.S. 1970. Dynamic Hydrology. McGraw-Hill Book Company, New York.

Garg, S.K. 1987. Irrigation Engineering and Hydraulic Structures. Khanna Publishers, Delhi.

Gupta, S.K., Ram Babu and Tejwani, K.G. 1970. Nomograph and important parameters for estimation of peak rate of runoff from small watershed. Indian Journal of Agricultural Engineering, ISAE, New Delhi, Vol. 8(3): 25-32.

Hann, C.T. , Johnson, H.P. and Brakensick, D.L. 9 Editors) 1982. Hydrologic Modelling of Small Watersheds. American Society of Agricultuural Engineers, St. Joseph.

Kirpich, Z.P. 1940. Time of concentration of small agricultural watersheds. Civil Engg., 10:362.

Murty, V.V.N. 1998. Land and Water Management Engineering. 2nd Edition, Kalyani Publishers, New Delhi, pp. 586.

Sharma, R.K. and Sharma, T.K. 2002. Irrigation Engineering including Hydrology, S. Chand and Company Ltd. New Delhi.

Sharma, R.K., 1987. A Text book of Hydrology and Water Resources. Dhanpat Rai and Sons, Delhi.

Sharpley, A.N. and Williams, J.R. (Edn.) 1990. EPIC-Erosion/Productivity Impact Calculator: (1) Model Documentation, pp.235.

Shaw, E.M. 1994. Hydrology in Practice. 3rd ed., Chapman and Hall.

Singh, J. 1964. Irrational use of the Rational formula. Er presented at the third annual meeting of the Indian Society of Agricultural Engineers, Allahabad.

Sriboonlue, V., Khetkratok, N and Polkern, K. 2003. Inflow calculation for on-farm ponds in northeast Thailand, Comprehensive Assessment of Water Management in Agriculture Series, Water Productivity in Agriculture: Limits and Opportunities for Improvement. Book Edited by J.W.Kijne, R. Barker and D. Molden. CABI Publishing, pp. 352.

Tejwani, K.G., Gupta, S. K. and Mathur, H.N. 1975. Soil and Water Conservation Research 1956-1971. Indian Council of Agricultural Research, New Delhi.

US Soil Conservation Service.1972. National Engineering Handbook, Section 4, Hydrology, Department of Agriculture, Washington, D.C.

van den Akker, C. and Boomgaard, M.E. 1996. Hydrologie. Technische Universiteit Delft.

Viessman, W. Jr., Knapp, J.W., Lewis, G.L. and Harbaugh, T.E. 1977. Introduction to Hydrology. Thomas Y. Crowell Harper and Row Publishers, New York.

3

Storage of Irrigation Water

3.1 Introduction

Water used for irrigation can be stored mainly in three sources i.e. soil reservoir, surface reservoir or groundwater reservoir. Accordingly the storage system can be defined as soil reservoir storage system, surface reservoir storage system or groundwater storage system. Each of these storage systems has its own merit and demerit. Some of the storage systems though economical but are not helpful to store the irrigation water for longer periods. On the other hand, some other storage systems are very costly, but the irrigation water can be stored there for a longer period. However, before adopting a particular storage system, economics of the system need to be calculated. The cost of installation of the system and the water stored in the storage system and its use as supplemental irrigation to the crops to increase the production in terms of monetary benefit need to be assessed and then the economics of the system should be computed. The storage system that gives maximum benefit-cost ratio and net present value should be adopted for storage of irrigation water. The different types of storage systems are discussed below.

3.2 Soil Reservoir

Soil itself acts as a natural reservoir to store some amount of water. When precipitation occurs, the initial process of hydrologic cycle becomes infiltration and then runoff occurs. The infiltrated water is distributed in the soil and sub soil mantle by many processes. The distribution of soil moisture in the soil profile depends on soil texture, soil structure, porosity and void ratio of the soil, packing factor of the soil, initial soil moisture content organic matter content of the soil etc. Storage in soil reservoir is cheaper than any other storage system. But it has the demerit that water stored in the soil reservoir cannot support irrigation to the plants for a longer period. In other words, the soil moisture support system to the plants depends on the soil moisture retention capacity of the soil which in turn depends on the soil texture. For example, the coarser soils like sandy soil has less water holding or water retention capacity than the fine textured soils like clay soil. Hence, plants grown in the sandy soil will need frequent irrigation than the plants grown in clay soil. In the absence of any supplemental irrigation, plants grown in sandy soil and extracting water from the stored water in this soil will suffer more compared to the plants grown in clay soil. The water holding capacity of the soil reservoir can be increased by soil manipulation, increasing the organic matter content of the soil, increasing the earthworm population in the soil, cultivation of green fodders including dhanicha etc. Details of soil water, water holding capacity of the soil, Soil-moisture –irrigation relationship, soil moisture measurement, dynamics of soil water etc. are discussed in Chapter 6.

3.3 Surface Reservoir

Surface reservoirs are built to store irrigation water for use as rainfall supplement in rainy season or as protective source of irrigation in other seasons including summer. The surface reservoirs can store in-situ rainwater as well as the runoff generated from a catchment. The runoff may be surface or sub-surface runoff. The runoff amount generated from a catchment can be computed by methods like runoff coefficient, runoff percentage, Strange's tables and curves, empirical formula or infiltration method as discussed in chapter 2. The peak runoff rate can be computed by rational method, Crook's method or by SCS curve number method which have been discussed in chapter 2. Surface reservoirs are constructed over the soil surface and the water stored in the reservoirs can be safely managed and used. The surface reservoirs are provided with spillway large enough to convey the maximum anticipated flood flows. The larger the structure and the reservoir, the greater the danger to life and property and safer must be the spillway.

The capacity of each reservoir must be fixed largely by the natural conditions of the site in which the reservoir is to be constructed. The design of the structure must take into account the water availability from different sources, water demand of the crops along with the irrigation and water management practice of the crops. The economics of the irrigation system should also be taken into account while deciding the optimum design of the structure (Replogle *et al.*, 1980). The capacity of the reservoirs may range from a few hundred hectare meters to a large one with values as high as millions of hectare meters. Likewise, dams constructed for storage of irrigation water vary from a few metres in height to over 200 metres built across large catchment areas with massive masonry structures. Generally small earthen dams, farm ponds, and other small water harvesting structures have low capacity and the height of these structures vary from 2 to 5 m. These small structures generally does not require high techniques in designs and the cost involved in construction of these structures are low compared to large masonry dams. The design life span of the small structures is in the range from 5 to 10 years, where as for large mason nary structures, it is more than 100 years. The different surface reservoirs used to store irrigation water are discussed as follows:

3.3.1 Dams

Dams are valuable for impounding water from springs and streams so that it can be more effectively utilized. Reservoirs built for irrigation frequently have multiple advantages. They are very often used for pisciculture, agriculture, along with domestic uses in various forms. The dams are constructed to create artificial lakes or reservoirs behind themselves. The purpose for which these reservoirs are meant to serve may be divided into two broad categories. They are (i) flood control and (ii) water conservation. Conservation of water may be required for any one or more of the following purposes – irrigation, hydro electric power generation, navigation, water supply to towns, industries, recreation etc.

3.3.1.1 Selection of Dam Site

Storage can be obtained by constructing a barrier across a depression receiving runoff from a large catchment or across a well defined stream. For economic feasibility of a storage project, it is necessary that the length of the barrier or dam should be as small as possible and for a given height, it should store the maximum volume of water. The ideal site for construction of the dam is the site of the river valley which is narrow. Other important requirements of a good dam site are:

(i) For a concrete or masonry dam, solid rock at the surface or within reasonable depth below is essential. In such formation, the dams can be constructed to a large height. However, if the foundation materials are relatively soft, then there is chance of undermining and in such case, the height of the dam is less being limited to about a height of 15 m.

(ii) A major portion of the dam should be on high ground as compared to the river basin. This will facilitate in saving cost and drainage of the dam section.

(iii) A suitable site for spillway is that at which the gorge (a narrow valley with steep high banks) portion is separated from the flank by a hillock with its surface above the dam. If such a site is available, the spillway section can be built independently of the dam. At such a site the flank may be used to provide a separate spillway and the main valley spanned by an earthen or concrete dam.

(iv) The dam should be constructed at such a site where the construction materials are available in the vicinity at large quantities with relatively cheaper rate. This will save the transportation cost and labour wages for handling the construction materials and hence the construction of the dam would be economically feasible.

(v) The rim of the dam should be water tight upto the proposed top elevation of the dam. The reservoir basin should also be reasonable water tight; the stored water should not be able to escape under the surrounding hills through continuous pervious strata.

(vi) The value of the property and land submerged by the construction of the proposed dam should be low compared to the benefits expected from the dam. It should be noted that less population is displaced and the displaced people should be properly rehabilitated.

(vii) The dam site and the reservoir should be eco-friendly without affecting the local flora and fauna. The dam should also be constructed on a earth quake free zone so that there is no danger to dam because of earthquake which otherwise will create environmental hazards besides causing heavy loss to property and human and other animals' life.

(viii) The dam should be constructed at the site which will have minimum submergence of cultivated land so that the land acquisition and compensation charges are to the minimum.

(ix) The dam should be constructed at a site such that the water coming from the catchment upstream of the dam site is relatively silt free. If the cathcment is prone to severe soil and water erosion, then the silt laden water will get deposited in the reservoir which will reduce the storage capacity of the reservoir besides endangering the stability and life span of the dam.

A reservoir created behind a dam is called as storage reservoir, It stores surplus water during the period of excess flow for a considerable time with a view to its subsequent utilization when the demand is keen but the supply is meager for the purpose of irrigation, hydel power generation, water supply etc. Such a reservoir is provided with gate controlled outlets. The reservoir is called as multipurpose reservoir when the storage is used to cater a number of purposes such as irrigation, flood control, navigation, fishing, hydel power generation, public water supply, recreation etc. A multipurpose reservoir has larger capacity. The multipurpose reservoirs are generally financially viable, since they serve a number of purposes simultaneously besides giving indirect benefits such as development of fishing, tourism etc. But they have the disadvantage that it needs huge cost for construction and there are chances of rapid silting up causing capacity reduction. The repair and maintenance of the multipurpose reservoirs are also very high.

3.3.1.2 Investigations and Studies

Thorough and systematic investigations are required before undertaking a project. Investigations include the study of various alternatives regarding the layout of the scheme as a whole and also details of alternatives considered for the types and location of various features of the project. Followings are the different types of surveys/investigations which need to be undertaken for preparation of a project.

3.3.1.3 Topographic Survey

Topographic survey may be preliminary and final. Preliminary survey includes reconnaissance of the area, collection of data to find out merits and demerits of the project at various sites, tentative economic analysis of the project considering costs and benefits of the project and undertaking geological and hydrological studies. Once the project site is selected by preliminary survey, final survey is conducted in respect of the finally selected site to collect more precise data for detailed designs and final economic study. The final survey includes.

(i) Topographical survey plan of the dam site,

(ii) Reservoir submergence plans,

(iii) Command area survey plans,

(iv) Detailed survey maps to cover the area under barrage,

(v) River surveys in a length of about 10 km upstream and downstream of the diversion structure,

(vi) Longitudinal section plan along the canal alignment,

(vii) Survey plan for location of power station,

(viii) Longitudinal and cross section plans covering power channel, penstock and tail race channels,

(ix) Longitudinal and contour plan for construction of tunnels,

(x) Contour plan of the surge tank area,

(xi) Contour plan of the underground power house,

(xii) Survey for cross drainage systems,

(xiii) Plane table survey for selection of suitable sites for residential colony, workshop complex,

(xiv) Computation of land acquisition and compensation for the same,

(xv) Areal survey to estimate the capacity of the reservoir at various levels,

(xvi) Determination of types and height of the dam,

(xvii) Implication of riparian laws to be observed.

3.3.1.4 Geological Investigations

Geological investigations include the study of the following data.

(i) Geological and structural features like folds, faults, fissures etc. their nature and extent,

(ii) Existing and potential slide areas,

(iii) Leaking of water basins,

(iv) Presence of sink holes, solution channels etc.,

(v) Assessment of valuable mineral deposits in the reservoir,

(vi) Groundwater conditions,

(vii) Seismic conditions of the region with reference to geological map of the vicinity.

3.3.1.5 Foundation Investigation

Foundation investigations for earth/rock fill/masonry concrete dams include investigations to determine properties of foundation soils shear zones, depth of over-burden etc. It also includes study of contours of bed rock strata, study of location and thickness of weathered/softened/altered zones and their characteristics and structural weakness etc.

3.3.1.6 Meteorological Studies

Meteorological studies are carried on with respect to the following points:

(i) Watershed plans must be prepared with representations of orographic features, normal annual isohyets, location of raingauge stations in and around the catchment, gauge and discharge sites and inter-state boundaries.

(ii) Assessment of weighted rainfall in the catchment,

(iii) Collection and evaluation of data and frequencies of heavy rainfall in the catchment

(iv) Determination of maximum probable storm rainfall

(v) Determination of mean monthly data of temperature including maximum and minimum monthly temperature, relative humidity, wind velocity, sunshine hours, pan evaporation etc.

3.3.1.7 Hydrological Investigations

Hydrological investigations include

(i) Gauge and discharge observation data of the project site,

(ii) Hourly gauge and discharge data during flood occurrence and survey maps of flood marks of past floods,

(iii) Determination of observed river hydrographs to fix the spillway capacity of the reservoir

(iv) Estimation of design flood,

(v) Study of rainfall-runoff relationships,

(vi) Estimation of monthly/seasonal/annual yields of runoff,

(vii) Estimated yields at reliability levels of 75, 90 and 100 per cent for irrigation, power and water supply, respectively.

3.3.1.8 Sediment Studies

(i) Sediment observations including suspended load, bed load, and catchment characteristics with reference to soil erosion should be carried out.

(i) Prioritization of the sub watersheds for the erosion control measures to be taken and taking up the control measures on the sub watersheds as per the prioritization.

(ii) Quantitative yield of each of the principal sediment sources.

In addition to the above mentioned survey/investigations, other studies required for taking up a project including construction of dam and creation of reservoir are construction material investigation study, environmental survey and survey for provision of communication facility for the project site.

3.3.2 Area and Capacity of Reservoir

Area of water spread and reservoir area is computed in million square metres or hectares by measuring the area at definite contours of the contour map by a planimeter. A curve showing the water spread at various elevations is prepared for reference and presented in Fig..3.1.

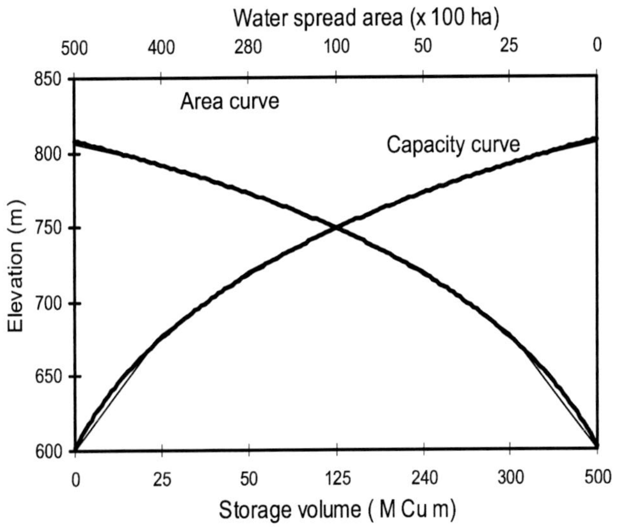

Fig. 3.1: Typical area-capacity curve

Capacity or volume of storage of the reservoir (V) at equal contour interval (H) is obtained from a contour plan. The contour interval required depends on the height of the dam and the side slopes in the basin. For the map of the reservoir area a contour interval of 2.5 m for dams of moderate height would normally be sufficient while for high dams and steep slopes, an interval of 5.0 to 7.5 m would be appropriate. For the work area in the vicinity of the proposed dam site, however, a very accurate triangulation survey with a contour interval of about 1.5 m and a scale of 1:1000, or, preferably, 1:500 is required. Volume, V is obtained by trapezoidal or prismdial formula is as follows (Sharma, 1987).

Trapezoidal Formula:

Volume between two successive contours is given as:

$$V = \frac{H}{3}\left(A_1 + A_2 + \sqrt{A_1 A_2}\right) \tag{3.1}$$

For even number of horizontal sections

$$V = \frac{H}{3}\left(0.5A_1 + A_2 + A_3 + ... + A_n\right) \tag{3.2}$$

Simpson's Formula:

Volume between any odd numbers of contours is given as:

$$V = \frac{H}{6}\begin{pmatrix} 2 \times area\ of\ each\ odd\ contour + 4 \times area\ of\ each\ even\ contour + \\ area\ of\ first\ and\ last\ contour \end{pmatrix} \tag{3.3}$$

Prismodial Formula:

$$V = \frac{H}{6}\left(A_1 + A_2 + 4A_m\right) \tag{3.4}$$

where, A_l and A_2 = areas at contour intervals and A_m = area within a contour line midway between the two adjacent contours and H is the contour interval.

The volumes upto various elevations having been determined as above, are plotted in the form of a capacity curve showing the capacities of the reservoir below different elevations (Fig. 3.1). The volume is generally measured in million cubic metres or hectare metres or millions of hectare metres for large reservoirs.

3.3.3 Storage Zones of Reservoir

The various storage zones of a reservoir are represented in Fig. 3.2. In general there are five types of storages. They are:

(i) Flood storage

(ii) Live storage

(iii) Dead storage

(iv) Valley storage

(v) Bank storage

Some of these storage zones are associated with certain reservoir levels. There are three types of reservoir levels i.e. normal reservoir level, maximum reservoir level and minimum reservoir level.

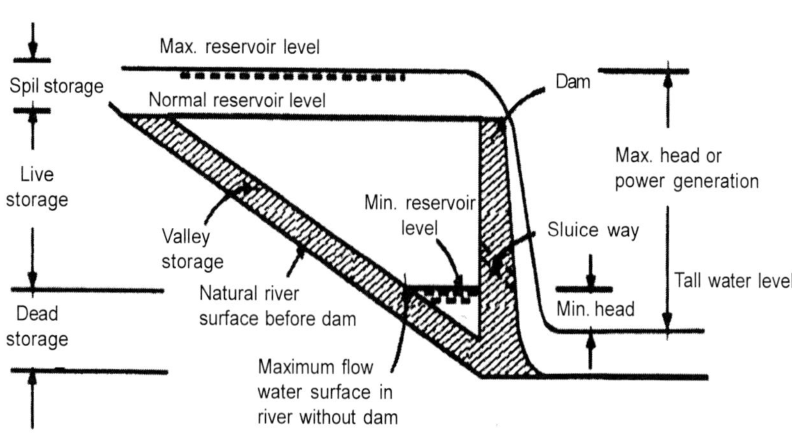

Fig. 3.2: Storage zones of a reservoir

Normal Reservoir Level

Normal reservoir level is also called as full reservoir level. It is the maximum level to which the water in the reservoir shall rise and is maintained without spillway discharging or without passing downstream through sluice except that under ordinary operating conditions to meet the downstream water demand.

Maximum Reservoir Level

It is the level to which the water in the reservoir shall rise during the design flood. In the dam, a free board of certain height is provided so that it will store the excess water entering the reservoir during flood and is helpful to maximize the storage as well as save the dam from flood damage.

Minimum Reservoir Level

Minimum reservoir level is also called as dead storage level. Reservoir storage capacity gradually decreases, when the water stored in the reservoir is used for irrigation. But some minimum storage level in the reservoir has to be maintained to trap the sediment deposits of the reservoir to be used for power generation as well as for satisfactory functioning of intake structures.

Flood Storage

The storage contained between the full reservoir level and maximum reservoir level is called as flood storage. It is the storage space provided in a reservoir for storing flood water temporarily to moderate the flow downstream. Flood storage varies with the spillway capacity for a given design flood. Effective flood storage is the difference between the amount of water stored in the reservoir during the period of flood peak reduction and the amount of natural valley storage during the same period.

Live storage

Storage capacity above the dead storage level is termed as live storage. This storage is important to meet the demand of irrigation, navigation, power generation or water supply. Live storage in an irrigation project should be sufficient so that the project is successful for 75 per cent of its life period, whereas in power generation and water supply, it is successful for 90 and 100 per cent of the life period, respectively. Live storage depends on the amount of inflow entering into the reservoir, the uses of water from the reservoir for various purposes as well as seepage and evaporation loss of the reservoir.

Dead Storage

Dead storage in a reservoir is used to trap the sediment present in the flowing water that enters into a reservoir. About 10 per cent of the gross storage capacity of a reservoir is provided for dead storage. Dead storage is equivalent to the volume of sediment expected to be deposited in the reservoir during the designed life of the reservoir which is usually taken as 100 years. A reservoir is not allowed to deplete below the dead storage level.

Valley storage

The storage in the river/stream after the flood water has over flown from it is called as valley storage.

Bank Storage

It is the storage absorbed and stored in the bed and banks of a stream in the voids in the soil surface in a reservoir and becomes available in whole or in part as seepage water when the water level drops down in the reservoir. It increases the reservoir capacity over and above that given by the elevation storage curve.

3.3.4 Design Capacity of Reservoir

The design capacity of a reservoir is fixed in considerations of the following factors:

(i) Long term precipitation records for the catchment,

(ii) Long-term runoff data at or near the site,

(iii) Sediment yield into the reservoir from the catchment,

(iv) Area and capacity curve,

(v) Sediment trap efficiency,

(vi) Seepage and evaporation losses of the reservoir,

(vii) Maximum requirement of water for various uses to be met from the reservoir and direct flows of the river irrespective of the availability of assured supplies,

(viii) Uses in the upper side of the catchment,

(ix) Density current aspects and location of outlets,

The storage capacity of a reservoir is determined by:

(i) Analytical method,

(ii) Mass curve method and

(iii) Bar graph method. Out of these, since, mass curve method is usually employed in determination of reservoir storage, this method is described in this book.

3.3.4.1 Mass Curve Method

It is the best method to compute the capacity of a reservoir. Mass curve is a curve of accumulated total flow against time. The curve rises continuously with time since the accumulated inflow goes on increasing with time. The rate of rise is steep if the rate of inflow is high. If there is no inflow at any time, then the curve is flat with no slope The relative dry periods are indicated as hollows on the curve. Any point on it represents the total inflow in million cubic metres from beginning of the period till that time. Slope of a tangent to the mass curve at this point represents the discharge in cumecs. Mass curve of demand representating accumulated outflows is also plotted in the same figure. The demand curve isa straight line, rising from the origin, if there is a constant rate of demand. In case of variable demand rate, the demand line is a curve. The mass curve thus gives the relationship between the accumulated inflow or outflow and the water available for storage at any given time from beginning of the year. Demand lines drawn tangent to the high points of inflow mass curve represents rates of withdrawal from the reservoir. The vertical distance between the successive tangents indicates the water spilled through spillway. When the demand curve coincides with the inflow curve, the reservoir is full. The maximum vertical ordinates between the two curves represent the reservoir storage capacity to meet the water demand during lean periods.

Procedure to determine the storage capacity

The step by step procedures employed are:

(i) Data of annual rainfall is collected for at least 35 years. From the annual rainfall values, inflows and outflows (demand) are calculated for at least three consecutive years of minimum inflow. The monthly rainfall observations during these periods are also noted.

(ii) For each month of the above period, runoff values of the catchment area are determined from the stream gauging records.

(iii) A mass inflow curve of cumulative runoff corresponding to each month (month taken in abscissa and cumulative runoff as ordinate) is developed.

(iv) Demand per year is obtained by adding the quantity of water actually consumed in each month of that year. This demand per year when multiplied by the number of years of data gives the cumulative demand for the period.

(v) The demand curve is also plotted as ordinate so that the starting point of this demand curve and that of the inflow curve are the same.

A mass curve (Fig. 3.3) is developed by using the data of sample calculation 1 (mentioned below) and employing the steps mentioned as above. Referring to this figure, the end P of the demand curve AP meets the inflow curve ABCD indicating that the total inflow for the period is equal to the demand. Point B is the lowest point and D is the highest point on the mass inflow curve. From O to B, the slope of inflow curve is less than that of the demand curve. This indicates that the inflow is less than the demand and the reservoir is emptying. From B to C, the slope of inflow curve is steep being more than that of the demand curve indicating that the inflow is more than the outflow and thus the reservoir is filling. Vertical intercepts BQ represents the initial storage to meet the demand from O to B. The point C in the mass curve represents the reservoir level same as it is in the beginning corresponding to point O. The slope of the mass inflow curve after the point C to D is more than that of the demand curve which indicates that the reservoir level is still in rising stage. Point D indicates the full reservoir level. The vertical intercept DR represents the storage between the initial water level corresponding to point O and full reservoir level. The vertical intercepts BQ and DR gives the minimum storage capacity of the reservoir. We can also draw lines DM and BN parallel to demand curve AP and tangential to the inflow curves at the lowest and highest points B and D, respectively. The vertical distance MB between the two tangential lines represents the required storage capacity (Fig. 3.3).

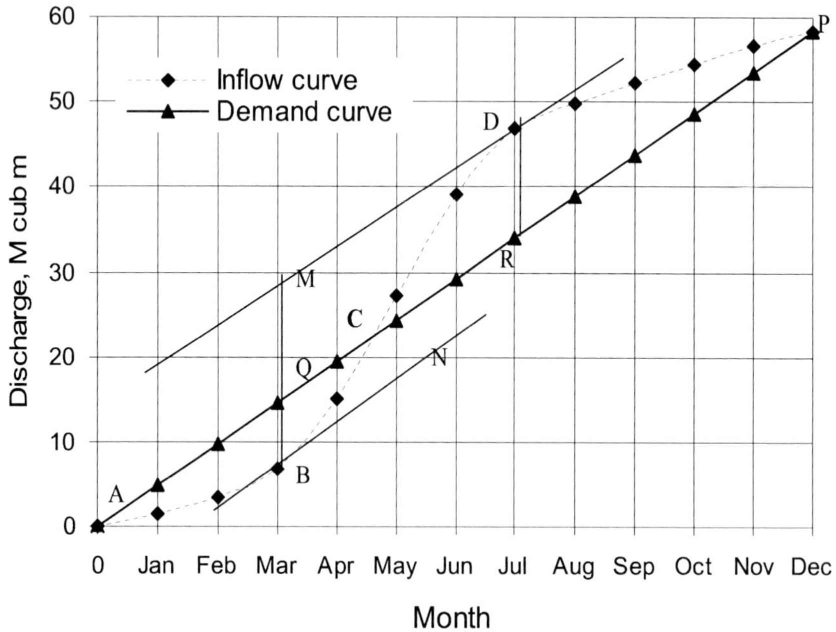

Fig. 3.3: Typical mass curve

Sample Calculation 1

The data of the table given below represents the water yield data of a catchment in different months. Determine the minimum storage capacity of the reservoir required to allow the above volume of water to be drawn off at a uniform rate assuming that there is no loss of water over the spillway.

Month	Jan	Feb	Mar	Apr	May	Jun	Jul	Aug	Sep	Oct	Nov	Dec
Inflow, M m^3	1.5	2.0	3.2	8.4	12.0	12.0	7.7	3.0	2.5	2.2	2.0	1.7

Solution

Total inflow = 1.5 + 2.0 + 3.2 + 8.4 + 12.0 + 12.0 + 7.7 + 3.0 + 2.5 + 2.2 + 2.0 + 1.7 = 58.2 M m^3.
Monthly demand = 58.2 /12 = 4.85 M m^3.

The inflow and outflow data are presented in tabular form as given below. The cumulative deficit and cumulative surplus are computed and both are observed to be same with a value of 20. 70 M m^3. The mass curve of the data is presented in Fig. 3.3 as given above. The vertical distance BQ which represents the cumulative deficit from January to March is calculated as 7.85 M m^3. Similarly, the vertical distance DR which represents the cumulative surplus from January to July is determined as 12.85 M m^3. Hence the storage capacity of the reservoir which is represented by vertical distance MB of mass curve (MB = BQ + DR) is obtained as 20.70 M m^3.

Table showing inflow versus demand of the data of sample calculation 1 is placed below:

Month	Inflow, M m^3	Outflow/demand, M m^3	Deficit, M m^3	Surplus, M m^3	Cumulative deficit, M m^3	Cumulative surplus, M m^3
Jan	1.5	4.85	3.35			
Feb	2.0	4.85	2.85			
Mar	3.2	4.85	1.65			
Apr	8.4	4.85		3.55		
May	12.0	4.85		7.15		
Jun	12.0	4.85		7.15		
Jul	7.7	4.85		2.85		
Aug	3.0	4.85	1.85			
Sep	2.5	4.85	2.35			
Oct	2.2	4.85	2.65			
Nov	2.0	4.85	2.85			
Dec	1.7	4.85	3.15		20.70	20.70

3.3.5 Life of a Reservoir

Water that enters into a reservoir is not silt free. The deposit of silts in the reservoir decreases its storage capacity as well as life span. The earlier belief was that the silt carried by water at first decreases the dead storage of the reservoir and then it encroaches the live storage. But in actual practice, because of back flow of water from the reservoir, some silts gets deposited in the live storage also and thus simultaneously decreasing the live storage capacity. Decrease of dead as well as live storage of the reservoir, affects in the irrigation, water supply and hydel power generation. As per Central Water Commission of India, a minimum draw down level, a little above the zero elevation is fixed which is likely to develop after 100 years of operation. Generally the life of a reservoir in India is taken as 100 years and to provide for silt storage space (dead storage) equal to the estimated total volume of silt deposition during its entire life. The reservoir is said to loose its useful life when its storage capacity gets reduced to 20% of its live storage capacity.

3.3.6 Reservoir Losses

The two major forms of losses in reservoir are (i) Evaporation losses and (ii) Seepage losses.

3.3.6.1 Evaporation Losses

Evaporation is the process of conversion of water from liquid state to gaseous state and its transfer into atmosphere in this form. It is a natural process and occurs when solar radiation falls on a water body and energy so transmitted is acquired by water molecules which in turn gets separated and move upward causing evaporation. In the hydrologic cycle, evaporation plays a vital role in causing precipitation. Evaporation losses depends on the water spread area of the reservoir and the rate of evaporation which depends on the atmospheric demand like temperature, wind velocity, relative humidity of air, atmospheric pressure as well as nature and size of the water body. Water spread area of a particular size and shape of storage structure with a given side slope varies with depth of water in the structure and hence the evaporation loss of the storage structure varies daily. The various methods of determining evaporation losses are:

- Analytical Method
- Water Budget Method
- Mass Transfer Method
- Empirical Method
- Energy Budget Method and
- Pan Evaporation Method

Pan Evaporation Method: In this method, the evaporation loss measured by an evaporimeter located near the reservoir is measured. The ratio of the evaporation loss measured by the evaporimeter to the reservoir evaporation loss on annual basis is a constant and hardly varies from region to region. However, on short term period basis, the ratio depends on a number of factors like heat storage of the evaporimeter/pan, size of the pan, water depth in the pan, colour of pan etc. The evaporation loss from a pan is greater than that of the reservoir. As the size of the pan increases, the difference of the evaporation loss of both pan and reservoir decreases. The rate of evaporation, in general, is observed to be more for deep buried pan with higher depth of water than shallow pans with less depth of water. Similarly, the evaporation from dark pans is greater than that from unpainted, galvanized pans. The white coloured pan shows the lowest evaporation. Covering the pan with a screen helps to reduce the pan evaporation to the equivalent of that from a reservoir. The screen helps to maintain the pan temperature uniform throughout day and night and reduces turbulence at the water surface so as to maintain uniform rate of evaporation (Lenka, 2001).

3.3.6.2 Reduction of Evaporation Losses

Storage of irrigation water in a reservoir is a very valuable commodity. Appreciable evaporation loss from large water spread area of a reservoir has a consequential loss on power generation, irrigation supply and many other associated activities. Some of the methods used for reducing evaporation losses are:

Reduction of Reservoir Surface Area

Evaporation is a surface phenomenon and hence if the surface area of the reservoir is reduced, then their evaporation loss is considerably reduced. A reservoir with greater depth of stored water having smaller surface area has less evaporation loss compared to a shallow reservoir with large surface area; both the reservoirs having same volume of stored water.

Covering the reservoir with shading materials

Shading materials like plastic films, thatches, paddy straw, sugarcane trash etc. reduces the radiation affects on water surface and decreases the evaporation loss. Biological shading like creepers over the small reservoir water surface can help to reduce the evaporation loss of the reservoir. Some other shading materials are dye mixed with pond water, coloured pan, plastic mesh and sheet, polystyrene beads and sheet, white spheres, white butyl sheet etc. Out of these measures, polystyrene raft, plastic sheet and foamed butyl rubber are the best measures since they are reported to reduce the evaporation loss by more than 90%

Monomolecular Film

Alcohol such as cetyl alcohol, also called as hexadecanol and stearyl alcohol (octo decanon) are known to form a monomolecular film on contact with water which is sufficiently enduring during field conditions. The invisible film is non-toxic and reduces evaporation by about 50-60% and at the same time helps as a free passage for rain, oxygen and sun light. (Cooley, 1974). However, the monomolecular films have the disadvantages that they get diluted in water quickly and then become ineffective to reduce the evaporation of water.

Wind Breaks

It is a well known fact that the turbulence affect of the water surface caused due to wind action increases the evaporation loss. Hence, wind breaks like growing tall trees around reservoir can minimize the evaporation loss to some extent. But, these wind breaks are useful only for small reservoirs. It is found out that a reduction of wind velocity by 25% reduces the reservoir evaporation loss by only 5% and thus is not significant on sustained basis.

3.3.6.3 Seepage Losses in Reservoir

Water stored in a storage structures like reservoir is a costly commodity and care should be taken to store it for the maximum period as possible with minimum loss. Most of the storage structures used for irrigation are unlined which cause huge losses due to seepage. This loss is significant when the reservoir is underlain by porous strata or when the bed material of the reservoir consists of coarse textured soil. Small farm reservoirs constructed in coarse textured soils; especially in arid and semi-arid regions are found to get dry completely just after the rainy season is over. However, reservoirs constructed in heavy soils are found to have less seepage losses. In general the seepage loss in unlined small farm reservoirs range from 40-60% depending on the water table position below the ground surface, types of soils in which the reservoir is constructed and hydraulic gradient available between the reservoir water level and water level of adjacent areas. It is observed that the seepage loss in newly constructed reservoir is low and as time progresses, silttaion of the reservoir reduces the seepage loss. Seepage loss in a reservoir depends on the wetted surface area of the reservoir which in turn depends on the depth of water in the reservoir.

3.3.6.4 Reduction of Seepage Losses

Seepage loss depends on the wetted surface area of the reservoir. By reducing the wetted surface area and depth of water in the reservoir, seepage losses can be reduced to a great extent. Cluff (1977) has suggested compartmentalized pond in which the pond is divided into two or more sections and as one section is emptied, the water is pumped out from another section. By doing so, the wetted area and top water surface area of the reservoir can be considerably reduced and so seepage and evaporation loss of water in the reservoir can be reduced to a great extent. Since, irrigation water is very costly, efforts should be made to reduce the seepage loss of the reservoir. Lining of the reservoirs though costly can reduce the seepage loss and improve the effective storage of reservoir. Different lining materials used to reduce the seepage losses are plastic film/polythene sheet, soil cement lining, bitumen lining, clay lining, cow dung lining, brick cement lining etc. Out of the various linings, plastic films are found to be very effective in reducing the seepage loss to the maximum extent. Lining for reduction of seepage losses is feasible only for small reservoirs. Effectiveness of different lining materials for seepage control in small farm reservoirs are presented in Table 3.1 (Murty, 1998).

Table 3.1: Effectiveness of different lining materials for seepage control

Sl. No.	Lining materials	Seepage loss, lit/hr/m^2
1.	Control (No lining)	18.56
2.	Cowdung + Paddy husk + Soil plaster (1:1:10)	16.98
3.	Cement plaster at bottom (1:6)	12.99
4.	Cement + Soil plaster (2:10)	0.85
5.	Polythene sheet	0.32
6.	Paddy husk ash plaster	11.60
7.	Coastal saline soil plaster	5.47
8.	Fly ash + Sand plaster (1:1)	2.5
9.	Clay	12.07

Before using any lining materials for seepage control in reservoir, the economic analysis relating to life of reservoir, reservoir lining materials etc. must be taken into account. At the same time, the amount of saving of irrigation water by reduction of seepage and its use in increasing the crop productivity, total crop production and other associated benefits must also be considered. Finally with present worth analysis the different economic indicators like net profit, benefit-cost ration, internal; rate of returns and pay back periods are to be worked out. The lining material that gives more net profit, benefit-cost ratio, and internal rate of returns and less pay back period is the best from seepage control point of view of the reservoir.

3.3.7 Reservoir Sedimentation

The runoff that moves over the soil surface carries some silt and when this silt laden water enters to a reservoir, it gets deposited in it. Fig. 3.4 shows a silted reservoir which has been badly affected by the silt laden water from the catchment. The fine sediment in the river water is carried in suspension, as suspended load, while the larger solids move on or near the bed as bed load. When the velocity of water decreases, the bed load gets deposited in the reservoir. The suspended sediments take more time to settle down the reservoir. The bed load and a part of the suspended load is deposited in the reservoir bed if the reservoir discharges the surplus freely for long time which is the case when runoff of the river is large compared to the reservoir capacity. On the other hand, if the runoff is less in comparison to the capacity of the reservoir, the sediment laden water of the runoff is deposited in the reservoir bed.

Fig. 3.4: Schematic view of a silted reservoir

3.3.7.1 Sediment Load

Sediment load consists of bed load and suspended load. Bed loads are generally bigger in size compared to the suspended loads. Because of gravity, they first get deposited in the bed of the reservoir and decrease the dead storage capacity of the reservoir. Immediately above the bed, a thin layer consisting of relatively coarser materials with high concentration exists. Above it and extending up to the water surface, a thick layer of relatively fine grained particles exist as suspended load. In case of bed load, movement of particles takes place by rolling, sliding and saltation depending on the velocity of flow. The bed load varies from 10 to 20 per cent of the total suspended load depending on the nature of bed material, its grade, and concentration of suspended materials. The suspended load constitutes the larger share of total sediment load. It consists of the finer materials and is carried away by the river to the delta or by the irrigation channels to the fields. Periodic sampling of river water is done at various discharges and the suspended sediment concentration is measured. A sediment rating curve relating the sediment concentration against various discharges is developed and is used to estimate the approximate suspended sediment load. The sediment distribution pattern of a reservoir depends on the following factors (Sharma, 1987):

(i) Slope of the river

(ii) Length of the reservoir

(iii) Constriction in the reservoir

(iv) Size of sediment

(v) Capacity-inflow ratio

(vi) Reservoir operation

(vii) River inflow pattern

(viii) Shape of reservoir

(ix) Sediment-reservoir volume ratio

(x) Outlets near the dam and

(xi) Suspended load

Factors Affecting Sedimentation of Reservoirs

Factors affecting the sedimentation of reservoirs are as given below (Sharma, 1987):

(i) Quantity, quality and concentration of sediment carried by river water into the reservoir

(ii) Percentage of sediments intercepted by the river that convey water to the reservoir

(iii) Percentage of sediments carried away by outflow from the reservoir through outlet

(iv) Degree of consolidation i.e. weight of silt per unit volume

(v) Length of reservoir

(vi) Runoff volume and peak discharge from the catchment and sub catchments

(vii) Ratio of reservoir capacity to annual runoff

(viii) Depth and age of sediment deposited

(ix) Shape and size of reservoir

(x) Slope of the river

(xi) Growth of vegetation at the upper end of the reservoir

(xii) Soil texture of the catchment area

(xiii) Precipitation and its characteristics in the catchment

(xiv) Temperature and wind velocity of the catchment area

(xv) Geological formation of the catchment

3.3.7.2 Rate of Sedimentation/Silting of Reservoir

Sedimentation rate varies from rivers to rivers. Even for the same river, it also varies from year to year depending on rainfall, topography, geology, soil types, types of the vegetation cover in the watershed, area of the watershed, relief and slope of the watershed, drainage density, drainage frequency, stream grade etc. Typical measured values of sedimentation of a few Indian rivers are given below in Table 3.2 (Singh, 1997).

Table 3.2: Typical measured values of sedimentation of a few Indian rivers

River/Dam site	Sediment production rate, ha m/100 km²/year
Beas	23.9
Ram Ganga	17.3
Damodar-Barakar	9.9-12.2
Hirakud	6.6
Sutlej	6.2
Narmada	5.3
Chambal	5.3

According to Dr. Khosla, the annual rate of silting of Indian reservoirs is estimated as 0.036 Mm^3 per 100 sq. km of catchment areas. He developed the enveloping curves considering data on 38 points. But since only 5 out of 38 points as considered by him pertains to the data of Indian reservoirs, the above mentioned recommendation has limitation to apply. According to Dr. Joglekar, the annual rate of silting of Indian reservoirs is given by equation:

$$Y = 0.597\ A^{-0.24} \tag{3.5}$$

where, Y = annual silting rate (M m^3 per 100 sq. km), and A = catchment area at dam site (sq. km).

Results of sediment survey of Indian reservoirs have revealed that the rate of sedimentation for major catchments vary between about 4.75 to 14.25 ha m annually per 100 sq. km and the reservoirs are losing a capacity of half a per cent to one per cent annually. IS: 6518-1972 has recommended provision of sedimentation at the rate of 0.1 to 0.2 ha m /year/sq. km of drainage area for the entire economic life of the reservoir (Sharma and Sharma, 2002). The economic importance of silting of a reservoir can be evident from the following hypothetical data.

Gross storage capacity of a reservoir = 500 M m^3

Dead storage capacity of the reservoir = 120 M m^3.

Average volume of the sediment inflow to the reservoir = 10 ha m/100 km^2/year

The area of the watershed conveying runoff to the reservoir = 6000 km^2.

From the above data it is obvious that the dead storage capacity of the reservoir will be silted up in 20 years and the entire storage capacity of the reservoir will be silted up in about 83 years.

3.3.7.3 Methods to Reduce Reservoir Sedimentation

Sedimentation decreases the dead as well as live storage of a reservoir and hence reduces the life span. Proper remedial measures must be taken up to prolong the life span of the reservoir. The method of reservoir sedimentation can be broadly classified into two categories i.e. preventive and curative.

3.3.7.4 Preventive Measures

Reservoir Site Selection

Reservoirs gets silted up due to the sediments coming to it with the runoff from the catchment area. The site where sediment inflow with runoff is high should be abandoned and preferably the catchment producing low sediment should be preferred for construction of reservoir. Further, the area having susceptibility to earthquake, especially in seismic zones, should be avoided since earthquake brings in more sediment to reservoir.

Reservoir off Main River Channel

Possibility of location of reservoir off the main river channel and creating an auxiliary reservoir on the main channel and transferring silt clear water to the main reservoir through a diversion channel may be fully explored.

Design of Reservoir

In designing a reservoir, ratio of storage capacity to catchment inflow, design and location of outlet works are crucial for estimating the rate of sedimentation. It is recommended that, if economics permit, the life of reservoir should be enhanced by increasing the storage capacity-catchment inflow ratio to such an extent as to have the annual loss of 0.25 per cent only. The ratio between the capacity of the reservoir and size of the drainage area is one of the most important factors governing the annual rate of sedimentation in the reservoir. Proper reservoir operation can also reduce the sedimentation of reservoirs. Some amount of sediment flows out with the discharge from the reservoir through the outlet including sluices. Concentration of sediment is more at some levels than at the others and if more water can be withdrawn from the former, silt deposition will be reduced considerably. Hence, adequate outlets must be provided in the reservoir and sluices must be placed at suitable elevations. The sediment content of a reservoir is relatively high after floods and if more water can be escaped at such times, the percentage of total sediments trapped in the reservoir will be reduced significantly.

Gradual Increase of Height of Dam

In the first instance, the height of the dam should not be erected very high. A high dam will create huge final storage and will eventually hold almost entire amount of silt laden runoff. This will ultimately decrease the dead storage capacity of the reservoir. Instead, raising dam height subsequently may be considered so as to adjust the capacity runoff ratio after some capacity has been lost by sedimentation. Raising dam height in stages may also reduce the evaporation loss of the reservoir (Singh, 1997).

Control of Sediment Inflow

Sediment inflow to a reservoir can be reduced by constructing check dams on the tributaries bringing highly silt charged waters, constructing contour trenches on the steep slopes of the hills in the catchment area, adopting suitable watershed management practices including biological and engineering measures in the catchment areas of the reservoir and undertaking massive afforestation in the upper side of the catchment areas of the reservoir.

Catchment Area

The catchment area conveying runoff to the reservoir should not be too large. As suggested by Brown (Sharma and Sharma, 2002), a reservoir can have a safe life of 100 years if it holds at least 3.6 ha m of water per sq. km of watershed area.

Suitable Soil and Water Conservation Measures

Assuming 40% of the rainfall contributing to runoff losses from a barren land, it is estimated that on an average about 4.012 ton of runoff is produced from one hectare of land in monsoon season in the country and assuming every ton of runoff water carries a minimum of 2 kg fertile top soil from barren land, than 4.012 ton runoff will wash away about 8.024 tons of fertile top soil per hectare per year. This continuous process of soil erosion makes the soil impoverished resulting in barren lands. Further, this silt laden water will decrease the dead storage capacity of the reservoirs and threaten to minimize the life span of the reservoirs. To conserve the reservoir storage capacity and make it more sustainable, the pre-requisite is to develop suitable soil and water conservation measures. The different soil and water conservation measures include construction of contour bunds and terraces, trenching, ditching, listering, provision of rock and loose brush wood check dams, afforestation, control of grazing, gully plugging, providing vegetation screens, contour cultivation, strip cropping, adopting suitable cropping patterns, buffer strip cropping, construction of v-ditch and miniature bund along the contours with and without vegetative filter strips etc.

3.3.7.5 Curative Measures

Silt which is deposited in the reservoir can be removed by excavation, dredging, scouring and sluicing with hydraulic and mechanical agitations. However, it is very costly to remove the silts from the reservoirs which have already been deposited by excavation as well as by dredging. Scouring of deposited sediment through sluices in the dam is also found to be rather ineffective. Loosening sediment and pushing it towards sluices by mechanical means simultaneously with scouring would increase the effectiveness of the scouring action to some extent. However for doing so, provision of large sluices near the bottom of a high dam is essential which limits the use of this measure in case of large reservoirs (Singh, 1997).

Question Banks

Q1. What are the different sources for storage of irrigation water? Of these which one is the cheapest method to store the irrigation water and why?

Q2. Describe the mechanisms by which soil reservoirs stores water. How can the storage of water in the soil reservoir be increased?

Q3. Why it is important to store water in the surface reservoir? What are the points need to be taken into account before designing a surface reservoir?

Q4. What is a multi-purpose reservoir? How is it different from ordinary small reservoirs?

Q5. What are the advantages and disadvantages of a multi-purpose reservoir?

Q6. What is the purpose of construction of dam in a reservoir? What are the points to be considered for selecting an ideal site for a dam?

Q7. What are the different types of surveys/investigations which need to be undertaken for preparation of a project?

Q8. What is a topographical survey? What are the various points need to be considered for topographical survey for preparation of a project?

Q9. Describe the various data essential to carry out the geological investigation for design of a reservoir and dam.

Q10. How are the hydrological investigations helpful in construction of a dam?

Q11. How are the sediment studies useful in design of a reservoir?

Q12. How is the contour map helpful in estimation of storage of a reservoir?

Q13. What are the different formulae used to find out the capacity or storage volume of a reservoir? Write these formulae and explain the various terms of these formulae.

Q14. Draw a typical area capacity curve and explain how it is prepared.

Q15. What are the uses of area capacity curve? How can you find out the storage of a reservoir at a particular elevation from this curve?

Q16. With a diagram describe the various storage zones of a reservoir.

Q17. What are the various reservoir levels? Enumerate each type of the reservoir level.

Q18. What are the differences between the maximum and minimum reservoir levels?

Q19. What is flood storage? How is it helpful in controlling flood?

Q20. What is the difference between the live and dead storage?

Q21. Write notes on the following types of reservoir storages.

(i) Flood storage (ii) Live storage (iii) Dead storage (iv) Valley storage (v) Bank storage

Q22. What are the various considerations need to be undertaken for deciding the design capacity of a reservoir?

Q23. What is a mass curve? How is it prepared?

Q24. Draw a typical mass curve and describe how it is useful in computing the storage of a reservoir?

Month	Jan	Feb	Mar	Apr	May	Jun	Jul	Aug	Sep	Oct	Nov	Dec
Inflow, M m^3	1.5	2.2	3.0	8.2	12.2	12.0	7.7	3.0	2.5	2.2	2.0	1.7

Q25. The water yield data of a catchment measured in different months are presented in table given below. Determine the minimum storage capacity of the reservoir required to allow the above volume of water to be drawn off at a uniform rate assuming that there is no loss of water over the spillway.

Q26. What are the two measures losses of water in reservoir? How does it vary with respect to the total out flow of water in the reservoir?

Q27. How does the evaporation loss of water bodies vary from place to place in India?

Q28. How does the evaporation loss occur in a reservoir?

Q29. How does the open water surface area of a reservoir affect the evaporation rate in it?

Q30. How does the depth of water in a reservoir influence the evaporation rate in it?

Q31. Write the water budget method to estimate the evaporation loss in a reservoir and describe how it is used to compute the evaporation rate.

Q32. Describe the mass transfer and energy budget method to compute the evaporation rate

Q33. Describe the combination equation for estimation of evaporation loss of a reservoir. Write the equation and explain the different terms of the equation.

Q34. Draw the diagram of a USDA Class A Pan evaporimeter and write about its working principle.

Q35. What are the criteria followed in selecting site for a pan evaporimeter?

Q36. What are the different types of evaporimeters? Write notes on construction and working of these evaporimeters with label diagram.

Q37. An evaporation pan is 1.6 m in size and is used to measure evaporation loss of a reservoir. The initial and final water level reading in the pan are found to be 8 cm and 9 cm, respectively. A rainfall of 4.1 cm has occurred during the period the observation was taken. To keep the water level same in the pan, 2.1 cm depth of water has been removed from the pan. Find out the pan evaporation. If the pan coefficient used to convert the pan evaporation to reservoir evaporation is 0.75, find out the reservoir evaporation also.

Q38. Briefly describe the various methods used to reduce the evaporation loss in a reservoir.

Q39. How are the shading materials useful in reducing the evaporation loss in a reservoir?

Q40. What do you mean by seepage loss in a reservoir? Why it is more in newly constructed reservoirs?

Q41. How does the seepage loss decreases in the reservoirs after some years of construction?

Q42. With a diagram write the water balance model of a reservoir and explain how this is used to estimate the seepage rate in a reservoir.

Q43. Enumerate the various methods employed to reduce seepage loss in a reservoir.

Q44. What do you mean by reservoir sedimentation? How does it decrease the storage capacity and hence life span of a reservoir?

Q45. What are the differences between bed load and suspended load?

Q46. What are the factors on which the sediment distribution pattern of a reservoir depends on?

Q47. What are the factors that affect the sedimentation of a reservoir?

Q48. How is the sedimentation rate determined? Why does the sedimentation rate of reservoirs vary from river to river in a country and also from year to year?

Q49. Briefly enumerate the different measures employed to reduce reservoir sedimentation.

Q50. How do the effective soil and water conservation measures help in reducing the sedimentation of a reservoir?

Q52. How does the construction of large reservoir affect the environment?

References

Belgaumi, M.I., Itnal, C.J. and Radder, C.D. 1997. Farm Ponds. Technical series 4, Publication Centrte, University of Agricultural Sciences, Dharwad, pp. 47.

Cluff, C.S (1977). The compartmented reservoir, a method of efficient water storage. Ph.D. dissertation, Colorado State University, Fort Collins.

Cooley, K.R (1974). Evaporation suppression for conserving water supplies. Proceedings of Water Harvesting Symposium, Phoenix, Arizona: 192-199.

Lenka, 2001. Irrigation and Draiange. Kalyani Publishers, New Delhi, pp. 397.

Murty, V.V.N. 1998. Land and Water Management Engineering. 2nd Edition, Kalyani Publishers, New Delhi, pp. 586.

Panigrahi, B. 2001. Water Balance Simulation for Optimum Design of On-Farm Reservoir in Rainfed Farming System. Ph.D. Thesis, Indian Institute of Technology, Kharagpur.

Penman, H. L. (1948). "Natural evaporation from open water, Bare Soil, and Grass." Proc. Roy. Soc. London, al93, 120-146.

Pochop, L.O., Borrelli, J. and Hasfurther, V. (1985). "Design Characteristics for Evaporation Ponds in Wyoming." Final Rep. submitted to the Wyoming Water Research Center, Laramie, WY, 1-13.

Replogie, J.A., Merriam, J.L. Swarner, L.R. and Phelan, J.T. 1980. Farm water delivery systems, In Design and Operation of Farm Irrigation Systems, M.E. Jensen (Ed.), ASAE Monograph 3, St.Joseph, MI.pp.317-343.

Sharma, R.K. and Sharma, T.K. 2002. Irrigation Engineering including Hydrology, S. Chand and Company Ltd. New Delhi.

Sharma, R.K., 1987. A Text book of Hydrology and Water Resources. Dhanpat Rai and Sons, Delhi.

Singh, B. 1997. Fundamentals of Irrigation Engineering, Nem Chand & Bros., Civil Lines, Roorke, pp. 616.

4

Discharge Measurement

4.1 Introduction

Discharge is also called as rate of flow. Discharge is defined as the volume of water flowing through a cross section in a unit time. It can be expressed as cubic metre per second (cumec), cubic feet per second (cusec), litre per sec (lps) etc. One cumec is equivalent to 35.5 cusecs or 1000 lps. The measurement of discharge is very important in everyday life starting from deciding the amount of irrigation to be supplied to a crop at a particular growth stage, or amount of water supply for domestic and industrial uses, for flood control and power generation etc. It plays a vital role in water resources planning. If a stage-discharge (also called as gauge-discharge) relationship for a particular gauging site is developed, then daily measurement of discharge in the gauging site is not required. Discharge can be estimated from the developed relationship of discharge versus stage if the measurement of stage is known.

4.2 Objectives of Discharge Measurement

The various objectives of discharge measurements are given as:

(i) For accurate assessment of available and dependable water resources of river basins,

(ii) For flood forecasting and flood warning

(iii) For comprehensive and coordinated planning for the optimum utilization of water resources in rivers and their tributaries,

(iv) For ascertaining annual and seasonal variation of runoff,

(v) For estimation of water losses and regeneration in the various reaches of the river and

(vi) For investigation and design of multipurpose river valley projects for development of irrigation and power potential, flood control, water supply etc.

4.3 Methods of Discharge Measurement

The various methods of discharge measurement can be broadly classified into the following types: (i) volume method of discharge measurement (ii) velocity-area methods (iii) slope-area method (iv) gauge-discharge relationship (v) flow measuring structures such as orifices, mouthpieces, flumes and weirs and (vi) chemical dilution method and (vii) radioisotope method.

4.3.1 Volume Method of Discharge Measurement

This is a simple method for discharge measurement used in small streams or irrigation channels. In this method, the flow of water of the stream or the small irrigation channel in a container such as bucket of known volume for a given time period is collected (Fig. 4.1). The rate of flow/discharge of the stream/small irrigation channel is computed by the formula (Lenka, 2001):

$$Discharge, lit/\sec = \frac{Volume\ of\ container, lit}{Time\ required\ to\ fill\ the\ container, \sec} \quad (4.1)$$

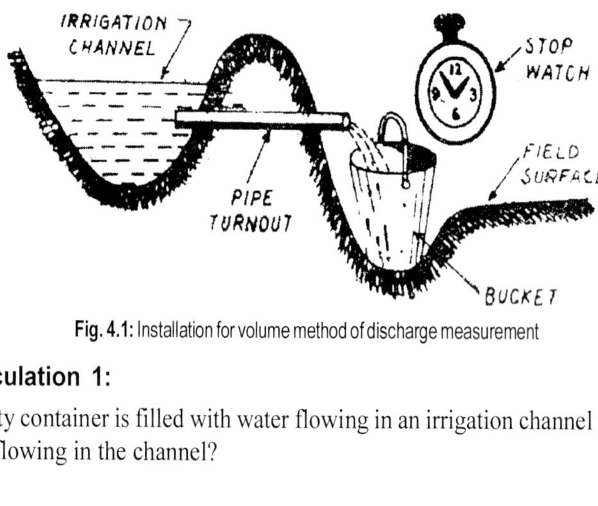

Fig. 4.1: Installation for volume method of discharge measurement

Sample Calculation 1:

A 20 lit capacity container is filled with water flowing in an irrigation channel in 5 sec. What is the discharge flowing in the channel?

Solution:

Volume of the container = 20 lit.

Time required to fill the container with water of the channel = 5 sec.

Hence, $discharge = \frac{20\ lit}{5\ \sec} = 4\frac{lit}{\sec}$ (4.2)

Volume method can be used to determine discharge of pumps and other water lifting device such as persian wheel, chain washer pump, indigenous water lifting devices such as tenda, doon, swing lift, wheel lift, wind mill etc. The irrigation water to a crop field can also be measured by a pump. By knowing the volume of water required by a crop and the pump discharge capacity, we can find out what a time the pump has to run to irrigate the field. In other words, the discharge capacity of a pump which is irrigating a crop field can also be calculated by volume method by knowing the irrigated area, depth of water irrigated and the time required to irrigate the field by the pump. Multiplying the irrigated area with depth of irrigation, volume of water is found out. Then dividing the volume of water with the time required to irrigate the field, discharge capacity of the pump is found out. A few important pumps generally used to irrigate the fields are (i) low lift hand pump (ii) centrifugal pump (iii) pedal operated pump (iv) krishak bandhu pump. In addition, many farmers irrigate their crop fields from the farm pond by indigenous water lifting devices such as tenda, doon, swing lift, wheel lift, wind mill etc. Details of various pumps and indigenous water lifting devices used for irrigation in fields are discussed in chapter 10.

Sample Calculation 2:

A pump is irrigating a field of 1 ha and the average depth of irrigation water is 5 cm. If the time required for irrigation is 5 hr, then find out the discharge of the pump.

Solution

Volume of irrigation water = Area x average depth of irrigation = 10000 m^2 x 0.05 m = 500 m^3.

Time of irrigation = 5 hr

So, discharge of pump = 500 m^3 /5 hr = 100 m^3/hr = 27.78 lit/sec.

Sample Calculation 3

Wheat crop require 40 cm of irrigation water during 120 days irrigating period. If area under cultivation is 10 ha and the pump used for irrigation is engaged every day for 10 hours each, then find out the discharge of the pump.

Solution

Total volume of water during the irrigation period of 120 days
= 0.40 m × 10 × 10000 m^2 = 40, 000 m^3.
Total time of irrigation in 120 days @10 hours per day = 120 × 10 = 1200 hr.
Discharge of the pump = Total volume of water/ total time of irrigation
= 40,000 /1200 = 33.33 m^3/hr = 9.26 lit/sec.

4.3.2 Velocity-Area Method

In this method the rate of flow passing in a pipe or open channel is determined by multiplying the cross sectional area of the flow section at right angles to the direction of flow by the velocity of water. The formula is:

Discharge (Q) = Area (A) x Velocity (V) (4.3)

The discharge is generally expressed in the unit of m^3/sec, area in m^2 and velocity in m/sec. The velocity is usually measured by float method, tracer method or by a current meter. If the small irrigation channel is of regular cross section in the form of a rectangle or trapezium, then cross sectional area of flow can be measured by measuring the depth of flow, base width and side slope (in case of trapezoidal section) of the channel and using the formula to compute the area of rectangle or trapezium as the case may be. In case of large irrigation channels, the cross section is divided into suitable compartments or vertical sections and total discharge is computed by adding the discharges in various sections on the basis of the depth and velocity observed in each vertical section. The method of dividing the large section in to several small vertical sections called as segments is termed as segmentation. In discharge computation by velocity-area method, 2 approaches are followed. They are mid-section method and mean section method.

4.3.3 Slope Area Method

This method of discharge measurement is approximate and is used when the actual measurement by velocity-area method is not possible. In this method, the average cross sectional area is multiplied with the velocity derived from the observed water surface slope. Cross sectional area is measured at three locations, two at the ends and one at the centre of a straight reach of section. The mean cross sectional area, A is given as (Sharma, 1987):

$$A = \frac{\left[A_1 + 2\,A_2 + ...(2\,A_{n-1}) + A_n\right]}{2\,(n-1)} \tag{4.4}$$

where, A_1, A_2..... A_n are the cross sectional areas measured at various locations.

Mean wetted perimeter, P is similarly calculated as (Sharma, 1987):

$$P = \frac{\left[P_1 + 2\,P_2 + ...(2\,P_{n-1}) + P_n\right]}{2\,(n-1)} \tag{4.5}$$

where, P_1, P_2.....P_n are the wetted perimeter measured at various locations.

Mean velocity is determined by from the observed water surface slope by either Chezy's formula or Manning's formula as written below.

Chezy's Formula: $V = C\sqrt{R\,S}$ (4.6)

Manning's Formula: $\mathrm{V} = \frac{1}{n} R^{2/3}\, S^{1/2}$ (4.7)

where, V = mean velocity, m/sec; C = Chezy's constant depending on shape and surface the channel; R = hydraulic mean radius, m; n = rugosity coefficient (value of n varies from 0.020 to 0.060 depending on types and sizes of bed materials of the channel) and S = water surface slope between two ends of a canal section or reach.

If the depth of water levels at two ends of a canal section of length, x m are Z_1 and Z_2, respectively, then water surface slope is given as, $S = (Z_1 - Z_2)/x$ (4.8)

4.3.4 Depth Measurement

The depth of flow is measured directly by (i) suspension rod (ii) sounding rod (iii) lead lines (iv) reel line and cranes and indirectly by (i) echo sounder and (ii) theodolite. Suspension rod is used when the depth of flow is less (up to 1 m) and velocity of flow is of the order of 1 m/sec. Sounding rods are used when the depth of flow is more (up to 6 m) and velocity is medium. The sounding rods may be made of wood or bamboo. The sounding rods are graduated in metres, decimeters and centimeters and are painted white, black or red. Lead lines also called as log lines are used when the depth of flow is more then 6 m but the velocity of flow is less. It consists of a cable of 2.5 mm diameter to the lower end of which is attached a hanger for placement of the current meter and sounding weight during observations. Weight of 7 to 14 kg is necessary to avoid deflection of cable from the true vertical. A core of insulated copper wire in the log line enables electrical circuit to be made for counting ticks of the revolution of the current meter when the current meter is operated by battery. When the depth of flow is high and velocity of flow is more, a cable line is lowered by means of a crane to measure the depth of flow. The arrangement, either mounted on a boat or used from a bridge parapet, consists of a hand-operated reel fixed to a pedestal and carrying 3 mm diameter hard steel flexible cable passing over pulley at the other end.

When the channel is of sufficient depth having high velocity of flow, echo sounder is used for accurate and rapid measurement of depth. A recording type echo sounder records a continuous graph of the stream bed during the traverse on moving chart drawn by a stylus. An indicator type echo sounder records a single reading for each depth measurement.

4.3.5 Velocity Measurement

Velocity can be measured by the following methods.

4.3.5.1 Velocity Measurement by Float Method

Approximate values of velocity of flow in a channel can be measured by float method. In this method the velocity is measured by noting the rate of movement of a floating body. A surface float usually a small piece of wooden block or a long necked bottle partly filled with water is used to travel a known length of a straight channel section having uniform cross section. The time taken for the surface float to cover the known length of the channel section is noted. Dividing the time with the known length of the channel section, surface velocity of water called as float velocity is obtained. Several measurements of time to cover the same known length of the channel section are taken from which average float velocity is determined. To convert the average float velocity to average velocity of flow of the channel/stream, a conversion factor 0.85 is selected (Michael, 1987). Several measurements of depth and width are made within the trial section of known length in order to find out the average cross sectional area. Multiplying the average velocity of flow of the channel/stream with its average cross sectional area, rate of flow is determined.

4.3.5.2 Velocity Measurement by Current Meter

The velocity in a channel varies approximately as a parabola, from zero at the bottom of the channel to a maximum at or near the water surface at the top. The maximum velocity occurs at a depth ranging from 0.10 to 0.20 of the depth below top water surface. The ratio of the maximum to average velocity is nearly equal to 1.2. The average velocity on any vertical section occurs at a depth 0.6 of the total depth. The average velocity at two depths i.e. 0.2 and 0.8 depth below the water surface equals the mean or average velocity whereas the velocity taken at 0.6 depths below the top water surface gives nearly the mean velocity. Velocity measurement at a single point i.e. 0.6 depth below top water surface is taken for stream with low depth of water. For streams with more depth of water, velocity measurement at two points i.e. 0.2 and 0.8 depths below top water surface is taken. Velocity measurement using two points method is more correct than that done by one point (velocity taken at 0.6 depth) method.

Current meter is of two types: (i) Vertical axis current meter and (ii) Horizontal axis current meter. Vertical axis current meter is also called as differential meter, direct meter, Price or Gurley current meter. Horizontal axis current meter is called as propeller type current meter. Sometimes to measure the low velocity of water (velocity up to 1 m/sec) in small shallow streams/irrigation channels such as minors and sub minors, a miniature type current meter called as Pigmi current meter is preferred.

4.3.5.3 Rating of Current Meter

The current meter has to be carefully rated in still water before using it to measure velocity of water in a stream. It is mounted on a carriage which moves on rails along a straight channel. The rating is done by moving the carriage through a still water tank. The speed of the carriage is estimated from the time required to traverse a known distance in a straight channel. Several readings of runs/revolutions of meter per second at various speeds of flow are taken and a curve is plotted called as rating curve. The rating curve is expressed as:

$$V = a N + b \tag{4.9}$$

where, V = velocity of flow, m/sec; N = number of revolutions of the meter per second; a and b = constants given by the manufacture or generally determined by the rating curve.

Sample Calculation 4:

Following velocities are recorded in irrigation channel with a current meter.

Depth above the bed of channel, m	0	1	2	3	4
Velocity, m/sec	0	0.4	0.75	0.8	1.0

Find the discharge per unit width of the channel near the point of measurement. Depth of flow at the point is 5 m.

Solution

Given depth of flow = 5 m. Velocity of flow at 0.2 depth below water surface (1m) i.e. 4 m above the channel bed is given as 1.0 m/sec. Similarly, the flow velocity at 0.8 depths below water surface (4 m) i.e. 1 m above the channel bed is given as 0.4 m/sec. Assuming velocity measurements by 2 points method i.e. at 0.2 and 0.8 depths below water surface, the mean velocity is estimated as (1.0 + 0.4) /2 = 0.70 m/sec.

Hence, discharge per unit width of the section is computed as Q = area x mean velocity = (5 × 1) × 0.7 = 3.5 m/sec.

Sample Calculation 5:

Calculate the discharge of a canal section from the following observations taken by the current meter whose rate curve is given by $V = 0.3\ N + 0.05$

Distance from bank, m	2	4	6	8	10	12	14	16
Depth (d), m	0.4	0.8	1.0	1.2	0.9	0.7	0.5	0.5
Current meter revolution/sec (N)	0.8	1.1	1.5	2.0	1.5	1.2	0.8	1.0

Solution:

Total discharge is computed in tabular form as given below.

Distance from bank, m	Segmented width (b), m	Depth (d), m	Current meter revolution (N)	Velocity ,m/sec $V = 0.3\ N + 0.05$	Discharge, m^3/sec $Q = b \times d \times V$
2	2	0.4	0.8	0.29	0.232
4	2	0.8	1.1	0.38	0.608
6	2	1.0	1.5	0.50	1.000
8	2	1.2	2.0	0.65	1.560
10	2	0.9	1.5	0.50	0.900
12	2	0.7	1.2	0.41	0.574
14	2	0.5	0.8	0.29	0.290
16	2	0.5	1.0	0.35	0.350
Total discharge					5.514

Sample Calculation 6:

Water is flowing in a river cross section. The velocity of a surface float of the river cross section as observed is given below. Estimate the total discharge of the river cross section passing at the gauging site. Assume the velocity factor to convert the surface float to mean velocity of water as 0.89.

Solution:

Estimation of total discharge of river cross section is given in tabular form as below.

Width of section (b), m	Mean depth of each section (d), m	Surface velocity observed (V_s), m/sec	Mean velocity (V_m),	Area (A), $m^2A = b \times d$ m/sec$V_m = 0.89 \times V_s$	Discharge (q), m^3/sec$q = A \times V_m$
25	1.5	0.8	0.712	37.50	26.70
30	1.5	1.0	0.890	45.00	40.05
30	2.0	1.0	0.890	60.00	53.40
28	2.5	0.8	0.712	70.00	49.84
25	2.0	0.8	0.712	50.00	35.60
25	2.2	1.2	1.068	55.00	58.74
30	1.8	1.1	0.979	54.00	52.87
30	1.8	1.2	1.068	54.00	57.67
30	2.0	0.8	0.712	60.00	42.72
30	2.0	0.8	0.712	60.00	42.72
Total Discharge					460.31

4.3.6 Gauge-Discharge Relationship

In a particular gauge site of a canal section or river section, numerous observations of the gauge height or called as stage above certain datum and the corresponding discharge passing the gauge site are measured. A relationship in graphical form called as gauge-discharge or stage-discharge curve is developed. A typical gauge-discharge curve is shown in Fig. 4.2 below (Sharma and Sharma, 2002).:

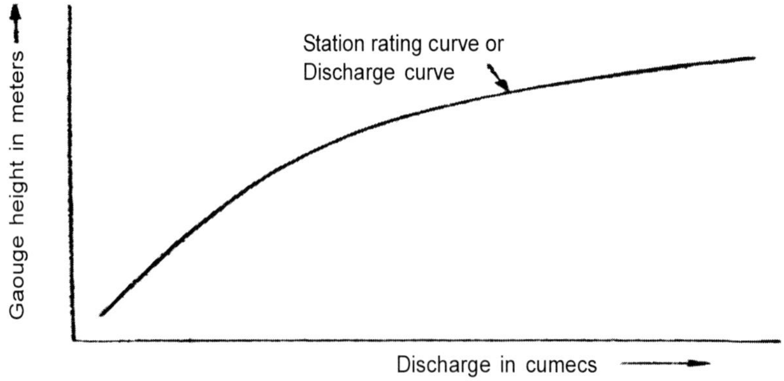

Fig. 4.2: A typical gauge-discharge relationship

This developed curve helps in predicting the discharge in periods when no discharge records are available but the stage records are available. The curve is prepared by taking a series of data of both stages and corresponding discharges and plotting the discharges in abscissa and the stages in ordinate. From this developed curve, discharge at any observed stage can be found out.

The stage-discharge curve in case of low or medium flood stage is a single curve. However, during the high flood, the rate of rise of the gauge/stage on a rising flood is generally higher than the rate of fall on a falling flood thereby passing a higher discharge at a particular gauge on the rising gauge than on the falling gauge (Fig. 4.3) (Sharma and Sharma, 2002).

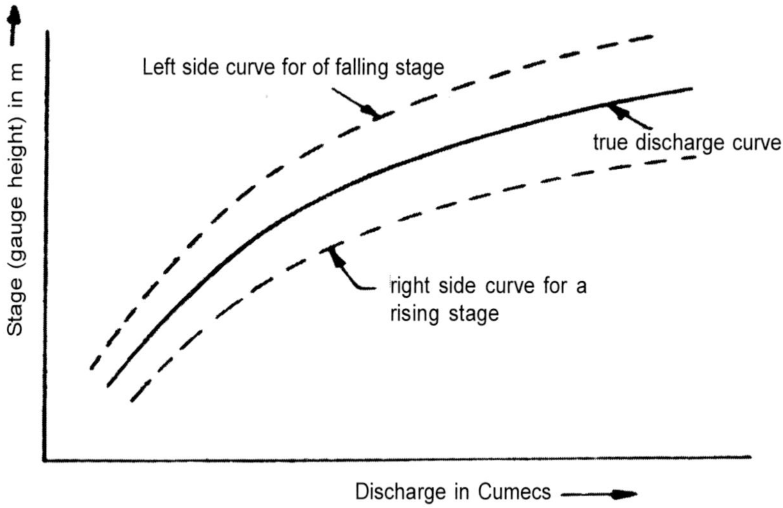

Fig. 4.3: Typical gauge-discharge relationship for high flood

4.3.6.1 Selection of Gauge Site

Selection of gauging site along with the essential requirement of gauge site are very important to know before data on stage and discharge are recorded and required relationship between the recorded data is developed.

A gauge site should be selected after careful study of different features of the available topographical map of the site. These sites are later inspected and surveyed to ensure that the physical and hydraulic features of the proposed site confirm to the requirements for the application of the method of flow measurement on a river/stream to furnish the gauge data.

4.3.6.2 Essential Requirements of a Gauge Site

The essential requirements of a gauging site are:

(i) The gauging site should be permanent and should not be shifted frequently because of silting or souring problem.

(ii) The site should be clearly visible and easily accessible.

(iii) The site should always register the same elevation of water surface corresponding to the given discharge.

(iv) The site should be so located that it can record the highest and lowest stage data corresponding to the high and low floods.

4.3.6.3 Types of Gauges

Gauges are generally divided into two types. They are direct gauges and indirect gauges. The direct gauges include staff gauges, hook gauges and float gauges. The indirect gauges include crest gauges and self record type gauges. Again the direct gauges may be permanent type or temporary type. A permanent type gauge has a fixed position to gauge the discharge of a site where as in an indirect type gauge, the gauge site changes time to time. The temporary gauges may be wooden or made of steel plates but the latter is preferred for its longevity. Followings are different types of gauges commonly used.

Staff Gauges

A staff gauge is a graduated scale on a staff or metal plate against which elevation of water surface is read. It is economical in construction and has long lasting in use. It resists to alternate wetting and drying. It is also simple in construction. Depending on the material of construction, it may be wooden type or vertical enameled type. Similarly depending on installation, it may be vertical staff gauge or inclined staff gauge.

Vertical Staff Gauge

Vertical staff gauge preferably of enameled plate gauge is graduated to millimeter. The graduation marks are clear and have bold line indications at every 20 mm interval. The graduations are made either by etching or by painting or enameling so as to read centimeters, decimeters, and metres clearly. The smallest graduation mark is the centimeter. If there are large fluctuations in flow, then several gauges running normal to the flow direction may be used. The staff gauge is generally fixed at or near the water in the river which facilitates in taking reading easily. It also helps in easy accessibility to the site for taking gauge treading.

Inclined Gauge

The inclined staff gauge is fixed on a sloping river bank. It is calibrated on the site by precise leveling. Inclined gauge may be constructed to one continuous slope or may be a compound of two or more slopes. It is convenient to construct a flight of steps along the side of inclined gauge to facilitate in taking readings.

Hook Gauge

It is a gauge made of brass rod pointing upwards with its top coinciding with the water surface in the river or stream. If the rod points downward, then it is called as a point gauge. Hook gauge is used in a vertical well.

Crest Stage Gauge

It is a gauge used to obtain the records of flood crests at sites where recording gauges are not installed. Different types of crest stage gauges are used. In one type, a small float is used which rises with the stage but are restrained at the maximum level. In another type, water soluble paints are used at bridge pier side shielded from rain. The portion of the paint coming in contact with water is washed out to indicate high water level.

Self Recording Gauge

Self recording gauge is also called as stage recorder. Like the self recording type rain gauge, it is an instrument which continuously records the rise and fall of water level of the stream or river in which water is flowing with respect to time on a graph paper. The recorder has two essential elements i.e. (i) time element consisting of a clock mechanism actuated by a spring, a weight and electricity and (ii) a gauge height element consisting of non-corrosive metallic floats, float cables, a counter weight and a gear reduction mechanism. There are various types of record. Commonly the time element operates parallel to the axis of the drum while the gauge height element rotates the drum in proportion to the rise or fall in water level. The recorder is usually operated by an 8-day spring driven clock. The height element activates a pen stylus which moves parallel to the axis of the paper roll so that its travel represents a change in water stage. Self recording gauge gives continuous record of gauges from which maximum and minimum gauges as also time of their occurrences can be known. From the continuous records, daily average gauge height can be estimated.

4.3.6.4 Discharge Site

Generally the site selected for gauge installation is the ideal site for discharge observations. If there are some limitations for the two sites to be distinct, then they should be as close as possible. The discharge measurement site should consider all the essential requirements of gauge site. In addition, it should consider additional requirements like (Sharma and Sharma, 2002):

- At or as close as possible to the gauge site in the case of velocity measurement by current meter or near the middle of a straight reach where float is used to measure the velocity,
- Located in a straight reach of length which is the least of 3 to 4 times the river width or 0.75 km during high flood both upstream and downstream of the normal cross section where velocity is to be measured,
- The flow in the straight reach is streamlined, steady and showing less departure from the normal flow at all stages of the rivers,
- The water flow is preferably in a single channel,
- Site is free from aggradations and degradation,
- Located where greatest range of fluctuations in river stage occurs,
- Site relatively free from any obstructions,
- Site is free from the tendency of formation of backward flow,
- Located preferably away from any hydraulic structures,
- Reasonably stable bed and banks for affording a regular gauge-discharge relationship, and
- Easily accessible all the time.

In case of a hilly area where discharge measurement site is selected, including the above requirements, some additional requirements are needed. They are

- Straight reach of minimum length of 150 m above section line,

- Free from projection in the beds and sides,
- Minimum 150 m removed from incoming current upstream and downstream from the section line,
- A bridge without piers and
- Nearest to a proposed dam site.

4.3.7 Weir

Weir is a notch of regular form through which irrigation streams are allowed to pass. It is built across the stream or channel. They may be built as stationary structures or may be portable. They consist of a weir wall of concrete, timber, strong tin plates, galvanized iron sheet or metal with a sheet metal weir plate fixed to it. They are generally divided into three types as (i) rectangular (ii) trapezoidal and (iii) triangular. The weir should be so placed across a channel section that the flow is under free flow condition. This necessitates the nappe of the freely falling water not to be submerged by tail end water of the channel section.

According to crest, the weir may be sharp crested weir or broad crested weir. In a sharp crested weir, water passing over the crest of the weir just touches only a line. In case of broad crested weir, there is a rounded upstream edge, or a crest so broad that water flowing over the weir touches the crest with a surface. In practice, the sharp crested weir is more accurate than a broad crested weir.

4.3.7.1 Rectangular Weir

It takes the name from the shape of the notch. It is used to measure large discharge. It has a horizontal crest and vertical side. As described above, it may be sharp crested or broad crested type. It is also dived into two more types i.e. contracted rectangular or suppressed rectangular weir. In a contracted weir, the crest length is less than the width of the up stream channel. In a suppressed rectangular weir, the crest length is equal to or more than the width of up stream channel or called as approach channel. The discharge formula through a rectangular weir can be calculated empirically by Francis' formula, Bazin's formula and Rehbock's formula as stated below:

Francis' Formula

It is one of the most commonly used formula for calculating discharge over sharp or narrow crested weirs with or without end contraction. The formula was proposed by J.B. Francis. In this formula, the length of the crest varied from 1.07 to 5.19 m, but in most of the experiments, the crest length was 3.05 m. The head varies from 0.18 m to 0.49 m. He conducted more than eighty experiments and from these experiments, he proposed the following formula for computation of discharge over rectangular weir. For a suppressed rectangular weir (Fig. 4.4), discharge is given by (Murty, 1998):

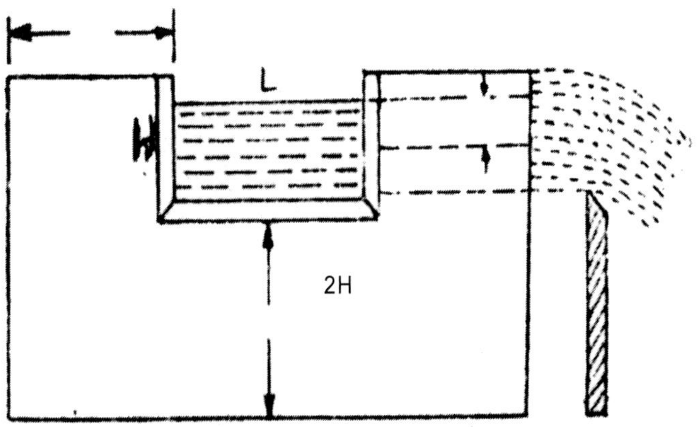

Fig. 4.4: Rectangular weir

$Q = C\,L\,H^{3/2}$ (4.10)

where Q = discharge in lit/sec; L = length of crest in cm; H = height of water called as head above the crest in cm and C = coefficient of discharge which is 0.0184.

In a contracted rectangular weir (both sides contracted), the discharge is given as:

$Q = C\,(L\text{-}0.2\,H)\,H^{3/2}$ (4.11)

For one side contraction, the formula is:

$Q = C\,(L\text{-}0.1\,H)\,H^{3/2}$ (4.12)

where the terms are defined as above.

The value of C is 0.0184. The terms inside the bracket i.e. $(L\text{-}\,0.1\,H)$ in Eq. (4.12) and $(L\text{-}0.2H)$ in Eqn. (4.11) are called as effective length of the weir. Fig. 4.4 (as given above) shows the view of a rectangular weir without end contraction.

It is to be noted that in above equations (Eqns. 4.10 to 4.12), the velocity of approach is neglected. When the velocity of approach is taken into account, the discharge formulae for Eqn. (4.10), Eqn. (4.11) and Eqn. (4.12) becomes as

$Q = C\,L\,(H_1^{3/2} - h_a^{3/2})$ (4.13)

Similarly for one side end contraction, discharge is given as:

$Q = C\,(L\text{-}0.1\,H_1)\,(H_1^{3/2} - h_a^{3/2})$ (4.14)

For both sides end contraction, discharge is:

$Q = C\,(L\text{-}0.2\,H_1)\,(H_1^{3/2} - h_a^{3/2})$ (4.15)

where, $H_1 = (H + h_a) = (H + V_a^2/2g)$; H = height of water surface above the crest of the weir and V_a = velocity of flow in the upstream side of the weir in the channel/stream.

Bazin's Formula

In 1886, H. Bazin of France conducted a number of experiments with rectangular weirs. From these experiments, he proposed the following formula for computation of discharge over the rectangular weir as follows:

Without velocity of approach, the discharge formula is:

$$Q = m\,(2\,g)^{1/2}\,L\,H^{3/2} \tag{4.16}$$

In which according to Bazin, the value of the coefficient m is:

$$m = (0.405 + 0.003/H) \tag{4.17}$$

With consideration of velocity of approach, discharge is given as:

$$Q = = m_1\,(2\,g)^{1/2}\,L\,H_1^{3/2} \tag{4.18}$$

In which the coefficient m_1 is:

$$m_1 = (0.405 + 0.003/H_1) \tag{4.19}$$

and H_1 *is* (H_1 is still water head) given as:

$$H_1 = (H + \alpha\,V_a^2/2g) \tag{4.20}$$

where α is a constant, the mean value of which is proposed by Bazin as 1.6.

Rehbock's Formula

In 1929, T. Rehbock on the basis of experiments with suppressed rectangular weir, proposed the following empirical formula for computation of discharge as:

$$Q = 0.667\,(0.605 + 0.08\,H/Z + 0.001/H)\,(2\,g)^{1/2}\,L\,H^{3/2} \tag{4.21}$$

where H = head, m and Z = crest height, m.

4.3.7.2 Trapezoidal Weir

It is also called as cipoletti weir. It is a contracted weir in which each side of the notch has a slope of 1 horizontal to 4 vertical. The weir has sharp crest and sharp sides which is beveled from downstream side only. It is commonly used to measure medium discharge.

The discharge through a trapezoidal weir is given as:

$$Q = C\,L\,H^{3/2} \tag{4.22}$$

where Q = discharge, lit/sec; L = crest length, cm; H = height of water called as head over the crest, cm and C = coefficient of discharge which is 0.0186.

The crest length is given as

$$L = (L_1 + L_2)/2 \tag{4.23}$$

where L_1 = length of crest at bottom and L_2 = length of crest at top. Fig. 4.5 represents a trapezoidal weir.

4.3.7.3 Triangular Weir

It is also called as *V* notch. The 90^0 triangular or V notch (Fig. 4.6) has greater practical utility than any other weir. It is ideal to measure discharge under medium or low flow condition. It is popular since it is easy construct and install. It has both sides sharp, beveled from downstream side only. The discharge equation is given as (Lenka, 2001):

$$Q = 0.0138\, H^{3/2} \tag{4.24}$$

where Q = discharge, lit/sec and H = height of water above the cres, cm.

In the above equations used to measure discharge, values of coefficient of discharge are used when they are not available. In practice the values of the coefficient of discharge for each type of weir should be found out by calibration in laboratory. The calibrated values of the coefficient can be used provided similar condition in the field exists. To get correct flow, the devices should be correctly designed and installed. The flow quantity should be checked from time to time.

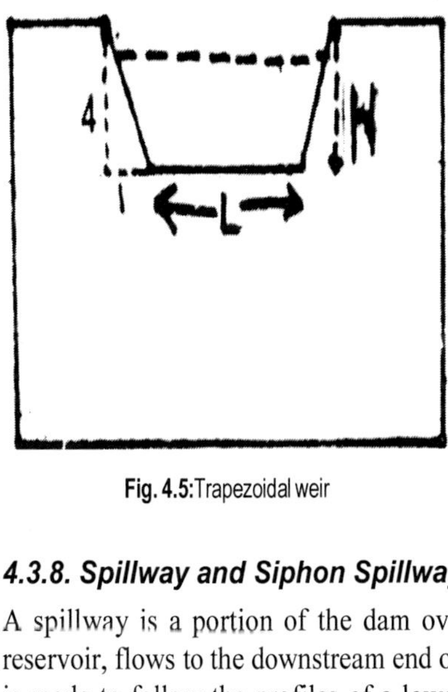

Fig. 4.5: Trapezoidal weir

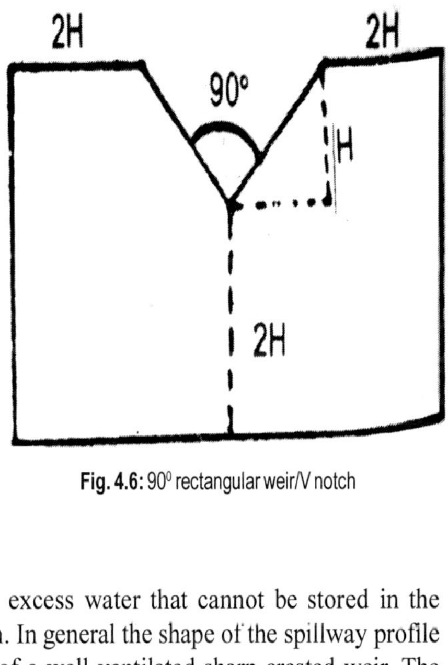

Fig. 4.6: 90^0 rectangular weir/V notch

4.3.8. Spillway and Siphon Spillway

A spillway is a portion of the dam over which excess water that cannot be stored in the reservoir, flows to the downstream end of the dam. In general the shape of the spillway profile is made to follow the profiles of a lower nappe of a well ventilated sharp crested weir. The spillway is formed by filling the space between the sharp crested weir and the lower nappe with concrete or masonry as shown in Fig. 4.7 (a). Such a spillway is called as *ogee spillway or ogee weir.* The discharge over an ogee spillway is given as (Modi and Seth, 2005):

$$Q = C\, L\, H^{3/2} \tag{4.25}$$

where, C = coefficient of the spillway, L = length of the spillway and H = head above the crest of the spillway. The value of the coefficient C is determined by calibration.

Siphon spillway consists of an ogee weir which is provided with an air-tight cover as shown in Fig. 4.7 (b). This converts the discharge face of the spillway into a large rectangular sectioned

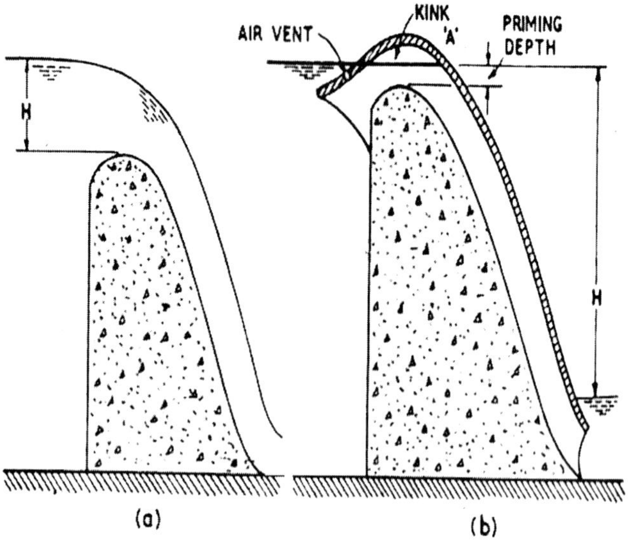

Fig. 4.7: (a) Ogee spillway (b) Siphon spillway

pipe connecting the upstream and downstream water surfaces. In the case of a siphon spillway the head H under which the water flows is equal to the difference between the water surface on the upstream and downstream sides. As such as compared with an ordinary spillway, the head in the case of a siphon spillway is more, on account of which the siphon spillway has a much greater discharge for a given length than an ordinary open spillway. The working of a siphon spillway is automatic.

When the water level in the upstream side of the reservoir rises above the crest level of the spillway, water starts flowing over the crest through the siphon. The jet of the flowing water strikes the inside of the cover, thus enclosing a small space *A* called kink in the upstream portion of the cover, in which air is trapped. As more and more water flows down, it sucks the trapped air from the kink, thus creating a partial vacuum in this portion which sucks up the water from the reservoir and completely fills the pipe. The symphonic action is started and the effective head causing the flow of water becomes equal to H, the difference between the water surface in the upstream and downstream sides. When the level of water in the reservoir decreases and becomes almost equal to the crest level of the spillway, the symphonic action stops. Air vents are provided in the cover at a level slightly above the crest of the spillway to facilitate this action (Garg, 1978).

The discharge through a siphon spillway is given same as ogee spillway which is

$$Q = C\,L\,H^{3/2} \qquad (4.26)$$

A siphon spillway has the following advantages over the ogee spillway.

(i) The operating head is increased considerably allowing higher discharge to flow.

(ii) The depth of water (priming depth) needed for the starting of the siphonic action is only a few centimeter above the crest of the spillway. As such the crest of the siphon spillway can be raised, thereby allowing more amount of water to get stored in the reservoir.

4.3.9. Proportional Weir (Sutro Weir)

In case of the rectangular or trapezoidal weir as discussed above, the discharge, Q varies directly with the head, H not linearly but the relation is in the form of power function i.e. $Q \,\alpha\, H^n$, where $n = 3/2$ for a rectangular or trapezoidal weir and $n = 5/2$ for a triangular weir. It is possible to design a proportional weir or called as Sutro weir in which the discharge, Q is linearly proportional to the head, H over the crest of the weir i.e. Q α H.

The analytical relationship for the shape of the proportional weir profile having Q α H, is

$$x \alpha y^{-\frac{1}{2}} \tag{4.27}$$

However, it may be noted from the above relationship that as $y \to 0, x \to \infty$ that means the width of the weir becomes infinitely small at the crest. The infinitely tending profile is however not practicable.

In order to overcome this problem, the shape of the weir profile was modified so that a finite width at the weir crest may be provided. Fig. 4.8 shows the view of the modified proportional weir profile that has its sides diverging downward in the form of hyperbolic curves having the equation as (Modi and Seth, 2005):

$$2\frac{x}{L} = \left[1 - \frac{2}{11}\tan^{-1}\left(\frac{y^{\frac{1}{2}}}{a}\right)\right] \tag{4.28}$$

where, a and L are respectively the height and width of the small rectangular shaped aperture that forms the base of the weir. The discharge through this weir is given by

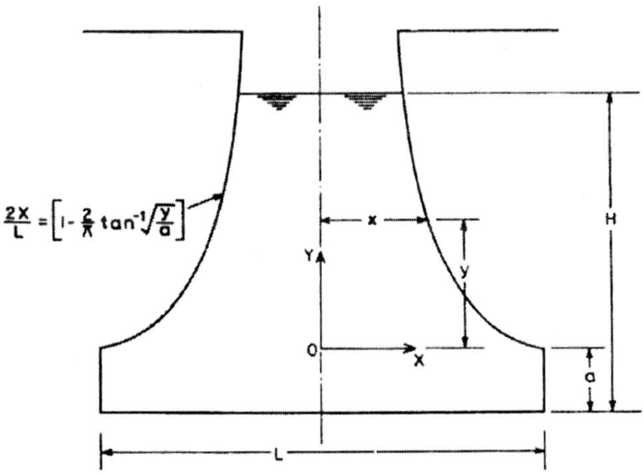

Fig. 4.8: Proportional weir

$$Q = k\ (H - a/3) \tag{4.29}$$

where k is given as:

$$k = C_d\ L\ (2\ g\ a)^{1/2} \tag{4.30}$$

In the above equation C_d is the coefficient of discharge which varies from 0.60 to 0.65.

Sample Calculation 7:

Find the discharge over a suppressed rectangular weir having length 5 m and head over the crest being 0.60 m.

Solution:

The discharge through a rectangular weir as given by Francis formula is:

$Q = 0.0184\ L\ H^{3/2}$

Given L = 5 m = 500 cm and H = 0.6 m = 60 cm

Hence Q = 0.0184 x 500 x $(60)^{3/2}$ = 4276 lit/sec.

4.3.10 Flow through Orifices and Mouthpieces

Water flows to the crop fields both through open channel as well as through pipes. The flow in pipes may be through orifices or mouthpieces. An orifice is an opening having a closed perimeter, made in the walls or bottom of a tank containing a fluid, through which fluid may be discharged. A mouthpiece is a short tube of length not more than two to three times its diameter, which is fitted to a circular opening or orifice of the same diameter, provided in a tank containing the fluid, such that it is an extension of the orifice and through which also the fluid may be discharged. According to the discharge condition, the orifice may be classified as orifice discharging free and submerged orifice. The submerged orifice may be further divided as completely or partially submerged orifice. The mouthpiece may be classified on the basis of their shape, position and discharge conditions. According to the shape the mouthpiece may be classified as cylindrical, convergent, divergent and convergent-divergent (Modi and Seth, 2005). According to the position, the mouthpiece may be classified as external and internal mouthpieces. According to the discharge condition, the mouthpiece may be classified as running full and running free mouthpieces. Formulae for calculation of discharge through the different types of orifices and mouthpieces are discussed in details by Modi and Seth (2005).

4.3.11 Meter Gate

A meter gate is basically a modified submerged orifice, so arranged that the orifice is adjustable in area. They are manufactured commercially and are used to control water flowing from one channel to another. They may serve the rate of flow, if the head and area of opening can be determined and the gate has been calibrated the head is same as that under submerged orifice.

Normally the gate and the opening in to outlet are circular and their area can be determined easily if the gate is fully open. However, under most conditions the gate is only partially open, leaving a crescent shaped area which is difficult to measure. Consequently, most gates are calibrated and tables are supplied, giving the rate of flow as a function of the head and the degree of gate opening as measured by the displacement of the gate stem.

4.3.12 Water Meters

Water meters utilizes a multi-blade propeller made of metal, plastic or rubber, rotating in a vertical or horizontal plane and geared to a totaliser in such a manner that a numerical counter can the flow in any volumetric units. Water meters are available for a range of sizes suiting the pipe size commonly used on the farm.

There are two basic requirements for accurate operation of water meter:

1. The pipe must flow full at all the times, and
2. The rate of flow must exceed the minimum for the rated range.

Meters are calibrated in the factory and field adjustments are usually not required. When water meters are installed in open channels, the flow must be brought through a pipe of known cross-sectional area. Care must be taken that no debris or other foreign materials obstruct the propeller.

4.3.13 Parshall Flume

Parshall flumes consists of (i) a converging inlet section (ii) a throat section and (iii) a diverging outlet section. The throat section lies inbetween the inlet and outlet sections. Fig. 4.9 shows the geometry of a Parshall flume. Parshall flumes may be built of wood, concrete, galvanized sheet metal or other materials. Large flumes are usually constructed on site, while smaller ones can be purchased as prefabricated units that are installed in the channel. Dimensions and ranges of flow for the various standard Parshall flumes as well as the design procedures can be found in Water Measurement Manual of the U.S. Bureau of Reclamation (1975).

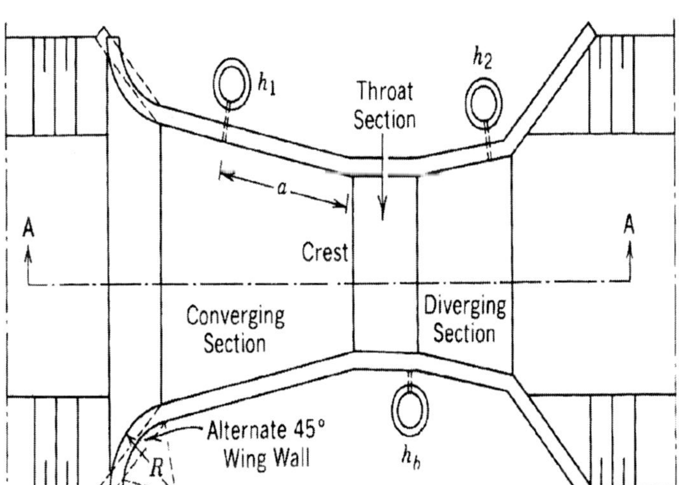

Fig. 4.9: Geometry of a Parshall flume

The primary advantage of Parshall flumes is their ability to provide accurate flow measurements over a wide range of flows with minimum of head loss (U.S. Bureau of Reclamation, 1975). This allows their uses in relatively shallow channels with flat grades. The main disadvantages of Parshall flumes are their relatively large size (Walker and Skogerboe, 1987).

The rate of flow through a Parshall flume may or may not be affected by the tail water depth in the downstream of the throat. Free flow condition exists when the tail water depth is not high enough to affect the flow. A flume is considered to be submerged when the tail water depth is sufficient to affect the flow and the submergence ratio, S defined as the ratio of head in the diverging section (H_b) to that in the converging section (H_a) exceeds the transition ratio S_t. Values of S_t for different sizes of Parshall flumes as decided by the throat widths are presented in Table 4.1 (James, 1988).

Table 4.1: Free flow and submerged flow coefficients and exponents for Parshall flume

Throat width, *W*, cm	C_f for *Q* in (lit/s)	C_s for *Q* in (lit/s)	n_f	n_s	S_t
2.54	9.57	8.47	1.55	1.00	0.56
5.08	19.14	17.33	1.55	1.00	0.61
7.62	28.09	25.91	1.55	1.00	0.64
15.24	58.34	47.01	1.58	1.00	0.55
22.86	86.94	71.08	1.53	1.08	0.63
30.48	113.28	88.08	1.52	1.06	0.62
45.72	169.92	125.17	1.54	1.08	0.64
60.96	226.56	168.22	1.55	1.115	0.66
76.20	283.2	204.47	1.555	1.14	0.67
91.44	339.84	243.55	1.56	1.15	0.68
121.92	453.12	314.35	1.57	1.16	0.70
152.40	566.40	383.74	1.58	1.185	0.72
182.88	679.68	448.87	1.59	1.205	0.74
213.36	792.96	514.01	1.60	1.230	0.76
243.84	906.24	577.73	1.60	1.250	0.78
304.8	1136.48	702.05	1.59	1.260	0.80

The following head discharge equation is used for free flow Parshall flumes (James, 1988).

$$Q = C_f \left(K\, H_a\right)^{n_f} \tag{4.31}$$

where, Q = discharge, lit/sec; C_f = free flow coefficient depending on throat width (Table 4.1); n_f = free flow exponent (Table 10.5); H_a = head measured in the converging section of the flume, m and K = unit constant taken as 3.28 when H_a is in m.

In submerged case, the ratio of head in the diverging section *(H_b)* and that in the converging section *(H_a) i.e* $S = H_b/H_a >$ transition submergence, S_t. Table 4.1 can be referred for deciding submergence case considering the value of S_t. In submergence case, discharge is given (James, 1988) as:

$$Q = \frac{C_s\left[K\left(H_a - H_b\right)\right]^{n_f}}{\left[-\left(\log S + C\right)\right]^{n_s}} \tag{4.32}$$

where, Q = discharge, lit/sec; C_s = submerged flow coefficient depending on throat width Table 4.1); n_f and n_s = free and submerged flow exponent (Table 4.1); H_a *and* H_b = heads measured in the converging and diverging sections of the flume, m, respectively and K = unit constant taken as 3.28 when H_a is in m, S = *ratio* H_b *to* H_a, *and* C = 0.0044.

Sample Calculation 8:

A Parshall flume of throat width 7.5 cm is used to measure discharge. If the head measured in the converging section of the flume is 30 cm and the flow is under free flow, find out the discharge measured by it.

Solution:

Given throat width, W = 7.5 cm; H_a = 30 cm = 0.3 m and free flow condition.

Referring Table 4.1 for W = 7.5 cm, C_f = 28.09; n_f = 1.55.

Using H_a = 0.30 m and K = 3.28 (when H_a is in m),

Discharge is obtained as

Q = 28.09 x (3.28 x 0.3) $^{1.55}$ = 27.4 lit/sec.

Sample Calculation 9:

The discharge is measured by a Parshall flume having the following data.

Throat width, W = 7.5 cm; head measured in the converging and diverging section of the flume are 45 and 30 cm, respectively. Find out the discharge measured by the flume.

Solution:

Given, throat width, W = 7.5 cm; H_a = 45 cm = 0.45 m, H_b = 0.35 m.

Value of $S = H_b/H_a$ = 0.30/0.45 = 0.67.

Referring Table (4.1) for W = 7.5 cm, we find S_t = 0.64

Since $S > St$, it is submerged case.

Now refereeing Table (4.1), for W = 7.5 cm and under submerged case, C_s = 25.91; n_f = 1.55, n_s = 1.00; C = 0.0044; S = 0.67; K = 3.28 (H_a is taken in m).

Using the above data and Eq. (4.32), discharge is obtained as = 50.90 lit/sec.

Sample Calculation 10:

A parshall flume is used to irrigate a field with tomato crop .The throat width of the flume is 15 cm.The head measured in the converging and diverging sections are 50 cm and 35 cm, respectively. Whether the flow is free or submerged? Find out the discharge measured by flume?

Solution:

Given data

Throat width, W=15 cm; H_a=30 cm =0.3 m and H_b = 50 cm = 0.5 m

Value of $S = H_a/H_b = 0.3/0.5 = 0.7$

Referring Table (4.1) for W=15 cm, we find S_t = 0.55 (by interpolating)

Since $S > S_t$, it is submerged case.

Now referring Table (4.1) and by interpolating, for W = 15 cm and under submerged condition, C_s = 46.35

$n_f = 1.58$, $n_s = 1$, $C = 0.0044$, $S = 0.7$, $K = 3.28$ (where H_a *and* H_b are taken in meter)

Discharged can be computed from the following formula

$Q = Cs\ [\ K\ (Ha- Hb)]^n{}_f\ /\ [-\ (log\ S + C\ )^n{}_s$

$= 46.34\ [3.28\ (0.5 - 0.35)]^{1.58}\ /\ [-(\log 0.7 + 0.0044)]^{1}$

= 100.00 lit/sec.

The discharge of the parshall flume is obtained as = 100.00 lit/sec.

4.3.14 Cutthroat Flume

A cutthroat flume consists of (i) level floor (ii) a uniformly converging inlet section (iii) a uniformly diverging outlet section and (iv) a throat occurring at the intersection of the inlet and outlet sections. A plane view of a cutthroat flume and construction details are shown in Fig. 4.10.

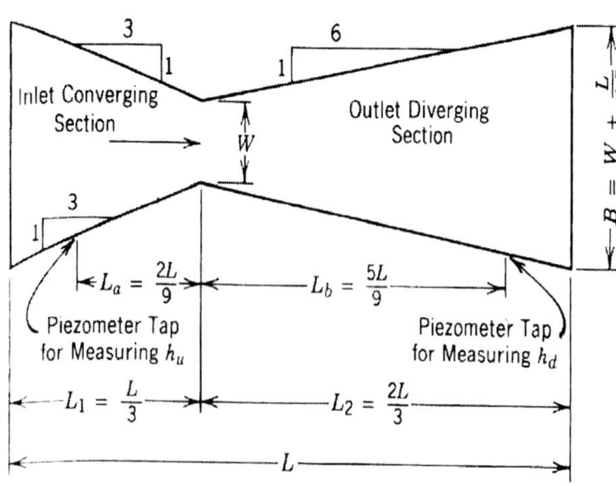

Fig. 4.10: Geometry of a cutthroat flume

Cutthroat flumes are easier and more economical to build and install than the Parshall flume. The same equations as used for discharge determination for free and submerged flow conditions in Parshall flume (Eqns. 4.31 and 4.32, respectively) are also used for discharge computation in cutthroat flume for free and submerged flow conditions, respectively. However, the parameters of these equations like C_f, C_s, n_f, n_s, S_t, K_f and K_s are different and are given in Fig.4.11 (James, 1988). Value of C is considered as zero. The following equations are used to calculate C_f and C_s.

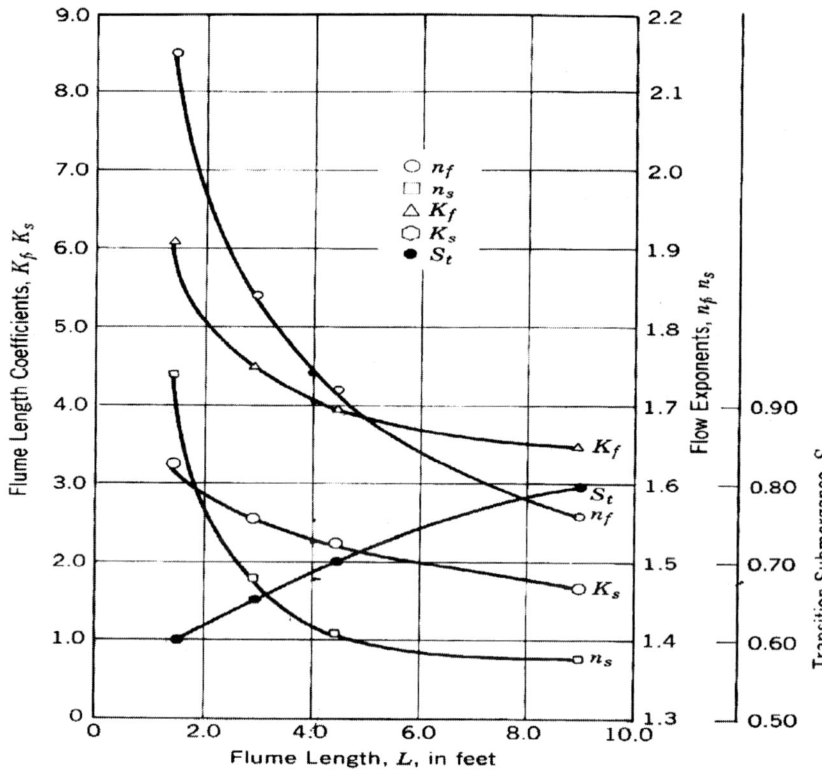

Fig. 4.11: Free flow and submerged flow coefficients (K_f and K_s), exponents (n_f and n_s) and transition submergence (S_t) for cutthroat flumes

$$C_f = K\,K_f\,W^{1.025} \tag{4.33}$$

$$C_s = K\,K_s\,W^{1.025} \tag{4.34}$$

where, W = throat width, ft; K_f *and* K_s are constants for free and submerged flow conditions, respectively (taken from Fig. 4.11); K = unit constant (1 for Q and H in Eqns. 4.31 and 4.32 in ft^3/sec and ft, respectively and also for H_a and H_b in Eq. 4.32 in ft).

Sample Calculation 11:

A cut throat flume is used to measure discharge under free flow condition for irrigating ground nut field. The flume has a length 4 ft. The throat width of the flume is 3 ft. The head measured in the converging inlet section of the flume is 2 ft.

Solution:

Given length of flume, L = 4 ft.

Width at throat, W = 3 ft.

Head at the converging inlet section of the flume, $H_a = 2$ ft.

From Fig. 4.11, when length, $L = 4$ ft, $n_f = 1.74$ and $K_f = 4.1$.

From Eq. (4.33),

Putting $K = 1$, $K_f = 4.1$ and $W = 3$, C_f is found as = 12.64

Under free flow case, discharge

Putting $C_f = 12.64$, $K = 1$, $H_a = 2$ and $n_f = 1.74$,

Discharge is obtained as 42.23 ft^3/sec which is equivalent to 1195.54 lit/sec.

Sample Calculation 12:

A cut throat flume is installed in an irrigation channel to measure the discharge. Length of flume is 6 ft. Width of flume at throat is 3.5 ft. Head measured at upstream and downstream end of the flume are 2.2 and 1.8 ft respectively. Decide the flow conditions and find out the discharge.

Solution:

Given data:

Throat width (W) = 3.5 ft, $H_a = 2.2$ ft, $H_b = 1.8$ ft, Length of flume (L) = 6 ft

From graph (Fig. 4.11), for flume length of 6 ft, the transition ratio (S_t)= 0.75

Again submergence ratio (S) = $H_b/H_a = 1.8/2.2 = 0.82$

As the submergence ratio is > transition ratio, so, the flow through the flume is submerged flow.

The discharge can be found out from the equation as:

$$Q = Cs\ [K\ (Ha - Hb)]^{n}{}_{f}\ /\ (-\log S\ )^{n}{}_{s}$$

In above equation H_a and H_b are in ft and Q is in cusec and so $K = 1$

Again, $Cs = K\ Ks\ W^{1.025}$

From graph (Fig. 4.11) for the throat length of 6 ft

$Ks = 2,\ n_f = 1.64,\ n_s = 1.38,$ and so, $Cs = 1{\times}2{\times}3.5^{1.025} = 7.22$

So, discharge = 7.22 $[1(2.2\text{-}1.8)]^{1.64}$ / $(\text{-log } 0.82)^{1.38}$ = 47.32 cusec which is equivalent to 1339.64 lit/sec.

4.3.15 Chemical Dilution Method

In this method of flow measurement, a relatively large quantity of dye, called tracer is dissolved in a small quantity of water and placed in a bottle so that the tracer solution can be discharged at a known rate into the water flowing in the stream. By measuring the concentration of tracer at upstream and downstream side of the stream, the discharge is computed as (Michael, 1987):

$$Q = q_1 \frac{(C_2 - C_1)}{(C_0 - C_2)} \qquad (4.35)$$

where, Q = discharge flowing in the stream, q_1 = quantity of tracer added, C_0 = original tracer concentration, C_1 = concentration of the tracer in the bottle, C_2 = concentration of the tracer at the downstream side of the stream.

4.3.16 Radioisotope Method

It may be used in place of chemical or dye tracer and the degree of dilution is determined by counting the gamma ray emissions from the diluted isotope solution using Geiger counters. In the total count method, a known amount of radioisotope is introduced into the flow in a relatively small time. At the downstream side measuring station where the isotope is thoroughly mixed, concentration of radioisotope tracer is determined from the gamma ray emissions detected and counted by the counter. The discharge is given as (Michael, 1987):

$$Q = \frac{F\,A}{N} \qquad (4.36)$$

where, Q = discharge flowing in the stream/channel, F = counts per unit of radioactivity per unit volume of water per unit of time, A = total units of radioactivity to be introduced for each discharge measurement and N = total counts.

Radioactive materials must be handled very carefully. Emissions of Gamma radiations are highly injurious. Only licensed personnel are allowed to handle it.

Question Banks

Q1. What is the definition of discharge? What are the different units used to find out discharge?

Q2. What are the objectives of discharge measurement?

Q3. Write the different methods used to estimate discharge.

Q4. How is the volume method of discharge measurement done?

Q5. Water is flowing in a channel. A container of 10 lit capacities is placed in the channel and it takes 10 sec to fill the container fully. Find out the discharge flowing in the channel.

Q6. You are given a container and a stop watch. How can you find out the discharge of a pump with using only these two equipments?

Q7. A pump is used to irrigate a crop field which has 5 ha area. The average depth of irrigation water supplied by the pump is 5 cm. If the time required to irrigate the crop field is 30 min, find out the discharge of the pump.

Q8. Describe in details the velocity-area method used to estimate the discharge.

Q9. What are the approaches followed for discharge computation by velocity-area method?

Q10. What are the differences between the mid section and mean section method for discharge measurement?

Q11. Explain with figures how the mid section and mean section methods are used to measure discharge in a channel.

Q12. Narrate how the slope-area method is used to estimate the discharge?

Q13. Write down the Manning's and Chezy's formula for estimation of velocity of flow.

Q14. Explain how the water surface slope and energy slope of a channel are determined.

Q15. What are different methods used to find out the depth of flow in a channel?

Q16. How the sounding rods are used to measure depth of flow?

Q17. How does the echo sounder help to measure the depth of flow in a channel?

Q18. Briefly describe how the float method is used to measure velocity of water.

Q19. What is a current meter? What are the various types of current meters?

Q20. With neat sketch describe the working and construction of vertical axis current meter.

Q21. What is a Pigmi current meter? Describe its advantages in measurement of velocity.

Q22. Describe how the current meter is calibrated to measure the velocity.

Q23. Following velocities are recorded in an irrigation channel by a current meter.

Depth above the channel bed, m	0	1.0	2.0	3.0	4.0
Velocity, m/sec	0	0.4	0.5	0.7	0.8

Find out the discharge per unit width of the channel near the point of measurement if the total depth at the point of measurement is 5 m.

Q24. Following readings are taken by a current meter in a main canal section.

Distance from one bank to the other, m	0	0.5	1.0	1.5	2.0	2.5	3.0	3.5	4.0
Velocity at 0.2 depth, m/s	0	0.4	0.6	0.7	0.9	1.2	0.9	0.6	0
Velocity at 0.8 depth, m/s	0	0.3	0.5	0.7	0.8	1.0	0.8	0.5	0
Depth, m	0	0.3	0.8	1.0	1.5	1.2	1.0	0.5	0

Compute the total discharge flowing in the canal by both mid section and mean section method.

Q25. Calculate the discharge in a channel section from the following observations taken by a current meter whose rating curve is $V = 0.04 + 0.3\,N$

Distance from bank, m	2	4	6	8	10	12	14
Depth, m	0.4	0.8	1.0	1.2	1.0	0.8	0.5
Revolution of current meter/sec	0.9	1.0	1.2	2.2	2.0	2.0	1.0

Q26. How a gauge-discharge curve is is prepared? Explain how this curve helps in estimation of discharge.

Q27. What are the essential requirements of a gauge site?

Q28. Write short notes on the followings.

(i) Staff gauge (ii) vertical staff gauge (iii) inclined gauge (iv) hook gauge (v) self recording gauge

Q29. Describe the construction and function of a stage recorder.

Q30. Write down the essential requirements of a discharge measuring site.

Q31. Write down the essential requirements of a discharge measuring site in case of a hilly area.

Q32. Draw a typical stage-discharge curve and explain how does this curve help in computation of discharge.

Q33. What are the differences between the weir and flume?

Q34. What is a weir? What are the different types of weirs?

Q35. With a neat sketch, explain the working of a rectangular weir for measurement of discharge.

Q36. Write short notes on the followings.

(i) Francis' formula (ii) Bazin's formula (iii) Rehbock's formula

Q37. Write down the formula to compute the discharge through trapezoidal and triangular weir.

Q38. What is a cipoletti weir? Draw a diagram of the cipoletti weir and describe how the discharge is measured by it.

Q39. Find the discharge over a suppressed rectangular weir having 4 m length and the height of water above the crest level is 60 cm under (i) no end contraction and (ii) with one side end contraction.

Q40. Find the discharge over a rectangular weir using the following data.

The length of the weir crest = 5 m. The crest level height above the channel bed = 1.2 m and the height of water flowing over the crest is 2.0 m above the channel bed.

Q41. What are the differences between ogee spillway and siphon spillway?

Q42. With a neat sketch describe the construction and function of an ogee spillway and a siphon spillway.

Q43. What is a proportional weir? Draw a diagram of proportional weir and describe the working of it.

Q44. How can you measure the discharge flowing in a pipe line? Discuss the use of orifices and mouthpieces for measurement of discharge in a pipe line.

Q45. Differentiate between a parshall flume and cutthroat flume.

Q46. With a neat label sketch, explain the construction and working a parshall flume.

Q47. How can you know the flow through a parshall flume and cutthroat flume are under free or submerged condition?

Q48. Draw a label sketch of a cutthroat flume and describe its principle to use to measure the discharge.

Q49. Write down the formula to measure the discharge through a cutthroat flume under submerged condition.

Q50. Compute the discharge passing through a parshall flume under the following condition.

(i) Throat width = 15 cm and (ii) Head measured in the inlet and out let sections of the flume are 50 and 40 cm, respectively.

Q51. A cutthroat flume is installed in a channel to measure the flow of water passing through it. Length of the flume = 6 ft, width of the flume at throat = 3.5 ft, Head of water measured at the upstream and down stream end of the flume are 2.0 and 1.7 ft, respectively. Find out the flow condition and accordingly find out the discharge.

Q52. Describe the chemical dilution method and radioisotope method for measurement of discharge in a stream/channel.

References

Garg, S.K. 1978. Irrigation Engineering and Hydraulic Structures. Khanna Publishers, Delhi.

James, L.G. 1988. Principles of Farm Irrigation System Design. John Wiley & Sons, New York, pp. 543.

Lenka, D. 2001. Irrigation and Drainage. Kalyani Publishers, New delh.

Michael, A.M. 1978. Irrigation Theory and Practices. Vikash Publishing House Pvt. Ltd., New Delhi.

Modi, P.N. and Seth, S.M. 2005. Hydraulics and Fluid Mechanics Including Hydraulic Machines. Standard Book house, Delhi.

Murty, V.V.N. 1998. Land and Water Management Engineering. 2nd Edition, Kalyani Publishers, New Delhi, pp. 586.

Sharma, R.K. and Sharma, T.K. 2002. Irrigation Engineering including Hydrology, S. Chand and Company Ltd. New Delhi.

Sharma, R.K., 1987. A Text book of Hydrology and Water Resources. Dhanpat Rai and Sons, Delhi.

U.S. Bureau of Reclamation 1975. Water Measurement Manual, U.S. Government Printing Office, Washington, D.C. pp. 327.

Walker, W.R and Skogerboe, G.V. 1987. The Theory and Practices of Surface Irrigation. Prentice-Hall, Inc., Englewood Cliffs, N.J.

5

Water Requirement of Crops

5.1 Introduction

A reservoir that collects and harvests *in-situ* rainfall as well as runoff from the catchment area can be used for many purposes: the most important of which is the supply of water to crops as supplemental irrigation. The water of the reservoir can be used for domestic and industrial uses. It is also a good source for recreational use. Total water demand of the reservoir including irrigation to different crops in different seasons, domestic and other non agricultural uses are to be assessed. While estimating the crop water requirement and irrigation requirement, various crops to be grown in the service area of the reservoir, their water requirements, crop duration, extent of coverage and water management practices followed are to be taken into consideration. Multiplying the irrigation requirement with crop coverage area, volumetric irrigation water requirement of crop can be found out. Summation of irrigation requirement of all crops grown in service area of the reservoir gives total irrigation demand of the reservoir.

Small farm reservoirs are normally used for irrigation to crops as well as to meet the live stock water demand. Normally the large live stocks like horse, cattle require 20 thousand litres of water per head per year whereas the small live stocks like sheep, goat, pigs require 2 thousand litres of water per head per year. Adding the irrigation demand to the livestock demand, total water requirement of the reservoir is calculated. In addition to this, water used for industrial, domestic and other sectoral purposes are to be added to irrigation and livestock demand to estimate the total water demand of the reservoir.

5.2 Water and Irrigation Demand of Crops

Of the various demand of water from the reservoir, irrigation demand of crops grown in the command area of the reservoir is very important. Most of the crop water demand is met from rainfall. However, when the rainfall is deficient to meet the crop water demand, then the deficit is met from the stored water of the reservoir in the form of irrigation, more commonly called as supplemental irrigation. The irrigation to the crops is required mostly to meet the evapotranspiration requirement of the crop. Before discussing the water and irrigation demand/requirement of crops, it is essential to know some commonly used terms frequently used.

5.3 Certain Important Definitions

Consumptive Use: It is the quantity of water lost in evaporation from the soil and water surface, transpiration by the vegetations grown on the soil and that used in metabolic activity. Water used in metabolic activity is very less (usually less than 1% of evapotranspiration). Thus, in usual sense, the term consumptive use is taken as evapotranspiration, *ETcrop*.

Water Requirement: It is the water needed for raising a crop in a given period starting from seedling to harvest. It includes consumptive use and other unavoidable losses and also the water required for special operations like land preparation, transplanting and leaching.

Evapotranspiration: It is the total amount of water lost by the evaporation from the soil and water surface and transpiration by the vegetations. It depends on the crop types, its coverage amount and management and cultural practice of the crop.

Irrigation Requirement: It refers to the quantity of water required to supply for successful growth of plants exclusive of precipitation, groundwater contribution and other natural resources.

Effective Rainfall: It is that part of rainfall which forms part of water requirement. Effective rainfall depends on a number of factors. It depends on the amount, intensity and distribution of rainfall event, antecedent soil moisture content, crop cover and management practice, and the season. A light intensity rainfall in winter may be 100% effective whereas a high intensity rainfall in rainy season occurring for a small duration may be 30-50% effective.

Gross Commanded Area (GCA): It is the total area, bounded within the irrigation boundary of a project. The boundary of GCA is usually defined by the drainage in either side across which irrigation cannot be expanded economically.

Cultivable Command Area (CCA): It is that part of the gross command area in which cultivation is possible. It is less than the gross command area. The area under small drainage, ponds, abadies, reserve forest roads, railways, hills etc. is excluded from the gross command area to get the cultivable/culturable command area.

Intensity of Irrigation: It is the percentage of CCA which is irrigated annually or seasonally. It varies with the type of irrigation, climate of the area, water management practice of the crops, rainfall and its distributions etc.

Area to be Irrigated: It is the area to which irrigation is provided. It is to be remembered that not all the CCA are the irrigated areas.

Paleo Irrigation: Sometimes, in the initial stage before the crop is sown, the land is very dry. This particularly happens in case of *rabi* crops when it is sown when the soil is too dry to be sown. In that context, the field is irrigated to moisten the soil so as to help in sowing the crops. This irrigation is called as Paleo irrigation.

Kor-watering: The first watering given to a crop when the crop is a few centimetres high is called as kor-watering. It is usually the maximum single watering followed by other waterings at usual intervals.

Crop Period: It is the time taken by a crop from sowing/planting till harvesting. The crop period varies with the types of crops and also for the same crop, it varies with the type of varieties.

Base Period: It is the time period between the first watering of a crop at the time of sowing to its last watering before harvest.

5.4 Duty

It is the relation between the area irrigated, or to be irrigated, and the amount of water used, or required to irrigate the crop for the purpose of its growing. When applied to a channel, it is the area irrigated during a base period divided by the mean supply used in cumecs. It may be defined as the area irrigated by unit discharge that means the number of hectares under a particular crop brought to maturity by a constant supply of one cumec flowing continuously for the base period. Duty based on discharge (discharge calculated at the head regulator) passed through the outlet and excluding all losses in the canal system is called the outlet discharge factor and is usually represented by symbol *D (*Sharme and Sharma, 2002*)*.

5.5 Delta

It is an expression used in the irrigation practice to mean the depth of water that would result over a given area for a given discharge for a certain length of time. It is a term equivalent to duty of irrigated water when the latter is expressed in water depth units. Delta is the total depth of water in cm needed by a crop for its growth. The average delta for rice (short duration) is 80-100 cm, for medium and long duration rice it is 100-120 cm and 120-150 cm, respectively. The delta of non-rice crops are generally less: pulses have values about 30 cm, oil seed crops, 40 cm and for most of the vegetables it is 25-35 cm. The delta of crops depends on the types of crops, crop varieties, duration of the crops, and climate of the areas where the crops are grown.

5.6 Relationship Between Duty and Delta

Let D be the duty of water in hectares/cumecs; B be the base period in days and Δ be the delta of water in metre.

Volume of one cubic metre of water flowing in one day

$$= 1 \times 24 \times 60 \times 60 = 86,400 \text{ m}^2 = 8.64 \text{ ha m.}$$

Volume of one cubic metre flowing for B days = 8.64 B ha m.

By definition of duty D, one cubic metre = 10,000 D m^2 of water supplied for B days matures D hectares of land.

Total depth of water applied in the land = volume/area = 86400 B/10000 D. = 8.64 B/D m.

Now delta is the total depth of water, Δ = 8.64 B/D m which gives

$D = 8.64\ B/\Delta$ m and $\Delta = 8.64\ B/D$ m = 864 B/D cm.

Sample Calculation 1:

Find out the delta of potato if the crop has a base period of 90 days. Irrigation is applied at an average interval of 10 days and at each irrigation, 5 cm depth of water is applied.

Solution:

Base period of the crop = 90 days

Average interval of irrigation = 10 days

Number of irrigations required = 90/10 = 9

Depth of irrigation at each interval = 5 cm

Hence, total delta of irrigation = Δ = 9 x 5 = 45 cm.

Sample Calculation 2:

Calculate the delta of a crop given its duty is 864 hectares/cumec on the field. Its base period is 100 days.

Solution:

Δ (cm) = 864 B/D where B (base period) in days and D (duty) is in hectares/ cumec.

It is given that B = 100 days and D = 864 hectares/cumec

Hence, Δ = 864 x 100/864 = 100 cm = 1 m.

5.7 Duty at Various Places

In a large canal irrigation system, water flows from the head regulator in to the crop field through main canal first; then it moves from the main canal to branch canal; from branch canal it moves to distributaries; then water moves to the minor; from minor to water courses (field channel) and finally to the crop field. The entire water distribution system from the head

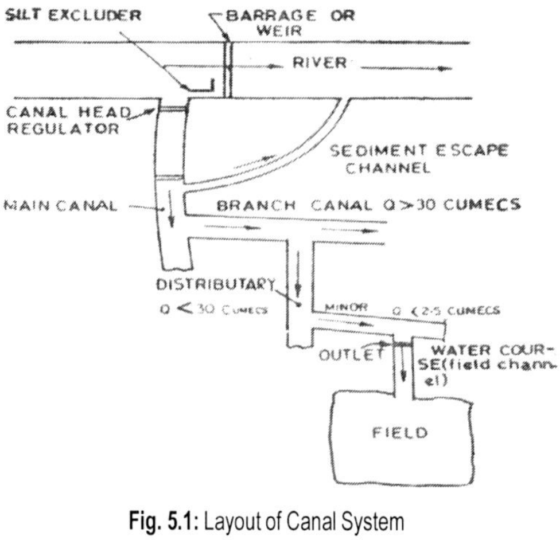

Fig. 5.1: Layout of Canal System

regular to the crop field is called as canal irrigation system (Fig. 5.1). During this passage of water in the canal irrigation system, some amount of water is lost due to the evaporation and seepage losses in the canal systems. These losses are called as transit or transmission or conveyance losses.

Duty of water for a crop is the number of hectares of land, which the unit flow of water for B days can irrigate. So, if the water requirement of a particular crop in a given location is more, then irrigation coverage will be less. In other words, if the water requirement of a crop is more, then its duty will be less. It, therefore, becomes clear that the duty of water at the head of the water course will be less than the duty of water at the field. The reason is because, when water flows from the head of the water course to the field, some amount is lost as seepage losses in the conveyance system. Duty of water varies from place to place in the canal system and in general it increases as one moves downstream from head of the regulator (main canal) to the crop field. The duty at the head of the water course (i.e. at the outlet point of the minor) is very important and is called as the outlet discharge factor or average outlet factor. The control of the Irrigation Department ends at the outlet point of the minor i.e. at the head of the water course and then water is carried into the crop field through the water courses or field channel and may be managed by the beneficiary cultivators.

5.8 Factors Affecting Duty of Crop

Duty of irrigation water depends on the following factors:

(i) **Type of crop:** Different crops require different amount of water, and thus, their duties are different. Duty will be more for a crop whose water requirement will be less and vice-versa.

(ii) **Climate and season:** Duty varies from season to season since the losses of water due to evaporation will be different for different seasons. Moreover, it also varies from day to day within the same season. In general the duty is less during the summer than the rainy or winter season. The figure for duties which is generally expressed is the average value considered over the entire season.

(iii) **Effective rainfall:** Effective rainfall is the rainfall which is directly useful to the crop. Rainfall excluding the percolation and surface runoff is termed as effective rainfall. If the effectively of the rainfall is more; in other words the effective rainfall is more, then the irrigation requirement will be less, and so, the duty will be more.

(iv) **Type of soil:** If the soil is light textured like sandy loam soil, then the percolation loss in it will be more which will necessitate more irrigation water. In such case, the duty of the water will be less. In other words, the heavy soil like clay has more duty.

(v) **Efficiency of cultivation method:** If the cultivation method including the irrigation and water management practice is faulty resulting more requirement of water, then the duty will be less. In other words, an economic irrigation management practice will enhance the duty.

5.9 Measures for Improving Duty of Water

Following measures are essential to increase the duty of water

(i) Land to be used for cultivation should be leveled.

(ii) Improved cultivation method should be practiced.

(iii) Soil porosity should be reduced to decrease the percolation.

(iv) Soil should be perfectly ploughed to the required depth.

(v) Alkaline soil should be leached well before sowing.

(vi) Organic matter content of the soil should be improved with addition of manures, green crops etc to enhance the water holding capacity of the soil.

(vii) Rotation of crops must be practiced to enhance the water-use-efficiency.

(viii) Water saving irrigation methods should be followed to reduce the water requirement of the crop and increase the water-use-efficiency of the crop.

(ix) Conveyance losses in the canal systems including the water courses should be reduced by lining the canals so that the losses will be less and duty will be more. If possible, the R.C.C. pipes may be used for conveyance system to reduce the conveyance loss.

(x) If surface irrigation is followed, then furrow irrigation should be used instead of free flooding.

(xi) It is better to irrigate the fields by sub-surface irrigation or presurised irrigation including drip or sprinkler method.

(xii) Volumetric measurements must be used to control the flow of water into the field.

(xiii) Field channels should be constructed and water through them should flow in to the field so that duty will be more.

(xiv) Rotational irrigation system will increase the duty of water.

5.10 Important Crop Seasons in Indian Agriculture

From the agricultural point of view, the Indian agriculture is mainly divided into two main seasons i.e. *Kharif* and *Rabi. Kharif* starts from April to September whereas *Rabi* starts from October to March. These dates are not fixed. They may vary from one to three months on either side. In some cases there are three crop seasons i.e. *Kharif, Rabi* and *Summer.* In such case, *Kharif* starts from June to September, *Rabi* starts from October to January and the *Summer* starts from February to May.

The important *Kharif* crops are *rice, bajra, jowar, maize, cotton, tobacco, groundnut* etc. The *Rabi* crops are *wheat, barly, gram, linseed, mustard, potatoes* etc. The *Rabi* crops are also called as *Winter* crops.

5.11 Discharge Computation at the Head Regulator

Discharge can be computed by average outlet factor as D = 864 B/Δ where D is duty in hectares/cumec, B is in days and Δ is delta of water in cm. However, a channel designed on this average outlet factor will prove to very in adequate, as it will fail to supply the required water to the crop at its peak demand. It is recommended that a channel be designed to carry maximum discharge which is estimated by taking into account the peak water demand in the peak period which will be clear in some of the worked out examples as mentioned below. Provision of channel losses should be there to decide the final channel capacity.

To obtain precise channel design capacity, one should find out the monthly or 15 days water requirement of various crops. The water depth required in this interval is multiplied by the crop area so as to give us the volume of water required in this interval. Dividing the volume by interval, we can find out the discharge required in each interval, by various crops; the summation of which for all the crops will give us the discharge required by all the crops in each interval. The channel may then be designed, for maximum of these values. It is evident that in order to be more precise, the interval should be as small as possible. Generally, monthly water requirement studies are conducted, and the channel capacity is increased by 20 to 25% to cater for peak demand in the month (Garg, 1987).

Sample Calculation 3

The gross commanded area for a distributary is 4000 hectares, 60% of which is cultivable. The intensity of irrigation for *rabi* season is 50% and that for *kharif* season is 30%. If the average duty at the head of the distributary is 1800 hectares/cumec for *rabi* season and 1000 hectares/cumec for *kharif* season, find out the discharge required at the head of the distributary from average demand considerations.

Solution:

G.C.A = 4000 hectares

C.C.A = 4000 × 60/100 = 2400 hectares.

Area to be irrigated in *rabi* season

= C.C.A × intensity of irrigation

= 2400 × 50/100 = 1200 hectares

Area to be irrigated in *kharif* season

= 2400 × 30/100 = 720 hectares.

Water required at the head of the distributary to irrigate *rabi* area

= 1200/1800 cumecs = 0.667 cumecs

Water required at the head of the distributary to irrigate *kharif* area,

= 720/1000 = 0.720 cumecs

Thus, the requirement in *kharif* season is 0.720 cumecs and that in *rabi* season is 0.667 cumecs. The required discharge is maximum of the two. i.e. 0.720 cumecs.

Hence, the distributary should be designed for 0.720 cumecs discharge at its head from average demand considerations.

Sample Calculation 4:

The gross command area for a distributary is 4000 hectares, 60% of which is cultivable. The intensity of irrigation for *rabi* season is 50% and that for *kharif* season is 30%. Wheat is grown in the entire command in *rabi* and paddy in *kharif*. Find out the discharge required at the head of the distributary considering the peak water requirement of *rabi* crop (wheat) 15 cm and the period of peak water requirement 4 weeks and that for *kharif* cop (paddy) 30 cm with peak period of water requirement 4 weeks.

Solution:

Outlet factor for *rabi* crop (wheat) = 864 × B/D= 864 x (4 × 7)/15.0

= 1613 hectares/cumecs and that for *kharif* crop (paddy)

= 864 × (4 × 7)/30

= 806 hectares/cumecs.

C.C.A. = 4000 x 60/100 = 2400 hectares

Area to be irrigated in *rabi* season

= C.C.A x intensity of irrigation

= 2400 x 50/100 = 1200 hectares

Area to be irrigated in *kharif* season

= 2400 x 30/100 =720 hectares.

Water required at the head of the distributary to irrigate *rabi* area
= 1200/1613 = 0.744 cumecs

Water required at the head of the distributary to irrigate *kharif* area

= 720/806 = 0.893 cumecs

The required discharge is maximum of the two i.e.0.893 cumecs.

Note: The required discharge from peak or maximum water requirement consideration have gone up to 0.893 cumecs from 0.720 cumecs.

Sample Calculation 5:

The cultivable commanded area of watercourse is 1000 hectares. Intensities of irrigation of sugarcane and groundnut crops (*rabi* season crop) are 30% and 40% respectively. The duties for the crops at the head of the watercourse are 750 hectares/cumec and 1400 hectares/ cumec, respectively. Compute the discharge required at the head of the watercourse.

Solution:

Cultivable command area. = 1000 hectares

Intensity of irrigation for sugarcane = 30%

Area to be irrigated under sugarcane

= 1000 × 30/100 = 300 hectares.

Intensity of irrigation for groundnut = 40%

Area to be irrigated under groundnut

= 1000 × 40/100 = 400 hectares.

Given duty for sugarcane = 750 hectares/cumec

Duty for groundnut = 1400 hectares/cumec

Hence discharge required for sugarcane = 300/750 cumec = 0.40 cumec

and discharge required for groundnut = 400/1400 cumec = 0.285 cumec

Now, sugarcane requires water throughout the year and groundnut for only rabi season. Hence the water requirement at the head of the watercourse at any time of the year will be the summation of the two, i.e. equal to 0.40 + 0.285 = 0.685 cumecs.

It is to be noted that the discharge during rabi season will be 0.685 cumec and for the rest of the year, it will be 0.40 cumec.

5.12 Methods to Estimate Crop Water Requirement

The most important component of water requirement of crop is its evapotranspiration. There are several methods used to determine the crop evapotranspiration. Broadly they can be classified as:

(i) Direct measurements and

(ii) Estimation method

The direct measurement method comprises the followings:

(i) Lysimeter experiment

(ii) Field experimental plots

(iii) Soil moisture depletion studies and

(iv) Water balance method.

Estimation of evapotranspiration can be achieved by:

(i) Aerodynamic method

(ii) Energy balance method

(iii) Combination method and

(iv) Empirical method

5.12.1 Lysimeter Method

Lysimeter studies involve the growing of crop in large containers (lysimeters) and measuring their water loss and gains. Mainly there are two types of lysimeters (Fig. 5.2). They are (i) weighing type and (ii) drainage type. Weighing type lysismetrs give more accurate results compared to the drainage type. To get more accurate and short period results, weighing type lysimeter can be used. The soil and crop conditions of the lysimeters should be close to the natural condition. From the irrigation point of view, weighing type lysimeter is set up to enable the operator to measure the water balances: water added, water retained by the soil, and water lost through all sources-evaporation, transpiration and deep percolation. These measurements require weighing that may be made with scales or by floating the lysimeters in water in which case the change of liquid di88splacement is computed against water loss from the tank.

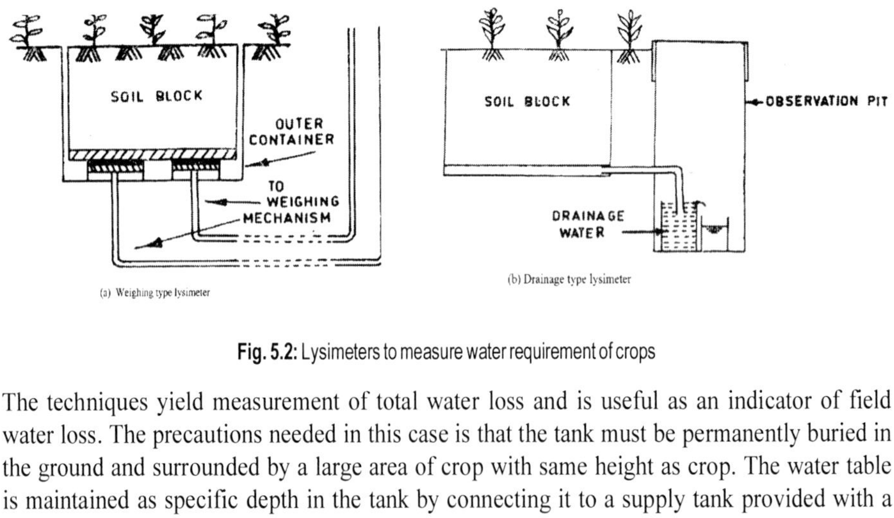

Fig. 5.2: Lysimeters to measure water requirement of crops

The techniques yield measurement of total water loss and is useful as an indicator of field water loss. The precautions needed in this case is that the tank must be permanently buried in the ground and surrounded by a large area of crop with same height as crop. The water table is maintained as specific depth in the tank by connecting it to a supply tank provided with a float mechanism which has an arrangement for receiving excess water that tends to build up in the tank. Water is applied in measured amount to the lysimeter as irrigation is applied to the crops in the fields where lysimeter is set up. The water received either from the reservoir or precipitation excluding the outflow, constitutes the water used by the crops (Lenka, 2001).

5.12.2 Soil Moisture Depletion Studies

This method is usually suitable for irrigated areas where soil is fairly uniform and the depth to the groundwater is such that it will not influence the soil moisture fluctuations within the crop root zone. In this method, the soil moisture within the crop effective root zone before and just after irrigation and also within each irrigation cycle is measured. The soil moisture content from different profile layers within the crop effective root zone are measured and then these values are used to compute the total soil moisture content in the entire crop root zone. There are a number of methods used to measure the soil moisture content i.e. gravimetric method, neutron probe method, infra red, tensiometer, gypsum block, pressure membrane and pressure plate technique.

The water use from the crop root zone is given (Mishra and Ahmed, 1987) as:

$$u = \sum_{i=1}^{n} \frac{(M_{1i} - M_{2i})}{100} . A_i . D_i + ER - Dp \qquad (5.1)$$

where, u = water use from the crop root zone for successive sampling periods or within one irrigation cycle, mm; n = number of soil profile layers in the crop effective root zone D; M_{li} = soil moisture percentage at the time of the first sampling in the ith layer; M_{2i} = soil moisture percentage at the time of the second sampling in the ith layer; A_i = apparent specific gravity of the ith layer of the soil and D_i = depth of the ith layer of the soil profile, mm, ER = effective rainfall, mm and Dp = percolation from the crop effective root zone, mm. The seasonal value of water use by the crop is calculated by adding the water use of each sampling interval. A correction is made by adding the potential evapotranspiration value(s) for the accelerated water loss for the interval(s) just after irrigation(s) and before soil moisture sampling.

5.12.3 Field Experimental Plots

In this method, measurement of water supplies to the fields and change in soil moisture contents of field plots are made. The seasonal water requirement is computed by adding measured quantities of irrigation water, effective rainfall received during the season and contribution of soil moisture form the crop effective root zone. This is expressed by the relationship (Mishra and Ahmed, 1987):

$$WR = \sum_{i=1}^{n} \frac{(M_{bi} - M_{ei})}{100} . A_i . D_i + IR + ER \qquad (5.2)$$

where, WR = seasonal water requirement, mm; IR = total irrigation water supplied, mm; ER = seasonal effective rainfall, mm; M_{bi} and M_{ei} = soil moisture percentage at the beginning and end of the season in the ith layer of the soil, respectively and other terms of equation are as discussed above. This method has the limitation that it does not provide information on intermediate soil moisture conditions, short term use, profile use, deep percolation and peak use rate of the crop.

Sample Calculation 6

Profile soil moisture content of a potato field in 60 cm effective root zone depth are measured as follows:

Profile layer, cm	Soil moisture content,%		Apparent Specific gravity, gm/cc
	Beginning of irrigation cycle	End of irrigation cycle	
0-15	18	10	1.60
15-30	20	11	1.58
30-45	25	15	1.59
45-60	25	15	1.60

If effective rainfall within the irrigation cycle is10 mm, and percolation from the effective root zone of the crop is 10 mm, estimate the water use from the crop root zone within the irrigation cycle.

Solution:

Water use of the crop is given by Eq. (5.1) as

$$u = \sum_{i=1}^{n} \frac{(M_{1i} - M_{2i})}{100} . A_i . D_i + ER - Dp$$

Given effective rainfall, ER = 10 mm and deep percolation loss Dp = 10 mm.

The soil has 4 layers of each 15 cm thick (Di = 15 cm).

The apparent specific gravity (Ai) is given in the question for each layer.

The values of soil moisture content at beginning and end of each irrigation cycle (M_{li} and M_{2i}) are also mentioned in the question for each layer.

Using the given data, water use is calculated as

= (18-10)/100 × 1.60 × 150 + (20-11)/100 × 1.58 x 150 + (25-15)/100 × 1.59 × 150 + (25-15)/100 × 1.60 × 150 + 10 -10 = 88.5 mm.

Sample Calculation 7

A field is sown with ground nut. The profile soil moisture content of the field at sowing and at harvest time are measured and presented in table as below.

Profile layer, cm	Soil moisture content,%		Apparent Specific gravity, gm/cc
	At sowing	At harvest	
0-15	20	10	1.62
15-30	22	11	1.60
30-45	25	13	1.59
45-60	25	15	1.60

If the crop is supplied with 10 cm irrigation, effective rainfall in the crop growing season is 12 cm and there is no percolation loss from the crop effective root zone in the entire season, then estimate the seasonal water requirement of the crop.

Solution:

Water requirement, WR of the crop is given by Eq. (5.2) as:

$$WR = IR + ER + \sum_{i=1}^{n} \frac{(M_{bi} - M_{ei})}{100} . A_i . D_i$$

Given effective rainfall, ER = 12 cm = 120 mm and irrigation = IR = 10 cm = 100 mm.

The soil has 4 layers of each 15 cm thick (D_i = 15 cm).

The apparent specific gravity (A_i) is given in the question for each layer.

The values of soil moisture content at beginning and end of crop growing season (M_{bi} and M_{ei}) are also mentioned in the question for each layer.

Using the given data, water requirement is calculated as:

= (20-10)/100 × 1.62 × 150 + (22-11)/100 × 1.60 × 150 + (25-13)/100 × 1.59 × 150 + (25-15)/100 × 1.60 × 150 + 100 -120 = 323 mm.

5.12.4 Water Balance Method

The method is suitable for large fields over long periods. It is based on inflow and outflow of water to the soil control volume. The water balance model of a crop field (generally non-paddy) under unsaturated phase is expressed (Panigrahi, 2001) as:

$$S_{w,i} = S_{w,i-1} + R_i + I_{s,i} - ETcrop_i - SR_i - S_i - D_i \tag{5.3}$$

where, R = rainfall; I_s = supplemental irrigation, $ETcrop_i$ = evapotranspiration of the crop; SR = runoff from the crop field; D = percolation loss in the crop field, S = seepage loss from the crop filed and S_{wi} and S_{wi-1} = soil moisture content on ith and (i-1)th day, respectively. The term i of the above equation refer to the day which may be taken as 1 day.

Water balance model for paddy field (which is generally under saturated/ponding phase) is given (Panigrahi, 2001) as follows:

$$D_{p,i} = D_{p,i-1} + R_i + I_{s,i} - ETcrop_i - SR_i - S_i - D_i \tag{5.4}$$

where, D_{pi} and D_{pi-1} are the ponding depths occurred at the end of day i *and* i-1, respectively. Other terms of the equation are described as above. The unit of various terms in the above equation is mm. It is to be remembered that the capillary rise due to groundwater table is ignored (Odhiambo and Murty, 1996).

Measuring all the parameters of the water balance model (Eqs.5.3 and 5.4) except crop evapotranpiration, we can calculate evapoptranspiration. The method has the advantage that we can compute the crop evapotranspiration on short term basis like daily basis. Details of estimation of the various parameters of the above water balance model are available in the book (Panigrahi, 2011).

5.12.5 Estimation of Evapotranspiration

All methods for computing evapotranspiration (*ETcrop*) involve the following equation:

$$ETcrop = Kc\ ET_o \tag{5.5}$$

where, *ETcrop* = crop evapotranspiration or simply evapotranspiratuion; *Kc* = crop coefficient and ET_o = reference crop evapotranspiration or potential evapotranspiration.

Potential evapotranspiration is the maximum rate at which water can be removed from the soil and plant surfaces. Potential evapotranspiration depends on the amount of energy available for evaporation and varies from day to day. Reference crop evapotranspiration, simply called as reference evapotranspiration, is the potential evapotranspiration for a specific crop (usually either grass or alfalfa) and set of surrounding (advective) conditions. Reference crop evapotranspiration is defined as the evapotranspiration from an extensive surface of 8 to 15 cm tall green grass cover of uniform height, actively growing, completely shading the ground and having no shortage of water (Doorenbos and Pruitt, 1977). Reference evapotranspiration is preferred over the potential evapotranspiration, since the later may vary from crop to crop due to the differences in aerodynamic roughness and surface reflectance (albedo) and from location to location. Reference evapotranspiration, on the other hand, is defined for a specific crop and set of advective conditions.

Many methods with differing data requirements and levels of sophistications have been developed for calculating ET_o. Some are physically based while others are determined empirically. Details of estimation of ET_o by various methods are discussed below.

5.12.5.1 Estimation by Aerodynamic Method

In the aerodynamic methods, vapour flux is proportional to mean wind speed and the vapour pressure difference between the evaporating surface and the surrounding air.

Dalton proposed the following equation for estimating evaporation from a water surface (James, 1988).

$$ET_o = (e_s - e)\, f(u) \tag{5.6}$$

where, e_s = vapour pressure at the plant surface (within the boundary layer surrounding the leaf); e = vapour pressure at some height above the plant and $f(u)$ = function of the horizontal wind velocity.

Dalton equation is not widely used to estimate ET_o because of the difficulty of determining e_s within the boundary layer surrounding the plant. This is due to primarily the very large vapour pressure and temperature gradients that often exist within the boundary layer.

A more common aerodynamic method of determining ET_o involves determining the vapour pressure and wind speed differences between measurement heights z_1 and z_2. The evaporation in this method is computed as:

$$ET_o = \frac{-10K^2\rho(q_2 - q_1)(u_2 - u_{1})}{\left(\ln\left\{\frac{z_2}{z_1}\right\}\right)^2} \tag{5.7}$$

where, ET_o = evaporation, mm; K = von Karman constant = 0.41; ρ = density of water vapour, (g/cc); q_2 and q_1 are specific humidity at heights z_2 and z_1, respectively that are taken as 2 m and 1 m, respectively above the water surface; and u_2 and u_1 are wind speeds (cm/s) at heights z_2 and z_1, respectively.

5.12.5.2 Estimation by Energy Balance Method

Based on the principles of conservation of energy and considering a soil column extending from the surface to some depth where vertical heat exchange is negligible, the energy balance can be written (Michael, 1987) as:

$$(Q+q)(1-a)+I\downarrow - I\uparrow + H + LE + G + M = 0 \tag{5.8}$$

where, Q = direct solar radiation incident on earth; q = diffuse solar radiation incident on the earth; a = surface resistance (albedo) which is a fraction; I ↓ = incoming long wave radiation; I ↑ = outgoing long wave radiation from the land; H = sensible heat transfer; LE = latent heat flux due to evaporation; G = net rate at which heat content of soil column is changing and M = miscellaneous heat terms.

Eq. (5.1) can be used for any time period. The units are generally expressed in cal/cm^2/unit time. It is to be noted that 1 cal/cm^2/min 1 mm of water evaporating per hour. In Eq. (5.8) we can use a term R_n = net solar radiation which is the difference between the total incoming and outgoing solar radiation.

Neglecting M, Eq. (5.8) can be written as:

$$R_n = H + LE + G \tag{5.9}$$

Based on Penman's equation (1948), the Kohler-Nordenson-Fox equation (Pochop *et al.*, 1985) describes evaporation from water bodies as the combination of water loss due to radiation heat energy and the aerodynamic removal of water vapur from a saturated surface.

5.12.5.3 Estimation by Combination Method

Based on Penman's equation (1948), the Kohler-Nordenson-Fox equation (Pochop *et al.*, 1985) describes evaporation from water bodies as the combination of water loss due to radiation heat energy and the aerodynamic removal of water vapour from a saturated surface.

The general form for the combination equation is:

$$ET_o = \frac{\Delta}{\Delta+\gamma}Rn + \frac{\gamma}{\gamma+\Delta}Ea \tag{5.10}$$

where, ET_o = evaporation in mm/day; Δ = slope of the saturation vapor pressure curve at air temperature T_a in mbar/^{0}C; γ = psychrometric constant in mbar/^{0}C; R_n = net radiation expressed in mm/day; and E_a = an empirically derived bulk transfer term (aerodynamic term) in mm/day. E_a is expressed as:

$$E_a = f(u)\,(e_s - e_d) \tag{5.11}$$

where, $f(u)$ = wind function; and $(e_s - e_d)$ = vapour pressure deficit.

Detail calculations of various terms of the above equation are available in the book (James, 1988)

5.12.5.4 Estimation by Empirical Method

Owing to the difficulty in measuring the different water balance parameters including the data required for computation of crop evapotranspiration by lysimeters, estimation methods by busing climatic data are preferred. There are several climatologically factors which will influence and decide the rate of evaporation. In fact when water is not a limiting factor, the rate of evaporation is decided by atmospheric demand of the water. Some of the important factors of climate influencing the evaporation are radiation, temperature, humidity, and wind speed. The estimation methods are empirical methods. Some of the estimation methods require only one or two climatic data and does not give accurate result of evapotranspiration. However, there are other estimation methods like FAO Penmain-Monteith model which require a lot of climatic data and give accurate result of evapotranspiration. The empirical method/estimation method involves estimating the reference crop evapotranspiration and then multiplying the reference crop evapotranspiration with crop coefficient we can estimate the crop evapotranspiration. Some of the more commonly used empirical methods to estimate reference crop evapotranspiration are given below. For detail estimation of various other empirical methods, readers are suggested to refer books (Doorenbos and Pruitt, 1977; Allen *et al.*, 1998).

Thornwaite Method

Thornwaite (1948) assumed that the amount of water lost through evapo-transpiration from soil surface covered with vegetation is governed by climatic factors and is independent of species when moisture supply is not limiting. He proposed the following formula:

$$e = 1.6\left(\frac{10\,t}{I}\right)^{a} \tag{.5.12}$$

where, e = unadjusted potential evapotranspiration, cm/month; t = mean air temperature, °C; I = annual or seasonal heat index, the summation of 12 values of monthly heat indices (i) where $I = (t/5)^{1.514}$ and a = an empirical exponent computed by the equation

$$a = 0.000000675\, I^3 - 0.0000771\, I^2 + 0.01792\, I + 0.49239 \tag{5.13}$$

The factor e is an unadjusted value based on a 12 hour day and 30 day month and is corrected by actual day length in hours and day in a month, to give the adjusted potential evapotranspiration

Table 5.1. Correction factors to be used in Thotnthwaite formula (Mean possible duration of sunlight expressed in units of 30 days of 12 hours each)

North Latitude	Jan	Feb	Mar	Apr	May	June	Jul	Aug	Sept	Oct	Nov	Dec
0	1.04	0.94	1.04	1.01	1.04	1.04	1.04	1.04	1.01	1.04	1.01	1.04
5	1.02	0.93	1.03	1.02	1.06	1.03	1.06	1.05	1.01	1.03	0.99	1.02
10	1.00	0.91	1.03	1.03	1.08	1.06	1.08	1.07	1.02	1.02	0.98	0.99
15	0.97	0.91	1.03	1.04	1.11	1.08	1.12	1.08	1.02	1.01	0.95	0.97
20	0.95	0.9	1.03	1.05	1.13	1.11	1.14	1.11	1.02	1.00	0.93	0.94
25	0.93	0.89	1.03	1.06	1.15	1.14	1.17	1.12	1.02	0.991	0.9	0.91
26	0.92	0.88	1.03	1.06	1.15	1.15	1.17	1.12	1.02	0.99	0.91	0.91
27	0.92	0.88	1.03	1.07	1.16	1.15	1.18	1.13	1.02	0.99	0.90	0.90
28	0.91	0.88	1.03	1.07	1.16	1.16	1.18	1.13	1.02	0.98	0.90	0.90
29	0.91	0.87	1.03	1.07	1.17	1.16	1.19	1.13	1.03	0.98	0.90	0.89
30	0.9	0.87	1.03	1.08	1.18	1.17	1.20	1.14	1.03	0.98	0.89	0.88
31	0.9	0.87	1.03	1.08	1.18	1.18	1.20	1.14	1.03	0.98	0.89	0.88
32	0.89	0.86	1.03	1.08	1.19	1.19	1.21	1.15	1.03	0.98	0.88	0.87
33	0.88	0.86	1.03	1.09	1.19	1.20	1.22	1.15	1.03	0.97	0.88	0.86
34	0.88	0.85	1.03	1.09	1.20	1.20	1.22	1.16	1.03	0.97	0.87	0.86
35	0.87	0.85	1.03	1.09	1.21	1.21	1.23	1.16	1.03	0.97	0.86	0.85
36	0.87	0.85	1.03	1.10	1.21	1.22	1.24	1.16	1.03	0.97	0.86	0.84
37	0.86	0.84	1.03	1.10	1.22	1.23	1.25	1.17	1.03	0.97	0.85	0.83
38	0.85	0.84	1.03	1.10	1.23	1.24	1.25	1.17	1.04	0.96	0.84	0.83
39	0.85	0.84	1.03	1.11	1.23	1.24	1.26	1.10	1.04	0.96	0.84	0.82
40	0.84	0.83	1.03	1.11	1.24	1.25	1.27	1.18	1.04	0.96	0.83	0.81
41	0.83	0.83	1.03	1.11	1.25	1.26	1.27	1.19	1.04	0.96	0.82	0.80
42	0.82	0.83	1.03	1.12	1.26	1.27	1.28	1.19	1.04	0.95	0.82	0.79
43	0.81	0.82	1.02	1.12	1.26	1.28	1.29	1.20	1.04	0.95	0.81	0.77
44	0.81	0.82	1.02	1.13	1.27	1.29	1.30	1.20	1.04	0.95	0.80	0.76
45	0.80	0.81	1.02	1.13	1.28	1.29	1.31	1.21	1.04	0.94	0.79	0.75
46	0.79	0.81	1.02	1.13	1.29	1.31	1.32	1.22	1.04	0.94	0.79	0.74
47	0.77	0.80	1.02	1.14	1.30	1.32	1.33	1.22	1.04	0.93	0.78	0.73
48	0.76	0.80	1.02	1.14	1.31	1.33	1.34	1.23	1.05	1.93	0.77	0.72
49	0.75	0.79	1.02	1.14	1.32	1.34	1.35	1.24	1.05	0.93	0.76	0.71
50	0.74	0.78	1.02	1.15	1.33	1.36	1.37	1.25	1.06	0.92	0.76	0.70
South Latitude												
5	1.06	0.95	1.04	1.00	1.02	0.99	1.02	1.03	1.00	1.05	1.03	1.06
10	1.08	0.97	1.05	0.99	1.01	0.96	1.00	1.01	1.00	1.06	1.05	1.10
15	1.12	0.98	1.05	0.98	0.98	0.94	0.97	1.00	1.00	1.07	1.07	1.12
20	1.14	1.00	1.05	0.97	0.96	0.91	0.95	0.99	1.00	1.08	1.09	1.15
25	1.17	1.01	1.05	0.96	0.94	0.88	0.93	0.98	1.00	1.10	1.11	1.18
30	1.20	1.03	1.06	0.95	0.92	0.85	0.90	0.96	1.00	1.12	1.14	1.21
35	1.23	1.01	1.06	0.94	0.89	0.82	0.87	0.94	1.00	1.13	1.17	1.25
40	1.27	1.06	1.07	0.93	0.86	0.78	0.84	0.92	1.00	1.15	1.20	1.29
42	1.28	1.07	1.07	0.92	0.85	0.76	0.82	0.92	1.00	1.16	1.22	1.31
44	1.30	1.06	1.07	0.92	0.83	0.74	0.81	0.91	0.99	1.17	1.23	1.33
46	1.32	1.10	1.07	0.91	0.82	0.72	0.79	0.90	0.99	1.17	1.25	1.35
48	1.34	1.11	1.08	0.90	0.80	0.70	0.76	0.89	0.99	1.18	1.27	1.37
50	1.37	1.12	1.08	0.89	0.77	0.67	0.74	0.88	0.99	1.19	1.29	1.41

PET (reference evapotranspiration) as.

PET = *e* x correction factor (5.14)

The correction factor for different latitudes and months are provided in Table 5.1.

Sample Calculation 8:

Kharagpur is situated at a latitude of 25^0N .Mean monthly temperature of Kharagpur for 12 months are given below:

Month	Jan	Feb	Mar	Apr	May	Jun	July	Aug	Sept	Oct	Nov	Dec
Mean monthly temp., °C	18	24	28	30	31	31	32	29	28	26	25	23

Find out the reference crop evapotranspiration/*PET* for the region for a crop grown during June to September.

Solution:

Thornwaite proposed the formula to compute reference crop evapotranspiration/*PET* as:

$$= 1.6\left(\frac{10t}{1}\right)^{\mathbf{a}}$$

where, e = unadjusted reference rop evapotranspiration/*PET*, cm/month; t = mean air temperature, °C; I = annual or seasonal heat index ,the summation of 12 values of monthly heat indices, where and a = an empirical exponent computed by the equation, $a = 0.000000675I^3 - 0.0000771I^2 + 0.01792I + 0.49239$

Reference crop evapotranspiration/*PET* = e × correction factor

Step 1: For determination of annual heat index I

Month	Mean temp °C	Annual heat index
Jan	18	6.954116887
Feb	24	10.74977151
Mar	28	13.57552622
Apr	30	15.07026861
May	31	15.83729542
Jun	31	15.83729542
July	32	16.61714772
Aug	29	14.31627336
Sept	28	13.57552622
Oct	26	12.13470051
Nov	25	11.43511632
Dec	23	10.07895154
Total		156.1819897

Step 2: For determination of a = (an empirical exponent)

$a = 0.000000675I^3 - 0.0000771\ I^2 + 0.01792I + 0.49239$

$= 0.000000675 \times (156.182)^3 - 0.0000771 \times (156.182)^2 + 0.01792 \times (156.182) + 0.49239$
$= 3.982045$

Step 3: For calculation of unadjusted reference crop evapotranspiration/*PET*, cm/month for four month from June to September:

$$e_{June} = 1.6\left(\frac{10\times 31}{156.182}\right)^{3.98} = 24.52997$$

$$e_{July} = 1.6\left(\frac{10\times 32}{156.182}\right)^{3.98} = 27.83573$$

$$e_{Aug} = 1.6\left(\frac{10\times 29}{156.182}\right)^{3.98} = 18.80885$$

$$e_{Sept} = 1.6\left(\frac{10\times 28}{156.182}\right)^{3.98} = 16.35596$$

Step 4: For calculating adjusted reference crop evapotranspiration/*PET* = e correction factor. Correction factor can be calculated from Table 5.1 for given month with known value of latitude

$PET_{jun} = e_{jun} \times correction\ factor$

$= 24.53 \times 1.14$
$= 27.96417$

$PET_{july} = e_{july} \times correction\ factor$

$= 27.83573 \times 1.17$
$= 32.56781$

$PET_{Aug} = e_{Aug} \times correction\ factor$

$= 18.80885\ \ 1.12$
$= 21.06592$

$PET_{Sept} = e_{Sept} \times correction\ factor$

$= 16.35596\ \ 1.02$
$= 16.35596$

Total PET from June to Sept = 27.96417+32.56781+21.06592+16.68308
= 98.28098 cm/month 98.3 cm/month

Modified Blaney-Criddle Method

In the modified Blaney-Criddle method the relationship recommended, representing mean value over the given month is expressed (Doorenbos and Pruitt, 1977) as:

$$ET_0 = C\left[P\left(0.46\,T + 8\right)\right] \tag{5.15}$$

where, ET_0 = reference crop evapotranspiration, mm/day for the month considered; T = mean daily temperature,^{0}C over the month considered; P = mean daily percentage of total annual daytime hours (Table 5.2) for a given month and latitude and C = adjustment factor which depends on minimum relative humidity, sunshine hours and day time wind estimates.

The adjustment factor is determined by using Fig. 5.3. Fig.5.3 can be used to estimate ET_0 graphically using the calculated values of [P+(0.46T+8)] given on X-axis and the value of ET_0 can be read directly from the Y axis. Relationships are presented in Fig. 5.3 for (i) minimum relative humidity (RH_{min}) (ii) three levels of the ratio of actual to maximum possible sun shine hours (*n/N*) and (iii) three levels of day time wind velocity at 2 m height (U_{day}). Values of maximum possible sunshine hours for different latitudes and months are given in Table 5.3 (Doorenbos and Pruitt, 1977). The limitations of use of the method are also mentioned in the above mentioned reference.

Calculation of mean daily ET_0 should be made for periods no shorter than one month. ET_0 should preferably be calculated for each calendar month for each year of record rather than by using mean temperature based on several years' record. Once ET_0 has been determined, the crop coefficients (*Kc*) can be used to determine *ETcrop*.

Table 5.2: Mean daily percentage (p) of annual daytime hours for different latitudes

Latitude	North South	Jan. -July	Feb. -Aug	March – Sept.	April- Oct.	May- Nov.	June- Dec.	July- Jan.	Aug.- Feb.	Sept.- March	Oct.- April	Nov.- May	Dec.- June
60°		0.15	0.20	0.26	0.32	0.38	0.41	0.40	0.34	0.28	0.22	0.17	0.13
58 °		0.16	0.21	0.26	0.32	0.37	0.40	0.39	0.34	0.28	0.23	0.18	0.15
56 °		0.17	0.21	0.26	0.32	0.36	0.39	0.38	0.33	0.28	0.23	0.18	0.16
54 °		0.18	0.22	0.26	0.31	0.36	0.38	0.37	0.33	0.28	0.23	0.19	0.17
52 °		0.19	0.22	0.27	0.31	0.35	0.37	0.36	0.33	0.28	0.24	0.20	0.17
50 °		0.19	0.23	0.27	0.31	0.34	0.36	0.35	0.32	0.28	0.24	0.20	0.18
48 °		0.20	0.23	0.27	0.31	0.34	0.36	0.35	0.32	0.28	0.24	0.21	0.19
46 °		0.20	0.23	0.27	0.30	0.34	0.35	0.34	0.32	0.28	0.24	0.21	0.20
44 °		0.21	0.24	0.27	0.30	0.33	0.35	0.34	0.31	0.28	0.25	0.22	0.20
42 °		0.21	0.24	0.27	0.30	0.33	0.34	0.33	0.31	0.28	0.25	0.22	0.21
40 °		0.22	0.24	0.27	0.30	0.32	0.34	0.33	0.31	0.28	0.25	0.22	0.21
35 °		0.23	0.25	0.27	0.29	0.31	0.32	0.32	0.30	0.28	0.25	0.23	0.22
30 °		0.24	0.25	0.27	0.29	0.31	0.32	0.31	0.30	0.28	0.26	0.24	0.23*
25 °		0.24	0.26	0.27	0.29	0.30	0.31	0.31	0.29	0.28	0.26	0.25	0.24
20 °		0.25	0.26	0.27	0.28	0.29	0.30	0.30	0.29	0.28	0.26	0.25	0.25
15 °		0.26	0.26	0.27	0.28	0.29	0.29	0.29	0.28	0.28	0.27	0.26	0.25
10 °		0.26	0.27	0.27	0.28	0.28	0.29	0.29	0.28	0.28	0.27	0.26	0.26
5 °		0.27	0.27	0.27	0.28	0.28	0.28	0.28	0.28	0.28	0.27	0.27	0.27
0 °		0.27	0.27	0.27	0.27	0.27	0.27	0.27	0.27	0.27	0.27	0.27	0.27

Table 5.3: Mean daily maximum duration of bright sunshine hours (*N*) for different months and latitudes

North South lats.	Jan-July	Feb.-Aug	March - Sept.	April-Oct.	May-. Nov.	June-Dec.	July-Jan.	Aug.-Feb.	Sept.-March	Oct.-April	Nov. -May	Dec.-June
50	8.5	10.1	11.8	13.8	15.4	16.3	15.9	14.5	12.7	10.8	9.1	8.1
48	8.8	10.2	11.8	13.6	15.2	16.0	15.6	14.3	12.6	10.9	9.3	8.3
46	9.1	10.4	11.9	13.5	14.9	15.7	15.4	14.2	12.6	10.9	9.5	8.7
44	9.3	10.5	11.9	13.4	14.7	15.4	15.2	14.0	12.6	11.0	9.7	8.9
42	9.4	10.6	11.9	13.4	14.6	15.2	14.9	13.9	12.6	11.1	9.8	9.1
40	9.6	10.7	11.9	13.3	14.4	15.0	14.7	13.7	12.5	11.2	10.0	9.3
35	10.1	11.0	11.9	13.1	14.0	14.5	14.3	13.5	12.4	11.3	10.3	9.8
30	10.4	11.1	12.0	12.9	13.6	14.0	13.9	13.2	12.4	11.5	10.6	10.2
25	10.7	11.3	12.0	12.7	13.3	13.7	13.5	13.0	12.3	11.6	10.9	10.6
20	11.0	11.5	12.0	12.6	13.1	13.3	13.2	12.8	12.3	11.7	11.2	10.9
15	11.3	11.6	12.0	12.5	12.8	13.0	12.9	12.6	12.2	11.8	11.4	11.2
10	11.6	11.8	12.0	12.3	12.6	12.7	12.6	12.4	12.1	11.8	11.6	11.5
5	11.8	11.9	12.0	12.2	12.3	12.4	12.3	12.3	12.1	12.0	11.9	11.8
0	12.1	12.1	12.1	12.1	12.1	12.1	12.1	12.1	12.1	12.1	12.1	12.1

Sample Calculation 9:

Calculate the reference evapotranspiration by modified Blaney-Criddle method using the following data for a place which has Latitude 29°N: Altitude 283.89 m. The month is December. Mean

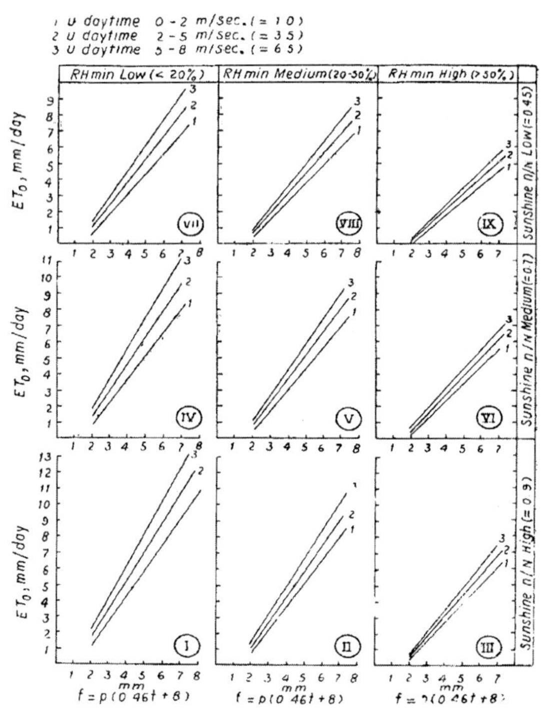

Fig.5.3: Correction factor to be used in adjustment of reference evapotranspiration by Blaney-Criddle method

daily maximum temperature = 22.96 °C and mean daily minimum temperature = 5.86 ^{0}C. *RHmin* = 65%, day time wind velocity at 2 m height = 4 m/sec and actual sunshine hour = 9.4 hours.

Solution:

T daily mean = (22. 96 + 5.86) /2 = 14.41 °C.

P from Table 5.2 for 29^{0}N = 0.23

$P\ (0.46\ T+8\ ) = 0.23\ (0.46 \times 14.41 + 8) = 3.36$ mm/day

RHmin = high

Value of *N* is taken from Table 5.3 and using given value of n (9.4 hr) we get,

n/N = high

u_2 day time at 2 m height = moderate

ET_0 from Fig.5.3 Block III (line 2) =2.3 mm/day

Pan Evaporation Method

This method is a relatively inexpensive and simple method of assessing the evaporative demand of the atmosphere. It has been observed that a close relationship exists between the rate of consumptive use by crops and the rate of evaporation from a properly located evaporation pan. There are different types evaporation pans in use which have been discussed earlier. The standard USDA class A open pan evaporimeter may be used for the estimation (Fig. 5.4). In this case daily reading of water depth in the pan is taken. The difference of the two consecutive day's water depth reading in the pan (assuming that water loss due to wind action, animals, birds etc has been prevented or is negligible) gives the daily pan evaporation. If in between the two consecutive readings of taking water depth in the pan, there is some rainfall, then this amount of rainfall is to be subtracted from the difference of the two consecutive days water depth reading of the pan so that the actual pan evaporation data is obtained. The relationship between reference evapotranspiration (*ETo*) and pan evaporation (*Epan*) is given by the pan coefficient (*Kp*) as:

$$ETo = Epan\ \ Kp \tag{5.16}$$

The value of *Kp* depends on environment and geographical location and also on the nature and characteristics of pan. The shape and colour of the pan also affects the value of *Kp*. For example, water loss from circular pans is independent of wind direction, while evaporation from square pans depends on wind direction. Colour difference between the pans affects the reflection of radiation and hence evaporation. Screens mounted above pans to prevent birds and animals to drink from the evaporation pan reduce pan evaporation by as much as 10 percent (Doorenbos and Pruitt, 1977). Conditions upwind of the evaporation pans also significantly affect *Kp*. Table 5.4 lists values of *Kp* for different winds, relative humidities and windward ground cover conditions as (illustrated in Fig.5.5) for class A evaporation galvanized pans annually painted with aluminum. There will be little difference in values of *Epan* when

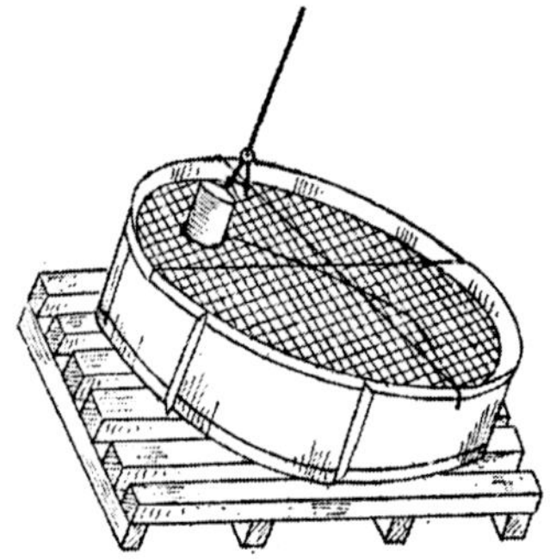

Fig. 5.4: Figure of U.S.D.A. class A open pan

inside and outside surfaces of the pan are painted white. An increase of *Epan* value up to 10 percent may occur when the pan is painted black.

Table 5.4 Values of *Kp* for class A open pan for different ground covers, 24 hour wind speed and levels of mean relative humidities *(RH)*.

Class A Pan		*Case A*[a] *Pan surrounded by Short Green Crop*				*Case B*[ab] *Pan surrounded by Dry-Fallow Land*		
RH mean, Percent		Low (< 40)	Medium (40-70)	High (> 70)		Low (< 40)	Medium (40-70)	High (> 70)
Average daily wind speed (m/sec)	Upwind distance of Green Crop (m)				Upwind distance of Dry Fallow (m)			
Light (< 2)	1	0.55	0.65	0.75	1	0.7	0.8	0.85
	10	0.65	0.75	0.85	10	0.6	0.7	0.8
	100	0.70	0.8	0.85	100	0.55	0.65	0.75
	1000	0.75	0.85	0.85	1000	0.5	0.6	0.7
Moderate (2-5)	1	0.5	0.6	0.65	1	0.65	0.75	0.8
	10	0.6	0.7	0.75	10	0.55	0.65	0.7
	100	0.65	0.75	0.8	100	0.5	0.6	0.65
	1000	0.7	0.8	0.8	1000	0.45	0.55	0.6
High (5-8)	1	0.45	0.5	0.6	1	0.6	0.65	0.7
	10	0.55	0.6	0.65	10	0.5	0.55	0.65
	100	0.6	0.65	0.7	100	0.45	0.5	0.6
	1000	0.65	0.7	0.75	1000	0.4	0.45	0.55
Very High (> 8)	1	0.4	0.45	0.5	1	0.5	0.6	0.65
	10	0.45	0.55	0.6	10	0.45	0.5	0.55
	100	0.5	0.6	0.65	100	0.4	0.45	0.5
	1000	0.55	0.6	0.65	1000	0.35	0.4	0.45

Source: Doorenbos and Pruitt (1977) [a]Cases A and B are defined in Fig. 5.5

[b]For extensive areas of bare fallow soils and areas without agricultural development, reduce *Kp* values by 20 percent under hot windy conditions and by 5 to 10 percent for moderate wind, temperature and humidity conditions.

Sample Calculation 10:

Calculate the reference crop evapotranspiration value (*ETo)* as measured by class A open galvanized pan in a given site under the following conditions with data measured as:

Water level in the pan on previous day at 7 A.M. is 18.5 cm and that measured after 24 hours at the same time is 20. 5 cm. There is 7.0 cm rainfall in between the 24 hours. The pan is surrounded by green crop and is situated upwind distance of green crop of 100 m. Mean relative humidity is 50 percent and wind speed is 4 m/sec.

Solution:

There is 7 cm rainfall within the two consecutive day's readings of water measurement.

Water level readings on previous day = 18.5 cm

Water level reading on day after 24 hours = 20.5 cm

Hence, the value of *Epan* = 18.5 +7.0 – 20.5 = 5 mm/day.

Kp = 0.75 (referring Table 5.4).

Hence, *ETo* = *Epan Kp*

= 5 x 0.75 = 3.85 mm/day.

Other methods used for computation of *ETo* are Penmen method, Modified Penmen method, Radiation method, Christiansen method, Energy Balance method, Oliver method, Van Bavel method, FAO Penmen-Monteith method etc. Out of these methods, FAO Penmen-Monteith method gives reasonable correct value of *ETo* since it considers a number of climatic data. The details of FAO Penmen-Monteith method is discussed in FAO Paper No. 56 (Allen *et al.*, 1998).

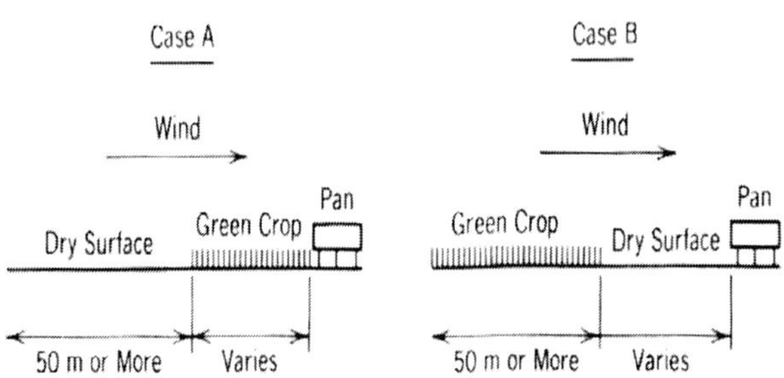

Fig. 5.5: Definition sketch for Cases A and B in Table 5.4 (from Doorenbos and Pruitt, 1977)

5.12.6. Crop Coefficient (Kc) for Estimating Actual Evapotranspiration

The values of ET_0 (reference evapotranspiration) estimated by using different methods as described above, need to be adjusted for actual evapotranspiration (*ETcrop*) under natural field conditions. ET_0 rarely occurs in most of the irrigated field crops, with an exception of lowland rice and probably for two to three days immediately after irrigation or rains. For converting ET_0 values into *ETcrop*, suitable crop coefficients (K_c) should be evolved for different crops, soils and climatic conditions and also for different stages of growth for the same crop. Conversion of *ETo* to *ETcrop* is given as:

$$ETcrop = K_c \times ET_0 \tag{5.16}$$

The range of K_c values during different growth stages of crop is different. The value of K_c largely depends on the level of ET_0 and frequency with which soil is wetted by rain and/or irrigation. The smooth curve in the figure of *Kc* presents average *Kc* values rather than actual sharp increase in *Kc* just following rain and irrigation, with a less sharp but marked decline afterwards, until the next rain or irrigation (discussed later on). The ratio may exceed 1.0 in areas prone to advection since the same is not reflected in *ETo* estimated by most of the formulae. The correct estimation of *ET* from ET_0, therefore, depends upon the accuracy of the *Kc* values adopted. Values of actual crop evapotranspiration are low during the .early stages of crop growth and increases as the plant approaches full growth period and remain fairly constant for some time. There is a general decline during the late/matured periods (Michael, 1978). The values of Kc_c for a particular crop at a location should be determined experimentally. If the values of Kc_c for particular crop and place are not available then the K_c values mentioned in Table 5.5 (Doorenbos and Pruitt, 1977) may be used.

The value of *Kc* depends on type of crop, its management practice, soil moisture condition, length of crop growing season, growth stages of the crop, crop sowing/planting date and climatic conditions. Particularly following sowing and during the early growth stage, the frequency of rain or irrigation is important. The values of crop coefficient are low during the initial and matured stage of the crop and are more during the crop development and mid season stage. During the mid season and crop development stage, the crop has more canopies, covers the soil to the maximum extent and has more developed and maximum root development. This necessitates the plants to require higher amount of water which is reflected in higher values of *Kc* . The crop growing season is generally divided into the four stages i.e. initial stage, crop development stage, mid-season stage and late season stage/maturity stage. Initial stage covers germination and early growth stage when the soil surface is covered up to 10 percent. Crop development stage starts from the end of initial stage to attainment of effective full ground cover of 70 to 80 percent. Mid season stage starts from the end of the crop development stage to time of start of maturing as indicated by discolouring of leaves whereas the late season stage covers the period from the end of mid-season stage to full maturity or harvest.

Table 5.5: Values of *Kc* for field and vegetable crops for different crop growth stages and prevailing climatic conditions

Crop	Humidity Wind speed, m/sec	RHmin 0-5	> 70% 5-8	RHmin 0-5	< 20% 5-8
	Crop stage				
All field crops	Initial 1 Use Fig. 5.6 & 5.7by interpolation				
	Crop development 2				
	Mid-season 3	0.95	0.95	1.0	1.05
	Late season/maturity 4	0.9	0.9	0.95	1.0
Barley	3	1.05	1.1	1.15	1.2

(Contd.)

Crop	Humidity Wind speed, m/sec	RHmin 0-5	> 70% 5-8	RHmin 0-5	< 20% 5-8
	4	0.25	0.25	0.2	0.2
Beans (green)	3	0.95	0.95	1.0	1.05
	4	0.85	0.85	0.9	0.9
Beans (dry)/pulses	3	1.05	1.1	1.15	1.2
	4	0.3	0.3	0.25	0.25
Beets	3	1.0	1.0	1.05	1.1
	4	0.9	0.9	0.95	1.0
Carrots	3	1.0	1.05	1.1	1.15
	4	0.7	0.75	0.8	0.85
Sweet corn (maize)	3	1.05	1.1	1.15	1.2
	4	0.95	1.0	1.05	1.1
Cotton	3	1.05	1.15	1.2	1.25
	4	0.65	0.65	0.65	0.7
Crucifers (cabbage,	3	0.95	1.0	1.05	1.1
cauliflower, broccoli)	4	0.80	0.85	0.9	0.95
Cucumber	3	0.9	0.9	0.95	1.0
	4	0.7	0.7	0.75	0.8
Lentil	3	1.05	1.1	1.15	1.2
	4	0.3	0.3	0.25	0.25
Melons	3	0.95	0.95	1.0	1.05
	4	0.65	0.65	0.75	0.75
Millet	3	1.0	1.05	1.1	1.15
	4	0.3	0.3	0.25	0.25
Oats	3	1.05	1.1	1.15	1.2
	4	0.25	0.25	0.2	0.2
Onion (dry)	3	0.95	0.95	1.05	1.1
	4	0.75	0.75	0.8	0.85
Onion (green)	3	0.95	0.95	1.0	1.05
	4	0.95	0.95	1.0	1.05
Peanuts (Groundnut)	3	0.95	1.0	1.05	1.1
	4	0.55	0.55	0.6	0.6
Peas	3	1.05	1.1	1.15	1.2
	4	0.95	1.0	1.05	1.1
Potato	3	1.05	1.1	1.15	1.2
	4	0.7	0.7	0.75	0.75
Radish	3	0.8	0.8	0.85	0.9
	4	0.75	0.75	0.8	0.85
Safflower	3	1.05	1.1	1.15	1.2
	4	0.25	0.25	0.2	0.2
Sorghum	3	1.0	1.05	1.1	1.15
	4	0.5	0.5	0.55	0.55
Soybeans	3	1.0	1.05	1.1	1.15
	4	0.45	0.45	0.45	0.45
Spinach	3	0.95	0.95	1.0	1.05
	4	0.9	0.9	0.95	1.0
Sugarbeet	3	1.05	1.1	1.15	1.2

(*Contd.*)

Crop	Humidity Wind speed, m/sec	RHmin 0-5	> 70% 5-8	RHmin 0-5	< 20% 5-8
	4	0.9	0.95	1.0	1.0
Sunflower	3	1.05	1.1	1.15	1.2
	4	0.4	0.4	0.35	0.35
Tomato	3	1.05	1.1	1.2	1.25
	4	0.6	0.6	0.65	0.65
Wheat	3	1.05	1.1	1.15	1.2
	4	0.25	0.25	0.2	0.2

N.B. Values of *Kc* mentioned as above are for field and vegetables, Values of *Kc* for other crops are reported by Doorenbos and Pruitt (1977).

5.12.6.1 Determination of Kc *Value in Initial Stage*

The crop coefficient for the initial growth stage (K_{cini}) can be derived from Figs. 5.6 and 5.7 which provide estimates for $K_{c\,ini}$ as a function of the average interval between wetting events, the evaporation power ET_o, and the importance of the wetting event.

Fig.5.6 is used for all soil types when wetting events are light. When wetting during the initial period is only by precipitation, one will usually use Fig.5.6 to determine $K_{c\,ini}$. The graph can also be used when irrigation is by high frequency systems such as micro irrigation and centre pivot and light applications of about 10 mm or less per wetting event are applied.

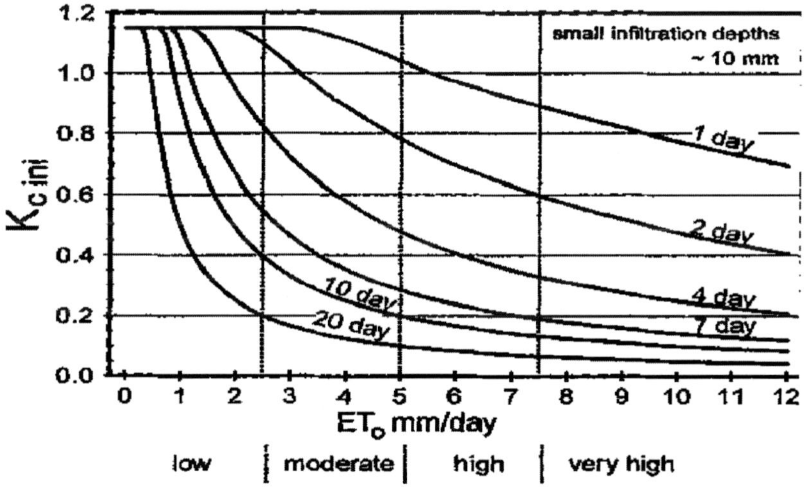

Fig. 5.6: Average K_{cini} as related to the level of ET_o and the interval between irrigations and/or significant rain during the initial growth stage for all soil types when wetting events are light to medium (3-10 mm per event)

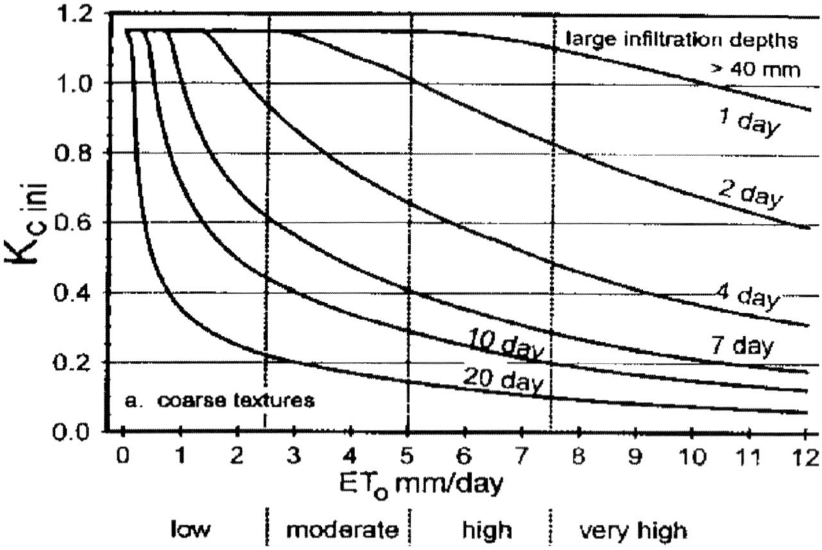

Fig.5.7(a): Average $K_{c\,ini}$ as related to the level of ET_o and the interval between irrigations greater than or equal to 40 mm per wetting event, during the initial growth stage for coarse textured soils

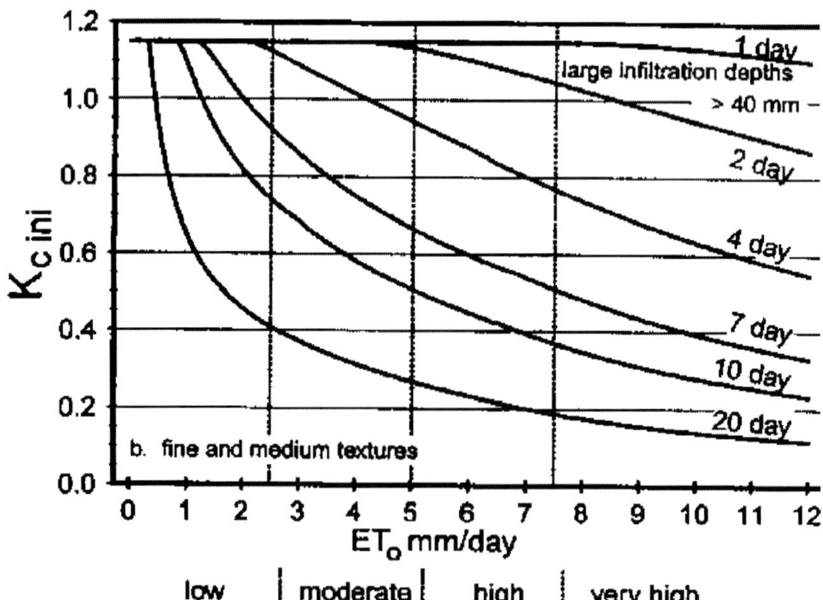

Fig. 5.7 (b): Average $K_{c\,ini}$ as related to the level of ET_o and the interval between irrigations greater than or equal to 40 mm per wetting event, during the initial growth stage for fine and medium textured soil

Fig. 5.7 is used for heavy wetting events when infiltration depths are greater than 40 mm, such as for when wetting is primarily by periodic irrigation such as by sprinkler or surface irrigation.

Fig.5.7(a) is used for coarse textured soils and Fig.5.7 (b) is used for fine and medium textured soils. Coarse textured soils include sands and loamy sand textured soils. Medium textured soils include sandy loam, loam, silt loam and silt textured soils. Fine textured soils include silty clay loam, silty clay and clay textured soils.

Where average infiltration depths are between 10 and 40 mm, the value for $K_{c\ ini}$ can be estimated from Figs.5.6 and 5.7 and using the Eqn. (5.17):

$$K_{c\ ini} = K_{c\ ini\ (Fig.\ 5.6)} + \frac{(I-10)}{(40-10)}\left[K_{c\ ini\ (Fig.\ 5.7)} - K_{c\ ini\ (Fig.5.6)}\right] \tag{5.17}$$

where, $K_{c\ ini\ (Fig5.6)}$ is the value for $K_{c\ ini}$ from Fig. 5.6, $K_{c\ ini\ (Fig.5.7)}$ is the value for $K_{c\ ini}$ from Fig. 5.7, I is the average infiltration depth, mm.

The values 10 and 40 in Eq. (5.17) are the average depths of infiltration (mm) upon which Figs. 5.6 and 5.7 are based.

The steps needed to arrive at the *Kc* values for the different stages are given below (Doorenbos and Pruitt, 1977).

(i) Establish planting/sowing date from local information.

(ii) Determine total growing season and length of crop development stages from local information.

(iii) Initial stage: Predict irrigation and/or rainfall frequency; for predetermined values of reference evapotranspiration and obtain *Kc* values from Figs. 5.6 or 5.7 (as the case may be) and plot *Kc* value as shown in Fig. 5.8 which shows the daily variation of *Kc* of a crop in a growing season.

(iv) Mid-season stage: For given climatic condition select *Kc* values from Table 5.5 (Doorenbos and Pruitt, 1977) and plot as straight line.

(v) Late season stage: For time of full maturity or harvest within a few days select *Kc* values from Table 5.5 for given climatic condition and plot values at the end of growing season or full maturity. Assume straight line between the *Kc* values at the end of mid-season period and at the end of growing season.

(vi) Crop development stage: Assume straight line between the *Kc* values at the end of initial to start of the mid-season stage.

Fig. 5.8 is prepared for corn planted in mid may with total growing period of 125 days (initial stage = 20 days; crop development stage = 35; mid-season stage = 40 days and late season stage = 30). For total growing season winds are light to moderate (0-5 m/sec); *RHmin* = 30 – 35 percent; *ETo* in initial stage = 8.4 mm/day and irrigation frequency in the initial stage = 7 days.

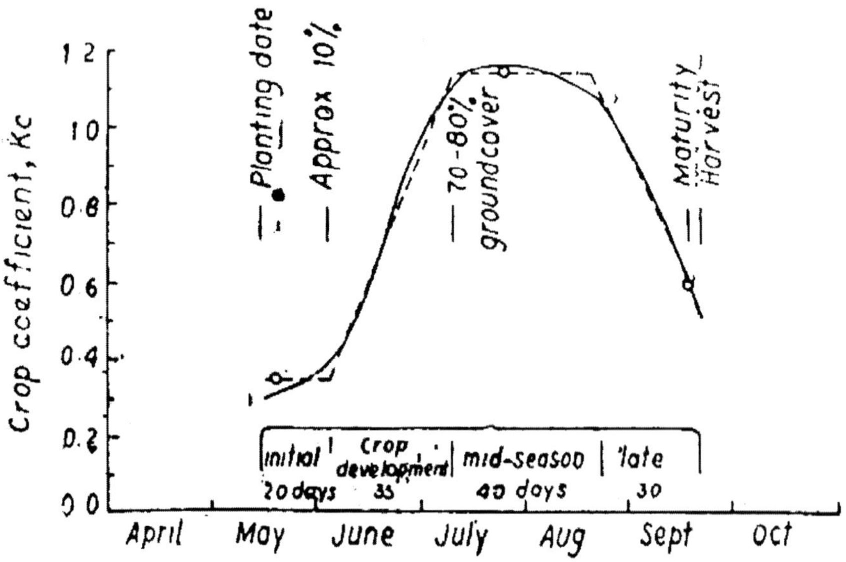

Fig. 5.8: Examples of crop coefficient, K_c

For simplification, the values of *Kc* for the different periods within the growing season are represented as straight lines. A smoothed curve may be drawn as indicated in Fig. 5.8, although this may have little effect in terms of accuracy added.

Rice is an important crop, especially in many Asian countries. The values of *Kc* for rice grown in humid Asian wet season (without any water stress) vary from 0.95 to 1.15 for various crop growth stages and for different climatic conditions. There is no difference in the values of *Kc* between the broadcast/sown and transplanted rice. Similarly, there is little difference in the values of *Kc* among the upland, medium and low land rice. Only during the initial crop stage, *Kc* need to be reduced by 15 to 20 percent (Doorenbos and Pruitt, 1977).

It is also to be remembered that the soil moisture content is a deciding factor for evaluation of K_c values. Under adequate soil moisture condition the values of K_c are more and as the soil moisture decreases and the plants remain under stress condition, values of K_c become less as a result of which, the actual crop evapotranspiration decreases from the potential value. Under such stress condition, the actual crop evapotranspiration (*ETcrop*) is given as:

$$ETcrop = K_s \cdot K_c \cdot ET_0 \tag{5.18}$$

where, K_s = crop stress factor and other terms are defined earlier.

The value of crop stress factor is equal 1.0 under adequate soil moisture condition (optimum level) and decreases from 1.0 as the soil moisture decreases from the optimum value. Some researchers report that there is a linear decrease of actual crop evapotranspiration from the potential values as the soil moisture decreases from the optimum value whereas others opine a logarithmic decline.

Paddy crop, in general has K_c values ranging from 0.95 to 1.15 under ponding (standing water condition) and at about saturation condition. However, as the ponding water from the crop field disappears such that the unsaturated condition occurs in the fields, the value of K_c linearly decreases. Jensen *et al.* (1993) reported that K_c values of paddy decreases by 10% every day from the day of disappearance of ponded water from the field. Panigrahi (2001) reported that in paddy field under unsaturated condition, the crop stress factor is equal to the relative available soil moisture which is the ratio of current available soil moisture in the crop effective root zone to the saturated moisture content in the crop effective root zone. The values of crop stress factor results in decline of actual crop evapotranspiration and hence yield. It is to be remembered that the stress condition in the crop during the critical growth stages has pronounced effect on decreasing the yield than that occurring in other stages. Detail discussions on effect of stress factor on crop at different growth stages along with the values at various stages of different crops are available in FAO Paper No. 33 (Doorenbos and Kassam, 1979).

5.12.6.2 Approximate Range of Seasonal Crop Water Requirement of Some Crops

As discussed above, the values of crop water requirements (*ETcrop*) is a variable factor. The values mainly depend on climatic condition of the place and crop type including sowing/planting time, crop growing period, crop growth stages, cultural and water management practices etc. Because of these reasons, the same crop may have wide variation in the value of crop water requirement. Table 5.6 given below reports the seasonal variation of crop water requirement of some important crops grown in different places (Doorenbos and Pruitt, 1977).

Table 5.6: Approximate range of seasonal *ETcrop* of some crops

Crop	Seasonal *ETcrop*, mm	Crop	Seasonal *ETcrop*, mm
Alfalfa	600-1500	Potato	350-625
Banana	700-1700	Rice	500-950
Beans	250-500	Sorghum	300-650
Coffee	800-1200	Soybeans	450-825
Cotton	550-950	Sugarbeets	450-850
Deciduous trees	700-1050	Sugarcane	1000-1500
Grains (small)	300-450	Sweet potato	400-675
Grape	650-1000	Tobacco	300-500
Maize	400-750	Tomato	300-600
Oil seeds	300-600	Vegetables	250-500
Onions	350-600	Vineyards	450-900
Oranges	600-950	Walnuts	700-1000

The above mentioned values of *ETcrop* may be used for design of irrigation projects. However, it is recommended that local available values of *ETcrop* should be used in irrigation project design.

Question Banks

Q1. What are the multiple uses of water of a farm pond?

Q2. What are the factors considered for computing the total crop water demand of a farm pond service area?

Q3. What is the normal water demand of large livestock like horse and cattle and small livestock like pig and sheep?

Q4. How can the total water demand of the farm pond be determined?

Q5. (a) Define the following terms.
(i) Consumptive use (ii) Water requirement (iii) Evapotranspiration (iv) Duty and (v) Delta
(b) Derive the relationship between duty and delta.

Q6. (a) Briefly explain the energy balance method and aero-dynamic method to estimate the crop water requirement.
(b) What is a combined method for estimation of reference crop evapotranspiration?

Q7. With label diagram, describe how the evapotranspiration of crop is measured by the lysimeters.

Q8. Which type of lysimeter gives more correct measured values of crop evapotranspiration? Describe how this lysimeter is used to measure the crop evapotranspiration.

Q9. Describe how the soil moisture depletion study is helpful to compute the crop evapotranspiration.

Q10. Profile soil moisture content of a potato field in 60 cm effective root zone depth are measured as follows:

Profile layer, cm	Soil moisture content,%		Apparent Specific gravity, gm/cc
	Beginning of irrigation cycle	End of irrigation cycle	
0-15	20	12	1.60
15-30	22	13	1.58
30-45	24	15	1.59
45-60	25	15	1.60

If effective rainfall within the irrigation cycle is 20 mm, and percolation from the effective root zone of the crop is 11 mm, estimate the water use from the crop root zone within the irrigation cycle.

Q11. A field is sown with ground nut. The profile soil moisture content of the field at sowing and at harvest time are measured and presented in table as below.

Profile layer, cm	Soil moisture content,%		Apparent Specific gravity, gm/cc
	At sowing	At harvest	
0-15	22	10	1.62
15-30	24	11	1.60
30-45	25	13	1.59
45-60	25	15	1.60

If the crop is supplied with 20 cm irrigation, effective rainfall in the crop growing season is 12 cm and there is no percolation loss from the crop effective root zone in the entire season, then estimate the seasonal water requirement of the crop.

Q12. Write the general water balance model of crop field (both paddy and non-paddy) and explain the different terms. How can the model be used to predict the soil moisture content in different soil layers of the crop filed?

Q13. Write down the different steps involved in estimating the reference crop evapotranspiration by Thronwaite method.

Q14. Following data are recorded for a particular place having latitude 29°N. Estimate the daily reference crop evapotranspiration for the month of January using the data and using Thornwaite method.

Month	Jan	Feb	Mar	Apr	May	Jun	Jul	Aug	Sep	Oct	Nov	Dec
Mean temperature, °C	13.0	15.9	16.5	23	29.8	30.4	29.0	28.6	25.7	20.6	18.6	12.1

Q15. Write down the different steps used to compute the reference evapotranspiration by the modified Blaney-Criddlee method.

Q16. Calculate the reference evapotranspiration with using the following data.

Place is Chiplima having 22°N latitude and altitude is 205 m period is month December. The mean maximum temperature is 23°C and mean minimum temperature is 6°C. The place has actual sunshine hour (average monthly) is 9 hr. The day wind velocity at 2 m height is 3.5 m/s and the mean minimum relative humidity is 54%.

If the crop coefficient value is 1.05 then find out the actual crop evapotranspiration.

Q17. (a) The water level in a pan evaporimeter on a day at 7 AM is 20 cm below the rim of the pan and that on the next day is 19.2 cm. If there is no rainfall within the time both readings are taken then find out the pan evaporation. Had there been 5 mm rainfall then what would have been the value of pan evaporation? Using pan coefficient value as 0.8, find out the reference evapotranspiration in both the cases.

(b) Briefly describe what the factors that influence the pan coefficient value.

Q18. What do you mean by crop coefficient and how it is useful in estimation of actual crop evapotyranspiration?

Q19. What are the factors that influence values of the crop coefficient?

Q20. What are the different crop growth stages and how can you decide these stages of a crop.

Q21. What do you mean by crop stress factor? How does it affect the actual crop evapotranspiration?

Q22. Find out the seasonal actual crop evapotranspiration of a crop using the data given in tabular form as below.

Crop growth stage	Values of reference evapotranspiration, cm	Values of crop coefficient	Values of crop stress factor
Initial	20	0.8	0.5
Crop development	25	1.0	0.9
Mid season	30	1.2	0.9
Maturity	18	0.9	1.0

Q23. What are the various measures to enhance the duty of water in a command?

Q24. Enumerate the different factors that affect the duty of water.

Q25. With a label diagram explain how the class A pan evaporimeter is used to measure the reference crop evapotranspiration.

Q26. Explain how can the crop coefficient in the initial growth stage of a crop can be determined.

References

Allen, R.G., Pereira , L.S., Raes, D. and Smith, M. 1998. Crop Evapotranspiration: Guidelines for Computing Water Requirements. Irrigation and Drainage Paper 56, FAO, Rome, pp 300.

Doorenbos, J. and Kassam, A.H. 1979. Yield Response to Water. Irrigation and Drainage Paper 33, FAO, Rome.

Doorenbos, J. and Pruitt, W.O. 1977. Guidelines for Predicting Crop water Requirements. Irrigation and Drainage Paper 24, FAO, Rome

Jensen, J.R., Mannan, S.M.A. and Uddin, S.M.N. 1993. Irrigation equirement of transplanted monsoon rice in Bangladesh. Agricultural Water Management, Vol. 23(3):199-212.

Lenka, D. 2001. Irrigation and Drainage. Kalyani Publishers, New Delhi.

Michael, A.M. 1978. Irrigation Theory and Practices. Vikash Publishing House Pvt. Ltd., New Delhi.

Misra, R.D. and Ahmed, M. 1987. Manual on Irrigation Agronomy, Oxford & IBH Publishing Co. Pvt. Ltd., New Delhi, pp. 412.

Odhiambo, L.O. and Murty, V.V.N. 1996. Modeling water balance components in relation to field layouts in low land paddy fields. I. model development, Agricultural Water Management, Vol. 30(2): 185-199.

Panigrahi, B. 2001. Water Balance Simulation for Optimum Design of On-Farm Reservoir in Rainfed Farming System. Ph.D. Thesis, Indian Institute of Technology, Kharagpur.

Panigrahi, B. 2011. Irrigation Systems Engineering. New India Publishing Agency, New Delhi.

Thornthwaite, C.W. 1948. An approach towards a rational classification of climate. Georgr. Rev.: 55-94.

6

Basic Soil Water Relationship

6.1 Soil

The term 'soil 'is defined in various ways depending on the general professional field in which it is used. However to an agriculturist, soil is the substance existing on the earth surface, which grows and develops plant life. For an engineer, soil is the unaggregated deposits of mineral and organic particles covering major part on the earth crust. It includes widely different materials like boulders, sand, gravel, clays and silts. Soil is produced by the weathering of solid rocks. The formation of soil is due to geologic cycle continuously taking place on surface of the earth. The geologic cycle comprises weathering or denudation, transportation, deposition and upheaval, again followed by the above mentioned processes. Weathering may be natural or artificial. In natural weathering, physical processes like temperature changes, flow of water, ice, wind and rain play the role. In artificial weathering animals especially human beings interference causes soil erosion and formation.

6.2 Soil as a Three Phase System

A soil mass is a three phase system consisting of soil particles called as soil grains, water and air. the void space between the soil grains is field partly with water and partly with air. In case of a dry soil mass, void spaces fully occupy air and water. In case of a fully saturated soil, voids are completely filled with water. In general the soil mass has three constituent parts which are blended together forming a complex material (Fig. 6.1 (a)); the properties of which depend upon the relative proportion of these constituent, their arrangement and a variety of other factors. For calculation purpose, it is more convenient to show these constituents occupying separate space (Fig 6.1(b)(i) and (ii)).

As shown in Fig.6.1 (b) (i), the total volume (V) of the soil mass consists of (i) volume of air (V_a)(ii) volume of water (V_w) and (iii) the volume of solids (Vs). The volume of voids (V_v), is, therefore, equal to volume of air plus the volume of water. Fig. (Fig. 6.1 (b) (ii) also represents

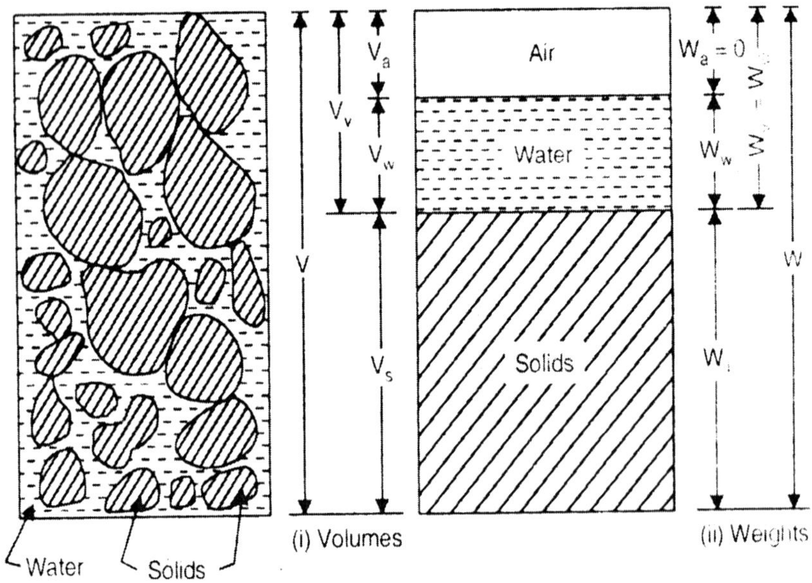

(a) Elements of natural soil (b) Elements separated into three phases

Fig. 6.1: Soil as a three phase system

the constituents occupying separate spaces in terms of weights. The weight of air is considered to be negligible. So the weight of total voids is equal to the weight of water (W_w).The weight of solids is represented by W_d *or* W_s, which is evidently equal to the dry weight of soil sample. The total weight (*W*) of the moist sample is, therefore, equal to ($W_w + W_d$)

6.3 Water Content, Density and Unit Weights

6.3.1 Water Content

The water content (*w*) also called the moisture content, is the ratio of weight of water (W_w) to the weight of solids (**W_s** *or* W_d) in a given mass of soil. It is generally expressed in percent and is given as:

$$w = \frac{W_w}{W_d} x\,100 \tag{6.1}$$

Eq. (6.1) can be written as:

$$w = \frac{W - W_d}{W_d} x\,100 = \left(\frac{W}{W_d} - 1\right) x\,100 \tag{6.2}$$

In the field of agriculture, water content (w) is defined as the ratio of weight of water (W_w) to the total weight of soil sample (W).When expressed in percentage from the water content /soil moisture content is given as

$$w = \frac{W_w}{W} x\,100 \tag{6.3}$$

Relationship between w and w′ is as follows

$$w = \frac{W_w}{W} = \frac{W_w}{W_w + W_d}$$

$$\frac{1}{W} = \frac{W_w + W_d}{W_w} = 1 + \frac{W_d}{W_w}$$

$$\frac{1}{W} = \frac{1}{W} + 1 \text{ (since } w = \frac{W_w}{W_d})$$

From which we get,

$$W = \frac{W}{1 - W} \tag{6.4}$$

When expressed in percent

$$W = \frac{W}{1 - W} x\,100 \tag{6.5}$$

From Eq (6.5) we find that w is always greater than w. Many a time the term w and w′ are called as soil moisture content in dry weight basis and wet weight basis, respectively.

6.3.2 Density of Soil

It is defined as the mass of soil per unit volume.

(i) Bulk Density (ρ)

It is also called as moist density. It is the ratio between the total mass of the soil (M) per unit of its volume (V).Thus

$$\rho = \frac{M}{V} \tag{6.6}$$

It is expressed in g/cm^3 or kg/m^3

(ii) Dry density (ρ_d)

It is the ratio of mass of solid (M_d) per unit of its total volume (V).Thus

$$\rho_d = \frac{M_d}{V} \tag{6.7}$$

(iii) Density of Solid (ρ_s)

It is expressed as the ratio between mass of soil solid (M_d) and volume of soil solid (V_s) and is given as

$$\rho_s = \frac{M_d}{V_s} \tag{6.8}$$

6.3.3 Unit Weight of Soil Mass

The unit weight of soil mass is defined as its weight per unit volume.

(i) Bulk Unit Weight (γ)

The bulk unit weight or moist unit weight (simply called as unit weight) is the total weight (W) of a given soil mass per unit of its total volume (V). Thus

$$\gamma = \frac{W}{V} = \frac{M\,g}{V} = \rho\,g \tag{6.9}$$

where, g = acceleration due to gravity.

(ii) Dry Unit Weight (γ_d)

It is the ratio between weight of solid (W_d) and total volume of soil mass (V).

$$\gamma_d = \frac{W_d}{V}$$

(iii) Unit Weight of Solids (γ_s)

The unit weight of solids is the weight of solids (W_d) per unit volume of solids (Vs).

$$\lambda_s = \frac{W_d}{V_s} \tag{6.10}$$

It is to be noted that when we want to convert the density into unit weight, we have to multiply the former by 'g'(acceleration due to gravity)

6.4 Void Ratio, Porosity and Degree of Saturation

6.4.1 Void Ratio

Void ratio (e) of a given soil sample is the ratio of the volume of voids (V_v) to the volume of soil solids (V_s) in the given soil mass. Thus

$$e = \frac{V_v}{V_s} \tag{6.11}$$

6.4.2 Porosity

Porosity (η) is defined as the ratio between the volume of voids (V_v) to the total volume of a given soil sample (V) and thus

$$\eta = \frac{V_v}{V} \tag{6.12}$$

Void ratio is generally expressed in fraction where as porosity is expressed in percent. Fig 6.2 (a) and (b) represent void ratio and porosity of a given soil sample respectively.

If the volume of voids is taken equal to e, then volume of solid (V_s) from (Eq 6.11) is equal to 1 and the total volume is equal to $(1+e)$ [Fig 6.2 (a)]. Similarly if the volume of voids is taken equal to η (Fig. 6.2 (b)), the total volume of the element will be 1 (unity) and hence the volume of solid would be equal to (1- η).

From Fig. 6.2 (a), we have by definition of porosity,

$$\eta = \frac{V_v}{V} = \frac{e}{1+e} \tag{6.13}$$

Similarly from Fig. 6.2 (b) and using definition of void ratio, we have

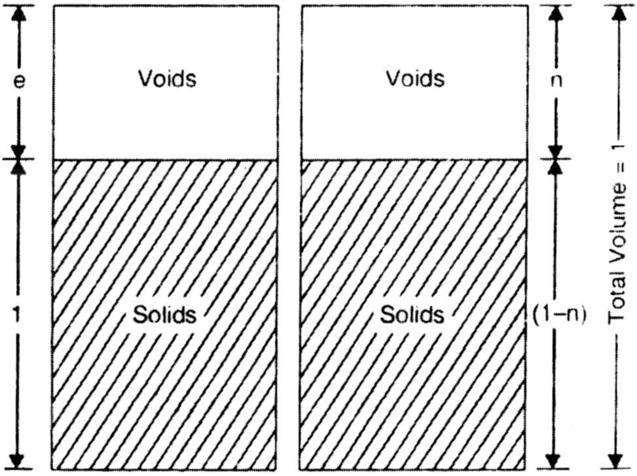

(a) Soil element in terms of e (b) Soil element in terms of η

Fig. 6.2: Void ratio and porosity

$$e = \frac{V_v}{V_s} = \frac{\eta}{1-\eta} \quad (6.14)$$

Combining Eqs. (6.13) and (6.14), we get

$$1-\eta = \frac{1}{1+e} \quad (6.15)$$

6.4.3 Degree of Saturation

In a given volume of soil, some space is occupied by water and rest is occupied by air. The degree of saturation (*S*) is defined as the ratio between volume of water (V_w) in a given soil sample to the total volume of voids (V_v) present in it.

$$S = \frac{V_w}{V_v} \quad (6.16)$$

Degree of saturation when expressed in percent, the term in Eq. (6.16) is multiplied by 100..In fully saturated soil (some times fully saturated soil is also called saturated soil where as unsaturated soil is the soil which partly occupied by water and rest are by air), all the voids are filled with water and there is no air i.e. $V_v = V_w$ and hence from Eq (6.16), degree of saturation (*S*) is 1.0 (100%). Thus, for a perfectly dry soil sample $V_w = 0$ and hence S=0 i.e. zero degree of saturation.

Depending upon degree of saturation, the soil mass may be classified as follows (Table 6.1).

Table 6.1: Classification of soil based on degree of saturation

Soil classification	Degree of saturation
Dry	0
Humid	$\geq 0 - \prec 0.25$
Damp	$\geq 0.25 - \prec 0.5$
Moist	$\geq 0.5 - \prec 0.75$
Wet	$\geq 0.75 - \prec 1.0$
Saturated	1.0

Detail functional relationship between various term like degree of saturation, moisture content, porosity, void ratio etc. are available in the book "Soil Mechanics and Foundations" written by Punmia *et al*. (2010).

Sample Calculation 1:

A soil sample has 50% porosity. Find its void ratio?

Solution:

Given porosity (η) = 50% = 0.50.

So void ratio e (From Eq 6.14) is

$$e = \frac{\eta}{1-\eta} = \frac{0.50}{1-0.50} = 1.0$$

Sample Calculation 2:

What will be the porosity of a soil sample having void ratio 0.4?

Solution:

From Eq (6.13), we have porosity (η)

$$\eta = \frac{e}{1+e} = \frac{0.4}{1+0.4} = 0.286 = 28.6\%$$

Sample Calculation 3:

An undisturbed soil sample has volume of 500 cm^3 and mass of 800 gm.When dried in oven for 24 hours, the soil mass comes out to be 600 gm. Calculate the water content, bulk density and its unit weight ?

Solution:

Given mass of wet soil = 800 gm and mass of dry soil mass M_d = 600 gm

So mass of water (M_w) = 800 – 600 = 200 gm

So percentage of moisture content from Eq (6.1) is

$$w = \frac{W_w}{W_d} x\,100 = \frac{M_w\, g}{M_d\, g} x\,100 = \frac{M_w}{M_d} x\,100 = \frac{200}{600} x\,100 = 33.3\%$$

Bulk density $\rho = \frac{M}{V} = \frac{800}{500} = 1.6\,\frac{gm}{cm^3}$

From Eq. (6.9), we have unit weight (γ) as:

$$\gamma = \rho\, g = 1.6\ x\ 9.81 = 15.70\,\frac{KN}{m^3}$$

Sample Calculation 4:

What will be the degree of saturation of a given mass of soil sample that contain volume of water half to volume of voids?

Solution:

$$\text{Given } V_w = \frac{V_v}{2}$$

$$S = \frac{V_w}{V_v} = \frac{V_v}{2 \, x \, V_v} = 0.5 = 50\%.$$

6.5 Physical Properties and Classification of Soil

Soil has two physical properties i.e. texture and structure. Texture relates to the fineness (size) of the particles whereas structure relates to the arrangements of the particles. Textural properties of the soil influence the water retention and their release capacity. Relative proportion of sand, silt and clay in a soil determines its textural class. The soil textural class can be determined by a triangular diagram (Fig. 6.3) in which percentage of sand, silt and clay are taken as three sides of the triangle taken in clockwise manner and several textural classes are formed by taking different possible combinations of proportion into account. Using the above triangular

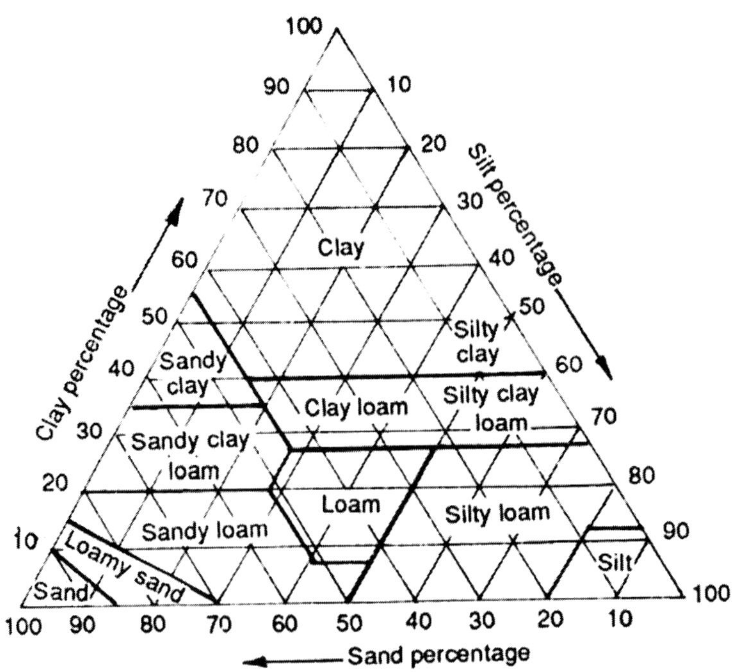

Fig. 6.3: Triangular diagram used to classify the soil texture

diagram (Fig. 6.3) data of Table 6.2 (data are arbitrary not measured in any experimental site) represents the soil textural class with various proportion of sand, silt and clay.

Table 6.2: Textural class according to the percentage of sand, silt and clay

Particles	Percentage						
Sand	85	84.4	57.4	8.9	60	20	25
Silt	10	55.6	20.2	21.4	28	60	30
Clay	5	1010.0	19.3	65.8	12	20	45
Textural class (Fig. 6.3)	Sandy	Sandy loam	Loam	Clay	Silt	Loam	Clay

Source: Lenka (2001)

6.6 Physical Characteristics of Soil Textural Class

Sandy soil contains more amount of sand and fewer amounts of silt and clay. This soil is open and friable. Its water holding capacity is very low and the crops in this soil needs frequent irrigation. Loamy soil contains almost equal proportion of sand, silt and clay. It gives presence of small grits when felt with fingers; slightly plastic when semi dry. Silt loam soil contains medium quantities of fine sand and relatively less clay. Percentage of silt is generally more than 50. This soil form hard clod when dried but the clods break easily. Clay loam soil contains more proportion of clay. The particles are very fine. The soils form hard clod on drying. The wet soil is quite plastic. In a clay soil, the clay content is more than 50%. The soil forms very hard clod on drying and do not break easily. Clods are such that they impede in farming operation.When become wet, stickiness and plasticity develop in the soil.

6.7 Soil Water/Soil Moisture

The soil profile lying below the soil surface up to the impervious strata/bed rock can be divided in to (i) zone of aeration (also called as unsaturated zone) and (ii) zone of saturation. Soil water is defined as the suspended water in the zone of aeration. It includes gravitational water, capillary water and hygroscopic water. Fig. 6.4 represents the classification of soil moisture.

Zone of aeration is important from the point of view of plant growth. Zone of aeration comprises of (i) soil water belt, (ii) intermediate belt and (iii) capillary zone. Soil water belt refers to the zone of water adjacent to the ground surface consisting of top soil and the sub soil. Water in the zone moves downward by gravity called as infiltration and upward by evapotranspiration by the plants. Water passed down this belt goes beyond the reach of the root zones of the plants and is not available for use by the plants. This water going down the plants root zone is called as percolation and percolation in the deeper zones is called as deep percolation. Percolated and deep percolated water adds to the zone of saturation containing the ground-water.

It is to be remembered that the soil water belt is the most important part of zone of aeration that helps in supporting the growth of the plants. Intermediate belt acts as passage for the downward movement of water through the soil under the action of gravity. The thickness may vary from zero (when the capillary fringe extends in to the soil) to several meters depending on the groundwater table position. Capillary zone is sandwiched between the intermediate zones lying at the top to the zone of saturation below. Water moves through this zone and adds to the zone

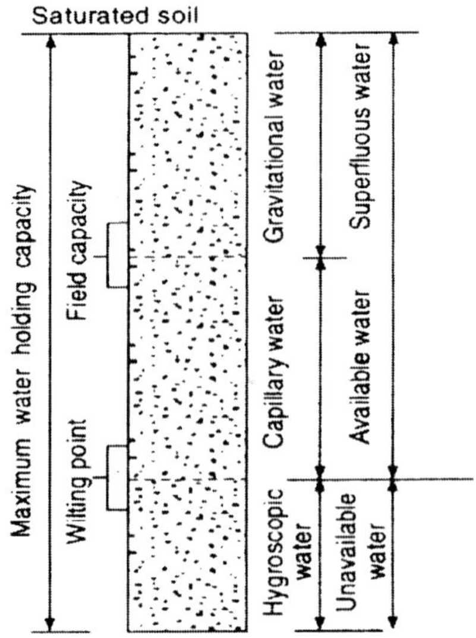

Fig. 6.4: Classification of soil moisture

of saturation that contains the groundwater. The thickness of this zone may vary from a few cm to about one meter or more depending on the type of soil. The capillary rise is about 20-50 cm for coarse sand, 50-80 cm in medium sand and over 80 cm in fine clay. Saturated zone has the upper boundary as groundwater table and lower boundary as the impervious/bed rock. It serves as a great reservoir of groundwater.

Gravitational water is the water in the unsaturated zone in excess of hygroscopic and capillary water which moves out of the soil under favourable drainage condition. Water held in the capillary spaces by surface tension is called as capillary water. This water moves freely under the influences of capillary forces and is available to the plants. Hence the capillary water is also called as available water. Hygroscopic water is the water held in static state with the atmospheric water vapour. It is not capable of any movement by gravity or capillary forces and hence, is not available for use to plants. In fact, there is no distinct boundary between these three states of soil water. The portion of each class depends in soil texture, structure, organic matter content, and temperature of the soil and depth of the soil column considered (Hansen *et al.*, 1979). Sometimes a term called as superfluous water is used which refers to the gravitational water that drains down so deeper that plants root cannot draw it.

6.7.1 Determination of Soil Moisture Content

The amount of water present in the soil is called as soil moisture content. It can be expressed as mm or cm of water depth present in 1 m depth of the soil or can be expressed in percentage. It is not same for all the soils and varied from time to time and space to space. The soil moisture content otherwise called as water content of the soil can be determined by the following methods:

(i) Oven dry method

(ii) Sand bath method

(iii) Calcium carbide method

(iv) Radiation method

(v) Alcohol method

(vi) Pycnometer method

(vii) Torsion balance method

(viii) Using Tensiometer

(ix) Using Neutron moisture meter

(x) Using gypsum resistance block

6.7.1.1 Oven Dry Method

This is the most accurate method to determine the soil moisture content. In this method, the moisture content of the soil is determined in the laboratory. Soil samples from a given layer of soil are collected by an auger in a clean dry non corrodible can. Before taking soil sample in it, its mass is found out with its lid on a balance accurate to 0.01 gm .Then the moist soil sample collected from a given layer of soil is kept in the can and its weight with moist soil and lid is taken. The can with lid open is kept inside a thermostatically controlled oven with interior of

°C for 2 hours .It should be remembered that if the soil is highly organic, such as peat, a lower temperature of about 60°C is preferable to prevent oxidation of organic matter. Similarly if soil contains gypsum the sample must be dried at not more than 80°C but for a longer time. After drying of the soil kept in the can, the can is removed from the oven and is allowed to cool in a desicator. The lid is then replaced and the can with dry soil is weighed. The water content is calculated as:

$$w = \frac{M_2 - M_3}{M_3 - M_1} x\,100 \tag{6.17}$$

where, M_1 – mass of empty can, M_2 – mass of can with moist soil and lid and M_3 = mass of dry soil in can with lid.

In the above expression (M_2 - M_3) stands for mass of water present in the soil and (M_3 – M_1) stands for mass of dry soil and w is the soil moisture content in percent in dry basis.

If soil moisture is to be calculated on the wet basis then water content (w) is given as

$$w' = \frac{M_2 - M_3}{M_2 - M_1} x\,100 \tag{6.18}$$

where as $M_2 - M_1$ stands for mass of moist/wet soil. It should be noted that soil moisture content on dry basis is more than that on wet basis. The relationship between them is:

$$w = \frac{W}{1-W} \tag{6.19}$$

It is to be noted that Eqns. (6.1) and (6.17) are same. Similarly, Eqns (6.2) and (6.18) are same. The only difference between them is that w and w' in Eqns. (6.1) and (6.2) are calculated based on weight basis whereas w and w' in Eqns (6.17) and (6.18) are calculated based on mass basis.

6.7.1.2 Sand Bath Method

By this method water content of the soil sample can be roughly determined. It is a field method for determining the soil moisture content. The can with the moist soil sample is placed in a sand bath which is heated over a kerosene stove. When the soil becomes dry, which generally takes half to 1 hour, the can is taken out and allowed to cool. The weights of empty can, can with moist soil and can with dry soil are taken. The water content is then determined from Eqns. (6.1) and (6.2) to find out the soil moisture content on dry and wet basis, respectively. This method has a limitation on temperature that the high temperature may break the crystalline structure of soil and so high temperature should be avoided. Another limitation is that determination of soil moisture content by this method should not be used for organic soils having higher percentage of gypsum.

6.7.1.3 Alcohol Method

This method is like sand bath method. This is a rough method to determine the soil moisture content used in the field. The wet soil sample is kept in an evaporating dish and is mixed with sufficient quantity of methylated spirit. The disc is then properly covered and the mixture is ignited. The mixture is kept stirred by a wire during ignition. As described above, three mass readings need to be taken before using Eqns. (6.1) and (6.2) to compute the soil moisture content. These three readings are (i) mass of empty dish, (ii) mass of dish with moist soil and (iii) mass of dish with dry soil. Then Eqns. (6.1) and (6.2) may be used to compute the soil moisture content on dry and wet basis, respectively. This method is most suitable to determine the water content of the soil containing large percentage of organic matter or gypsum.

6.7.1.4 Neutron Probe (NP) Method

The technique is based on the measurement of fast moving neutrons that are slowed (thermalised) in the soil by an elastic collision with existing Hydrogen particles in the soil. Hydrogen (H^+) is present in the soil as a constituent of:

(i) Soil organic matter

(ii) Soil clay minerals

(iii) Water

Water is the only form of H^+ that will change from measurement to measurement. Therefore any change in the counts recorded by the NP is due to a change in the moisture with an increase in counts relating to an increase in moisture content

The neutron moisture meter consist of two principal parts (i) a probe, which is lowered in access tube inserted vertically in the soil and which contains a source of fast neutron and detector of slow neutron ;(ii) a scalar or rate meter to monitor the flux of slow neutron, which is proportional to the soil water content (Fig. 6.5). In use, the probe is lowered into the access tube, and counting rates are determined at the desired depth. The count rates adjusted for background and standardized to counts made in tank of pure water are then calibrated against volumetric determination of soil moisture content (Fig 6.6).The effective volume of soil in which the water content is measured depends on the concentration of hydrogen nuclei.

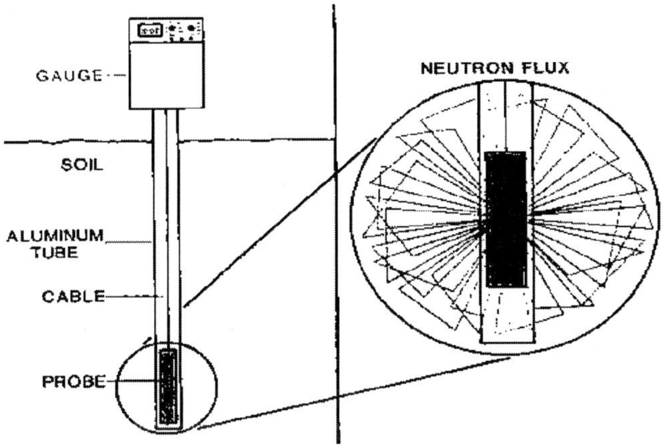

Fig 6.5: Neutron probe method

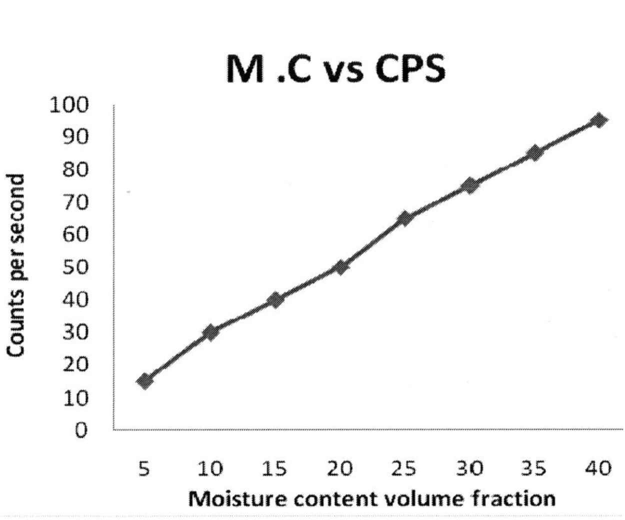

Fig 6.6: A typical calibration curve of a neutron moisture probe

This method has the following limitations:

(i) Improper use of the neutron moisture meter can be hazardous.

(ii) A protective shield (generally consist of lead and paraffin or polythene in a cylindrically shaped unit with a central hole to accommodate the probe) is an essential component of the equipment. It also serves as a standard absorber for checking the readings.

(iii) This method is not used to measure the soil moisture content in the first 15 cm soil layer.

6.7.1.5 Electrical Resistance Blocks

Electrical resistance blocks consist of two electrodes enclosed in a block of porous material. The block is often made of gypsum, although fiberglass or nylon is sometimes used. Electrical resistance blocks are often referred to as *gypsum blocks* and sometimes just *moisture blocks*. The electrodes are connected to insulated lead wires that extend upward to the soil surface.

Resistance blocks work on the principle that water conducts electricity. When properly installed, the water suction of the porous block is in equilibrium with the soil-water suction of the surrounding soil. As the soil moisture changes, the water content of the porous block also changes. The electrical resistance between the two electrodes increases as the water content of the porous block decreases. The block's resistance can be related to the water content of the soil by a calibration curve. Many designs have been used successfully. There have been considerable alternations in shape of the block as well as in the size, shape and configuration of the electrodes. In general, nylon and fiber-glass units are more sensitive than gypsum blocks at high moisture content and low tension condition. Gypsum blocks operate best at tension between 1 to 15 atm whereas nylon units are most sensitive at tensions lower than 2 atm.

Limitations of Use of Resistance Blocks

(i) Resistance blocks are not suitable for saline soil, since the resistance reading is affected by salt concentration.

(ii) Generally the resistance block should not be used for wet soil moisture range.

(iii) Resistance measurement changes somewhat with temperature and thermisters are installed with the block to measue the temperature.

6.7.1.6 Tensiometer Method

Tensiometer consists of a porus ceramic cup connected by a metallic or P.V.C. pipe to a manometer or a vacuum gauge. The manometer may be filled with mercury or may be simply a vacuum gauge (Fig. 6.7). Distilled water is filled in the ceramic cup as well as in the pipe of the tensiometer which is then placed in the soil layer where the potential is to be measured. Water in the cup is generally at atmospheric pressure and in the soil at sub-atmospheric pressure. After installation, if the soil is dry water from the cup moves out into the soil to establish equilibrium and pressure in the manometer falls, which is indicated by depression of mercury in the mercury type manometer or by rise of indicator in the vacuum gauge. If there is some rainfall or if some irrigation water is applied to the soil in the field, soil moisture content increases. Under such condition, water from outside enters into the cup and consequently the

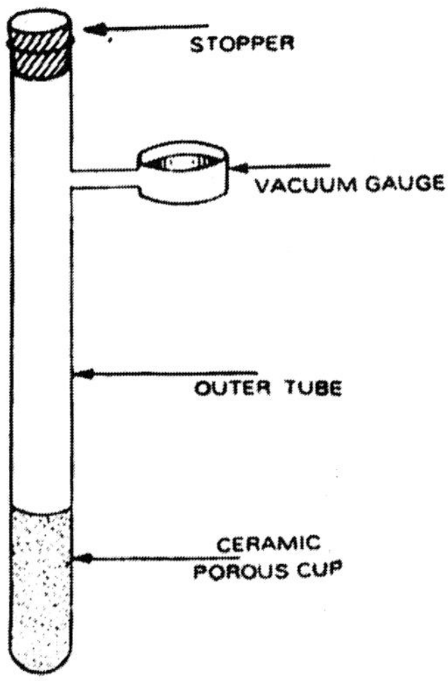

Fig. 6.7: Tensiometer used to measure the soil moisture content

pressure increases. The mercury column rises up or the indicator falls down as the case may be. Most tensiometer gauges are calibrated with readings from 0 to 100, called centribars. A reading of 0 indicates saturated soil in which plant roots may suffer from lack of oxygen. Readings from 0 to 5 in tensiometer (vacuum type) represent a condition too wet for most crops. Readings from 10-25 represent ideal water and aeration conditions. The ceramic porous cup is permeable to both water and solute and after equilibrium the cup-water attains the composition and concentration of solute as in soil water.

Limitations of Use of Tensiometer

Tensiometer is the most widely used instrument to measure the soil moisture content in wet range i.e. within 1 bar. However, it works most satisfactorilly at or below 0.85 bar. Another limitation for use of tensiometer is that it cannot be used to measure the moisture content of the soil which is at saturation (0 bar) or contains water at more than saturation condition (at more than saturation condition, there is standing water in the field). Tensiometers are insensitive to the osmotic effects of dissolved salts in soil solution. Hence, tensiometer only measures matric potential.

6.7.1.7 Calcium Carbide Method

In this method a moisture tester, which is also called as air tight container is taken and 6 gm of soil sample is taken in it. Sufficient quantity of calcium carbide is mixed with it. The mixture is shaken vigrously. The acetylene gas, which is produced by the reaction of the moisture of the soil and the calcium carbide, exerts pressure on a sensitive diaphragm placed at the end of the container. The dial gauge located at the diaphragm reads the moisture content directly. However, the calibration of the dial gauge is such that it gives the water content based on wet weight of the sample. Then the relation (Eq. 6.4) is used to convert this wet weight basis moisture content into dry weight basis.

6.7.1 8 Pycnometer Method

This is also a quick method of determining the water content of those soils whose specific gravity is known accurately. Pycnometer is a large size density bottle of about 900 ml capacity. A conical brass cap, having a 6 mm diameter hole at its top is screwed to the open end of the pycnometer. A rubber washer is placed between conical cap and the rim of the bottle so that there is no leakage of water. Detail methodologies for determining the soil moisture content by pycnometer method is available in the book "Soil Mechanics and Foundations" written by Punmia *et al.* (2010).

6.7.1.9 Radiation Method

This method is very much useful for the determination of water content of soil deposit in the *in-situ* condition. It uses two steel casings-casing A and casing B which is placed in two bore holes at some distance apart, in the soil deposit; the field moisture content of which is to be determined. A device containing some radio-active isotope material (such as cobalt-16) is placed in a capsule which in turn is lowered into casing A.

Similarly, a detector unit is lowered in steel casing B. Small openings are made in both casing A and B, facing each other. When the radio active device is activated, it emits neutrons. When these neutrons strike with the hydrogen atoms of water in the sub-soil, they loose energy. The loss of energy is evidently equal to water content in the soil. The detector device is calibrated to given directly the water content of the subsoil, at that level of emission. However proper shielding precautions should be taken to avoid radiation problems.

6.7.1.10 Torsion Balance Method

Torsion balance has two main parts: (i) Infraed lamp and (ii) Torsion balance. The infra red radiation is provided by 250 watt lamp built in the balance for use with alternating current 220-230 V, 50 cycle, and single phase mains supply. The weighing mechanism is provided by a torsion balance, which has a built in magnetic damper to reduce pan vibrations during quick drying. The balance scale is divided in terms of water percentage from 1to 100 water content in 0.2 percent division. The moisture meter is generally calibrated to use 25 gram of soil, and hence the maximum size of particle present in the specimen should be less than 2 mm.

The test specimen is kept in a suitable container so that the water content to be determined is not affected by ambient condition. Torque is applied to on end of the torsion wire by means of a calibrated drum to balance the loss of weight of the sample as it dries out under infra red lamp. To determine the percent reduction of mass at any instant the drum scale is rotated by turning the drum drive knob until the point returns to the index. The percent is read directly from the scale. However this percent (w') is the percent of water based upon the initial mass (i.e. wet mass) of the sample. The water content (w) based on the dry mass is computed from Eq. (6.4).

6.7.2 Soil Moisture Tension (Suction)

A wet soil contains some amount of water that is called as soil moisture. If some weight is applied to the wet soil, some amount of water will come out. This water is the quantity of water that comes out which was held by the soil with a force equivalent to that of weight. Such work done to remove the water from the soil has been termed in various ways such as suction, tension (negative pressure) and stress and each one is an expression of free energy. In a soil profile in the field that is in contact with water table underneath, water is held at different points with different potentials. At the surface it is zero; it increases positively downward (hydrostatic pressure) and decreases upward. Hence, the use of the term potential combines all the terms involved and can express uniquely the soil-moisture relationship in the profile soil moisture continuum. Energy levels of soil water can be expressed as soil water potential per unit mass *viz.*, erg/g, joules/kg per unit volume i.e. bars, cm of water height or soil water suction *viz.*, pressure (bars) and head (cm of water height). The relationship is shown in Table 6.3.

From the energy relations (Table 6.3) it would be seen that soil water potential can be expressed in terms of height of water column. Since soil water potential expressed thus can range from very small to very high values, use of such values in practice was felt very cumbersome. To avoid this Schofield (1935) suggested using logarithmic scale of such values, which he termed *pF.* The relationship between *pF* and free energy of water (*h*) expressed in terms of cm of water is given as:

$$pF = log_{10} h \tag{6.20}$$

Table 6.3 Equivalent expression of soil water energy

Soil water potentia per unit mass		Soil water suction Per unit volume		Head pressure		Vapour pressure	
erg/g	joules/ Kg	bars	cm of water	Bars	Atmos	cm of water	Torr 20^0 C
0	0	0	0	0	0	0	17.5350
-1×10^4	-1	-0.01	-10.2	0.01		10.2	17.5349
-1×10^5	-10	-0.1	-102.0	0.1		102.0	17.5337
-3×10^5	-30	-0.3	-306.0	0.3		306.0	17.5312
-6×10^5	-60	-0.6	-612	0.6		612	17.5273
-1×10^6	-100	-1.0	-1020	1.0	0.98692	1020	17.5222
-2×10^6	-200	-2	-2040	2	1.9738	2040	17.5089
-3×10^6	-300	-3	-3060	3	2.9607	3060	17.4961
-4×10^6	-400	-4	-4080	4	3.9476	4080	17.4833
-6×10^6	-600	-6	-6120	6	5.9241	6120	17.4572

1 bar = 1×10^6 dynes/ cm^2 = 0.98692 atmospheres
1 millibar = 1×10^3 dynes/ cm^2 = 1.0133 bars = 1 atmospheres
1 bar = 1033.26 g / cm^2 = 10227.7 cm H_2O at 25 ^{0} C
1 atmosphere = 76.39 Hg Hg Sp. Gr = 13

Due to expression of energy in terms of bar, centibar (1/100 bar), millibar (1/1000 bar) or atmosphere, the use of *pF* is not in vogue. Height of water column of 1, 10, 100, 1000, 10,000, 1,00,000 and 1,000,000 cm has the *pF* values of 0, 1, 2, 3, 4, 5 and 6, respectively). The *pF* at saturation, field capacity, permanent wilting point, hygroscopic coefficient and in oven dry soil are 0, 2.00-2.253, 4.18, 4.5 and 7, respectively (Lenka, 2001).

6.7.3 Soil Moisture Potential Curve

Each soil holds water with some potential or tension. More the water content greater the potential (lesser the tension). Each soil can be characterised with respect to how much moisture it can hold against how much of potential. Water or moisture content held by different soils is different at a given potential. Similarly, a particular soil holds different amount of water at different potentials. Table 6.4 gives a relationship between potential and soil moisture/water content in different soils collected in a particular place (Bargarh in Odisha). The relationship will be different for other soils at any other regions. The relationship between the soil moisture content in percent and potential can also be represented in a graphical form which is called as soil moisture potential curve/soil moisture tension curve/soil moisture release curve or simply soil moisture characteristics curve.

Table 6.4: Relationship between soil moisture potential and moisture content in various soils

Potential (- bars)	Soil Texture		
	Sandy	Loam	Clay
		Moisture content in percent	
0.33	16.5	32.2	45.4
1	12.2	26.4	31.7
2	8.0	18.2	27.0
4	5.1	14.5	20.4
15	3.1	7.2	12.1

It will be clear from Table 6.4 that sandy, loam and clayey soils hold different quantities of moisture at same potential. In other words for same quantities of moisture, the potential ($Ø_m$) value is different. At -0.33 bars, the moisture content in these soils are 17, 32, and 45%, respectively. At $Ø_m$ -2 bars, these soils holds 8, 18, and 27% moisture, respectively and at $Ø_m$ -15 bars, respective moisture contents are 3.4, 7.2 and 12%.

Soil structure determines pore size, and distribution and hence water holding characteristics of the soil. At higher soil water potential (low matric potential of 0 to 1 bar) moisture retention depends on its pore size, degree of packing and aggregation. As the soil moisture potential decreases (tension increases), the texture plays a greater role. Due to increase in clay and organic matter content, moisture content of heavy textured soil at any matric potential is more. The soil moisture potential curve indicates the nature of soil, how much water it can hold within the practical range and how quickly it releases water for plant purposes. It helps in deciding the irrigation frequency of crops.

6.7.4 Specific Water Holding Capacity

In irrigation, sometimes a term called as specific water holding capacity is used. This is given as:

$$W_e = \frac{d\theta}{d\psi} \tag{6.21}$$

where, W_e = specific water holding capacity; $d\theta$ = change in soil moisture content and $d\psi$ = change in potential.

6.7.5 Measurement of Soil Water Potential

As discussed above, there is a clear relationship between the soil water potential and soil moisture content of the soil held at this potential. Soil water potential is influenced by the energy of its retention. The relationship between the quantity of moisture held by the soil and the energy of its retention is continuous from wet to dry range without any sharp break. At each range the energy can be measured by different ways. In wet range, the soil moisture content is within 1 bar soil water potential. Tensiometers are the most widely used instruments to measure the soil water potential in wet range. Details of measurement of soil moisture content/soil water potential by tensiometers are discussed earlier. In moist and dry range, pressure plate and pressure membrane apparatuses are used to measure the soil water potential.

6.7.6 Saturation and Maximum Water Holding Capacity

In a soil sample, some space is occupied by water and rest is occupied by air. When the soil gets wet either because of rainfall or irrigation, the macro and micro pores gradually get filled with water and the volume of air in the soil decreases. When all the pores are filled with water, the air content in the void space becomes zero. Under such cases, the soil is said to be fully saturated (sometimes simply called as saturated). Degree of saturation of a fully saturated soil is 1.0 (100%). Thus, for a perfectly dry soil sample there is. zero degree of saturation. At saturation, soil moisture potential is maximum i.e. zero and the moisture held by the soil is also maximum. Capacity of the soil to hold moisture at saturation is called as maximum water holding capacity.

6.7.7 Field Capacity

Field capacity is defined as the amount of water held by the soil after excess water has been drained away by gravity. At field capacity, the rate of downward movement of water has been materially decreased to negligible value which takes place two to three days after irrigation in a pervious soil of uniform texture (Veihmeyer and Hendrickson, 1949). At field capacity, the macro pores are filled with air whereas the micro pores are filled with water. Field capacity is the upper limit of available water. When crops are to be irrigated, care should be taken such that the soil moisture content comes to the field capacity level and any addition of water beyond field capacity is a loss since it is drained out to the layer below the crop effective root zone and adds to percolation. Field capacity is not same for all soils. It depends on a number of factors like texture of the soil, its structure, depth to groundwater table, organic matter content of the soil, layering of the profile and rate and pattern of evaporation. The average matric potential of field capacity varies from -0.1 bars in sandy soil to -0.33 bars in clayey soil.

However, for all practical purposes, moisture at field capacity is at the matric potential of 0.33 bars.

Field capacity water is divided into two parts i.e. capillary water and hygroscopic water. One part is that which is attached to the soil molecules by surface tension against gravity and can be extracted by plants by capillarity is called as capillary water. The other part is that which is attached to the soil molecules by loose chemical bonds and cannot be removed by capillarity is called as hygroscopic water. This water is not available for the plants. Field capacity is defined as (Garg, 1987):

$$Field\ Capacity = \frac{Weight\ of\ water\ retained\ in\ a\ certain\ volume\ of\ soil}{Weight\ of\ same\ volume\ of\ soil} \tag{6.22}$$

6.7.8 Permanent Wilting Point

The roots of the plants absorb water from the soil which results in gradual decrease of soil moisture content around the plant roots. Roots exert more and more pressure to extract water from the soil to meet the crop water demand. If this process continues, then there will be deficiency of soil moisture and the plants will start wilting. This type of wilting is termed as temporary wilting. However, if this process continues for longer period then there will be excess deficiency of soil moisture in the root zone. The rate of absorption remains too low to counter balance the loss and the wilted leaves do not regain turgidity. Briggs and Shantz (1912) termed this state of the plant as permanent wilting. Permanent wilting point refers to the lowest limit of available soil moisture content below which water extraction by a given plant is not sufficient to meet the evapotranspiration requirement of the plants. Irrigation to the crops is to be provided before this stage reaches. It is to be remembered that permanent wilting point is not a fixed point on soil moisture potential curve; rather it refers to a range. Crops start wilting at matric potential of wilting point varying from of -12 to -23 bars. For most of the mesophytes (considering their growth and development), the permanent wilting point refers to the matric potential of -15 bars.

6.7.9 Available Soil Moisture

As stated earlier, soil water at field capacity (potential = 0.33 bars) is the upper limit of available soil moisture for the plants. Similarly, soil water at permanent wilting point (potential = -15 bars) is the lower limit of available soil moisture. The soil moisture between these two values is called as available soil moisture. The water left in the soil after permanent wilting point cannot be used by the plants and so is called as unavailable water or hygroscopic water. Available soil moisture has greater importance in growth and development of plants. It is to be remembered that the available soil moisture depends on texture of the soil. Soils containing higher proportion of silt and clay have more amount of soil moisture than the soils containing more amount of sand. For example, in Chiplima (Odisha), the sandy loam soil at 0-15 cm layer has field capacity 17.3% whereas the permanent wilting point is 4.6% and thus the available soil moisture is 12.7%. In Karnal (Haryana) the clay loam soil (0-15 cm soil layer) has field capacity 23.0% whereas the permanent wilting point is 7.0% and thus the available soil moisture is 16.0% (Lenka, 2001). Available moisture can be expressed as percentage moisture or as equivalent depth which ever is convenient. Table 6.5 gives the values of available soil moisture content for different soils in the Hirakud Command of Odisha (Misra and Nanda, 1985).

6.7.10 Readily Available Soil Moisture

Readily available soil moisture sometimes also called as relatively available moisture is that portion of available soil moisture that can be easily extracted by the plants. It varies from 75-80% of the available moisture.

Table 6.5: Available soil moisture content for different types of soil

Soil types	Mechanical composition			Available soil moisture (%)
	Sand (%)	Silt (%)	Clay (%)	
Coarse sand	94	4	2	3
Sand	89	5	6	3
Loamy sand	82	10	8	6
Sandy loam	63	22	15	10
Fine sandy loam	59	24	17	12
Loam	40	36	24	17
Sandy clay loam	50	21	29	14
Clay loam	37	28	33	16
Clay	31	23	46	16

6.7.11 Root Zone Depth

Depth below the ground surface in which the root systems of the plant develop to derive water and minerals from the underneath to prepare food and support growth is called as root zone depth. The root systems of the plants in the root zone may consist of (i) tap roots, (ii) shallow primary roots and (iii) lateral roots. The depth of the root zone depends on the type of the crop, soil texture where they grow, irrigated or non-irrigated condition of growth, depth of water table, subsoil formation etc. There is considerable variation of root growth systems among the various plant species and varieties. Various root growth simulation models are available to find out the daily increment of root length of plants. Crops, in general, develop most of their root systems in the upper most portion of the root zone. Pattern of moisture absorption of the plants varies from layer to layer in a given root zone. The first quarter of this depth uses 40% of the total soil moisture in the root zone whereas the second, third and the fourth quarter uses 30, 20 and 10% of the soil moisture, respectively (Fig. 6.8).

6.7.12 Infiltration

Infiltration is defined as the entry of water into the soil through the interstices or pores of the soil. Infiltration occurs both prior to and during the occurrence of surface runoff but decreases with the time until a minimum rate of infiltration is reached. The rate at which infiltration takes place during a rainfall or when irrigation is applied to crop field, expressed in terms of depth of water per unit time (usually mm/hr, cm/hr) is called as infiltration rate. It is equal to the rainfall rate or infiltration capacity which ever is less. Infiltration rate is maximum at first and gradually decreases with time. The infiltration capacity curve which is a graphical representation of

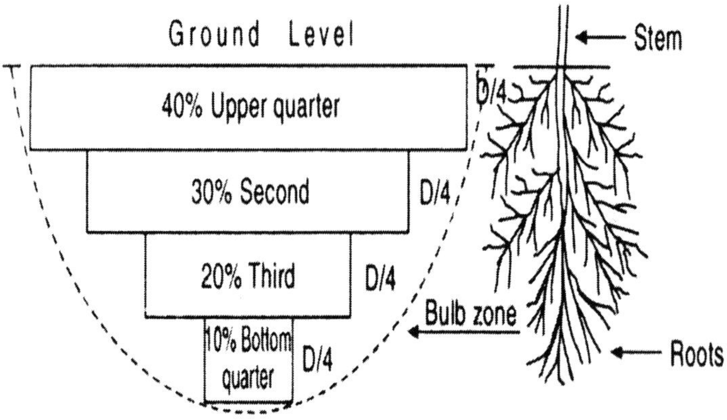

Fig. 6.8: Patterns of moisture absorption by roots

infiltration rate vs. time is an exponential decay curve and indicates that the rate may be zero at infinite time. A number of factors influences the infiltration capacity namely soil texture, soil structure, vegetation condition, slope of the soil, cultivation practices, antecedent moisture condition of the soil etc. A number of models in hydrology are available to predict the infiltration rate: most of them are empirical. Details of models used to predict the infiltration with their limitations are available in many Hydrology books.

6.7.13 Permeability

Permeability is one of the important properties of the soil which refers to the ability to convey flow of water through the pore spaces caused by a given force. It is defined as the velocity of flow caused by a unit gradient. Permeability is not influenced by the gradient and this is an important point of difference between permeability and infiltration. However, permeability is influenced most by soil texture and soil physical condition. It is a very widely varying factor varying from a few centimeters per hour to several meters per hour depending on soil types and soil physical conditions. For unsaturated soil, the moisture content is one of the dominant factors influencing permeability. It has the physical dimensions of length divided by time. Details of permeability are discussed later on.

6.7.14 Surface Tension

Water molecules are held together by cohesive force. If the water molecules are broken by a needle then the water molecules at the surface are pulled apart and the needle sinks down: otherwise the needle floats at the surface. At the middle of the water body contained in a container, any molecule is attracted equally in all directions. However at the surface, the molecule is pulled more in downward and side ward directions than upward. This causes the surface of the liquid to look like a stretched membrane. Tension exists exist along the surface of the stretched membrane. This tension is called as surface tension. Its unit is force/unit length (dynes/cm or newtons/m). It depends on temperature being linearly inversely proportional with temperature.

Since surface tension of water causes suction on water within the soil, the term "suction head" is being used by many researchers instead of "tension head". Suction head is generally expressed in terms of equivalent length of water column.

6.7.15 Tension Head

If a capillary tube is dipped in water in a container, then water will rise in the tube to a height more than the level of water in the container. This rise is due to the combined action of force of cohesion and adhesion-surface tension. The height of rise of water in the tube depends on surface tension and weight of liquid column. Water will rise in the capillary tube till the force (*F*) due to the surface tension (*F*) is equal to the weight of the liquid (*W*) being lifted up in the tube. The force is equal to the product of length of the film in the circular tube (length = 2 π r where *r* is radius of the tube) and surface tension is *T*. If α is the angle of contact between the vertical column of the tube and meniscus of the water level in the tube and resolving the tension in vertical direction, one get force as:

$$F=\prod\pi \text{ T} \cos\alpha \tag{6.23}$$

The weight of the water lifted in the tube is equal to the product of the volume of liquid lifted up and its unit weight. The volume is given as r^2h and unit weight is *g*. Thus, the weight of liquid lifted up is given as:

$$W = \pi\, r^2\, h\, \rho\, g \tag{6.24}$$

where, *r* = radius of the tube; *h* = height of water column in the tube; = density of water and *g* = acceleration due to gravity.

Under equilibrium condition,

$$2\,\pi\, r\, T \cos\alpha = \pi\, r^2\, h\, \rho\, g \tag{6.25}$$

$$h = \frac{2\,T \cos\alpha}{\rho\, r\, g}$$

If liquid taken in the tube is water, then angle is very small and then is nearly 1. Assuming *T* for water at 20^0C as 72.75 dynes/cm, equal to 1g cm^{-3} and *g* equal to 981 ergs/cm, *h* of Eq. (6.25) comes as:

$$h = \frac{0.148}{r} \tag{6.26}$$

In the above equation (Eq.6.26), both *h* and *r* are in same unit say if *h* is in cm then *r* is also in cm. The height of rise of water column in a tube (in a capillary tube it is called as capillary rise) is inversely proportional to the size of the tube (size is the diameter of the tube). In the soil-water-plant relationship, rise of water in the soil (called as capillary rise) is calculated if the size of the largest effective pore is known and vice versa. It is to be remembered that the capillary rise depends on the temperature, surface tension and density of water. Capillary rise also depends on the soil texture and pore diameter. In an experiment, the capillary rise of water in sandy soil was observed to be 155 cm whereas in the loamy sand it was 137 cm. (Premi and Oswal, 1989; Premi *et al.*, 1989). Further, capillary rise depends on the height of groundwater table. If the water table is near to the ground surface, then there will be more rise of capillary water than when the water table is far away from the ground surface.

6.8 Flow of Water into and Through Soils

The flow of water in the soil is governed by the fundamental property of inter conversion of mechanical work and energy. The total energy of a flowing fluid comprises potential energy, kinetic energy and pressure energy. This is given by Bernoulli's Eq (6.27) as:

$$H = \frac{V^2}{2g} + \frac{p}{\gamma} + Z \tag{6.27}$$

where, H = total energy head, = kinetic energy head, = pressure head and Z = potential energy head expressed as elevation head over some datum. In the above Eq (6.27), V = velocity, g = acceleration due to gravity, p = pressure, = unit weight of fluid and Z = elevation head. All the terms of the above equation are expressed in terms of dimensions of length, which is extensively used in flow of fluid in pipes and open channels including flow through the soils. Since the velocity of flow in soil is very small, the kinetic energy head term of Eq. (6.27) can be ignored and hence the total energy head (Eq. 6.27) is simplified as:

$$H = \frac{p}{\gamma} + Z \tag{6.28}$$

The head, H in Eq (6.28) can be used as the piezometric head or sometimes called as hydraulic head. This quantity is measured in saturated soil by the simple piezometer and in the unsaturated soil by a tensiometer.

Piezometer readings can be used to ascertain the direction of flow of water in soil and can also be used to measure the hydraulic gradient. Knowing the hydraulic gradient and permeability of the soil, the flow velocity in the soil can be determined applying the Darcy's law (explained later). It is to be remembered that flow always occurs in the direction of decreasing piezometric head. The piezometric head in two piezometers put into the soil near to each other can be used to find out the flow in upward and/or downward direction. If the piezometer readings of the first piezometer are more than the second one, then it indicates that flow of water in the soil is in downward direction and vice versa. The piezometric readings can be helpful to calculate the percolation rate in the soil in a given layer. This is explained as follows:

Suppose two piezometers A and B are put in an unsaturated soil at a given location at 1.5 and 2.0 m depths below the ground level. Let the piezometer heads of the two piezometers A and B be 1.8 and 1.7 m, respectively above some datum. The datum here is not the ground level but it is below the piezometer level in the ground surface. Then the hydraulic gradient of flow (flow is in downward direction) will be equal to the ratio of the difference of piezometric heads

(1.8 – 1.7 = 0.1 m) and the elevation head difference of the two piezometers (2.0 – 1.5 = 0.5 m) i.e. the hydraulic gradient (i) will be 0.2. Knowing the permeability of the soil under unsaturated condition (K), one can determine the percolation rate of the soil (V) by applying Darcy's law ($V = K\ i$). Piezometers can also be used to measure the groundwater table position in the soil. Knowing the permeability of the soil and the piezometric heads at two positions in a given soil stratum, one can determine the direction and velocity of groundwater flow in the soil stratum. Darcy's law states that the velocity of the flow is proportional to the hydraulic gradient sometimes called as the slope of the piezometric head line as long as the flow is laminar (Reynold's number is less than 1). Velocity of flow is then written as:

$$V = -K\,i = -K\,\frac{H_L}{L} \tag{6.29}$$

where, V= velocity of flow; K = permeability of soil in saturated condition; i = hydraulic gradient ($i = \frac{H_L}{L}$); H_L = difference of piezometric head and L = difference of elevation head of the two piezometers put in the soil.

Negative sign in the above equation infers movement of water in the direction of decreasing potential. However, in the subsequent equations involving Darcy's law only the absolute value is considered to determine the magnitude of flow (the negative sign is kept silent). The term is a head loss term in the direction of flow. It is also called as the slope of the piezometric head line. Hydraulic gradient is a dimensional less term and hence the unit of velocity depends on the unit of soil permeability. Soil permeability is a combined measure of soil and fluid properties. Details of determination of soil permeability under both saturated and unsaturated conditions are beyond the scope of the book. Readers can refer any book on Groundwater Hydrology for further readings on hydraulic conductivity.

Using the continuity equation, one can write the discharge, Q as:

$$Q = A\,V = A\,K\,i = A\,K\,\frac{H_L}{L} \tag{6.30}$$

where, A = cross sectional area of flow at right angle to the flow direction and other terms are defined as above.

In groundwater study, Darcy's law in addition to the permeability plays a vital role. It can be used both for unconfined as well as confined aquifer. Following example illustrates the use of Darcy's law and use of permeability to determine the quantity of groundwater flow.

Sample Calculation 5

An unconfined aquifer comprising sandy soil over lays a compact clay layer. Two piezometers A and B are put in the aquifer at a horizontal distance of 100 m. The piezomeric head readings of A and B are 50 and 40 m, respectively measured above some datum (Datum is below the ground level). Value of permeability of the soil in the layer where two piezometers are put is measured to be 1.1 m/day. Find out the velocity of flow. Determine the quantity of flow if the section of flow is 1000 m wide and the thickness of the layer is 20 m.

Solution:

Given permeability, $K = 1.1$ m/sec.

Piezometric head readings at A (H_1) and B (H_2) are 50 and 40 m, respectively. So, the piezomtric head difference is 50-40 = 10 m.

Since, distance between the two piezometers are 100 m, the hydraulic gradient, i = 10/100 = 0.1.

Using Darcy's law, velocity of flow is given as, $V = K\, i$ = 1.1 x 0.1 = 0.11 m/day.

Cross sectional area of flow = A = 1000 x 20 = 20000 m^2.

Hence, the quantity of flow is, Q = A V = 0.11 x 20000 = 2200 m^3/day = 91.7 m^3/hr = 25.5 l/sec.

6.9 Flow of Water in Horizontal and Vertical Soil Column in Unsaturated Condition

Flow of liquid occurs in soil column when there is some difference in piezometric /hydraulic head causing hydraulic gradient. As stated earlier, the piezometric head comprises the pressure and gravitational head. The elevation of the point relative to some reference point called as datum determines the gravitational head whereas the pressure head is determined by the height of water column in the piezometer put at that point. In case of horizontal soil column (Fig. 6.6), the reference point is same and so the gravitational head is ignored. The difference of the pressure head/hydraulic head determines the flow. If the pressure head is same at two points, then the hydraulic gradient will be zero and under such case, there will be no flow (static flow). Refering Fig (6.9), the hydraulic gradient is given as:

$$i = \frac{h}{L} \tag{6.31}$$

Flow velocity is (by Darcy's law):

$$V = K\frac{h}{L} \tag{6.32}$$

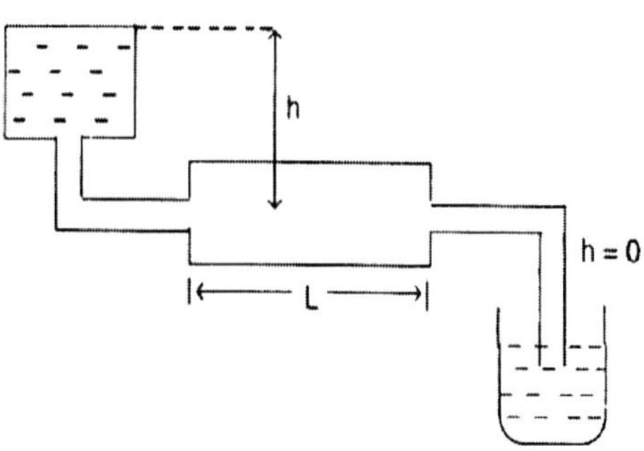

Fig. 6.9: Flow through horizontal soil column

In case of vertical column (Fig. 6.10), piezometric head at the outlet point is the sum of the hydraulic head and the length of the soil column. If the length of the soil column is L and the hydraulic head difference is h, then the hydraulic gradient in a vertical column of length L is given as:

$$i = \frac{(h+L)}{L} \text{ and} \tag{6.33}$$

Flow velocity is (by Darcy's law):

$$V = K\,i = K\,\frac{(h+L)}{L}$$

$$V = K + K\frac{h}{L} \tag{6.34}$$

It is observed from Eqn. (6.32 and 6.34) that the flow velocity in case of a vertical column is more than that of a horizontal soil column.

It is to be remembered that the soil permeability is not same in all the layers. So, in case of layered soil with varying values of soil permeability, the equivalent value should be considered to determine the flow velocity and flow quantity. Determination of equivalent soil permeability in layered soil with flow direction in both horizontal and vertical direction of soil layers is available in any standard Hydrology Book.

In an unsaturated soil, contribution of hydraulic and gravitational head is of less importance and the matric potential plays a crucial role. So, the hydraulic gradient is given as:

$$i = \frac{\Delta\psi_m}{L} \tag{6.35}$$

In unsaturated case, the soil permeability depends on the moisture content of the soil. In the wet range up to -1 to -1.5 bar matric potential, sandy loam soil has greater soil permeability

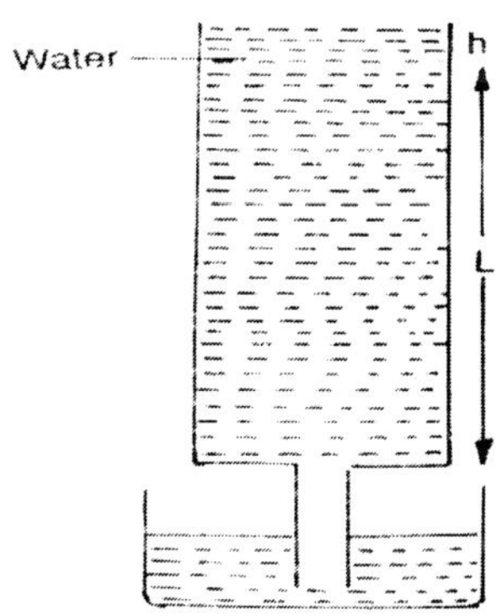

Fig. 6.10: Flow through vertical soil column

than the silty clay loam. In the moist range, when the matric potential is less than this, the relationship is reversed. Movement of unsaturated flow ceases in sand at a higher potential than in fine textured soil In moist range, the values of unsaturated permeability in *sand*.

6.10 Entry of Water into Dry Soil

When rain falls on the dry soil, there is discontinuous moisture content distribution in the soil which influences the movement of water. There is a sharp demarcation between the dry and wet soil causing wetting front. The region behind the wetting front is called as transmission zone having fairly uniform moisture content and pressure. Water intake decreases with time and reaches almost a constant value. The water intake is governed by a power form equation of the type as given below.

$$i = a\, t^{b} \tag{6.36}$$

where, i = water intake rate into the soil (sometimes it is also called as infiltration rate though in actual sense both terms are not same); t = time; a = a constant and b = exponent which is always –ve with values ranging from 0 to -1.

It is to be marked by the above equation that, the infiltration rate is never zero even at infinite time though it approaches zero. This equation of the infiltration rate was proposed by Kostiakov and so is called as Kostiakov equation. Many such empirical equations for determination of infiltration rate are available. Sum of the equations are Philip's equation, Lewi's equation, Lewi's and Green Ampt equation etc. Details of these equations can be read by referring any standard Hydrology book.

6.11 Water and Plant

Just like all other living beings, plants also need water to prepare food by the process called as photosynthesis. The roots of the plants draws water from the soil and convey the same to the leaves through various plant parts. This water helps in preparation of food. Moreover, water is an universal solvent. It is the medium through which, all substances are carried from the region of supply to the region of use. Water is essential to keep the plant cells turgid so that the stomata remain open to carry the photosynthesis uninterruptedly. Water also acts as a buffer against high and low temperature injury and keeps the leaves cool through transpiration. Further, it is an essential reagent in activating enzymes and in hydrolyzing starch to sugar.

Water movement from soil to the atmosphere through various parts of the plants takes place along the gradient of decreasing water potential. As in the soil, water potential in the cell is the sum of (i) osmotic potential, (ii) matric potential and (iii) pressure potential. There exists a wide difference in water potential in plant parts due to variations in resistances in the plant parts. The rate of movement (flux) varies directly with the potential and inversely with the resistance. There exists wide variation among the water potential in soil, plant and the atmosphere in the water path way from soil to the atmosphere in the soil-plant-atmospheric continuum.

6.12 Causes of Water Stress in Plants

Major amount of water in plants' cell is consumed by evapotranspiration (evaporation from the soil surface around the plant roots and transpiration by the leaves). Evaporation from the soil surface depends on the moisture condition of the soil. If the soil moisture around the root systems is adequate, then there will be more evaporation loss from the soil as a result, the water stress in the plants will be more. On the other hand, under dry soil condition, the flux (evaporation component) will be reduced as a result there will be less water stress in the plants. Transpiration depends on the leaf size, leaf area index, orientation of the leaves, size and distribution of stomata in the leaves, atmospheric humidity, temperature, wind velocity, and percentage of day light hours. The water uptake by the roots depends on the length of the roots and its distribution systems, density of roots, and soil moisture condition. High atmospheric temperature and intense sunlight due to increased transpiration causes closure of stomata and wilting of leaves even if soil moisture content is not a limiting factor. This deficit is made up during the night when there is no transpiration On the other hand, when there is more humidity in the atmosphere, then plants may not wilt even if there is limiting soil moisture. It is to be remembered that the roots are more sensitive to decrease ion soil water potential than the leaves.

Available soil moisture, soil solution concentration and root growth determines the absorption of soil moisture and thus affects the water stress. Water logging causing to decreased soil aeration and other associated problems like development of soil salinity and alkalinity causes dwindling growth of roots of the plants in non paddy crops as a result of which the roots absorb less water from the soil. This causes more water stress in the plants. Similarly, if the soil temperature falls below 20^{0}C, then it will affect the water absorption. Fall of temperature decreases absorption due to reduction in root-cell permeability. Further, fall of temperature increases the viscosity of water which increases the resistance to passive movement of water through roots there by the absorption of water is reduced. The geometry of plants root system plays a vital role in absorption of soil moisture. Long roots with branched fibrous root systems absorb more water from the deeper soil layer than shallow and less fibrous roots. Details about the root zone depth and pattern of soil moisture absorption from different soil layers by the roots are discussed earlier.

6.13 Soil Water and Crop Yield Response to Soil Water

The reduction of soil moisture on growth of plants is a complex phenomenon. The reduction of soil moisture affects a large number of physio-chemical activities which ultimately influences the growth of plants. Reduction of moisture reduces the leaf size and leaf area index, cell size, hampers root growth and water uptake is reduced. The process of photosynthesis is affected and hence, all these factors hamper the plant growth and its metabolic activities consequently reduce the yield. Chemical composition and quality of the crops also are affected. Water stress decreases sugar concentration in leaves of several crops (bean, sugarbeet etc.) while it increases sugar concentration in other crops (cotton, tobacco etc.). Moisture stress decreases both RNA and DNA.

Crop response to available soil moisture at different growth stages is different. During some particular stages, the plants are more sensitive to moisture stress than at others and this moisture – sensitive (critical) periods differ with crops and varieties. In some cases these critical periods may also differ under different climatic and soil conditions for the same crops. These critical moisture-sensitive periods, sometimes called as critical growth stage for various crops has been described in detail by Hukkeri and Pandey (1977) and by Hiler and Howell (1983). Application of water at these critical growth stages is more productive than when applied at other stages because moisture stress during these periods will considerably reduce crop yield which can not be recovered by subsequent application of water of later stages.

There are mainly two crop water production functions models available for assessment of soil water deficit affect on yield of dry land crops at different growth stages. They are (i) multiplicative model and (ii) additive model. Again each of the above two mentioned models may be seasonal or dated. Out of the various types, additive dated model is important. An additive model of dated crop-water production function derived from sensitivity factors for water stress in physiological growth stages as proposed by (Rao *et al.*, 1990) is given as:

$$\frac{Y_a}{Y_p} = 1 - \sum_{tg=1}^{NP} K_{tg}\left(1 - \frac{ETcrop}{PET}\right)_{tg} \qquad (6.37)$$

where, Y_a and Y_p = actual and potential yield of the crops, respectively, kg ha^{-1}; K_{tg} = yield response factor for growth stage, tg; *ETcrop and PET* = actual and potential evapotranspiration, respectively, mm; *tg* = growth stage index and *NP* = total number of periods for the crop (generally for all crops it is four). Values of K_{tg} for different crops for different growth stages are reported by Doorenbos and Kassam (1979). Potential yield of the crops is the maximum yield harvested under a set of ideal condition and to be obtained for a particular region from the experimental field observations.

Dated water production function of rice can be used for predicting the yield, but the yield response factors are not available for various stages of its growth. However, the physiological based rice growth and yield model *(RICEMOD)* developed by McMennamy and O' Toole (1983) can be used to predict the rice yield under practical field conditions.

Sample Calculation 6

Following table gives data of actual and potential evapotranspiration and yield response factor of a crop at different growth stages. Find out the percentage reduction of yield due to water stress. If the potential crop yield is 10 q/ha, then find out the actual crop yield?

Growth stages	Potential evapotranspiration, mm	Actual evapotranspiration, mm	Yield response factor
Initial	22.0	20.0	0.5
Crop development	30.0	25.0	0.8
Mid season	35.0	32.0	0.9
Maturity	20.0	15.0	0.5

Solution:

Solution of the above problem is illustrated with the calculated data in the tabular form:

Growth stages	Potential evapotranspiration (*PET*), mm	Actual evapotranspiration (*ETcrop*), mm	Yield response factor, *Ktg*	*Ktg (1-ETcrop/PET)*
Initial	22.0	20.0	0.5	0.045
Crop development	30.0	25.0	0.8	0.133
Mid season	35.0	32.0	0.9	0.077
Maturity	20.0	15.0	0.5	0.125
Total			0.380	

$$\frac{Y_a}{Y_p} = 1-\sum_{tg=1}^{NP} K_{tg}\left(1-\frac{ETcrop}{PET}\right)_{tg}$$

= 1-0.380 = 0.620

Hence, the percentage reduction of yield = *(1-Ya/Yp)* 100 = 38.0

If potential yield is 10 q/ha, then the actual yield will be 0.620 x 10 = 6.20 q/ha.

Question Banks

Q1. What is soil? How is it formed?

Q2. With labeled diagram explain the soil as a three phase system.

Q3. Define the following terms and write formulae to find them:

(i) Soil moisture (ii) Density (iii) Unit weight (iv) Bulk density
(v) Dry density (vi) Solid density (vii) Void ratio (viii) Porosity
(ix) Maximum water holding capacity (x) Saturation moisture content

Q4. Differentiate between the followings:
(i) Porosity and void ratio (ii) Bulk density and dry density
(iii) Saturated and unsaturated moisture content (iv) Field capacity and permanent wilting point
(v) Zone of aeration and zone of saturation (vi) Soil texture and soil structure
(vii) Gravitational and capillary water

Q5. How can you calculate degree of saturation of a soil sample? Write the degree of saturation for various types of soils.

Q6. Find the void ratio of a given soil sample that contains 40% porosity.

Q7. An undisturbed soil sample has volume of 800 cm^3 and mass of 1000 gm. When dried in oven for 24 hours, the soil mass comes out to be 500 gm. Calculate the water content, bulk density and its unit weight ?

Q8. What will be the degree of saturation of a given mass of soil sample that contain volume of water one fourth to volume of voids?

Q9. Draw a neat sketch of triangular diagram and explain how does it help to classify the textural classes?

Q10. Draw a figure to represent the different types of soil moisture and discuss the various types of soil moisture wi9th the help of this figure.

Q11. What is hygroscopic water? Why it is not helpful to support growth of plants?

Q12. Define soil moisture content. Why it is important to measure it in the field of irrigation?

Q13. Discuss the following methods employed to measure the soil moisture content.

(i) Sand bath method
(ii) Calcium carbide method
(iii) Radiation method
(iv) Torsion balance method
(v) Using neutron moisture meter
(vi) Using gypsum resistance block

Q14. What are the limitations of use of neutron moisture meter and resistance block for the measurement of soil moisture content?

Q15. What is a tensiometer? How does it help in determining the moisture content of the soil? Also write the limitations of its use.

Q16. What is soil moisture tension? How does it vary with the types of soil?

Q17. What is *pF* and give a relationship between the pF and free energy of water.

Q18. A soil sample has tension (*pF*) 2. Find out the free energy of water.

Q19. Find out the *pF* of if the free energy of water n terms of height of water column are 1, 10, 100, 1000, 10,000, 1, 00,000 and 1,000,000 cm, respectively.

Q20. Draw a typical soil moisture retention curve and discuss how it helps in irrigation to crops.

Q21. How the soil texture and the soil structure do affect the soil water holding capacity?

Q22. At a particular soil suction, which of the following two types of soil will retain more soil moisture and why?

(i) Sandy soil (ii) Clay soil

Q23. Differentiate between saturation and maximum water holding capacity?

Q24. What is field capacity moisture content? How does it play a great role in determining the irrigation to the crops?

Q25. How can the filed capacity moisture content of a soil be determined? Briefly explain the factors that influence it.

Q26. Why should not the plant be irrigated after permanent wilting point moisture level has reached?

Q27. What is available moisture content? How is it different from readily available soil moisture content?

Q28. Enumerate the factors that affect the available moisture content of a given soil.

Q29. How does the root zone depth play a vital role in irrigation scheduling?

Q30. Do the infiltration rate and intake rate are same? If not then explain how are they different?

Q31. How the infiltration capacity curve is prepared? What are the factors that affect the infiltration?

Q32. Write notes on the followings:

(i) Permeability
(ii) Darcy's law
(iii) Surface tension
(iv) Tension head
(v) Infiltration
(vi) Capillary tension.

Q33. Discuss how the flow occurs through the porous medium of the soil. What are the governing factors that influence the flow through the porous media?

Q34. Derive the expression *for tension head, h as* where h is the tension head.

Q35. Briefly narrate how the water enters into the soil in dry phase.

Q36. The flow of water in the horizontal soil column will be more or flow will be more in vertical soil column. Discuss the phenomenon with detail derivation of velocity under both conditions.

Q37. A sandy soil aquifer over lays a compact clay layer. Two piezometers A and B are put in the aquifer at a horizontal distance of 120 m. The piezomeric head readings of A and B are 60 and 50 m, respectively measured above some datum (Datum is below the ground level). Value of permeability of the soil in the layer where two piezometers are put is measured to be 1.0 m/day. Find out the velocity of flow. Determine the quantity of flow if the section of flow is 2000 m wide and the thickness of the layer is 22 m.

Q38. What the causes of water stress in plants? Hoe does the water stress affect the growth and yield of crops?

Q39. Write the formula of additive dated water production function and explain the various terms. How does this production function affect the crop yield?

Q40. What is RICEMOD? What for it is used?

Q41. How is infiltration measured in the field? What is the difference between the single and double ring infiltrometer?

Q42. The water level in two piezometers put at a horizontal distance of 100 m are 10 and 8 m, respectively. The elevation heads of these two piezometers above some reference level below the ground surface are 30 and 25 m, respectively. The permeability of the soil in between the two piezimeters is 1.2 m/day. Calculate the velocity of flow by using Darcy's formula. Estimate the quantity of water flowing in a section of 1 km wide with 20 m thickness of the soil layer.

Q43. Following table gives data of actual potential evapotranspiration and yield response factor of a crop at different growth stages. The potential evapotranspiration during the same growth stages of the crop is also given in the same table. Find out the ratio of actual and potential yield of the crop. If the potential crop yield is 10 q/ha, then find out the actual crop yield?

Growth stages	Potential evapotranspiration, mm	Actual evapotranspiration, mm	Yield response factor
Initial	20.5	16.7	0.5
Crop development	25.9	22.3	0.65
Mid season	35.5	30.3	0.88
Maturity	17.0	13.5	0.55

References

Belgaumi, M.I., Itnal, C.J. and Radder, C.D. 1997. Farm Ponds. Technical series 4, Publication Centrte, University of Agricultural Sciences, Dharwad, pp. 47.

Briggs, J. and Shantz, H.L. 1912. U.S.D.A. Bur. Plant Ind. Bull. 285.

Doorenbos, J. and Kassam, A.H. 1979. Yield Response to Water. Irrigation and Drainage Paper 33, FAO, Rome.

Garg, S.K. 1987. Irrigation Engineering and Hydraulic Structures. Khanna Publishers, New Delhi.

Hansen, V.E., Israelsen, O.W. and Stringham, G.E. 1979. Irrigation Principles and Practices. John Wiley and Sons, New York.

Hiler, E.A. and Howell, T.A. 1983. Irrigation options to avoid critical stress. An overview in limitations to efficient water use in crop production. Madison, Wis. Soil Sc. Soc. Am.

Hukkeri, S.B and Pandey, S.L.1977. Water requirement and irrigation of crops. In: Water Requirement and Irrigation Management of Crops in India. Monograph No. 4, New Series Water Technology Centre, IARI, New Delhi: 163-273.

Lenka, D. 2001. Irrigation and Drainage. Kalyani Publishers, New Delhi.

McMennamy, J.A and O' Toole. J.C. 1983. RICEMOD: A Physiologically Based Rice Growth and Yield Model. International Rice Research Institute, Manila.

Misra, D. and Nanda, S.S.K. 1985. Available water of different textural classes in Ayacuts of Hirakud Command Areas. Journal of Indian Society of Soil Science, Vol. (33): 372-379.

Premi, S.C. and Oswal, M.C. 1989. Comparison of observed and predicted maximum rise of water in soils. Journal of Indian Society of Soil Science, Vol. (32): 654-657.

Premi, S.C., Oswal, M.C., Nagpal, P.K. and Gulati, K.L. 1989. Dependence of inter-layer contact impendence on pore size and angle of contact. Journal of Indian Society of Soil Science, Vol. (37): 152-154.

Punmia, B.C., Jain, A.K. and Jain, A.K. 2010. Soil Mechanics and Foundations. Laxmi Publications (P) Ltd., New Delhi.

Rao, N.H., Sarma, P.B.S. and Chander, S. 1990. Optimal multicrop allocation of seasonal and intraseasonal irrigation water. Water Resour. Res., Vol. 26(4): 551-559.

Schofield, R.K. 1935. The *pF* of water in soils. Trans Intern. Cong. Soil. Scil. 3rd, 2: 37-48.

Sharma, R.K. and Sharma, T.K. 2002. Irrigation Engineering including Hydrology. S. Chand and Company Ltd., New Delhi.

Singh, R.P. 1983. Farm Pond. Project Bulletin No. 6, All India Co-ordinated Research Project on Dryland Agriculture, Hyderabad.

Veihmeyer, F.J. and Hendrickson, A.H. 1949. Methods of measuring field capacity and wilting percentages of soils. Soil Sci. Vol (68): 75-94.

7

Irrigation Scheduling

7.1 Introduction

For maximization of yield of crops, it is essential to provide irrigation to crops. When either the natural sources like rainfall, dew, snowfall and other forms of precipitation or when the capillary rise of groundwater is not sufficient to meet the crop water demand, then there is need to supply irrigation. This irrigation water is a precious resource and need to be used cautiously. There are three important things need to be understood for optimum utilization of irrigation water with basic purpose to maximize the yield and returns. These three things are:

(i) When to irrigate?

(ii) How much to irrigate? and

(iii) How to irrigate?

In fact these three things form the core concept of irrigation scheduling.

7.2 Benefits of Irrigation Scheduling

Proper scheduling is essential for the efficient use of water, energy and other inputs required to maximize the yield including fertilizer. It allows irrigation to be coordinated with other farming activities including cultivation methods and fertilizers-chemicals-pesticide applications. In general the benefits of irrigation scheduling are:

(i) It enables the farmers to schedule water rotation amount among the various fields to minimize crop water stress and maximize the yields.

(ii) It reduces the farmers' cost of water and labour to less irrigation, thereby making maximum use of soil moisture storage.

(iii) It lowers fertilizer cost by holding surface runoff and deep percolation (leaching) to a minimum.

(iv) It increases net returns by increasing crop yields and crop quality.

(v) It minimizes waterlogging problem by reducing the drainage requirements.

(vi) It assists in control of root zone salinity problems through controlled leaching.

(vii) It results in saving a lot of precious irrigation water which can cover more area under irrigation to non-cash crops that otherwise would not be irrigated during water-short periods. This fetches the farmers to get additional returns and enhances his livelihood status.

Irrigation scheduling is the main part of study of irrigation engineering. Since it is a very big chapter, for the ease and convenience of understanding the concept, the chapter has been divided into two parts. Chapter 7 with title "Irrigation Scheduling" covers the scope of studies of "when to irrigate" and "how much to irrigate" and Chapter 8 with title "Methods of Irrigation" constitutes the scope of study of "how to irrigate".

7.3 When to Irrigate?

Scheduling irrigation is based on crops, soil, atmosphere and plant water relations. The timing of irrigation is a complex phenomenon depending on a host of factors. It varies from region to region depending on the climatologically conditions of the region too. There are several approaches to decide when to irrigate and the most important approaches are described below.

(i) Plant indicators

(ii) Physiological growth stage

(iii) Soil moisture regime concept

(iv) Water balance technique and

(v) Climatological approach based on *IW:CPE* method

Plant and soil indicators involve monitoring the plant and soil, respectively to determine when to irrigate. The water balance approach involves using a water balance model for the crop field and knowing the management allowable deficit, one can ascertain when to irrigate the crop field (discussed later). On the other hand climatological approach makes use of some formula like *IW:CPE (IW* = irrigation water and *CPE* = cumulative pan evaporation) and knowing the value of *IW:CPE* and the value of *IW, CPE* can be found out which will help when to irrigate (*IW:CPE* approach is discussed in details later on).

7.3.1 Plant Indicators

Plant indicators are the most direct method of deciding when to irrigate, since the primary aim of irrigation is to supply the plants with the water they need when it is needed. Normally the plant parameters are related to soil water content to calculate the amount of irrigation. The various plant indicators are described as follows:

7.3.1.1 Appearance and Growth

Appearance and growth indicators of the plant can be visualized by eye. Leaf and shoot wilting and leaf colour are the several appearance parameters to decide whether it is time for irrigation or not. Similarly, measurement of stem diameter and height can be made regularly to find out the growth rate. If the low growth rate indicates the time for irrigation. Though it is a simple method but this technique is not accurate to ascertain the exact timing of irrigation. In fact the time of irrigation might have been called for before hand when there is wilting of leaf/shoot or discolourance of leaf has occurred or detection of low growth rate has been observed. Sometimes along with the water stress, disease and improper nutrient application cause the change in appearance and cause low growth rate.

7.3.1.2 Leaf Temperature

Rise in leaf temperature is associated with reduced transpiration rates resulting from partial or full stomata closure. A hand-held infrared thermometer is used to measure the difference between the plant canopy and ambient air temperature each day near the time of maximum surface temperatures. The number of degrees by which canopy temperature exceeds the air temperature for each day is accumulated till a certain critical level is reached. This critical level depends on the crop and soil parameters. When the critical level reaches, it is the time for irrigation. Days when the canopy temperature is less than the air temperature are neglected.

7.3.1.3 Leaf Water Potential

As the plant removes water from the soil, the water content of the soil decreases resulting total water potential of the soil (ψ_{soil}) to get reduced. A decreased value of soil water potential causes the low flow of water in the plant body. As this situation continues, plants begin to dehydrate and leaf water potential (ψ_{leaf}) decreases. When the ψ_{leaf} is sufficiently low, the leaf will be dehydrated to such an extent that stomata begins to close. This results in lower rate of transpiration and reduced photosynthesis rate causing a reduction of yield of crops (Hansen *et al.*, 1979). It is to be remembered that a lower or more negative leaf potential is an indicator of more water need and vice-versa. A pressure chamber or thermocouple psychrometer is used to measure the leaf water potential. The disadvantages of this method to decide the timing of irrigation are (i) method is time consuming, (ii) method requires sampling skill and (iii) data recorded is not easily interpreted.

7.3.1.4 Stomatal Resistance

Stomatal resistance is an indicator to the need for water, since it is related to the degree of stomatal opening and the rate of transpiration. In general, high resistance indicates significant stomatal closure, reduced transpiration rate and the need for water. Diffusion parameters are used to measure the stomatal resistance. The disadvantages of this method are same as described for the leaf water potential.

7.3.2 Physiological Growth Stages

In this method, the approach is to mark the important crop physiological stages and when there

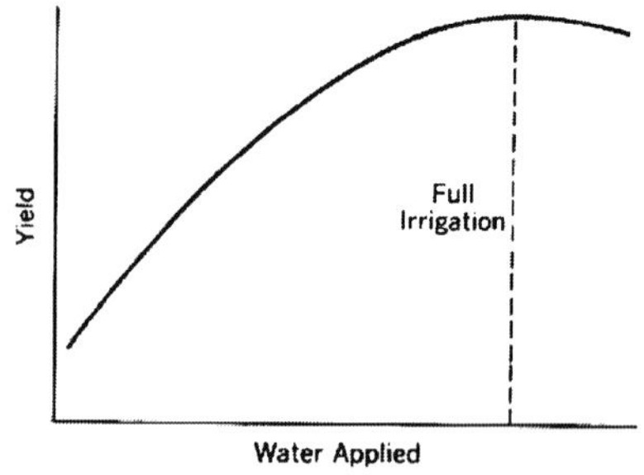

Fig. 7.1: A crop production function

is deficit irrigation supply, irrigate the crop at these stages only. Deficit irrigation refers to partial application of irrigation water to meet the crop water requirement. Deficit irrigation reduces the crop yield but saves adequate amount of irrigation water and is practiced when there is shortage in supply. It is economically justified when reducing water applications below full irrigation causes production costs to decrease faster than the revenues decline. Application levels can be reduced below full irrigation till the slope of production function (Fig. 7.1) is such that the decrease in revenue due to an incremental reduction in water application equals the accompanying decline in production costs. Irrigations should be scheduled to apply the seasonal application at this point on the production function, since the net benefit of irrigations are maximized.

Since water supply is limited, irrigation cannot be applied to the crops at all the growth stages. In other wards it involves applying water at some important physiological growth stages. Table 7.1 will helps in taking decision at which irrigations are to be given to some of the important field crops based on growth stages after giving the due considerations for the rainfall, especially for *kharif* crops. The real number of irrigations will depend upon agro-climatic conditions, rainfall in crop growth season, ground water contribution and soil types. If irrigation water is not adequately available, then limit the frequency and amount of irrigation at critical growth stages (*CGS*) only (Table 7.1).

Table 7.1 Physiological stages of crops with respect of irrigations

Sl. No.	Crops	Physiological stages	Critical stages
1.	Wheat	Crown root initiation (CRI), late tillering, late jointing, flowering, milk and dough	CRI, flowering
2.	Barley	Early vegetative stage, flowering and dough	30 days after sowing, pre-flowering
3.	Potato	Sprouting, stolonisation, stolon development, early tuberisation, 20% of tuber weight, 40% of tuber weight, 60% of tuber weight and 80% tuber weight	Sprouting, early tuberisation
4.	Gram	Pre-flowering and pod formation	Pre-flowering
5.	Soybean	Early seedling, pre-flowering, pod development	Pre-flowering, pod development
6.	Rice	Early tillering, panicle initiation, flowering, milk, dough	Early tillering, flowering
7.	Sorghum	Flowering, dough	Flowering
8.	Maize	Early vegetative growth, tasseling and silking, dough	Tasseling and silking
9.	Cotton	Branching, pre-flowering, boll formation	Pre-flowering, boll formation
10.	Sugarcane	Sprouting, tiller initiation, tillering	Sprouting
11.	Groundnut	Emergence, flowering, pod formation, pod development	Flowering, pod f ormation
12.	Mustard	Vegetative growth, flowering	Flowering

7.3.3 Soil Moisture Regime Concept

Water is consumed by the plants through their roots. It, hence, is necessary to ensure that sufficient moisture is made available to the plants in the root zone depth. Soil moisture in the root zone can vary between the field capacity (*FC*) which is the lower limit of available soil water to permanent wilting point (*PWP*) which is the lower limit of available soil water. As discussed earlier, the total maximum available water (*FC* – *PWP*) is not allowed to deplete i.e. plant is not allowed to get irrigated at *PWP* as it would result in considerable loss in crop yield. The optimum level up to which the soil moisture may be allowed to be depleted in the root zone without fall in crop yield has to be worked out for each crop from experimental data. Irrigation should be provided when this optimum level of soil moisture content reaches and its quantity should be just sufficient to replenish the soil moisture in the root zone up to field capacity level. Any addition of water content after field capacity level moisture content is a loss for use of the plants since gravitational drainage would occur (Garg, 1987).

When irrigation is applied, the soil moisture content would increase up to field capacity Water will be utilized by the plants after irrigation is provided and then there would be fall in the soil moisture content provided there is no rainfall or there is no other source to contribute water to the plants for use. It would increase again after the subsequent irrigation is provided when the soil moisture in the root zone reaches an optimum level (Fig. 7.2). The optimum level of soil moisture content depends on the management allowable deficit or available soil water depletions at certain depth (effective root zone), or soil water potential of the same depths at which irrigation is to be scheduled. The magnitude of available soil water depletions at which irrigation is to be scheduled may be a value ranging from 25 to 75 percent while in case of soil water potential, the values may range from 0.1 to 0.7 bars at some specified depth (Punmia and Lal, 1992). Table 7.2 shows the available soil water depletion and soil suction at specified depth for some important crops at which water should be applied for maximum yield. Table 7.3 represents the effective rooting depth of some crops and the available soil water of different types of soils (Doorenbos and Pruitt, 1977).

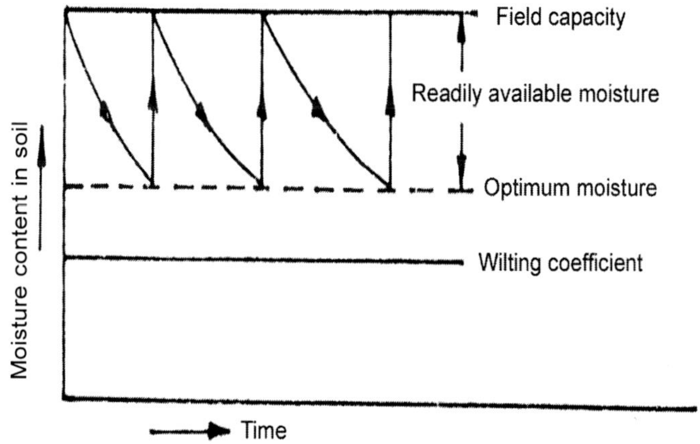

Fig. 7.2: Depth and frequency of irrigation

Table 7.2: Soil water depletion and suctions in the zone of maximum root activity at which water should be applied for maximum yields of some crops

Sl. No.	Crops	Available soil water depletion (%) at 0-30 cm soil depth	Soil water (suction bar)	Suction at depth, cm
1.	Wheat	50	0.5	30
2.	Potato	25	0.3	30
3.	Sugarbeet	50	0.5	30
4.	Sugarcane	65	0.7	30
5.	Soybean	60	0.6	30
6.	Tabacco	25	0.3	30
7.	Maize	50	0.5	30
8.	Sorghum	75	1.0	60
9.	Cotton	50	0.5	30
10.	Peas (grains)	65	0.7	30
11.	Berseem (fodder)	No depletion	0.3	30

In general soil based irrigation scheduling involves calculating the current available soil water. Soil water content may be measured directly or inferences from measurements of other soil parameters such as soil water potential or electrical conductivity can be helpful to determine the soil moisture content. Several common methods of estimating the soil water contents are listed in Table 7.4 (James, 1988) along with the advantages and disadvantages of various soil indicators. Details of measurement of soil moisture content by different methods/instruments have been discussed earlier.

Table 7.3: Generalised data on rooting depth of full grown crops, fraction of available soil water, and readily available soil water for different soil types in mm/m depth of soil.

Crop	Effective root depth, m	Fraction of available soil water	Readily available soil water Fine	Medium	Coarse
Alfalfa	1.0– 2.0	0.55	110	75	35
Banana	0.5 - 0.9	0.35	70	50	20
Barley	1.0 - 1.5	0.55	110	75	35
Beans	0.5 – 0.7	0.45	90	65	30
Beets	0.6 – 1.0	0.50	100	70	35
Cabbage	0.4 – 0.5	0.45	90	65	30
Carrots	0.5 – 1.0	0.35	70	50	20
Cotton	1.0 – 1.7	0.65	130	90	15
Cucumber	0.7 – 1.2	0.50	100	70	40
Dates	1.5 – 2.5	0.50	100	70	30
Grains (small)	0.9 – 1.5	0.60	120	80	30
Grains (winter)	1.5 – 2.0	0.60	120	80	40
Grapes	1.0 – 2.0	0.35	70	50	40
Grass	0.5 – 1.5	0.50	100	70	20
Groundnut	0.5 – 1.0	0.40	80	55	30
Lettuce	0.3 – 0.5	0.30	60	40	25

(Contd.)

Crop	Effective root depth, m	Fraction of available soil water	Readily available soil water Fine	Medium	Coarse
Maize	1.0 – 1.7	0.60	120	80	20
Melons	1.0 – 1.5	0.35	70	50	30
Olives	1.2 - 1.7	0.65	130	95	25
Onions	0.3 – 0.5	0.25	50	35	45
Peas	0.6 – 1.0	0.35	70	50	40
Potatoes	0.4 - 0.6	0.25	50	30	30
Safflower	1.0-2.0	0.60	120	80	15
Sorghum	1.0 – 2.0	0.55	110	75	50
Soybeans	0. 6 – 1.3	0.50	100	75	35
Spinach	0.3 - 0.5	0.20	40	30	35
Strawberries	0.2 – 0.3	0.15	30	20	15
Sugar beet	0.7 – 1.2	0.50	100	70	10
Sugarcane	1.2 – 2.0	0.65	130	90	30
Sunflower	0.8 – 1.5	0.45	90	60	40
Sweet potato	1.0 – 1.5	0.65	130	90	30
Tomatoes	0.7 – 1.5	0.40	180	60	40
Vegetables	0.3 – 0.6	0.20	40	30	25
Wheat	1.0 – 1.5	0.55	105	70	15

N.B. The above mentioned values of fraction of available soil water is meant when crop evapotranspiration (*ETcrop*) = 5-6 mm/day. When *ETcrop* is 3 mm/day or less, increase the value by 30% and when *ETcrop* = 8 mm/day or more, decrease the value by 30% (Doorenbos and Pruitt, 1977).

Table 7.4: Soil based indicators to decide when to irrigate

Observed/ measured parameters	Procedures/ required instrument	Advantages	Disadvantages
Appearance and feel	Hand probe	Simple	Time consuming, approximate and requires interpretative skill
Gravimetric sampling	Cans, soil augers, oven and balance	Simple and accurate	It may cause root zone damage if frequent samplings are collected from different layers, time consuming
Electrical resistance	Porous block	Provide indirect measure of soil moisture content	Requires careful installation, calibration and frequent readings, not sufficiently sensitive in coarse-textured soil
Soil matric potential	Tensiometer	Measures fundamental parameters affecting soil water flux	Requires preinstallation preparation, careful installation, frequent readings, need multiple site
Soil matric potential	Porous (ceramic) block	Accurate over a wide range of soil contents, facilitates automatic scheduling of block irrigation	Limited experience in use, calibration can deteriorate with time
Neutron scattering	Neutron probe block	Successive measurements at same location and soil depth are possible, quick and accurate	Equipment is costly, needs special skill to handle the equipment, dangerous to use, cannot measure soil moisture content at upper soil layers

For calculation of irrigation interval, allowable/management soil water depletion value and daily water use of the plants need to be found out. The irrigation interval is then given as (Michael, 1978):

$$Irrigation\ interval = \frac{Allowable\ soil\ water\ depletion}{daily\ water\ use} \tag{7.1}$$

Sample Calculation 1

Calculate the irrigation interval for a crop of tomato with the help of the following data.

Field capacity, $F\,C$ = 22.0% dry weight basis

Permanent wilting point, PWP = 10% dry weight basis

Bulk density, BD = 1.40 g/cc

Effective root zone depth, RD = 60 cm

Daily water use, $ETcrop$ = 0.25 cm/day

Allowable soil water depletion = 25 percent of available soil water.

Solution:

$$Available\ soil\ water = \frac{(FC - PWP)}{100} \times BD \times RD$$

$$\text{So, available water} = \frac{(22.0 - 10.0) \times 1.40}{100} \times 60 = 10.08\ cm$$

$$\text{Allowable soil-water depletion} = \frac{25 \times 10.08}{100} = 2.52\ cm$$

$$\text{Interval of irrigation} = \frac{2.52}{0.25} = 10.08 = 10\ days$$

Sample Calculation 2

After how many days will you irrigate a tomato crop if?

Field capacity of the soil = 26% (dry weight basis)

Permanent wilting point = 11% (dry weight basis)

Bulk density of soil = 1.3 g/cc

Effective depth of the root zone = 70 cm

Daily consumptive use of water for the crop = 12 mm

Allowable soil water depletion = 100 percent of available soil water

Solution

Available depth of water or deficit d in the root zone depth D is :

$$Available\ soil\ water\ = \frac{FC\ -\ PWP}{100}.\times BD \times RD$$

where FC = field capacity = 26%

PWP = permanent wilting point =11%

BD = Apparent specific gravity =1.3 g/cc

RD = Effective depth of the root zone = 70 cm

$$Available\ soil\ water\ = \frac{(26\ -\ 11)}{100} \times 1.3 \times 70\ = 13.65\ cm$$

$$\text{Allowable soil-water depletion} = \frac{100\ \times 13.65}{100}\ = 13.65\ cm$$

$$\text{Therefore, number of days between watering} = \frac{13.65}{1.2} = 11.34 \text{ days (11 days say).}$$

Sample Calculation 3

Wheat grown in loamy soil, having available soil water of 10 cm and allowable soil water depletion is 50 percent. Daily water use is 1 cm. Calculate the irrigation interval.

Solution:

Available soil water = 10 cm

Soil water depletion level = 50 percent

Allowable soil water depletion based on equivalent depth = 10 x 0.5 = 5 cm

Since daily water use is 1 cm, the irrigation interval is computed from Eq. (7.1) as 5/1 = 5 days.

Therefore, irrigation should be given at 5 days interval.

Sample Calculation 4

After how many days will you supply water to soil in order to ensure sufficient irrigation of the given crop, if

Field capacity = 28 percent

Permanent wilting point of the soil =15 percent

Bulk density of the soil = 1.5 gm/cc

Effective rooting depth = 50 cm.

Daily consumptive use of water for the given crop = 10 mm and

Optimum soil moisture level = 80 percent of available soil moisture.

Solution:

Given Field Capacity (*FC*) = 28 percent and permanent wilting point (*PWP*) = 15 per cent

Available soil moisture content = $FC - PWP$ = 28 -15 = 13 percent

Optimum soil moisture level = 0.80 x 13 = 10.4 percent

Irrigation is given when the soil moisture content is 28.0 - 10.4 = 17.6 percent i.e. soil moisture content varies from 17.6 to 28 percent.

Depth of water should be supplied such that the optimum soil moisture content rises from 17.6 to 28.0 percent which is estimated as:

$$d = \frac{1.5 \, x \, 50 \, x \, (28.0 - 17.6)}{100} = 7.8 \, cm$$

Daily consumptive use = 10 mm = 1 cm

$$\text{Hence the irrigation interval} = \frac{7.8}{1.0} = 7.8 = 7 \; days$$

Sample Calculation 5

Wheat is to be grown in a field having field capacity equal to 25 percent and permanent wilting point is equal to 11 percent. Find the maximum storage capacity in 80 cm depth of the soil if the bulk density is equal to 1.5 gm/cc. If irrigation is supplied when the average soil moisture falls to 18 percent, find the depth of water required to be given to the field if application efficiency is equal to 80 percent. Also find out the depth of water to be supplied at the canal head end considering transit loss as equal to 20 percent.

Solution:

Maximum storage capacity of the soil is given as:

$$Maximum \;\; storage \;\; capacity = \frac{Bulk \;\; density \;\; x \; depth \;\; of \;\; soil \;\; x \; (FC - PWP)}{100}$$

$$= \frac{1.5 \, x \, 80 \, (25.0 - 11.0)}{100} = 16.8 \, cm$$

Irrigation is supplied when the soil moisture falls from 25 to 18 percent.

Depth of water required to be supplied to the field to raise the moisture content from 18 to 25 percent is given as:

$$d = \frac{1.5\ x\ 80\left(25.0 - 18.0\right)}{100} = 8.4\ cm$$

Hence net irrigation requirement = 8.4 cm

Since the application efficiency is 80 percent, the quantity of water required at filed level

$$= \frac{8.4\ cm}{application\ efficiency} = \frac{8.4}{0.80} = 10.5\ cm$$

Considering transit loss (conveyance loss) of 20 percent, we get conveyance efficiency as 100 – 20 = 80 percent.

So the water to be delivered at the head end of the canal is given as:

$$= \frac{10.5}{0.8} = 13.125 = 13.13\ cm$$

7.3.4 Water Balance Technique

In this method a water balance model for crop is used to find out the amount of irrigation to be given to a crop. Irrigation management practice followed initial soil moisture content in the root zone and daily measurement of various water balance model parameters except irrigation helps to find out the soil moisture content in the root zone on daily basis. When the running soil moisture content in the root zone of any day reaches the pre-assigned optimum moisture content level as discussed above, then irrigation on that day is given to the crop. Irrigation is given to raise the optimum soil moisture content to field capacity level. A daily book keeping procedure to maintain all the records of the water balance parameters would help to find out the day of irrigation.

In the development of water balance model, rainfall (*R*) and irrigation (*I*) are inputs to the model whereas actual evapotranspiration (*ETcrop*), surface runoff (*SR)* from the field, seepage and percolation losses (*S*) and (*D*), respectively of the crop field form the outputs. Considering the effective root zone as a single layer, and neglecting the capillary flux of groundwater with assumption that groundwater lies at 2 m or more below the crop effective root, the general water balance model of crop field under unsaturated phase is given (Panigrahi, 2011) as:

$$S_{w,i} = S_{w,i-1} + R_i + I_i - ETcrop_i - SR_i - S_i - D_i \tag{7.2}$$

where $S_{w,i}$ is soil water content in the root zone of rice at the end of day *i* (*i* is the time index taken as 1 day in the study) and other parameters of the model are defined as above.

The general water balance model of rice field where standing water is maintained almost in all the growth periods is given (Panigrahi, 2011) as:

$$D_{p,i} = D_{p,i-1} + R_i + I_i - ETcrop_i - SR_i - S_i - D_i \tag{7.3}$$

In Eq. (7.3), D stands for depth of standing water in the field and subscript i refers to the present day whereas $(i-1)$ refers to the previous day and all other parameters are defined as above. The units of all the parameters of the models (Eqs. 7.2 and 7.3) are considered in mm. Measuring all the parameters of the above water balance models except, irrigation (I), one can estimate the amount of irrigation and day to irrigate to the crop.

Timing and frequency of irrigation to rice using the above water balance model (Eq. 7.3) requires a given water/irrigation management practice to be followed. The water balance model of rice is run on daily basis and various inputs and outputs parameters (except irrigation) are fed to the model and depth of standing water in the field is calculated on daily basis. When the running depth of standing water in the field reaches the lower limit of the pre-assigned value of the depth to be maintained, irrigation is supplied and the depth of irrigation is given till the depth reaches the upper limit of the is pre-assigned value of the depth to be maintained. Like non-rice crops, a daily book keeping procedure to maintain all the records of the water balance parameters would help to find out the day of irrigation.

Sample Calculation 6

Wheat is grown in sandy loam soil in Chiplima. The following daily water balance model parameters were recorded from July 5th to July 10th. The initial soil moisture content (in the morning of 5th July) in the effective root zone of the crop is 25.0 percent. Bulk density of the soil is 1.3 gm/cc. Assume the effective root zone depth of the crop as 60 cm. Irrigation is to be given to the crop when the available soil moisture content depletes by 50 percent. Assume field capacity and wilting point of the soil as 28 and 15 percent, respectively. Find the day(s) when irrigation is given to the crop.

Date	Rainfall, mm	Surface runoff, mm	Seepage, mm	Percolation, mm	*ETcrop*, mm
5th July	10.5	0	2.0	5.0	6.0
6th July	0	0	1.0	0.0	5.0
7th July	0	0	0.0	0.0	4.5
8th July	0	0	5.0	2.5	6.0
9th July	5.5	0	8.0	5.0	5.5
10th July	0	0	0.0	0.0	5.0

Solution

Initial soil moisture content in the morning of 5th July = 25 per cent.

Using the data of bulk density and effective root zone depth, the above initial soil moisture content of 25 percent can be converted to equivalent depth of

$$d = \frac{1.3 \; x \; 60 \; x \; 25}{100} = 19.5 \; cm = 195 \; mm$$

Available soil moisture content is:

$ASM = FC - PWP = 28 - 15 = 13$ percent

Optimum soil moisture content allowing 50 percent depletion at which irrigation is to be given is calculated as 0.5 x 13 = 6.5 percent i.e. irrigation is given when soil moisture content reaches a value of 28-6.5 = 21.5 percent.

The equivalent depth of soil moisture content of 21.5 percent in the effective root zone is:

$$d = \frac{1.3 \times 60 \times 21.5}{100} = 16.77\ cm = 167.7\ mm$$

Using the data given in tabular form in the question and using the water balance model (Eq. 7.2), soil moisture on daily basis in the effective root zone is calculated. The day in which the equivalent depth of soil moisture conent in the effective root zone reaches 16.77 cm, irrigation is given. The solution is given in tabular form as:

Date	Initial moisture content, mm	Rainfall mm	Surface runoff, mm	Seepage mm	Percolation mm	*ETcrop* mm	Final moisture content, mm
5th July	195.0	10.5	0	2.0	5.0	6.0	192.5
6th July	192.5	0	0	1.0	0.0	5.0	186.5
7th July	186.5	0	0	0.0	0.0	4.5	182.0
8th July	182.0	0	0	4.5	2.5	6.0	169.0
9th July	169.0	5.5	0	8.0	5.0	5.5	156.0
10th July	156.0	0	0	0.0	0.0	5.0	

Since the soil moisture content in the effective root zone in the morning of 10th July is 156.0 mm which is less than 166.7 mm, irrigation is to be given in the morning of 10th July. Irrigation is to be given to raise the soil moisture content in the effective root zone from 156.0 mm to the equivalent depth of root zone soil moisture at field capacity i.e.

$$d = \frac{1.3 \times 60 \times 28}{100} - 156.0 = 62.4\ mm$$

It is to be remembered that if the water balance model parameters of the entire crop growing period is available, then the above procedure can be run from sowing to harvest of the crop to decide total number of irrigations and total depth of irrigation water.

7.3.4.1 Water Management Practice Influencing Irrigation to Rice

There are many approaches of water management practice that influences irrigation to rice. In one approaches, shallow submergence of 3 to 7 cm of depth of water is maintained in the rice field whereas in other approaches medium to deep submergence varying from 10 to 15 cm is maintained in the rice field. The rice field water balance model (Eq. 7.3) is run on daily basis and feeding the various input and output parameters to the model, the depth of standing water in the field is calculated on daily basis. Irrigation to rice is to be given when the lower limit of depth of the standing water is reached in the field. In a recent water saving technique, irrigation to rice is recommended to be given at 3 days after disappearance of ponded water (*ADP*) in *kharif* and *for rabi* rice, irrigation is to be given at 1 *ADP* and during each irrigation 7 cm water may be supplied (Anonymous, 1997).

Sample Calculation 7:

The following daily water balance model parameters were recorded from July 5th to July 10th. The initial depth of standing water (in the morning of 5th July) in rice field is 5 cm. Find the day(s) when irrigation is given to the crop. Irrigation management practice followed is shallow submergence of 3 to 7 cm and 10 cm excess of standing depth of water is drained out as surface runoff.

Date	Rainfall, mm	Seepage, mm	Percolation, mm	*ETcrop*, mm
5th July	10.5	2.0	5.0	6.0
6th July	0	1.0	0.0	5.0
7th July	0	0.0	0.0	4.5
8th July	0	5.0	2.5	6.0
9th July	5.5	8.0	5.0	5.5
10th July	0	0.0	0.0	5.0

Solution

Initial standing depth of water (in the morning of 5th July) in rice field = 5 cm

The irrigation management practice followed is to irrigate the rice when the lower limit of the standing depth is 3 cm.

Now the water balance model of rice field is run starting from the 5th July to 10th July with the consideration of the various input and output parameters as given in the question in tabular form and the final standing water depth in the field is calculated at the end of each day. The day when the final standing depth of water is less than 3 cm (30 mm), irrigation is to be given. The solution is given in tabular form as follows:

Date	Initial standing water, mm	Rainfall, mm	Seepage, mm	Percolation, mm	*ETcrop*, mm	Final standing water, mm
5th July	50 .0	10.5	2.0	5.0	6.0	47.5
6th July	47.5	0	1.0	0.0	5.0	41.5
7th July	41.5	0	0.0	0.0	4.5	37.0
8th July	37.0	17.0	5.0	2.0	6.0	41.0
9th July	41.0	5.5	8.0	5.0	5.5	28.0
10th July	28.0	0	0.0	0.0	5.0	

Since the depth of standing water in the field in the morning of 10th July is 28.0 mm which is less than 30 mm, irrigation is to be given on this day.

7.3.5 Climatological Approach Based on IW:CPE Method

In this method, pan evaporation data on daily basis is calculated by a Pan evaporimeter. For a crop the ratio of irrigation water (*IW*) and cumulative pan evaporation (*CPE*) (*IW:CPE*) is decided and knowing the depth of irrigation to be given to the crop at each irrigation, the value of *CPE* is estimated. The time of irrigation is the day when the so calculated value of *CPE* is reached. It is to be marked that if there is any rainfall during the days CPE is calculated, then it has to be considered and irrigation date is accordingly extended. The sample calculation given below illustrates this method. For example if irrigation is scheduled at 1.2 *IW:CPE* and the depth of irrigation is 6 cm, then irrigation would be given when the *CPE* reaches 5 cm. The values of *IW:CPE* varies from crop to crop and depends on climatic condition and ground water table position of the crop field where irrigation by *IW:CPE* technique is to be adopted.

Results of extensive field trials of irrigation scheduling by *IW:CPE* for various crops by the scientists of All India Coordinated Research Project on Water Management in Chiplima centre of Orissa are compiled and presented in Table 7.5 (Anonymous, 1995). Data of the Table 7.5 also reveals how many irrigations and how much total irrigation has to be supplied by each irrigation schedule to each crop. One important drawbacks of this approach is that a constant value of *IW:CPE* is considered for all the crop growth stages for irrigation which is not correct. A crop in the initial stage may need a lower value of IW:CPE such as 0.8 whereas in the crop development and mid season stage it may require a somewhat higher value such as 1.2. A constant value of IW:CPE in the growth stages may cause mismatch of irrigation requirement with the crop water requirement causing water stress in crop development and mid season stage and water excess in the initial and maturity stages of the crop. Consequently it affects the growth and yield of the crop. But the advantages of this method are that no sophisticated instruments are required to monitor soil moisture content to plan irrigation except Class A pan evaporimeter which is of low cost and can be afforded by any user. Moreover no high technology is required to use this instrument to measure pan evaporation. It also saves considerable time which is needed in case of soil moisture measurement to monitor irrigation.

Table 7.5: Values of *IW: CPE* at which irrigation is scheduled and irrigation requirement of some crops in Chiplima, Orissa

Crop	Duration (days)	IW:CPE	Water required at each irrigation (cm)	Number of irrigation	Total amount ofi rrigation (cm)
Ground nut (Summer)	115	1.2	6	8	48
Mustard	75	0.9	6	3	18
Potato	90	1.2	5	4	20
Sugarcane	360	1.2	6	15	90
Wheat	95	1.0	6	5	30
Green gram	60	0.8	6	3	18
Black gram	60	0.8	6	3	18
Sesamum (Summer)	90	0.9	6	6	36
Sunflower	90	1.0	6	7	42
Maize	90	0.9	6	5	30
Chilli	150	1.0	6	6	36
Onion	105	1.2	5	8	40

It is important to remember that the above mentioned *IW:CPE* irrigation schedule suggested for various crops is to be followed when there is no deficit of irrigation water. However, if there is dearth of supply, we have to choose to supply water to the crop at the most responsive stage i.e. critical growth stage.

Sample Calculation 8

Sugarcane is grown in sandy loam soil in Bargarh. The crop requires to be irrigated by *IW:CPE* = 1.2. At each irrigation, a depth of irrigation (*IW*) of 5 cm is required. First irrigation is given on 10th June. The values of pan evaporation recorded by Class A pan evaporimeter of the site where the crop is grown starting from 10th June are given as follows: Find out the day when the next irrigation is required t be given to the crop.

Date	Pan evaporation, mm	Rainfall, mm
10th Jun	6.0	0
11th Jun	6.6	0
12th Jun	6.4	3.0
13th Jun	5.8	0
14th Jun	7.0	0
15th Jun	6.5	0
16th Jun	6.0	0
17th Jun	6.2	0

Solution

Irrigation is given at *IW:CPE* = 1.2

Given *IW* = 5 cm = 50 mm

So irrigation is to be given when cumulative pan evaporation (*CPE*) = 50/1.2 = 41.7 mm

Now *CPE* is calculated from 10th Jan till the day 41.7 mm is reached taking into consideration rainfall to find out the day of irrigation. The solution is given in the table as follows:

Date	Pan evaporation, mm	Rainfall, mm	*CPE*, mm
10th Jun	6.0	0	6.0
11th Jun	6.6	0	12.6
12th Jun	6.4	3.0	16.0
13th Jun	5.8	0	21.8
14th Jun	7.0	0	28.8
15th Jun	6.5	0	35.3
16th Jun	6.0	0	41.3
17th Jun	6.2	0	47.5

Since the *CPE* on 16th Jun is 41.3 mm which is slightly less than 41.7 mm, irrigation is to be given to the crop in the morning of 17th Jun. If one waits for irrigation on 18th Jun, the *CPE* would be 47.5 mm which is more than the prescribed value of 41.7 mm and then crop would suffer from water stress.

7.4 How Much to Irrigate? (Net Quantity of Water to be Applied)

After deciding when to irrigate a crop, next question comes how much water to be irrigated to a crop at each irrigation. Much of the things to decide how much to irrigate are in close association with when to irrigate a crop. When to irrigate a crop has been discussed in details earlier. In short, it may be summarized that net quantity of water to be applied depends on magnitude of moisture deficit in the soil, leaching requirement and expectancy of rainfall. When no rainfall is likely to be received and soil is not saline, net quantity of water to be applied is equal to the moisture deficit in the soil, i.e. the quantity required to fill the root zone to field capacity. Water will gradually be utilized consumptively by the plants after irrigation and the soil moisture will start gradually falling. When the depleting soil moisture reaches the optimum value, fresh dose of irrigation may be provided so that water content is again raised to field capacity level. Fig. 7.2 helps in estimating the amount of irrigation. The moisture deficit (*d*) in the effective root zone that is to be filled up by irrigation is found out by determining the field capacity moisture content, actual available soil moisture content and bulk density of each layer and is given as (Misra and Ahmed, 1987):

$$d = \sum_{i=1}^{n} \frac{(FCi - PWi)}{100} \; x \; BD_i \; x \; RDi \tag{7.4}$$

where, $\sum_{i=1}^{n}$ = summation of n-number of layers in the root zone; FCi = field capacity of the ith layer on oven dry weight basis; PWi = actual moisture content of the ith layer on oven dry weight basis; BD_i = apparent specific gravity/bulk density of *i*th layer, and RDi = depth of *i*th layer.

Sample Calculation 9

Find out the net quantity of irrigation water to be applied to ground nut with the following values of soil moisture content and other data.

Depth of soil layer in (cm)	Moisture % on oven-dry basis		Apparent specific gravity, gm/cc
	Field capacity	Actual	
0 – 15	26.0	17.4	1.39
15 – 30	24.0	17.8	1.47
30 – 60	22.3	19.2	1.51
60 – 90	22.2	20.5	1.53

Solutions

Moisture deficits in the different layers will be as follows:

$$\text{1st layer} = \frac{26.0 - 17.4}{100} \times 1.39 \times 15 = 1.79 \text{ cm}$$

$$\text{2nd layer} = \frac{24.0 - 17.8}{100} \times 1.47 \times 15 = 1.36 \text{ cm}$$

$$\text{3rd layer} = \frac{22.3 - 19.2}{100} \times 1.51 \times 30 = 1.40 \text{ cm}$$

$$\text{4th layer} = \frac{22.2 - 20.5}{100} \times 1.53 \times 30 = 0.78 \text{ cm}$$

So, net quantity of water to be applied is 5.33 cm.

Sample Calculation 10

A loam soil has field capacity of 22 percent, and wilting coefficient of 10 percent. The dry unit weight of soil is 1.5 gm/cc. Root zone depth is equal to 70 cm. Irrigation is applied when the soil moisture content falls to 14 percent. If water application efficiency is 70 percent, determine the depth of water required to be applied in the field.

Solution

Given Field Capacity (FC) = 22 percent and permanent wilting point (PWP) = 10 percent

Actual available soil moisture content at which irrigation is applied = 14 percent.

Actual available soil moisture content here is the optimum soil moisture level which is 14 percent.

Irrigation is given when the soil moisture content varies from 14 to 22 percent.

Depth of water should be supplied such that the optimum soil moisture content rises from 14 to 22 percent which is estimated as:

$$d = \frac{1.5 \; x \; 70 \; x \; (22.0 - 14)}{100} = 8.4 \; cm$$

Application efficiency = 70 percent.

$$\text{Hence, field irrigation requirement} = \frac{Net \;\; irrigation \;\; requiremen \;\; t}{applicatio \;\; n \;\; efficiency} = \frac{8.4}{0.7} = 12.0 \; cm$$

It is the usual practice to fill the soil to field capacity under full irrigation. However, there are several factors to be considered before irrigating. For example, if one wants to take the benefit of rainfall and there is some chance to have rainfall, then irrigation can be delayed for a day or two. In other words, irrigation is supplied to the crops when the soil moisture content comes below limiting value of optimum soil moisture content as discussed earlier. In another approach, soil may be filled with irrigation somewhat below field capacity level. This approach may be termed as deficit irrigation management.

Water balance models (Eq. 7.2 for non rice and Eq. 7.3 for rice) can also be used to find out the depth of irrigation. Rearranging Eq. 7.2, irrigation component of the water balance model is given (excluding groundwater contribution under the assumption that groundwater table lies at a depth of more than 1.5 m below the effective root zone of the crop) as:

$$I_i = (S_{w,i} - S_{w,i-1}) - (R_i - SR_i - S_i - D_i) + ETcrop_i \quad (7.5)$$

where all the terms have already been defined earlier in Eq. (7.2). Since the purpose of irrigation scheduling is to irrigate the crop with an amount such that the soil moisture content on the day of irrigation is brought back to the initial soil moisture content i.e. the moisture content just preceding to the present irrigation day, we can write the term $(S_{w,i} - S_{w,i-1}) = 0$. The term $(R_i - SR_i - S_i - D_i)$ refers to effective rainfall. Hence, Eq. (7.5) can be written as:

$$I_i = ETcrop_i - Effective\ Rainfall \quad (7.6)$$

Eq. (7.6) indicates that irrigation amount can be determined if the crop evapotranspiration (sometimes simply called as consumptive use since metabolic requirement of the crop is very less usually less than 1 percent of crop evapotranspiration) requirement and moistureffective rainfall amount is known.

Effective rainfall is that amount of rainfall which is directly used by the plants for meeting the evapotranspiration requirement. It excludes runoff, seepage and percolation from the given amount of rainfall. It depends on a number of factors like season (*kharif* or *rabi)*, amount, distribution and intensity of rainfall, antecedent moisture condition, crop cover and management practice, crop evapotranspiration and water holding capacity of the soil. Rearranging Eq. (7.3) to get the irrigation component of water balance model of rice and applying the same hypothesis as described above for non-rice crop, we can obtain same model i.e. Eq. (7.6) for calculating irrigation requirement of rice. It is to be remembered here that effective rainfall of rice grown in *kharif* is not same as that of other crops grown in *rabi.* In general effective rainfall (percentage of rainfall) in *kharif* is less than that in *rabi.* Details of calculation of effective rainfall from rainfall are available in FAO publication number 24 (Doorenbos and Pruitt, 1977).

Based on the studies at multilocational trials conducted at various places in Odisa, Lenka (2001) has reported the water requirement of several major crops along with the duration which is placed at Table 7.6.

Table 7.6 Water requirement of crops

Crop	Duration	Water requirement, cm
Rice (Transplanted)	90-150	70-120
Rice (Direct sown)	92-110	37-46
Wheat	95-115	35-40
Maize	95-120	50-60
Ragi	85-130	40-65
Jowar	110-125	55-65
Bajra	115-120	45-50
Greengram (Winter)	75	15
Greengram (Summer)	70	25
Blackgram	90	20
Horsegram	90	15
Gram	100	15-20
Redgram	160	40
Peas	138	15
Groundnut *(kharif)*	110	48
Groundnut (summer)	120	55
Sesamum	90	25
Mustard	100	30
Castor	130	50
Linseed	85	35
Jute	120-130	48
Sugarcane	300	170
Cotton	120	62
Potato	75-110	47-65
Tomato	100	50
Lady's finger	90	45
Chillies	120	52
Tobacco	110	36
Onion	115	70
Cabbage	70	60
Castor (*rabi*)	145	48

Question Banks

Q1. What do you mean by irrigation scheduling? What are the parameters used to define the irrigation scheduling?

Q2. Write about the different advantages of irrigation scheduling.

Q3. What are the main approaches to decide when to irrigate.

Q4. What are the various plant indicators to determine the time of irrigation?

Q5. How does the leaf water potential help in determining the time of irrigation?

Q6. Write short notes on the following plant indicators for knowing the time of irrigation:

(i) Leaf temperature (ii) Leaf water potential (iii) Stomatal resistance

Q7. Write the physiological and critical growth stages of crops mentioned below:

(i) Rice (ii) Wheat (iii) Maize (iv) Potato (v) Tomato (vi) Sorghum (vii) Ground nut and (viii) Mustard.

Q8. How is the irrigation timing decided based on the criteria of soil moisture level?

Q9. What is deficit irrigation management? What are its advantages and disadvantages?

Q10. Draw the figure of crop water production function and explain its use in the field of irrigation.

Q11. What is *IW:CPE* approach in irrigation scheduling? How does the irrigation timing get decided by this approach?

Q12. What is optimum moisture content and how does it help in deciding the irrigation timing?

Q13. Differentiate between available soil moisture and readily available soil moisture

Q14. What is moisture deficiency and how can you calculate it?

Q15. A crop of chilli is irrigated based on the approach of *IW: CPE*. Irrigation is given with fixed value of *IW: CPE* of 1.2. If 6 cm of water is supplied at each irrigation, then find out at what value of *CPE*, the irrigation is to be applied to the crop? If there is a constant rate of pan evaporation of all the days, then also find out after how many days, irrigation has to be supplied to the crop?

Q16. A crop of potato is grown in Sambalpur and the following data were recorded. Determine the time at which irrigation has to be supplied to the crop based on available soil moisture depletion study.

Field capacity, FC = 22.0% dry weight basis; permanent wilting point, PWP = 10% dry weight basis; bulk density, BD = 1.40 g/cc; effective root zone depth, RD = 60 cm; daily water use, $ETcrop$ = 0.25 cm/day and allowable soil water depletion = 25 percent of available soil water.

Q17. After how many days will you supply water to soil in order to ensure sufficient irrigation of the given crop, if Field capacity = 28 percent, permanent wilting point of the soil =15 percent, bulk density of the soil = 1.5 gm/cc, effective rooting depth = 50 cm, daily consumptive use of water for the given crop = 10 mm and optimum soil moisture level = 80 percent of available soil moisture.

Q18. Write the formula to compute the irrigation amount to be given to a crop based on moisture depletion level?

Q19. Find out the net quantity of irrigation water to be applied to a crop of potato with the following values of soil moisture content and other data.

Depth of soil layer in (cm)	Moisture % on oven-dry basis Field capacity	Actual	Apparent specific gravity, g/cc
0-15	28.0	19.4	1.39
15-30	26.0	17.8	1.47
30-45	22.3	19.2	1.51
45-60	22.2	20.5	1.50

Q20. Find out the gross irrigation amount of a crop if the net amount of irrigation is 5 cm and the field efficiency is 50 per cent.

Q21. Write down the water balance model of rice crop and non-rice crop and explain the various terms of these models.

Q22. How does the water balance model help in finding out the depth of irrigation?

Q23. Briefly explain the water management practices followed for rice.

Q24. What is water saving irrigation techniques? Write down the methods of water saving irrigation technique used for rice.

Q25. A sandy loam soil has field capacity of 25 percent, and wilting coefficient of 10 percent. The dry unit weight of soil is 1.5 gm/cc. Root zone depth is equal to 80 cm. Irrigation is applied when the soil moisture content falls to 14 percent. If water application efficiency is 75 percent, determine the depth of water required to be applied in the field.

References

Anonymous. 1995. Annual Report, AICRP on Water Management, Chiplima Centre, Orissa University of Agriculture and Technology, Bhubaneswar.

Anonymous. 1997. Status of Research on Agricultural Water Management (Eastern Region), AICRP on Water Management, Chiplima Centre, Orissa University of Agriculture and Technology, Bhubaneswar.

Doorenbos, J. and Pruitt, W.O. 1977. Guidelines for Predicting Crop water Requirements. Irrigation and Drainage Paper 24, FAO, Rome

Garg, S.K. 1987. Irrigation Engineering and Hydraulic Structures. Khanna Publishers, New Delhi.

Hansen, V.E., Israelsen, O.W. and Stringham, G.E. 1979. Irrigation Principles and Practices. John Wiley and Sons Inc., New York.

James, L.G. 1988. Principles of Farm Irrigation System Design. John Wiley & Sons, New York, pp. 543.

Lenka, D. 2001. Irrigation and Drainage. Kalyani Publishers, New Delhi.

Michael, A.M. 1978. Irrigation Theory and Practices. Vikash Publishing House Pvt. Ltd., New Delhi.

Misra, R.D. and Ahmed, M. 1987. Manual on Irrigation Agronomy, Oxford & IBH Publishing Co. Pvt. Ltd., New Delhi, pp. 412.

Panigrahi, B. 2011. Irrigation Systems Enginering. New India Publishinh Agency, New Delhi.

Punmia, B.C. and Lal, P.B.B. 1992. Irrigation and Water Power Engineering. Laxmi Publications Pvt. Ltd., New Delhi.

8

Methods of Irrigation

8.1 Introduction

After knowing when to irrigate and how much to irrigate, next important thing is how to irrigate the crop field. Methods of irrigation play a very crucial role in deciding how to irrigate the crop fields since all the crops do not require same methods of irrigation. Depending on the sources of water available for irrigation, type of crop to be irrigated, quantity and quality of water to be applied and other economic considerations, methods of irrigation will vary. Irrigation water may be applied to the crops by flooding it on the field surfaces as surface irrigation, by applying it beneath the soil surface as subirrigation, by spraying it under pressure as sprinkler or overhead irrigation or by applying it in drops around the root zone of the crop as trickle irrigation. The commonly used methods of irrigation are described in schematic diagram as follows:

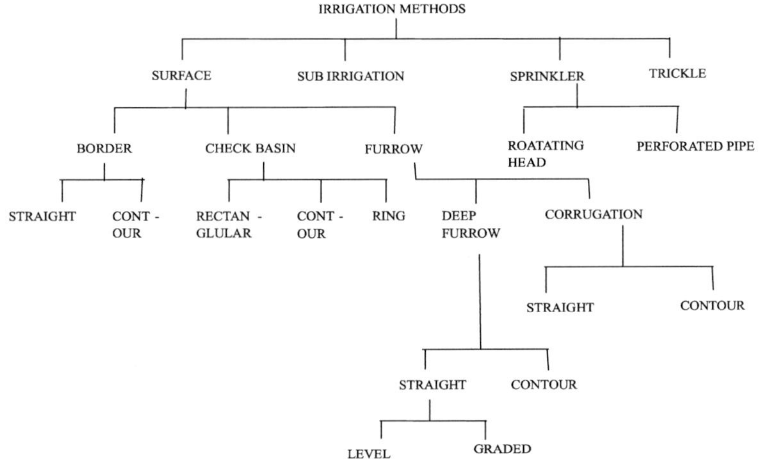

8.2 Surface Irrigation

Surface irrigation method is most widely practiced. This system is suitable for soil having low to medium infiltration rate and works well for soil having relatively uniform terrain. It also works better for land having less than 2 percent slopes. In this system water is conveyed to the point of infiltration directly on to the soil surface in channels that vary in shape, size and hydraulic characteristics. The methods may vary from corrugation to long narrow strips or large fields where water is impounded. Surface irrigation, in general, can be practiced as uncontrolled or wild flooding and controlled flooding. Controlled flooding can be achieved by (i) Free flooding (ii) Contour laterals (iii) Border strip (iv) Checks basin and (v) Furrow. Each type of controlled flooding is further divided into several types which are shown in the schematic diagram as presented above.

Different crops need different methods of irrigation and each method has some limitations. Table 8.1 represents the various methods of surface irrigation including free flooding as used for various crops (Lenka, 2001).

Table 8.1: Methods of surface irrigation for different crops and their limitations

Method	Crops to be irrigated	Topography	Stream size	Remarks
Free flooding	Rice, wheat, broadcast crop, fodder, millets	Sloppy and irregular up to 10%	Large	Inefficient method, causes unequal distribution of water
Boarder strips	Wheat, pulses, sorghum, forage crops	No cross slope, slope along length	Small to medium	Sandy loam to loam. graded, low layout cost, easy for intercultural works
Furrow (ridge and furrow)	Sugarcane, potato, cotton, Vegetables	Flat and gentle slopes.	Small to large supply	Economic water use, facilitates intercultural works
Corrugated method	Vegetables, onion garlic, spinach	Slopes perpendicular to furrow	Small to large supply	Economic water supply, interculture becomes difficult
Broad ridge and narrow furrow	Turmeric, ginger, yam, sweet potato	-	8-12m long, breadth depends on crops	Almost flat gentle slope in both directions
Basin or ring	Orchard crops, cucurbits	Flat	1-3 m size ring, 40-45 cm deep	Economic water use increases efficiency.
Check basin	Grain crops orchard crops	-	Small to medium	Low maintenance cost; erosion control

8.3 Uncontrolled or Wild Flooding

In this method water is spread or flooded on a smooth flat land. The method is easy and inexpensive. Crops like rice, wheat, oilseeds and fodder crops and other close growing crops are generally irrigated by this method (Fig. 8.1). In this method water flows from the ditch directly to the field without much control on either side of the flow. It covers the entire field and moves almost unguided. The rate of advancing front is controlled by the topography of the field. This method is generally used in the inundation irrigation system. This method is employed where there is ample scope of availability of irrigation water and irrigation water is inexpensive. Uneven distribution of water and low water application efficiency are the common drawback of this method.

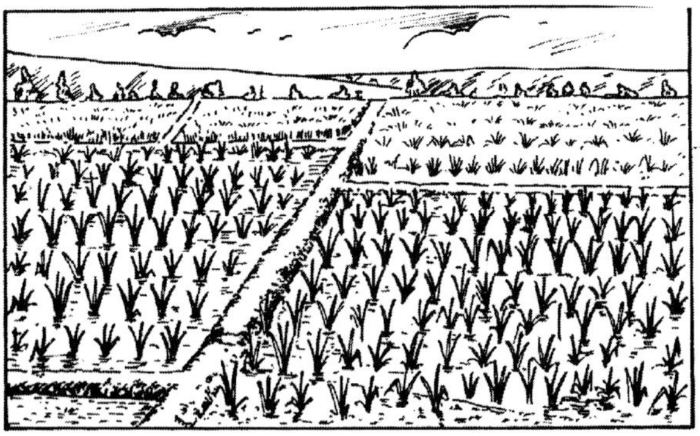

Fig. 8.1: Wild flooding

8.4 Controlled Flooding

In the controlled flooding, water is spread over the land, with controlled system such that depth of flow is controlled. Controlled flooding can be done by free flooding, contour laterals, border strips, check basin and furrow.

8.4.1 Free Flooding

Free flooding method is also known as irrigation by plots and is commonly used in India. In this method, the field is divided into a number of small plots which are generally level. Water enters into these plots at the upstream end of the plot and moves downward. When the downstream end attains sufficient depth, irrigation at the upstream end is cut off. In this method, rectangular plots are preferable to square plots (Fig. 8.2). The size of the plots depends on the porosity of the soil (Punmia and Lal, 1992).

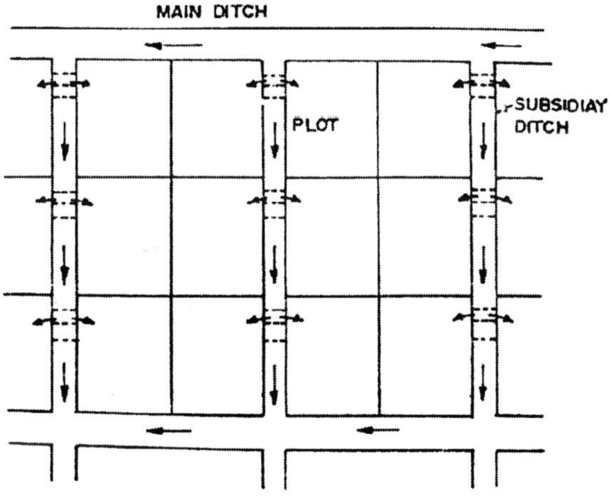

Fig. 8.2: Free Flooding method

8.4.2 Contour Laterals

This method is applicable for fields with steep gradient. The field is irrigated by a number of laterals laid on contours. The spacing of the contour laterals depends on the prominent grade of the field between two adjacent ditches or laterals, uniformity of slope and type of soil. Fig. 8.3 represents the contour laterals laid in a field.

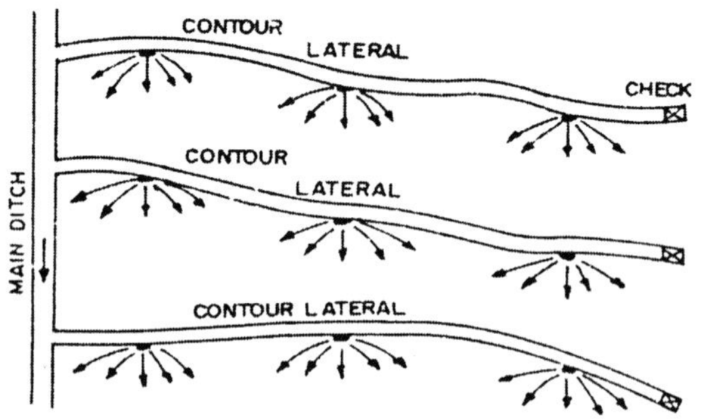

Fig. 8.3: Contour lateral

8.4.3 Border Strip Method

The border method of irrigation makes use of parallel ridges to guide a sheet of flowing water as it moves down the slope. The land is divided in to a number of long parallel strips called borders. These strips are separated by low ridges or levees. Irrigation water is supplied to the border by a discharge of 0.014 to 0.028 cumecs which is allowed to enter from a supply ditch to the head reach of the strip. Water advances in a form of thin sheet towards the downward direction of the strip along the longitudinal slope (Fig. 8.4). When the advancing water front either reaches the lower end, or a few minutes before that the stream is cutoff. Water temporarily stored in the border moves down the strip and infiltrates, thus completing the irrigation. Close growing crops like wheat, pulses, sorghum and forage crops are irrigated by this method.

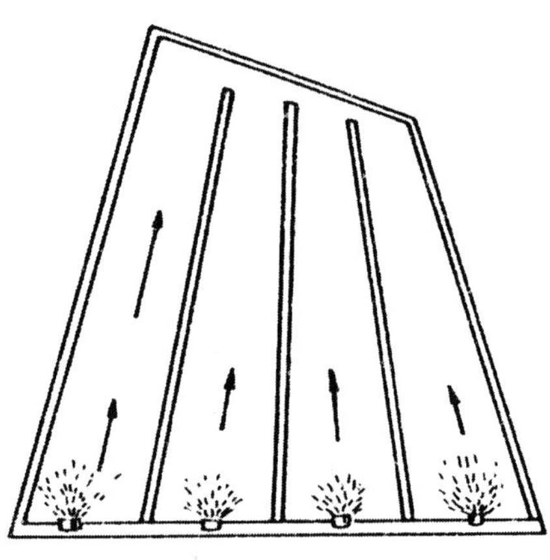

Fig. 8.4: Water movement along a border strip

Advantages of Border Irrigation

Border method has a number of advantages as follows:

(i) Border ridges can be constructed easily and economically with simple farm implements like a bullock drawn A-frame ridger or bund former or tractor drawn disc ridger.

(ii) Labour requirement in irrigation is greatly reduced as compared to the conventional check basin method of irrigation.

(iii) Uniform distribution and high water application efficiencies are possible if the system is designed properly.

(iv) Large irrigation streams can be efficiently used.

(v) Operation of the system is easy and simple.

Border method of irrigation is more adapted to soils having moderately low to high infiltration rates. It is not usually suitable to soils having very high infiltration rate like coarse sandy soils. Similarly, it is not well suitable to soils having very low infiltration rate.

8.4.3.1 Straight and Contour Borders

Borders may be laid along the general slope of the field (straight border) or may be laid across the general slope of the field (contour border). When the land slopes exceed safe limits, fields are undulating and leveling is not feasible, borders may be laid across the slope called as contour border. Design criteria of a contour border are same as that of a straight border. Just like straight border, contour border is leveled crosswise and has a gentle slope in longitudinal direction. For a given set of conditions, width and length of a contour border are identical to that of a straight border.

8.4.3.2 Border Specifications and Stream Size

Width of a border usually varies from 3 to 15 m depending on the size of the irrigation stream available and degree of land leveling permissible. If the size of the irrigation stream is small, width can be reduced. Reduction of width less than 3 m is not economical since a number of ridges will be required to separate so many strips per unit area of the field surface. Length of the border strip depends on how quickly it can be wetted over its entire length. This, however, depends on (i) infiltration of rate of the soil (ii) longitudinal slope of the field and (iii) available size of the stream. In general, length of the border strip varies from 60 to 120 m for sandy or sandy loam soil, 100 to 180 m for medium loam soil and 150 to 300 m for clay or clay loam soil (Michael, 1978). Borders usually have too little or no cross slope. However, they have some longitudinal slopes. High longitudinal slope will cause greater advancement of water front movement in the border causing insufficient irrigation in the head reach and deep percolation at the lower reach of the strip. They also cause soil erosion in the strip. On the other hand too flat slope cause very slow water front advancement causing deep percolation in the head reach and insufficient irrigation in the lower reach of the strip. Michael (1978) has recommended the following slopes for different types of soils. They are:

0.25 to 0.60 percent for sandy loam to sandy soils.

0.20 to 0.40 percent for medium loam soil and

0.05 to 0.20 percent for clay to clay loam soil.

Size of the irrigation stream depends on infiltration rate of the soil and width of the border strip. Large streams are required for coarse textured soils with high infiltration rate. Smaller streams are required for fine textured soils with low infiltration rates. It is to be remembered that size of the irrigation stream delivered into each border is a governing factor in border irrigation which can be varied after the irrigation systems have been installed. It is often convenient to express the requirement of the irrigation stream in terms of the rate of water flow per unit width of the border such as litres per second per metre of border width. When this value is multiplied with the width of the border strip, size of the irrigation stream that should be delivered into each border strip is found out. Table 8.2 gives some typical values of stream sizes for different soil types, infiltration rates and slopes (Michael, 1978).

Table 8.2: Typical values of stream sizes for various soil types, infiltration rates and border slopes

Soil type	Infiltration rate (cm/hr)	Border slopes (percent)	Flow per metre width of the border strip (lit/sec)
Sandy soil	2.5	0.20 – 0.40	10 – 15
		0.40 – 0.65	7 - 10
Loamy sand	1.8 – 2.5	0.20 – 0.40	7 – 10
		0.40 – 0.60	5 – 8
Sandy loam	1.2 – 1.8	0.20 – 0.40	5 – 7
		0.40 – 0.60	4 – 6
Clay loam	0.60 – 0.80	0.15 – 0.30	3 – 4
		0.30 – 0.40	2 - 3
Clay	0.20 – 0.60	0.10 – 0.20	2 - 4

8.4.3.3 Hydraulics of Border Irrigation

Efficient water application by border irrigation depends on the knowledge of hydraulic characteristic of irrigation stream. The flow in a border strip is a case of varied unsteady open cannel flow with decreasing discharge. The discharge rate decreases downstream due to infiltration. The hydraulic of border strip can be studied by classical open channel hydraulics approach. The prominent variables that influence the hydraulics of the border irrigation are the stream sizes, slope of the land, infiltration characteristics of the soil, and resistance to flow offered by the soil and vegetative cover. Advancement of water front in a border strip is the most important hydraulic parameter in designing an efficient border irrigation system. Advancement of water front varies with time and other parameters as described above for hydraulics of border strip.

As irrigation water flows into a field and progress down its length, an ever decreasing portion of the total volume of water applied flows above the ground, while the remainder infiltrates into the soil and composes the subsurface storage. Any rational approach to predict surface

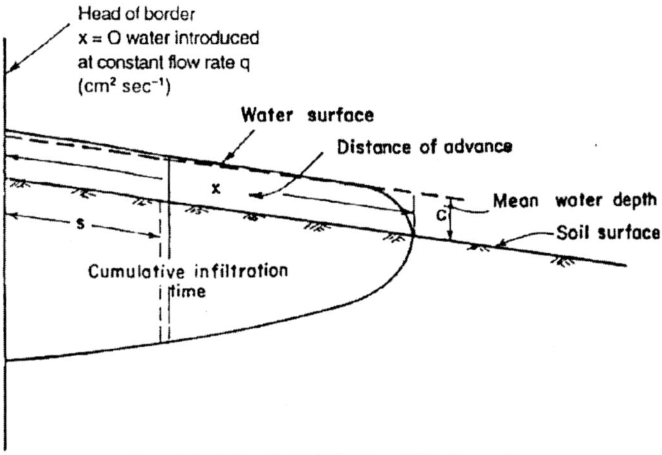

Fig. 8.5: Definition sketch for Lewis and Milne's equation

irrigation flows must equate the total volume of water discharge at the supply channel to the sum of surface storage and subsurface storage; Moreover this volume balance is obtained at every instant of time, subsequent to the initial turning of water on to the land. Fig. 8.5 shows the infiltration-advance problem in border irrigation.

Equation for water advance in a border strip using volume balance method as proposed by Lewis and Milne (1938) is as follows:

Let,

q = Constant rate of flow per unit width at the upstream end of the border, cm^3/min/cm of border strip

t = Time for which water is applied, min

d = Average depth of water applied, cm

x = Distance, the water front has advanced in time, cm

t_s = Time when distance covered x = s, min

y = Accumulated infiltration depth at the point x = s, when $t = t_s$

Now at the time $t = t_s$ the distance of water advancement is $\frac{dx}{dt}$

The total water admitted in t per unit width of border strip = $q.t$

Volume of water stored on surface of border strip per unit width = $d.x$

Volume of water infiltrated per unit width is $y.ds$

The total volume of water infiltrated over distance x is = $\int_0^x y.ds$

Using volume balance approach, we get $q.t = d.x + \int_0^x y.ds$ (8.1)

The accumulated infiltration at any time t is found to be as:

$$y = at^{\alpha} + b \tag{8.2}$$

where, y is accumulated infiltration at time t, cm; t is elapsed time, min and a, b and α are characteristics constants. Value of α is $0 \leq \alpha \leq 1$ and $t = 0$. Values of a and b are found out from experiments.

Putting the value of y in Eq.8.2 in Eq. 8.1,

$$q.t = d.x + \int_0^x (at^{\alpha} + b)ds \tag{8.3}$$

Phillip and Farrell (1964), using convolution theorem of Laplace transformation and gamma function, gave solution of water front advancement x as reported by Murty (2004) as:

$$x = \frac{q.t}{b+d}\left[\frac{1}{\Gamma 2} - \frac{\beta^1 t^{\alpha}}{\Gamma(2+\alpha)} + \frac{\beta^2 t^{2\alpha}}{\Gamma(2+2\alpha)} - \frac{\beta^3 t^{3\alpha}}{\Gamma(2+3\alpha)} + \cdots\right] \tag{8.4}$$

when $\dfrac{\beta^1 t^{\alpha}}{\Gamma(2+\alpha)} < 1$ and

$$X = \frac{q.t}{b+d}\left[\frac{1}{\beta t^{\alpha}\Gamma(2-\alpha)} - \frac{1}{(\beta t^{\alpha})^2\Gamma(2-2\alpha)} + \frac{1}{(\beta t^{\alpha})^3\Gamma(2-3\alpha)} + \cdots\right] \tag{8.5}$$

when $\dfrac{\beta^1 t^{\alpha}}{\Gamma(2+\alpha)} > 1$. In Eqns. (8.4 and 8.5),

$$\beta = \frac{\alpha\,\Gamma(\alpha+1)}{b+d}$$

To find out the gamma function for any constant quantity lying between 1.0 to 2.0, Table 8.3 can be used.

Table 8.3 Values of gamma function

K	√K	K	√K	K	√K	K	√K	K	√K
1.00	1.000000	1.205	0.916857	1.405	0.8870028	1.605	0.894088	1.805	0.932720
1.01	0.994325	1.21	0.915576	1.41	0.886764	1.61	0.894680	1.81	0.934076
1.02	0.988844	1.22	0.913106	1.42	0.886356	1.62	0.895924	1.82	0.936845
1.03	0.983549	1.23	0.910754	1.43	0.886036	1.63	0.897244	1.83	0.939690
1.04	0.978438	1.24	0.908521	1.44	0.885805	1.64	0.898642	1.84	0.942612

(Contd.)

K	$\sqrt{K}$	K	$\sqrt{K}$	K	$\sqrt{K}$	K	$\sqrt{K}$	K	$\sqrt{K}$
1.05	0.973504	1.25	0.906402	1.45	0.885661	1.65	0.900116	1.85	0.945611
1.06	0.968744	1.26	0.904397	1.46	0.885694	1.66	0.901668	1.86	0.948687
1.07	0.964152	1.27	0.902503	1.47	0.885633	1.67	0.903296	1.87	0.951840
1.08	0.959725	1.28	0.900718	1.48	0.885747	1.68	0.905001	1.88	0.955071
1.09	0.955459	1.29	0.899041	1.49	0.885945	1.69	0.906781	1.89	0.958379
1.10	0.951351	1.30	0.897471	1.50	0.886227	1.70	0.908639	1.90	0.961765
1.11	0.947395	1.31	0.896004	1.51	0.886591	1.71	0.910571	1.00	0.961766
1.12	0.943590	1.32	0.894640	1.52	0.887039	1.72	0.912581	1.92	0.968774
1.13	0.939931	1.33	0.893378	1.53	0.887567	1.73	0.914665	1.93	0.972376
1.14	0.936416	1.34	0.892216	1.54	0.888178	1.74	0.916826	1.94	0.976099
1.15	0.933040	1.35	0.891151	1.55	0.888868	1.75	0.919062	1.95	0.979880
1.16	0.929803	1.36	0.890185	1.56	0.889639	1.76	0.912375	1.96	0.983743
1.17	0.926699	1.37	0.889313	1.57	0.890489	1.77	0.923763	1.97	0.987684
1.18	0.923728	1.38	0.888537	1.58	0.891420	1.78	0.926227	1.98	0.991708
1.19	0.920885	1.39	0.887854	1.59	0.892428	1.79	0.928767	1.99	0.995813
1.20	0.918169	1.40	0.887264	1.60	0.893515	1.80	0.931384	2.00	1.000000

The following relationships of the gamma function, ($\sqrt{n}$), in which n is real number and positive are used in solution of the problems.

$\Gamma(n+1) = n\ \Gamma(n),\ \Gamma(1) = 1,\ \Gamma(2) = 1,\ \Gamma(n) = (n-1)!,\ \Gamma(4) = 3! = 6$ etc.

It is to be remembered that in Eqns. (8.4) and (8.5), water front advance is given in cm, stream size is given in cm^3/min/cm of border width, time is in min, and depth of water surface is in cm. For calculation of water front advance by using Eqns. (8.4) or (8.5) one can take first three terms inside the square bracket for simplicity and easy in calculation as higher terms have very small values being nearly equal to zero. It is to be noted that as the infiltration characteristic values of a, b and α changes, the distance of water front advance also changes. Other factors like stream size, depth of water applied, time of irrigation also influences the distance of water front movement. Using Eq. (8.4) or Eq. (8.5), we can get various values of water front movement at various values of elapsed time and then a water front advance curve can be drawn. Such a curve has been drawn taking the computed data of a sample calculation as given below.

Sample Calculation 1:

In a border strip experiment the infiltration characteristic is given by the equation $y = 0.75\ T^{0.5} + 0.7$.The irrigation applied for 2 hr with constant rate of flow per unit width of the channel at the upstream side is q = 1.5 lpm/cm of border strip. The mean depth of f low over the soil surface is 8 cm. Find out the distance of water front advance.

Solution:

Given that q = 1.5 lpm/cm = 1500 cm^3/cm/min,
Mean depth of flow d = 8 cm and
Time of irrigation t =2 hr = 120 min
Accumulated infiltration-time relationship is
$y = 0.75\ t^{0.5} + 0.7$

Which gives values of $a = 0.75$, $b = 0.7$,

$$\beta = \frac{\alpha \, \Gamma(\alpha+1)}{b+d} = \frac{0.75 \times 0.5 + 1}{0.7+8}$$

$$= \frac{0.75 \times 0.5 + 1}{0.7+8} \text{ (Value of 1.5 from Table 8.3 is 0.88622)}$$

$= 0.076$

$$= \frac{\beta^1 t^\alpha}{(2+\alpha)} = \frac{0.076 \, x \, 120^{0.5}}{\Gamma(2+0.5)} = \frac{0.076 \, x \, 10.954}{1.5 \, x \, 0.88622} = 0.626 = <1$$

Using Eq. (8.4), water front advance is given as

$$X = \frac{q.t}{b+d}\left[\frac{1}{\Gamma 2} - \frac{\beta^1 t^\alpha}{\Gamma(2+\alpha)} + \frac{\beta^2 t^{2\alpha}}{\Gamma(2+2\alpha)}\right]$$ neglecting values of higher exponents of the terms since the higher exponent values gives very small values nearly equal to zero.

$$= \frac{1500 \, x \, 120}{0.7+8}\left[\frac{1}{1} - \frac{0.076 \, x \, 120^{0.5}}{\Gamma(2+0.5)} + \frac{(0.076^2 \, x \, 120}{\Gamma(2+1)}\right]$$

$= 20689.66 \, (1\text{-}0.626 + 0.347) = 14902.26 \text{ cm} \approx 149 \text{ m}.$

Sample Calculation 2

In a border strip experiment the infiltration characteristic is given by the equation $y = 0.75 \, T^{0.5} + 0.7$. The irrigation applied for 1hr with constant rate of flow per unit width of the channel at the upstream side is q = 1.6 lpm/cm of border strip. The mean depth of f low over the soil surface is 8 cm. Find out the distance of water front advance in 1 hr. Find out also the water front advance in 10, 20, 30, 40 and 50 min. Draw water front advance curves. How much time will be required for the water front advance to cover a distance of 100 m?

Solution

Given that q = 1.6 lpm/cm = 1600 cm^3/cm/min,

Mean depth of flow d = 8cm,

Time of irrigation t =1 hr = 60 min

Accumulated infiltration-time characteristics is

$y = 0.75 \, t^{0.5} + 0.7$

Which gives values of $a = 0.75$, $b = 0.7$ and $\alpha = 0.5$

$$\beta = \frac{\alpha\,\Gamma(\alpha+1)}{b+d} = \frac{0.75 \times 0.88622}{0.7+8}$$

$$= \frac{0.75 \times 0.88622}{0.7+8} \quad \text{(Value of} \sqrt{1.5} \text{ from Table 8.3 is 0.88622)}$$

$= 0.076$

$$\frac{\beta^1 t^\alpha}{\Gamma(2+\alpha)} = \frac{0.076 \times 60^{0.5}}{\Gamma(2+0.5)} = \frac{0.076 \times 60^{0.5}}{1.5 \times 0.88622} = 0.45 < 1$$

Using Eq. (8.4), water front advance is given as

$$x = \frac{q.t}{b+d}\left[\frac{1}{\Gamma 2} - \frac{\beta^1 t^\alpha}{\Gamma(2+\alpha)} + \frac{\beta^2 t^{2\alpha}}{\Gamma(2+2\alpha)}\right]$$ neglecting values of higher exponents of the terms since the higher exponent values gives very small values nearly equal to zero.

$$x = \frac{q.t}{b+d}\left[\frac{1}{\Gamma 2} - \frac{\beta^1 t^\alpha}{\Gamma(2+\alpha)} + \frac{\beta^2 t^{2\alpha}}{\Gamma(2+2\alpha)}\right]$$

$$= \frac{1600 \times 60}{0.7+8}\left(1 - \frac{0.076 \times 60^{0.5}}{\Gamma(2+0.5)} + \frac{(0.076)^2 \times 60}{\Gamma(2+1)}\right)$$

$= 11034.5[1 - 0.443 + 0.17328] = 8058.26 \text{ cm} \approx 80.58$ m.

Values of water front advance at time, $t = 10$ min is:

$$= \frac{1600\,x\,10}{0.7+8}\left[\frac{1}{1} - \frac{0.076\,x\,10^{0.5}}{\Gamma(2+0.5)} + \frac{(0.076^2\,x\,10}{\Gamma(2+1)}\right] = 1559.70 \text{ cm} \approx 15.60 \text{ m}$$

Values of water front advance at time, $t = 20$ min is:

$$= \frac{1600\,x\,20}{0.7+8}\left[\frac{1}{1} - \frac{0.076\,x\,20^{0.5}}{\Gamma(2+0.5)} + \frac{(0.076^2\,x\,20}{\Gamma(2+1)}\right] = 2950.18 \text{ cm} \approx 29.50 \text{ m}$$

Values of water front advance at time, $t = 30$ min is:

$$= \frac{1600\,x\,30}{0.7+8}\left[\frac{1}{1} - \frac{0.076\,x\,30^{0.5}}{\Gamma(2+0.5)} + \frac{(0.076^2\,x\,30}{\Gamma(2+1)}\right] = 4267.58 \text{ cm} \approx 42.68 \text{ m}$$

Values of water front advance at time, t = 40 min is:

$$= \frac{1600\ x\ 40}{0.7+8}\left[\frac{1}{1} - \frac{0.076\ x\ 40^{0.5}}{\Gamma(2+0.5)} + \frac{(0.076^2\ x\ 40}{\Gamma(2+1)}\right] = 5546.19 \text{ cm} \approx 55.46 \text{ m}$$

Values of water front advance at time, t = 50 min is:

$$= \frac{1600\ x\ 50}{0.7+8}\left[\frac{1}{1} - \frac{0.076\ x\ 50^{0.5}}{\Gamma(2+0.5)} + \frac{(0.076^2\ x\ 50}{\Gamma(2+1)}\right] = 6805.84 \text{ cm} \approx 68.06 \text{ m}$$

Water front advance curves is drawn for various values of x and time, t as below (Fig. 8.6).

Finally time to cover a distance of 100 m is calculated as:
Putting x = 100 m, q = 1600 cm^3/min/cm of border width, b = 0.7, d = 8 cm, α = 0.5 and β = 0.076 in the above equation and with trial and error, value of time, t is computed as 75.5 min.

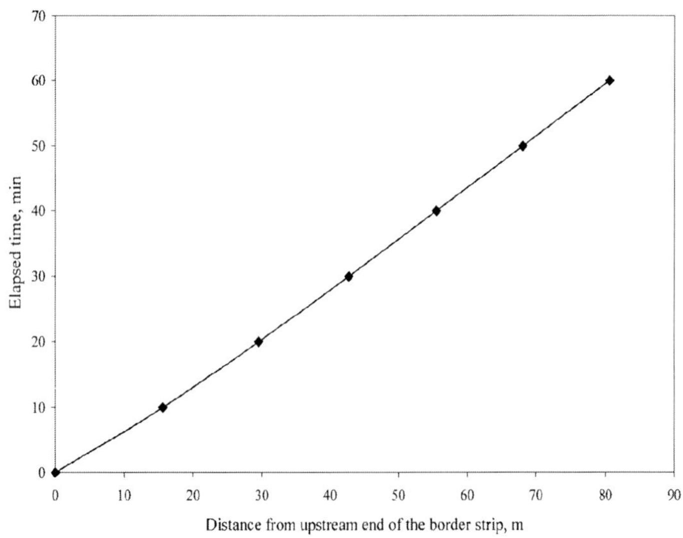

Fig. 8.6: Water front advance curve of border strip irrigation

8.4.3.4 Recession Flow

After the irrigation stream is cutoff, the tail water recedes downstream. The rate of recession of tail water is determined by noting the time at which water just disappears from the upstream end and recedes downstream past the border strip. In plotting the advance and recession curves, distance down the stream is plotted on the x-axis and elapsed time is plotted in y-axis. In the same graph, both the advance and recession curves are drawn with same scale in an arithmetic graph paper and the difference of time of recession and advance at any distance from the upstream end of the border gives infiltration opportunity time or time of ponding.

Parallelism of advance and recession curves ensures uniform distribution of water throughout the border.

In computation of water front advance by study of hydraulics of border irrigation, we assume that volume of water entering into the strip is equal to sum of water standing over the soil surface and that infiltrating into the soil surface. The infiltrated amount is a spatially varied unsteady flow component which is difficult to predict since it varies with both time and also along the distance of water front movement in a strip. This makes use of convolution theorem of Laplace transformation along with application of gamma function to find out the water front movement in the form of Eqns. (8.4 and 8.5). Application of Eqns. (8.4 and 8.5) are tedious and cumbersome as felt by some readers and needs simple alternate approach.

We can apply the same volume balance approach to a border strip with assumption that infiltration rate is constant over a distance x of water front movement and proceed as follows to arrive at simple models to find out the water front movement. Use Fig. (8.7) as a definition sketch for volume balance approach.

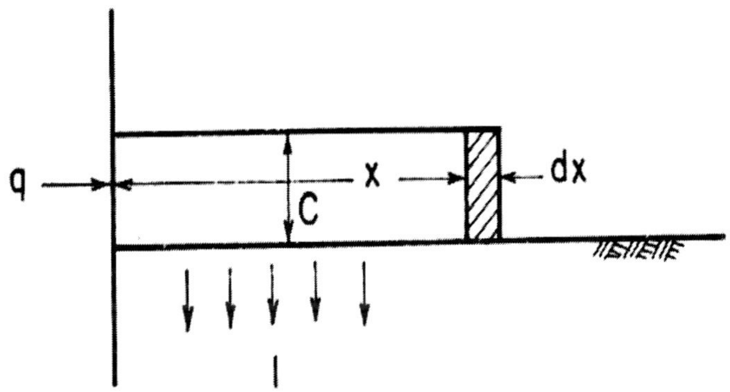

Fig. 8.7: Definition sketch for volume balance approach

Let

d = Average depth of water on the soil surface, cm

q = Inflow per unit width of border strip, cm^3/min/cm of border width

I_{av} = Average infiltration rate, cm/min

x = Distance of water front advance, cm

Now let in a small time interval, dt, water front moves over a distance dx as shown in Fig. (8.7). Using volume balance approach,

$$q.\ dt = d.\ dx + I_{av.} x.\ dt \tag{8.6}$$

Eq. (8.6) can be rearranged as

$$q.dt - I_{av.} x.dt = d.\ dx$$

$$(q - I_{av.} x)dt = d.\ dx$$

$$dt = \frac{d.\,dx}{(q - I_{av}\,x)}$$

$$\int_0^t dt = \int_0^x \frac{d}{(q - I_{av}\,x)} dx$$

$$t = -\frac{d}{I_{av}} \ln\left(\frac{q - I_{av}\,x}{q}\right)$$

$$\text{or, } \left(\frac{q - I_{av}\,x}{q}\right) = e^{-\frac{t\,I_{av}}{d}}$$

$$x = \frac{q}{I_{av}}\left(1 - e^{-\frac{t\,I_{av}}{d}}\right) \tag{8.7}$$

Let $D =$ where D = Total depth of water infiltrating into the soil

Putting $D =$ in Eq. (8.7) we get,

$$x = \frac{q\,t}{D}\left(1 - e^{-\frac{D}{d}}\right) \tag{8.8}$$

Average infiltration rate of a particular site can be predicted by infiltration models such as Kostiakov's model as

$I_{av} = c_1.t^n$

where, is the average infiltration rate, cm/min; , n are the empirical constants ; t = time of irrigation, min. The empirical constants are found out from a set of observations of a given experimental site. A perusal of Eq. (8.8) reveals that water front advance depends on infiltration characteristics of the soil and hence on soil types, discharge per unit width of the border, depth of water over the soil surface and time of irrigation application. Once all these values are known, then we can use Eq. (8.8) to predict the water front movement.

From experiment, it is found that $d = 0.665$

where, = depth of water at the upstream end of the border strip which can be calculated by simplifying Manning's equation as:

$$do = \left(\frac{nq}{so^{1/2}}\right)^{0.6} \tag{8.9}$$

where,

n = Manning's roughness coefficient ,

s_0 = Bed slope in m/m,

q = Inflow per unit width of the border in m^2/sec/m and

d_0 = Normal depth of surface in m..

Michael and Pandya (1971) from experimental studies in borders proposed an empirical equation for estimation of n, as:

$$n = 0.55425 - 0.0001386\, q$$

where, q = inflow per unit width of border in cm^3/min/cm

We can calculate the water front movement by using Eq. (8.8) with the same data of the sample calculation present above and compare the result with that predicted by using Eq. (8.4).

Sample Calculation 3:

In a border strip experiment the infiltration characteristic is given by the equation $y = 0.75\, t^{0.5} + 0.7$. The irrigation applied for 1hr with constant rate of flow per unit width of the channel at the upstream side is q = 1.6 lpm/cm of border strip. The mean depth of f low over the soil surface is 8 cm. Find out the distance of water front advance in 1 hr.

Solution

Given the accumulated infiltration-time characteristic as $y = 0.75\, t^{0.5} + 0.7$

Putting time, t = 60 min, y is calculated as 6.51. Here $y = D$ = accumulated or total infiltrated water entering into the soil.

Given mean depth of flow over soil = d = 8 cm.

Discharge per unit width of border strip, q = 1.6 lpm/cm = 1600 cm^3/min/cm

Using these values of q, d, D *and* t and putting them in Eq. (8.8) we get,

$$x = \frac{60\, x\, 1600}{6.51}\left(1 - e^{\frac{-6.51}{8}}\right)$$

= 8209 cm ≈ 82 m.

The above predicted value of 82 m is nearly same as that predicted by Eq. (8.4).

8.4.3.5 Design of Border Irrigation Systems

The border irrigation system should be so designed that it provides uniform distribution of water over the entire field. To achieve maximum water application efficiency, water should remain on the surface sufficiently long to allow just the desired quantity of water to infiltrate into the soil. The required infiltration opportunity time is obtained using the accumulated infiltration-time relationship for the soil in question. Following procedures are suggested (Michael, 1978) for the design of an efficient border irrigation system.

(i) Length, width and slope of the border should be determined.

(ii) The depth of water required to replenish the root zone soil moisture upto field capacity is estimated. This can be accomplished by determining the root zone soil moisture content before irrigation or predicted from evapotranspiration demand of the crop under the prevailing agro-climatic conditions.

(iii) The accumulated infiltration-time relationship of the soil under the existing soil conditions, and vegetation is determined. This can be done by measuring infiltration amount before each irrigation.

(iv) The desired infiltration opportunity time is determined. It is the time required for the soil to absorb the estimated depth of water, It is obtained from the accumulated infiltration-time relationship of the soil.

(v) Hydraulic resistance is estimated on the basis of the soil surface roughness and the hydraulic characteristics of the crop. Hydraulic resistance, expressed as Manning n, may be estimated by using Manning's formula. An approximate value of stream size is chosen first and the value of n is calculated. More accurate values of n are obtained by successive approximations on the basis of the stream size ultimately selected.

(vi) Average depth of flow is calculated. For a given value of q, n, bed slope, depth of water at the upstream end of the border is calculated by Eq. (8.10). Then the average water depth on the soil surface is calculated from the depth of water at the upstream end by multiplying the upstream depth with 0.655.

(vii) Water front advance is predicted by using Eqns. (8.4 or 8.5). Then a water front advance curve is drawn for various values of water front with elapsed time. To arrive at an efficient irrigation system design, it is necessary to calculate a family of advance curves using different values of q and possible border lengths.

(viii) Finally, the irrigation system is designed to obtain the optimum water application efficiency and border length. The procedure proposed by Hall (1960) and reported by Michael (1978) can be followed to arrive at a suitable design.

8.4.4 Check Basin Irrigation

It is the most commonly used method of irrigation in India and in many other countries. This is the simplest method of irrigation and is so popular. In this method, the entire field is divided into a number of small plots so that each plot has a nearly leveled surface. If the field is very large and it is not easy to level the small plots to one grade, then the field is divided into several small strips and then each strip is divided into several small plots by putting bunds. Then these bunded small plots are leveled. A common irrigation or drainage line runs in between two strips of such small plots. Such a leveled small plot is called as basin (Fig. 8.8).

Fig. 8.8: Basin methods of irrigation

The basins are filled to the desired depth and the water is retained till it infiltrates into the soil. The basins may be irrigated intermittently or kept under flooding. The bunds or ridges formed around a basin may be temporary for single irrigation as in case of pre-sowing irrigation of seasonal crops or may be for a crop season as in case of post-emergence irrigations. This may also be semi-permanently constructed for repeated use as in case of rice fields. Water enters into the basins through supply channels and lateral field channels. The supply channel is aligned at the upper side of the areas of the basins and there is a common field lateral in between two rows of check basins. When sufficient water has entered into the basins, the supply is cut off. The size of the basin depends on the infiltration rate of the soil in it. If the infiltration rate is high, the size of the basin must be reduced for a given value of stream size that irrigates it in order to cover the whole basin in a small time. On the other hand, a large stream size will be needed to irrigate a given size of the basin if the infiltration rate of the soil in it is high. In general sandy or sandy loam soil with high infiltration rate permit small size basins while clay soil having low infiltration rate permits large basins.

8.4.4.1 Types of Check Basins

Based on size, the check basin may be small comprising one square metre area to to a large one of about one or two hectares or more. Small basins are cultivated with growing vegetables whereas large ones are used to grow rice, wheat, oilseeds etc. The basins may be rectangular in shape. In rolling topography, the ridges follow the contour of the land surface. The contour ridges are connected by cross ridges at intervals. The vertical interval between the two contour ridges ranges from 6 to 12 cm in case of upland irrigated crops like wheat and 15 to 30 cm in case of low land irrigated crops like rice.

In case of orchard crops, ring method of basins may be used. The rings are generally circular in shape and surround the tree. Fruit trees are irrigated by ring basin (Fig. 8.9).

Initially when the tree is small, the ring basin is small in size and as the tree grows the size increase. Usually there is one tree in one basin. Sometimes two or more trees are planted in one basin. The depth of the ring depends on the quantity of water to be irrigated. In many cases the

Fig. 8.9: Ring method of irrigation

size of the ring and its distance from the tree depends on the crown of the tree. In this method, the whole field need not be irrigated. Only two-third of the ring depth becomes enough to irrigate the ring (Lenka, 2001). Thus, this method of irrigation saves a lot of water and hence, increases water use efficiency of the crop. .

8.4.4.2 Adaptability of the System

Check basin irrigation is suited to smooth, gentle and uniform land slopes and for soils having moderate to slow infiltration rates. Bo0th row and close growing crops are adapted to be used with basins. The method is especially suitable for irrigation to grains or fodder crops in heavy soils. Many different kinds of crops can be grown in sequence in the same field without making major changes in the design. Practically all irrigated rice is grown in basins. Check basins are helpful to leach out the salts from the fields. They also help in conserving rainfall and reduce soil and water erosion as the basins hold in-situ rainfall because of the levees around it.

Limitations

Followings are the limitations of the system:

(i) The ridges of the basins interfere with the movement of farm machineries.

(ii) There is considerable land lost to formation of ridges and field laterals.

(iii) The method impedes surface drainage.

(iv) System requires precise land grading and shaping.

(v) Labour requirements in land preparation and irrigation are more as compared to other method of irrigation except when the basins are large.

8.4.4.3 Hydraulics of Check Basin Irrigation

The hydraulics of flow in check basins may be considered to consist of the following stages.

(i) Initial spreading of the entrance stream to cover the full width of the basin and simultaneous advance of the irrigation stream

(ii) Advance of the water front after the initial spreading

(iii) Rise of water level after the advancing stream reaches the downstream end and

(iv) Subsidence of water after the irrigation stream is stopped.

The variables influencing the hydraulics of check basin irrigation are, in general, same as that of border irrigation. However, there are differences in the initial spreading of the entrance streams to cover the full width of the basin and in the characteristics of recession flow.

After the initial spreading, water front advance in the check basins may be predicted by using Eqns. (8.4) and/or (8.5) as in borders. The dominant variables are the same like entrance stream size, infiltration characteristics of the soil, slope of the land, elapsed time, hydraulic resistance offered by the soil and vegetation. After water front reaches the downstream end of the check basin, ponding occurs. Depth of ponding in a check basin can be calculated by using volume balance approach i.e. amount of water contributing to ponding is the difference between the amount of water entered into check and that infiltrated into the soil in a given time of water application.

8.4.4.4 Design of Check Basin Irrigation

A thorough study of hydraulics of flow of the basin is useful in efficient irrigation by check basin irrigation system. In check basins, it is possible to apply accurate amount of water required to replenish the depleting moisture in the root zone of the crop to field capacity if provisions are made to measure the rate of flow into the basin and control the duration of flow of the irrigation stream. This will increase the water application and water storage efficiency. A useful thumb rule in check basin design is that the water spread in the entire basin should be covered in one-fourth the time requirement to infiltrate the net depth of irrigation water. By careful selection of the size of the check basin and the stream size to suit the infiltration characteristics of the soil, it is possible to arrive at design of check basins that gives high value

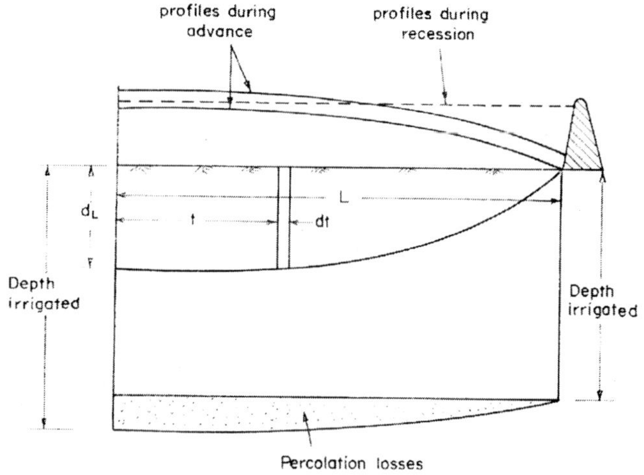

Fig. 8.10: Water surface and infiltration profile of check method

of water distribution efficiency. The time required to irrigate a basin of given size can be found out by dividing the stream size with the volume of irrigation water required to bring the soil moisture content before irrigation to field capacity. Fig.8.10 shows the principle of operation of the check system. Water surface and infiltration profiles at different stages are shown in this figure.

The amount of water required to irrigate a particular size of the check can be obtained using the formula (neglecting percolation):

$Q\,t = A\,d$

where, Q = stream size, t = time for which irrigation is applied, A = area of the check and d = depth of irrigation.

The quantity of water calculated by the above formula will irrigate the entire check if the opportunity time is same all over the check. In a check as some time is needed to spread, the opportunity time and consequently the depth absorbed will not be same over the entire area. It can be said that at the tail end of the check basin, the opportunity time available is less to an extent of the time required for the water to spread all over the check. As such, if the area is to be fully irrigated, an additional opportunity time equal to the time of spread should be allowed at the end. It is assumed that after the water has spread all over the check, rest of the water introduced into the check is uniformly available over the entire area. Fig. 8.11 represents the advance and recession curves.

Let in a time of irrigation application, t_L water spreads over a check of length, L. Consider water spread over a small elemental length dl in the check in a small time interval, dt at certain advance distance l from the upstream end of the check when the opportunity time is t_d and advance time is t_l. (Total time of irrigation application, $T = t_d + t_l$) .

Consider the infiltration rate defined by Kostiakov's equation as:

$$i = K\,t_d^{\,n} \tag{8.10}$$

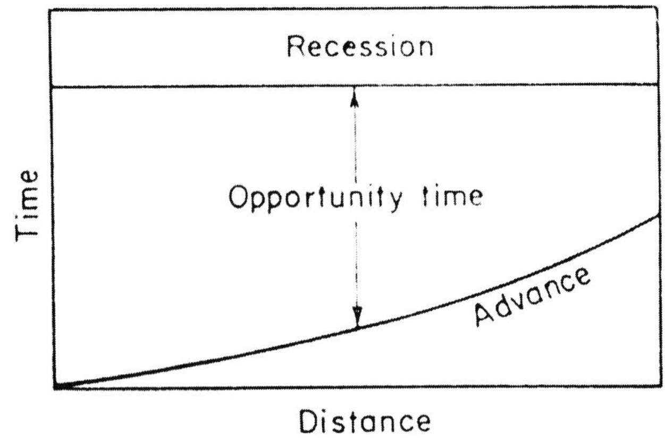

Fig. 8.11: Advance and recession curve of check method

Integrating above Eq. (8.10) with respect to time *dt*, we get the accumulated depth of infiltration, *d* as:

$$d = \frac{K}{n+1} t_d^{\,n+1} \tag{8.11}$$

Considering unit width of the check and a small elemental length dl of the water spread along the longitudinal section of the check, volume of water infiltrating is given as

$$v = \frac{K}{n+1} t_d^{\,n+1} . dl \tag{8.12}$$

Over total length of the check, *L*, the total volume of water infiltrating is given as:

$$V = \frac{K}{n+1} \int_O^L t_d^{\,n+1} . dl \tag{8.13}$$

The opportunity time, t_d can be written as $t_d = T - t_l$

Substituting the opportunity time in Eq. (8.13),

$$V = \frac{K}{n+1} \int_0^L (T - t_l)^{n+1} dl \tag{8.14}$$

The water front movement in a check is in the from of power function as:

$$l = a\, t_l^{\,b} \tag{8.15}$$

where, l = length of water front advance in time t_l and a and b are the characteristic constant.

From Eq. (8.15), $$t_l = \left(\frac{l}{a}\right)^{\frac{1}{b}} \tag{8.16}$$

Substituting the value of t_l from Eq. (8.16) in Eq. (8.14) and simplifying:

$$V = \frac{K\,T^{n+1}}{n+1} \int_0^L \left[1 - \frac{1}{T}\left(\frac{l}{a}\right)^{\frac{1}{b}}\right]^{n+1} dl \tag{8.17}$$

Expanding the term under integral through binomial expression and retaining the first two terms (since subsequent terms are negligible) we get:

$$V = \frac{K\,T^{n+1}}{n+1}\int_0^L \left[1 - \left(\frac{n+1}{T}\right)\left(\frac{l}{a}\right)^{\frac{1}{b}}\right] dl \qquad (8.18)$$

Integrating right hand side of Eq. (8.18) and simplifying we get.

$$V = \frac{K\,T^{n+1}}{n+1}\left[L - \frac{n+1}{T}\,a^{\frac{-1}{b}}\,\frac{L^{\frac{1}{b}+1}}{\frac{1}{b}+1}\right]$$

$$= \frac{K\,L}{n+1}\left[T^{n+1} - T^{n}\left(\frac{(n+1)\,b}{1+b}\right)\left(\frac{L}{a}\right)^{\frac{1}{b}}\right] \qquad (8.19)$$

Considering the length of advance as L, time of advance t_L from Eq. (8.16) can be written as:

$$t_L = \left(\frac{L}{a}\right)^{\frac{1}{b}} \qquad (8.20)$$

Assume $\theta = \dfrac{(n+1)\,b}{1+b}$ (8.21)

Using Eqns. (8.20) and (8.21), Eq. (8.19) can be written as:

$$V = \left(\frac{K\,L}{n+1}\right) T^{n+1} - T^{n}\,\theta\,t_L \qquad (8.22)$$

Let out of "*V*" volume of water that enters into the check, "*d*" be the depth of water that infiltrates and remains within the crop effective root zone over a length of the check". Hence volume of water adding to percolation is given as

$P = (V- d\ x\ L)$.

Percentage of percolation is given as:

$$P = \left(\frac{V - d\,x\,L}{V}\right) x\,100 \qquad (8.23)$$

Using Eq. (8.22) and Eq. (8.23) and with simplification, we get percentage of percolation:

$$P = \left(\frac{T^{n+1} - T^n t_L \theta - t_d^{\,n+1}}{T^{n+1} - T^n t_L \theta}\right) x\, 100 \qquad (8.24)$$

From Eq. (8.24), it is seen that if time $t_L = 0$, $T = t_d$, and then Eq. (8.24) gives percolation zero. This case occurs when the water could be spread instantaneously over the check and allowed to stay over the time.

In actual practice, water is allowed to flow over the entire basin and allowed to stand to infiltrate the desired depth and there is some percolation in the upper reach of the check. The application efficiency is hence reduced and is given as Ea which is given as:

$$Ea = (1-P)\ x\ 100 \qquad (8.25)$$

Using Eqns. (8.24) and (8.25) and on simplification we get,

$$E_a = \left[\frac{t_d^{\,n+1}}{T^{n+1} - T^n t_L \theta}\right] x\, 100 \qquad (8.26)$$

Since exponent n is negative and T is large, we can neglect the term $(T^n t_l \theta)$ in Eq. (8.26) and then application efficiency (Eq. 8.26) is:

$$E_a = \left(\frac{t_d}{T}\right)^{n+1} x\, 100 \qquad (8.27)$$

The expressions derived for percolation losses can be used in design of the check basin system. For a given stream size and plot size, estimate the percolation losses and if these losses are high, either reduce the size of the plot or increase the stream size. Knowing the percolation losses, size of the plot and depth of water to be applied, total amount of water to be applied can be estimated. Knowing the total volume of water and stream size, total time of irrigation can be estimated.

The following sample calculation illustrates the estimation of time required to irrigate a basin.

Sample Calculation 4:

A check basin of size 10 m x 10 m is to be irrigated by a stream of size 10 lit/sec. The depth of the effective root zone is 90 cm. This 90 cm depth of the root zone is divided into 4 profile layers and the following values of soil moisture content and other data of the basin in these four layers are recorded as below.

Depth of profile layer in (cm)	Moisture % on oven-dry basis		Apparent specific gravity, gm/cc
	Field capacity	Actual	
0 – 15	26.0	17.4	1.39
15 – 30	24.0	17.8	1.47
30 – 60	22.3	19.2	1.51
60 – 90	22.2	20.5	1.53

How long should the irrigation stream be applied to the basin to replenish the soil moisture to field capacity. Assume no deep percolation or surface runoff.

Solutions:

Moisture deficits in the different layers are as follows:

$$\text{1st layer} = \frac{26.0 - 17.4}{100} \times 1.39 \times 15 = 1.79 \text{ cm}$$

$$\text{2nd layer} = \frac{24.0 - 17.8}{100} \times 1.47 \times 15 = 1.36 \text{ cm}$$

$$\text{3rd layer} = \frac{22.3 - 19.2}{100} \times 1.51 \times 30 = 1.40 \text{ cm}$$

$$\text{4th layer} = \frac{22.2 - 20.5}{100} \times 1.53 \times 30 = 0.78 \text{ cm}$$

So, net quantity of water to be applied is 5.33 cm.

Area of the check basin = 10 x 10 = 100 m^2.

Volume of irrigation water required = 100 x 5.33 / 100 = 5. 33 m^3 = 5330 lit.

Stream size = 10 lit/sec

Time required for irrigation = 5330/10 = 533 sec ≈ 9 min.

Sample Calculation 5:

A check basin of size 10 m x 40 m is to be irrigated with total depth of irrigation of 7.5 cm. Accumulated infiltration time relationship is given as $d = 9.95\, t_d^{0.43}$ where d is accumulated infiltration, mm and t_d is time, min. Water front advance equation is expressed as $l = 5.6\, t_l^{0.63}$ for a stream size of 1.7 lit/sec/m width of the check basin. Find out the water application efficiency. Find out also the time required to flow into the check basin with the given stream size.

Solution:

Given size of the basin is 10 m x 40m from which maximum length of the basin is 40 m.

Water front advance equation is given as:

$$l = 5.6\, t_l^{\,0.63}$$

If the maximum length of the basin is 40 m (L = 40 m), then time of advancement (t_L) is:

$$t_L = \left(\frac{40}{5.6}\right)^{\frac{1}{0.63}} = 22.66 \text{ min}$$

Accumulated depth of infiltration, d = 7.5 cm = 75 mm

Accumulated infiltration- time relationship is given as:

$$d = 9.95 t_d^{0.43}$$

Hence, time to infiltrate 75 mm depth of water is:

$$t_d = \left(\frac{75}{9.95}\right)^{\frac{1}{0.43}} = 109.67 \text{ min}$$

Total time of irrigation, $T = t_d + t_L$ = 109.67 +22.66 = 132.33 min

$$\theta = \frac{(n+1)\,b}{1+b} = \frac{(0.43+1)\,x\,0.63}{1+0.63} = 0.553$$

Using Eq. (8.26) and putting the values of t_d = 109.67, n = 0.43, T = 132.33, t_L = 22.66 and θ = 0.553, we get application efficiency = 84.4 percent.

Time for which water is to flow into the basin with the given stream size is obtained as:

$$T = \frac{d\,(mm)\,x\,L\,(m)}{60\,x\,E_a(fraction)\,x\,q\,(lit/\sec/m)} = \frac{75\,x\,40}{60\,x\,0.844\,x\,1.7} = 34.6 \text{ min}$$

8.4.5 Furrow Irrigation

Furrow method of irrigation is used to irrigate row crops with furrows between the rows of crops. In this method, small furrows are used to convey water into the soil surface in small individual parallel streams. Water infiltrates through the sides and bottom of the furrows. Water moves both laterally as well as vertically downward and wet the root zone (Fig. 8.12).

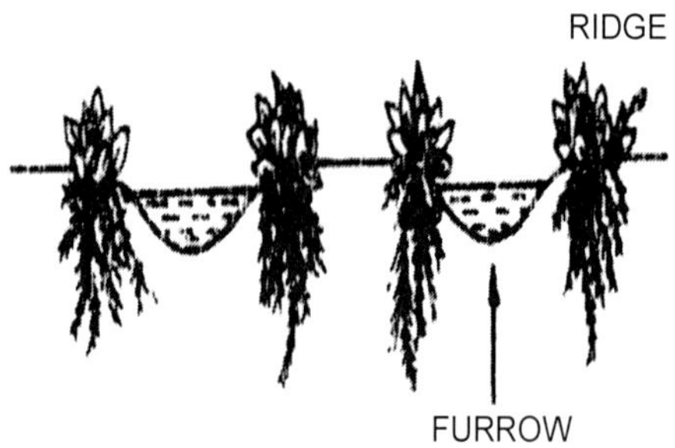

Fig. 8.12: Movement of water in furrow

Length of time the water is to flow in the furrow depends on the volume of water needed to replenish the root zone, infiltration rate of the soil and rate of the lateral spread of water in the soil. Size and shape of the furrow depends on crop grown, equipment used, soil type and spacing between the rows of the crops. Furrows have an added advantage that in rainy season, they can be used as surface drainage channel to drain out excess rainwater from the filed. (Hansen *et al.,* 1979)

8.4.5.1 Adaptability of Furrow Irrigation System

Furrow irrigation can be used to irrigate all cultivated crops planted in rows, including orchards and vegetables like tomato, potato, brinjal, Ladies finger, chillies, tobacco, groundnut, sugarcane, maize etc. Furrow irrigation is suitable to most soils except sandy soil that have high infiltration rate and has less lateral distribution of water than vertical distribution between the furrows. A clay soil has more lateral distribution than vertical one (Fig. 8.13). If sandy soil is to be used for furrow irrigation, then it should have closer furrow spacing than the clayey soil.

The advantages of furrow irrigation are (Lenka, 2001):

(i) Economic use of water and increased water-use-efficiency is possible due to lateral and downward flow of water in the root zone.

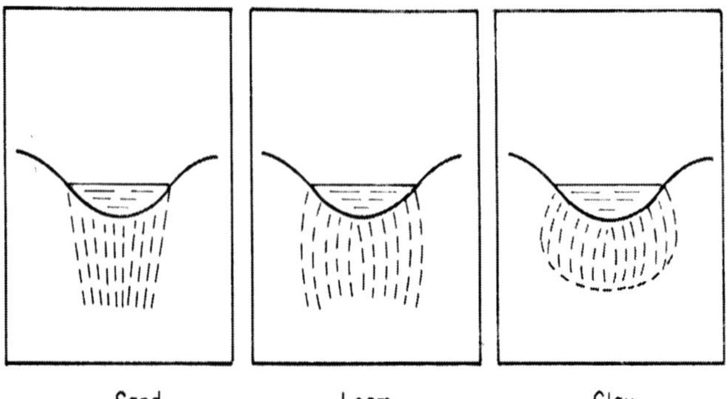

Fig. 8.13: Wetting pattern in furrows of various soils

(ii) Furrows can be used both as irrigation and drainage channel.

(iii) Cultural operation in furrows is easy.

(iv) Evaporation loss is minimized due to lesser contact zone in the furrow.

(v) Reduces ponding and crusting of the ridge since zone of wetting is only one-third to one-fifth of the total furrow depth.

(vi) Irrigation can be practiced in alternate furrows or paired rows to save water and increase water-use-efficiency.

(vii) The method reduces labour requirement in land preparation and irrigation.

However, furrow irrigation requires careful land grading so that water can easily flow in each furrow without causing any ponding.

8.4.5.2 Types of Furrow Irrigation

Depending on the method of irrigation and cultivation, furrow irrigation may be

(i) Irrigation in each furrow

(ii) Irrigation in alternate furrow

(iii) Alternate irrigation in alternate furrow and

(iv) Irrigation in furrow in paired rows.

Crops like potato, tomato, maize, cotton etc. need irrigation in each furrow whereas crops like lettuce, carrots, onions etc. often need two rows between the furrows which is called as irrigation in paired rows (Michael, 1978). Furrows of depth 7.5 to 12.5 cm are suitable for vegetables while some row crops and orchards require much deeper furrows. But it is advantages to use wide and shallow furrows for a wide range of crops.

Based on their alignment, irrigation furrows can be classified into two types i.e. (i) straight furrow and (ii) contour furrow. Again based on the size and spacing, they can be (i) deep furrows and (ii) corrugations.

Straight Furrows

Straight furrows are laid down the prevailing land slope and are best suited to sites where the land slope does not exceed 0.75 percent. In areas of high rainfall, furrow grade should not exceed 0.5 percent so as to minimize the soil erosion. The range of furrow slopes for efficient irrigation in various soil types are the same as those recommended for the borders.

Contour Furrows

Contour furrow method is similar to the graded and level furrow methods in that irrigation water is applied in furrows; but the furrows carry water across a slopping land rather down the slope. Contour furrows are curved to fit the topography of the land. The furrows are given gentle slope along its length as in case of a graded furrow. Light soils can be irrigated across slope up to 5 percent by contour furrow. For stable soils, slope up to 8 to 10 percent can be irrigated by contour furrows.

Corrugation Irrigation

Corrugations are small furrows often used to irrigate non-cultivated close growing crops in rolling lands that may cause some intermittent flooding. Crops like small grains and pastures are irrigated by this method. Corrugations are generally used in soils of very low infiltration capacity or for the soils that disperse and puddle on flooding and crusts are formed after drying. Corrugations may be used in conjunction with border irrigation on lands with relatively flat slopes. In this method, the land is prepared and the crop is sown before any layout of irrigation method is adopted. After sowing the seed and before germination of the seeds, a corrugator is used, making small furrows or corrugations to aid in controlling irrigation water. Corrugations are V or U-shaped channels about 6 to 10 cm deep, spaced 40 to 75 cm apart. The entire soil surface is wetted slowly by the capillary movement of water which flows in the corrugations. This system of irrigation is most suitable in loamy soils in which lateral movement of water takes place readily. Clay soils with low infiltration rate are difficult to be irrigated by this method unless the slopes are quite flat. This method is also not suitable to deep sandy soil. Corrugation methods are not recommended for saline soils. The corrugations are usually made in the direction of the steepest slope. They may also be laid across the slope as in contour furrow.

8.4.5.3 Furrow Irrigation Hydraulics

Study of hydraulics of furrow irrigation helps in obtaining efficient water application. Like border, flow in a furrow is an unsteady and spatial varied open channel flow. The dominant variables influencing the rate of flow in furrows are the entrance stream size, infiltration rate of the soil, size and shape of the wetted perimeter of the furrow, furrow slope and hydraulic resistance. Both the roughness of the wetted perimeter of the furrow and resistance offered by the crop influence the hydraulic resistance to flow. The time of ponding or infiltration opportunity time in a furrow should be the time period required for the net depth of water to infiltrate into the soil to wet the root zone to field capacity. The time of ponding should be uniform throughout the length of the furrow.

8.4.5.4 Design of Furrow Irrigation

The design of furrow irrigation system comprises in deciding (i) spacing of the furrows, (ii) shape and size of the furrows, (iii) stream size, (iv) furrow slope and (v) furrow length.

Furrow Spacing

Spacing of the furrows should fit the crops grown and type of machine used for planting and cultivation. Crops like potato, tomato, maize and cotton are planted at 60 to 90 cm apart and are irrigated by furrows between all the rows. Vegetables like lettuce, carrots and onions are spaced 30 to 40 cm apart and often have a paired row between furrows. Wide spaced crops like melon, fruit trees and berries, generally, require more than one furrow between crop rows. Furrows should be spaced close enough to ensure that water spreads to the sides into the ridge and root zone of the crop before it moves down below the root zone. In general, small plants need small furrows whereas large plants need large furrows.

Length of Furrow

The optimum length of the furrow is usually the longest furrow that can be safely and efficiently irrigated. The length of the furrow should not be too large. If furrow is of large length, then there will be more accumulation of infiltrated water at the upstream end than the downstream end of the furrow that will cause non uniform storage of water in the crop root zone. This means, it can be said that the upstream end becomes over irrigated whereas the downstream end becomes under irrigated. On the other hand short furrows require more area to be lost to field supply channels causing loss of more valuable crop land and increase in labour cost. Proper furrow length depends on hydraulic conductivity of the soil. Furrows must be shorter in porous sandy soil than on light clay soil. The length of the furrows may vary from 45 to 300 m or longer depending on the infiltration rate. Length of the furrows may also be limited by the size and shape of the field,. Furrow length depends on the slope and net depth of water application. Longer furrows can be used for deep rooted crops grown on clay soil than for shallow rooted crops grown in sandy soil. Table 8.4 represents the recommended maximum length of the furrows for various types of soil, land slope and depth of irrigation (Michael, 1978).

Table 8.4 : Recommended length of furrows for various types of soil, land slope and depth of irrigation

Furrow slope, percent	Furrow length, m						
	Soil type						
	Clay		Loam Net depth of water application, cm		Sand		
	7.5	15	5	10	5	7.5	10
0.05	300	400	120	270	60	90	150
0.10	350	440	180	330	90	120	190
0.20	370	470	220	370	120	190	250
0.30	390	500	280	400	150	225	280
0.40	380	500	280	370	120	190	250
0.50	270	400	250	300	90	150	220

Furrow Slope

Slope or grade of the furrow controls the speed of flow of water in the furrow in downward direction. A minimum slope of 0.05 percent is needed to enable the furrow to act as drainage channel. The slope of the furrow should be such that the flow of water in it is non-erosive and at the same time helps in efficient irrigation. Table 8.5 gives some general values of furrow slopes depending on type of the soil (Murty, 2004). When the land slope exceeds the safe limits for furrows, the furrows can be laid across the slope on the contour with desired slope.

Table 8.5: Furrow slopes for various types of soils

Soil type	Maximum furrow slope, percent
Sand	0.25
Sandy loam	0.40
Fine sandy loam	0.50
Clay	1.50

Furrow Stream

Size of the furrow stream varies from 0.5 to 2.5 lit/sec. In order to achieve the most uniform irrigation, the largest stream that causes no erosion, no overtopping of the furrows and no runoff at the tail end of the furrow is used to irrigate at the beginning in each furrow so that the entire length of the furrow is irrigated quickly thus obtaining the soil to absorb water evenly through the entire furrow length. This maximum non-erosive stream is given by Eq. (8.28).

$$q_{\max} = \frac{0.60}{s} \tag{8.28}$$

where,

q_{max} = maximum non-erosive, lit/sec and s = slope of furrow, percent.

Average depth of water applied during irrigation is given as:

$$d = \frac{q \; x \; 360 \; x \; t}{w \; x \; L} \tag{8.29}$$

where,

d = average depth of water applied, cm; q = stream size lit/sec; w = furrow spacing, m; L = length of furrow, m and t = duration of irrigation (cumulative), hr.

After the flow reaches the lower end of the furrow, the stream size is reduced or cut back so that it will keep the furrow wet throughout its length with a minimum waste at the end. This reduced stream is called as *cut back stream* which continues to flow till the required amount of water has been applied. The water is ponded at the tail end by closing the furrow so that the entire furrow absorbs the water. Following sample calculation will help understand the average flow depth governed by the cut back stream approach.

Sample Calculation 6:

A stream of size 2 lit/sec is initially used to irrigate a furrow of length 100 m and furrow spacing is 60 cm. When the initial furrow stream reached the tail end of the furrow, it was cut back and the cut back stream was 0.5 lit/sec. The initial furrow stream took 50 min to reach the tail end of the furrow. The cut back stream continued to irrigate the furrow for one hour.

Estimate the average depth of flow.

Solution

Given, $q = 2.0$ lit/sec; $t = 50$ min = 50/60 hr; $L = 100$ m and $w = 60$ cm = 0.60 m

Average depth of flow by the initial stream is estimated by Eq. (8.29) as:

$$d = \frac{2\,x\,360\,x\,50}{0.60\,x\,100\,x\,60} = 10\,cm$$

Average depth of flow by the cut back stream is also given by same Eq. (8.29) as:

$$d = \frac{0.5\,x\,360\,x\,1}{0.60\,x\,100} = 3.0\,cm$$

Hence, total depth of irrigation = 10 + 3 = 13 cm.

8.4.5.5 Evaluation of Furrow Irrigation

Infiltration of water into the furrow is the most important variable affecting the characteristics of flow in furrows. The infiltration rate in furrows can be better measured by inflow-outflow method or called as volume balance method. In the inflow and outflow method, the furrow is divided into a number of sections. Water measuring devices are installed at each section to measure the inflow and out flow in the section. The furrow cross-sectional profile is determined at the representative sites in the test section with a point gauge before admitting water. Furrow spacing is measured by taking distance between centre of two consecutive furrows. The rate of advance of water in the test section and the depth of flow at different points at definite time intervals are also measured. From the above measured parameters, furrow cross-sectional area and wetted perimeter are found out. The average values of the wetted perimeter when multiplied by the length of the test section give the wetted area. Volume of the accumulated values of infiltration is calculated by subtracting the volume of the accumulated storage from volume of the accumulated inflow. The accumulated depth of infiltration is given dividing the accumulated wetted area of the test section with the volume of the accumulated infiltration. Following sample calculation will help to evaluate the furrow infiltration.

Sample Calculation 7

Following data were recorded from a test furrow in a sandy loam soil.

Stream size, lit/min	Distance, m	Advance time, min	Wetted perimeter, cm	Furrow cross sectional area at corresponding depth of flow, cm^2
100.0	20	1.80	25.5	60.0
	40	5.80	25.8	90.0
	60	11.00	26.4	103.0
	80	18.00	26.7	108.5
	100	23.50	27.0	112.0

Compute furrow infiltration.

Solution:

Step I: Calculate the volume of the accumulated inflow by multiplying the stream size and advance time.

Step II: Calculate the accumulated storage volume by multiplying the furrow cross-sectional area with distance traveled by the stream.

Step III: Compute the accumulated wetted area by multiplying the wetted perimeter with the distance traveled by the stream.

A sample calculation for distance 60 m is given below.

Volume of accumulated inflow = 100 x 11 = 1100.00 lit

Volume of accumulated storage = 6000 x 103 = 618000 cm^3 = 618 lit

Accumulated wetted area = 6000 x 26.4 = 158400 cm^2

Similar calculations are done for other distances and the calculations are shown in table below.

Volume of accumulated inflow, lit	Volume of accumulated storage, lit	Accumulated wetted area, cm^2
180	120	51000
580	360	103200
1100	618	158400
1800	868	213600
2350	1120	270000

In the next stage volume of accumulated infiltration and depth of accumulated infiltration are computed as discussed above.

For a distance of 60 m, volume of the accumulated infiltration is given by subtracting 618 from 1100 and the result is 482 lit = 482000 cm^3.

Accumulated wetted surface area at 60 m is 158400 cm^2.

Hence, depth of accumulated infiltration = 482000/158400 = 3.04 cm.

Similar calculations for other distances are done and are presented in table below.

Distance, m	Advance time, min	Volume of accumulated infiltration, lit	Depth of accumulated infiltration, cm
20	1.80	60	1.18
40	5.80	220	2.13
60	11.00	482	3.04
80	18.00	932	4.36
100	23.50	1230	4.56

8.5 Subirrigation

Subirrigation method consists of supplying water directly below the ground to the root zone of the crop. In this case, an artificial water table is maintained at some depth below ground surface depending on soil texture and depth of the plant roots. Water reaches the plant roots through capillary action. Water may be introduced either by open ditches or by underground pipe lines such as mole drains or tile drains. Depth of open ditches vary from 30 to 100 cm and spaced at distance of 15 to 30 m. Water supply system consists of supply channels, ditches or trenches suitably spaced to cover the field adequately. This irrigation is suitable for nearly smooth and level topography or for topography with gentle and uniform slopes. It is ideal for soils having a low water holding capacity and high infiltration rate where surface method cannot be used and sprinkler irrigation is costly. The advantage of this method is that in this irrigation, it is possible to maintain the water level at optimum depths for crop needs at different growth stages. Evaporation losses from the ground surface are minimized by this method. However, water having high salt content cannot be used by this irrigation method. It also requires special site conditions since it is necessary to have complete control of the water table through controlled water application and drainage.

8.6 Sprinkler Irrigation

In sprinkler method of irrigation, water is spread into the air and allowed to fall on the ground surface somewhat resembling rainfall. The spray is developed by the flow of water under pressure with the help of a pump. This method of irrigation is also called as overhead irrigation. In some cases where the source of water is high enough above the area to be irrigated, the required pressure may be developed by gravity alone. Nozzles or orifices help in spray of water. With careful selection of nozzle sizes, operating pressure and sprinkler spacing, amount of irrigation water required to refill the root zone soil moisture depletion can be met nearly uniformly at a rate to suit the infiltration rate of the soil. The system reduces runoff and deep percolation losses and there by enhances water application efficiency.

8.6.1 Adaptability of Sprinkler Irrigation

Sprinkler can be used for all most all crops except rice, jute etc which are high water requiring crops. It can be used in almost all types of soils. However, it is not usually suitable for very fine textured soils like heavy clay soils where infiltration rate is less than 4 mm/hr. It is particularly suitable to sandy soils that have high infiltration rate. It can be used in almost all types of topographic situations with little land preparation. Land leveling is not usually essential for irrigation with sprinklers. Small streams f irrigation water can be used efficiently. They are now being used increasingly in regions of water scarcity areas. They are used to irrigate all vegetables, orchard crops, and high valued plantations like tea, coffee, cardamom and flower and fruit trees.

8.6.2 Advantages of Sprinkler Irrigation

Followings are the various advantages of sprinkler irrigation (Panigrahi, 2011).

(i) Can be used for land unsuitable or uneconomical for leveling

(ii) Soils too porous, highly erodible or relatively impermeable which causes other methods of irrigation difficult

(iii) Uniform application of water is possible

(iv) Small streams of irrigation water can be used efficiently

(v) Can be used in almost all types of soils but less suitable for heavy clay soil

(vi) Can be used where frequent light applications are needed like the case of seed germination, when seedlings are small

(vii) Land preparation is not required and so the labour cost for installation is reduced

(viii) Soluble fertilizers, herbicides and fungicides can be applied in the irrigation water economically and with little extra equipment. Thus saves labour cost to supply them to the crops

(ix) Elimination of channels for conveyance and there by saving land lost to channel conveyance system

(x) Areas located at a higher elevation than the source can be irrigated

(xi) Crop damage from frost can be reduced

(xii) Land too sloppy can be used for irrigation by this method where other methods are not efficiently used

(xiii) Irrigation method does not interfere with the movement of farm machinery

(xiv) Surface runoff of irrigation water can be eliminated

8.6.3 Disadvantages of Sprinkler Irrigation

Followings are the disadvantages.

(i) Uneven distribution of water due to high velocity of winds

(ii) More evaporation losses when operating under high temperature conditions. This is more important in case of irrigation water that has higher amount of dissolved salts

(iii) Initial investment cost is high

(iv) Ripening of soft fruits must be protected from spray

(v) The water used for spraying must be free from sand, debris and large amount of dissolved salts

(vi) Theft problem of sprinklers in the fields are frequent. If the system has to be dismantled after irrigation is over then carrying them to the house is problematic. Again reinstallation in the field for next irrigation is needed to be done thus making the system uneasy.

(vii) Power requirements are usually high since sprinklers operate with a water pressure of 0.5 to more than 10 kg/cm^2.

(viii) Cannot be used efficiently for fine textured soils that have low infiltration rate especially in hot windy areas

(ix) Sprinkling low quality irrigation water damages the citrus crops.

(x) Free water on crop surfaces induces growth of fungi and bacteria and multiplies diseases

(xi) Small, slender and tender seedlings break due to sprinkler force.

8.6.4 Types of Sprinkler Systems

On the basis of portability of sprinkler units, these are classified as: (i) portable, (ii) semi-portable, (iii) semi permanent, (iv) solid set system, (v) permanent system, and (vi) set-move irrigation system. A portable system has portable main lines, and laterals and a portable pumping unit. It is designed to move from field to field. Portable and semi-portable types are most commonly used in agriculture. This system can be moved manually or mechanically. In mechanical arrangement, a wheel move system is used to carry the pipes and sprinklers from field to field.

On the basis of arrangement of spraying of water, sprinklers may be fixed nozzle type, perforated type and rooting type. In the fixed nozzle type, the sprinklers are fixed in a row parallel and water sprinklers out at right angle from the pipes. By changing the pipes the plots are irrigated. In perforated pipe system, there are perforation in the lateral irrigation pipes in a specially design pattern to distribute water almost uniformly and the system works with low operating pressure of about 0.5-2.5 kg/ cm^2. The pressure is so low that the system can be equipped to an overhead tank to obtain the necessary pressure head. The sprays can be on either side of pipe and can cover a strip of land from 6 to 15 m wide. The rate of spray is high and so the system is suitable for high infiltrating soils. System is suitable for irrigation to lawns, gardens and small vegetable fields when the height of the crops does not exceed 40 to 60 cm (Patnaik and Mohanty, 2010).

In rooting head system, nozzles are placed in risers fixed at uniform interval along the length of lateral pipe (Fig. 8.14). Crops like tomato, potato, sugarbeet and many flowering plants are best irrigated by this system. Soil with low in-take rate and high water holding capacity is generally irrigated by rotating sprinkler systems. Most agricultural sprinklers are rotating types with either one or two nozzles that vary in size from 1.5 mm to 5 cm; the discharge from such nozzles varies from 4 to 4000 lpm and covers circular area from 6 to 180 m in diameter. The most commonly used ones are with capacities ranging from 7.5 to 75 lpm that operate at pressure from 1.4 to 4.2 kg/cm^2. The rotation is caused by a torque produced by a reaction of water leaving the nozzle or due to impact of a spring loaded arm. Some sprinkler is gear driven by a small turbine located at the base of the sprinkler or externally on one side of the nozzles.

The soil puddles and runoff begins if the drop becomes very large. Drop size depends on types and size of nozzle and operating pressure. For small drop sizes, nozzle sizes may be 3-5, 5-6 and 6-9 mm and the pressure for these nozzle sizes may be respectively 6-9, 8-11 and 9-13 kg/ cm^2. Water distribution depends on wind velocity and is affected if the velocity becomes more than 16-17 km/hr. If wind velocity is high the lateral and risers spacing should be closer.

Fig. 8.14: Rotating head sprinkler in operation

8.6.5 Components of Sprinkler System

A sprinkler irrigation system consists of a pump to provide necessary pressure (ii) main pipe lines (iii) laterals (iv) risers and (v) sprinkler head. Sometimes sub main are connected to main line from which laterals arise. Fig. 8.15 represents a typical sprinkler irrigation system with various components.

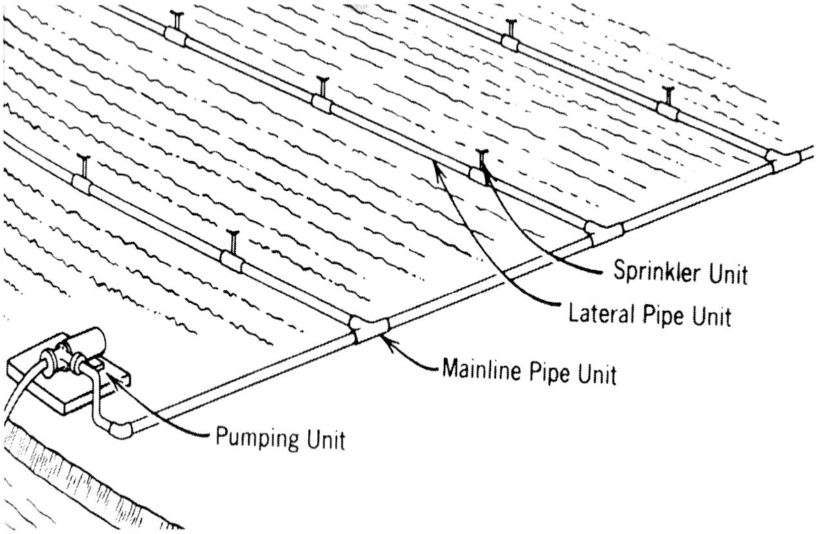

Fig. 8.15: Layout of a typical sprinkler irrigation system

The pump lifts water from the source and pushes it through the distribution systems and the sprinklers. The pump should be so designed that it lifts water from the source and pushes it to the highest point in the crop field and maintains adequate operating pressure in the distribution system. Volute type centrifugal pump or deep well turbine pump can be used to lift water and feed to the sprinkling system. Volute type centrifugal pump is used when the water level in the source is at lower level i.e. nearer to the ground. Otherwise turbine pump will be needed. Pumps can be installed permanently at the water source or may be portable type. Pump can run either by electric motors or by internal combustion engine. Electric motors have low initial and operating costs and are easier to maintain.

Main lines may be permanent or portable. Permanent main lines are used in situations where there is proper facility of watch and ward and where the crop needs full season irrigation. Asbestos cement and PVC pipes are used as main lines. In case of permanent mains, they can be buried underground at appropriate locations along the farm boundaries. Underground concrete pipelines on the farm can act as supply points to the sprinkler systems rather than as main lines because of the pressure limitations. Light weight aluminum pipes with quick couplers are used for most portable main lines. Water is taken from the main either through a valve placed at each point of junction with a lateral or in some cases through an L or T section that has been supplied in place of one of the couplings on the main.

Lateral lines are usually portable. Quick-coupled aluminum pipes are best for most of the portable laterals. Lateral pipes are available in length of 5, 6 or 12 m. Sprinkler head is the most important component of sprinkler irrigation system. Most of the sprinklers used for field crops are rotating type with either one or two nozzles. Most commonly used sprinklers are small to medium type with capacities varying from 7.5 to 75 lpm and which operate at pressure ranging from 1.4 to 4.2 kg/cm^2. They cover an area of 10 to 40 m in diameter. Rotating sprinklers rotate around a vertical axis. Rotation is caused either by the torque produced by the reaction of the water leaving a nozzle or by the impact of spring-loaded arm that periodically interrupts the jet from one of the nozzles. Sprinklers commonly used have two nozzles, one to cover the farther area from the sprinkler and the other to cover the area near the sprinkler. There are several variations in the design and operating characteristics of the rotating sprinklers to suit different conditions. Important among them are (i) popup sprinkler head used to irrigate lawns (ii) part-circle medium sized agricultural sprinkler and (iii) large sized sprinklers (Suresh, 2010). To remove the debris and sediments from entering into the sprinkler heads, debris basins and desilting basins required. Soluble fertilisers can be added to irrigation water and can be used by the system. For this case, a Venturi is fixed in the sprinkler line creating differential pressure and allowing the fertilizer solutions to flow into the water line. Details of classifications of rotating head sprinklers, their characteristics and adaptability are discussed by Michael (1978).

8.6.6 Design of Sprinkler Systems

Design of sprinkler system for a given situation requires consideration of soil, crop, climate and topography besides the equipment availability. The different steps involved are described below.

1. Inventory of Resources and Conditions

Following items are required to be collected for inventory of resources.

(i) Map of the area, (ii)Water supply, (iii)Climatic conditions (iv) Depth of irrigation, (v) Irrigation interval, (vi) Water application rate and (vii) Power source

A map of the area showing all relevant elevations with respect to water supply, pump location, and critical elevations in the fields to be irrigated should be prepared first with a proper scale before sprinkler system is to be installed at a site. The sources, availability and dependability of water source should be identified to meet the seasonal, actual and peak water demand of the crops in the site where irrigation system is to be installed. Water source should be relatively clean and free from suspended impurities and should not cause corrosion to the equipment. Water should also be physio-chemically suitable for irrigation to crops. Water requirement of the crops depend on climatic conditions. The system is so designed that it meets the peak water demand of the crops which in turn depends on the climatic data of the study site. A peak demand in the range of 2 to 10 mm depth per day is equivalent to a continuous flow of 0.23 to 1.16 lit/sec per hectare.

An accurate irrigation scheduling in respect of estimating the depth of irrigation and frequency of irrigation during the period of peak consumptive use of the crops are to be calculated for facilitating in design of sprinkler irrigation. Details of estimating the depth and frequency of irrigation are discussed in Chapter 7 and estimation of consumptive use of the crops is described in Chapter 5. Details of computing the depth and frequency irrigation intervals can be varied by changing the hours of the set to suit the irrigation requirement at different stages of the crops.

The rate of water application by sprinklers is limited by the infiltration capacity of the soil. Application less than the infiltration capacity will result deficient supply and cannot meet the consumptive use demand of the crop. Thus, any application in excess of the infiltration capacity will cause runoff and deep percolation loss and also causes soil erosion. Table 8.6 presents some suggested values of maximum rate of water application for different soil soils ands slopes (Michael, 1978).

Table 8.6: Maximum rate of water application for various soil and slope

Soil texture	Maximum application rate, cm/hr at slope of percent			
	0–5	5 – 8	8 -12	12 – 16
Coarse sand up to 2 m	5.0	3.7	2.5	1.3
Coarse sand over more compact soils	3.7	2.5	2.0	1.0
Light sandy loam up to 2 m	2.5	2.0	1.5	1.0
Light sandy loam over more compact soils	2.0	1.3	1.0	0.8
Silt loam up to 2 m	1.3	1.0	0.8	0.5
Silt loam over more compact soils	0.8	0.6	0.4	0.3
Heavy textured clays or clay loams	0.4	0.3	0.2	0.1

Finally, the sources of power to operate the pump, whether electricity or internal combustion engine is to be ascertained before installation of the system. When the installation site in permanent electric run motors will be better since it is cheaper in installation and maintenance. For portable system, pumps run by diesel engine can be selected.

2. Selection of Sprinkler Systems and Layout

The selection should fit with the sprinkler spacing for a particular operating pressure and nozzle size so that it provides best uniformity of water application under prevailing wind conditions without causing runoff and deep percolation. The selection should satisfy the irrigation requirement and is done on the basis of

(i) Operating pressure, (ii) Nozzle discharge rate, (iii) Spacing of sprinkler heads, (iv) Infiltration rate of soil, (v) Crop requirement, (vi) Cost of power requirement and (vii) Availability of labour.

Usually sprinklers are designed to work under varying pressures. At high operating pressure, coverage area of the sprinkler is more and cost of the system is more. At low pressure, the case is reverse. Hence, intermediate operating pressure is required for most of the field crops. The pressure variation in the system should not exceed 20 percent of the operating pressure. Nozzle discharge rate varies with pressure and at the same pressure, a bigger size nozzle results higher discharge. Sprinkler spacing is decided based on pressure, nozzle discharge, wind speed etc. Soil infiltration rate is determined by infiltrometers and water application rate should be less than that otherwise runoff will occur. The labour/energy cost should be less for a sprinkler to be selected.

Layout of the system should be such that the pumping unit is placed at the central part of the field which will reduce the conveyance distance and hence friction loss in the distribution system. Sometimes it is necessary to adopt a sprinkler irrigation system with an already laid underground pipe line water distribution system or field channels. In case of underground pipe lines, a portable pumping plant can be used with suction attached to the hydrants mounted on the pipe outlets. In case of field channels running along one edge of the farm, a portable pumping unit along with the sprinkler unit with lateral extending to the field may be used to draw water directly from the channel and distribute it through the sprinklers. Orientations of laterals should be such that the main line passes along the steepest slope and the laterals move at right angles to it and should pass on contours as close as possible.

3. Capacity of Sprinkler System

The capacity of the sprinkler system/capacity of the pump to be used depends on the area to be irrigated (design area), depth of water applied at each irrigation, time of this depth of irrigation and application efficiency. The capacity is given by the formula:

$$Q = 2780 \frac{A \, x \, d}{F \, x \, H \, x \, E} \tag{8.30}$$

where,

Q = discharge of the sprinkler system/capacity of the pump, lit/sec; A = area to be irrigated, ha; d = depth of application of irrigation water, cm; F = irrigation interval, days; H = operating hours per day; and E = application efficiency, percent.

4. Discharge of a Sprinkler

The discharge rate of an individual sprinkler depends on the spacing of laterals, spacing of sprinklers in each lateral as well as application rate. It can be computed as:

$$q = \frac{S_m \, x \, S_l \, x \, I}{360} \tag{8.31}$$

where,

q = discharge of individual sprinkler, lit/sec; S_l = sprinkler spacing along the lateral, m; S_m = lateral pipe spacing along the main, m and I = water application rate, cm/hr.

By multiplying the discharge of an individual sprinkler with total number of sprinklers used in the system, sprinkler system capacity or capacity of the pump to be used can be calculated.

5. Hydraulic Design of Sprinkler Systems

Hydraulic design of sprinklers is aimed at obtaining a uniform irrigation coverage, the desired rate of application, maximize area coverage and efficiency desired to reduce the energy requirements in operating the system. The main hydraulic principles are placed below.

(i) Discharge of Sprinkler Nozzle

The discharge of a sprinkler nozzle is given by the formula:

$$q = C \, x \, A \sqrt{2 \, g \, h} \tag{8.32}$$

where,

q = nozzle discharge, m^3/sec; A = area of nozzle or orifice, m^2; g = acceleration due to gravity, 9.81 m/sec^2; h = pressure head at the nozzle, m and C = coefficient of discharge varying from 0.80 to 0.95.

(ii) Water Spread Area of the Sprinkler

For rotating head sprinkler, area covered by sprinkler is given (Pillsbury, 1968) as:

$$R = 1.35 \sqrt{d \, x \, h} \tag{8.33}$$

where,

R = radius of wetted area covered by sprinkler, m; d = nozzle diameter, m and h = pressure head at nozzle, m.

Maximum coverage is obtained when the jet emerges from the sprinkler at an angle of 30 to 32 °C above the horizontal.

(iii) Break-up of Jet

Some break-up of the jet of water is required to attain uniformity of coverage and to minimize the size of big drops which will reduce the chance of soil erosion while irrigating in the field. The following empirical formula suggested by Pillsbury (1968) provides an index of jet break-up.

$$P_d = \frac{h}{(10\,q)^{0.4}} \tag{8.34}$$

where,

P_d = index of jet break-up; h = pressure head at nozzle, m and q = discharge of individual sprinkler, lit/sec.

If P_d is greater than 2, the condition of drop size is good; if 4, condition is best and if greater than 4, pressure is wasted (Pillsbury, 1968).

(iv) Rate of Application

The average rate of application for a single sprinkler can be computed as:

$$R_a = \frac{q}{360\ x\ A} \tag{8.35}$$

where,

Ra = water application rate, cm/hr; q = discharge of individual sprinkler, lit/sec and A = wetted area of the sprinkler, m^2.

Some suggested values of maximum rate of water application for different soil soils ands slopes have been presented in Table 8.6.

6. Design of Sprinkler Laterals

A sprinkler lateral has equally spaced sprinklers along its length. In the beginning the capacity of the system is equal to the sum of discharge of each sprinkler. This will decrease along the line until, at the fartherest end, there will be only the flow of the last sprinkler. Hence, friction loss in a lateral will be much less than if the total flows were carried through the entire length of the line. It is recommended that the total pressure variation in the laterals should not be more than 20 percent of the operating pressure. If the lateral runs uphill or downhill, allowance for this difference in elevation should be made in determining the variation in head. If the lateral carrying water runs uphill, less pressure will be available at the nozzle. On the other hand, if it runs downhill, there will be tendency to balance the loss of head due to friction. It is recommended that a uniform diameter of the lateral instead of tapering section must be adopted. The rate of flow entering a lateral is calculated and a trial diameter of the pipe is calculated. Using this diameter and assuming that the flow is through the entire length without sprinklers, head loss due to friction is computed using Scobey's formula as:

$$H_f = 4.10 \, x \, 10^6 \, \frac{K_s \, L \, Q^{1.9}}{D^{4.9}} \quad (8.36)$$

where,

H_f = head loss due to friction in lateral, m; K_s = coefficient of friction; L = length of pipe, m; Q = discharge, lit/sec and D = inside diameter of the pipe, mm.

Friction loss can be read from the standard tables based on Scobey's formula. Table 8.7 presents the friction loss in main pipe lines whereas Table 8.8 presents the friction loss in lateral pipe lines for portable aluminum pipes with couplings based on Scobey's formula. Friction losses of other pipe lines Scobey's formula are available in Michael (1978). The friction loss is multiplied with a correction factor (F) as given in Table 8.9 corresponding to the number of sprinklers on the laterals. The factor F accounts for the decreasing flow rate along the sprinkler lateral and is called as Christiansen friction factor. When the first sprinkler is located at the same distance as the sprinkler spacing, F can be calculated by Eqn. (8.37) (Murty, 2004) as:

Table 8.7: Friction loss in meters per 100 meters in main lines of portable aluminum pipe with couplings. (Based on Scobey's formula K_S 10 m pipe length

Flow Litres/s	GPM	7.5cm	10 cm	12.5 cm	15 cm	17.5 cm	20 cm	25 cm
2.52	40	0.658	0.157					
3.15	50	1.006	0.239					
3.79	60	1.423	0.339					
4.42	70	1.906	0.449	0.150				
5.05	80	2.457	0.584	0.193				
5.68	90	3.073	0.731	0.242				
6.31	100	3.754	0.893	0.295	0.120			
7.57	120	5.307	1.263	0.413	0.170			
8.83	140	7.113	1.693	0.560	0.227			
10.10	160	9.169	2.182	0.721	0.293			
11.36	180	11.47	2.729	0.967	0.366			
12.62	200	14.01	3.333	1.102	0.448	0.209		
13.88	220	16.79	3.996	1.321	0.537	0.251		
15.14	240	19.81	4.713	1.558	0.633	0.296		
16.41	260	23.06	5.448	1.814	0.737	0.344		
17.67	280	26.55	6.316	2.089	0.849	0.397		
18.93	300	30.27	7.203	2.381	0.967	0.452	0.235	
20.19	320	34.22	8.142	2.692	1.094	0.511	0.265	
21.45	340	38.39	9.137	3.020	1.227	0.573	0.298	
22.72	360	42.80	10.18	3.366	1.368	0.639	0.332	
23.98	380	47.43	11.29	3.731	1.516	0.708	0.368	
25.24	400	52.28	12.44	4.113	1.671	0.781	0.399	0.136
26.50	420		13.95	4.513	1.833	0.857	0.445	0.149
27.76	440		14.57	4.930	1.988	0.936	0.486	0.163
29.03	460		16.23	5.364	2.179	1.019	0.529	0.177

(Contd.)

Flow Litres/s	GPM	7.5cm	10 cm	12.5 cm	15 cm	17.5 cm	20 cm	25 cm
30.29	480		17.59	5.815	2.363	1.104	0.573	0.192
31.55	500		19.01	6.284	2.554	1.193	0.620	0.208
34.70	550		22.79	7.532	3.060	1.430	0.742	0.249
37.86	600		26.88	9.886	3.611	1.687	0.876	0.294
41.01	650		31.30	10.35	4.204	1.965	1.020	0.342
44.17	700		36.03	11.91	4.839	2.262	1.174	0.394
47.32	750		41.08	13.58	5.517	2.520	1.339	0.449
50.48	800			15.35	6.237	2.915	1.513	0.507
53.60	850			17.32	6.999	3.71	1.698	0.569
56.79	900			19.20	7.801	3.646	1.893	0.635
59.94	950			21.28	8.645	4.041	2.097	0.703
63.10	1000			23.45	9.530	4.454	2.312	0.775
69.49	1100			28.11	11.42	5.338	2.771	0.929
75.72	1200			31.75	13.58	6.298	3.269	1.096
82.03	1300				15.69	7.333	3.806	1.277
88.34	1400				18.06	8.441	4.382	1.470
94.65	1500				20.59	9.624	4.996	1.675
101.0	1600				23.28	10.88	5.648	1.894
107.3	1700				26.12	21.21	6.337	2.125
14.0	1800					13.61	7.064	2.369
120.0	1900					15.08	7.829	2.625
126.0	2000					16.62	8.630	2.894

Table 8.8: Friction loss in meters per 100 meters in lateral lines of portable aluminum pipe with couplings. (Based on Scobey's formula)

Flow, Lit/sec	5.0 cm $K_s = 0.34$	7.5 cm $K_s = 0.33$	10 cm $K_s = 0.32$	12.5 cm $K_s = 0.32$	15 cm $K_s = 0.32$
1.26					
1.89	0.32				
2.52	2.53				
3.15	4.40	0.565	0.130		
3.79	6.85	0.858	0.198		
4.42	9.67	1.21	0.280		
5.05	12.9	1.63	0.376	0.122	
5.68	16.7	2.10	0.484	0.157	
6.31	20.8	2.63	0.605	0.196	
7.57	25.4	3.20	0.738	0.240	0.099
8.83		4.54	1.04	0.339	0.140
10.10		6.09	1.40	0.454	0.188
11.36		7.85	1.80	0.590	0.242
12.62		9.82	2.26	0.733	0.302
13.88		12.0	2.76	0.896	0.370
15.14		14.4	3.30	1.07	0.443
16.41		16.9	3.90	1.26	0.522
17.67		19.7	4.54	1.47	0.608
18.93		22.8	5.22	1.70	0.700

(Contd.)

Flow, Lit/sec	5.0 cm K_S = 0.34	7.5 cm K_S = 0.33	10 cm K_S = 0.32	12.5 cm K_S = 0.32	15 cm K_S = 0.32
20.19		25.9	5.96	1.93	0.798
21.45		29.3	6.74	2.18	0.904
22.72		32.8	7.56	2.45	1.02
23.98		36.6	8.40	2.74	1.13
25.24		40.6	9.36	3.03	1.26
26.50		44.7	10.3	3.34	1.38
27.76			11.3	3.66	1.521
29.03			12.3	4.00	1.66
30.29			13.4	4.35	1.80
31.55			14.6	4.72	1.95
34.70			15.8	5.10	2.12
37.86			18.9	6.12	2.52
41.01			22.2	7.22	2.98
44.17			25.9	8.40	3.46
47.32			29.8	9.68	3.99
50.48			33.8	11.0	4.54
53.63				12.5	5.15
56.79				14.0	5.78
59.94				15.6	6.44
63.10				17.3	7.14
				19.0	7.86

Table 8.9: Correction factor, *F* for friction loss in aluminum pipes with multiple outlets

Number of sprinklers	First sprinkler one sprinkler interval from main	First Sprinkler one-half sprinkler Interval from main
1.	1.000	1.000
2.	0.630	0.510
3.	0.550	0.460
4.	0.480	0.410
5.	0.460	0.395
6.	0.430	0.380
7.	0.420	0.375
8.	0.410	0.370
10.	0.400	0.365
12.	0.390	0.360
14.	0.385	0.360
16.	0.380	0.360
18.	0.375	0.355
20.	0.370	0.350
25.	0.365	0.350
30.	0.360	0.350

$$F = \frac{1}{m+1} + \frac{1}{2N} + \frac{\sqrt{(m-1)}}{6N^2} \tag{8.37}$$

where,

F = friction correction factor; N = number of sprinklers along the lateral; m = exponent which is equal to 1.9.

To the friction loss thus calculated, the elevation is added if the lateral goes uphill or the drop is subtracted if the lateral goes downhill. It is recommended that the total pressure variation in the lateral should not be more than 20 percent of the average pressure.

Making allowances for the differences in elevation along the lateral and for the riser height, the head at the main line due to lateral can be computed as:

$$H_n = H_0 + H_f \pm H_e + H_r + H_{bm} \tag{8.38}$$

where,

H_n = head at the main line; H_0 = sprinkler operating pressure; H_f = friction loss in sprinkler lateral; H_e = maximum difference in elevation between the main and the last sprinkler on the lateral (*+ve* for lateral going uphill and *–ve* for lateral going downhill), H_r = riser height and H_{bm} = losses due to bend and other minor losses.

7. Design of Main Line

Diameter of the main line is selected based on the discharge it carries and the friction losses. Selection of main line size should take into account the economics. Main line friction loss of about 3 m for small system and 12 m for large systems may be allowed. Friction loss in main line ($H_{f,main}$) with portable aluminum pipe based on Scobey's formula can be computed by Eq. (8.36) or Table 8.7 can be used to find out the value of friction loss. The friction loss in main line ($H_{f,main}$) when added to the head at the main line due to lateral (H_n), total head requirement of the system (H_{Sys}), is calculated i.e.

$$H_{Sys} = H_0 + H_f \pm H_e + H_r + H_{bm} + H_{f,main} \tag{8.39}$$

8. Selection of Pump and Power Unit

For selecting the pump capacity to supply water to the system, the total head against which the units are to operate is to be determined. The total head, H_t is given as:

$$H_t = H_{Sys} + H_m + H_i + H_s + \frac{V^2}{2g} \tag{8.40}$$

where,

H_t = total design head against which the pump has to work; H_{Sys} = total head requirement of the system, H_m = maximum friction loss in suction and delivery pipe (Table 8.10) represents the values of friction loss in galvanized iron (G I) pipe for various discharges and pipe sizes); H_i = elevation difference between the pump and the junction of the lateral and the main (called as delivery head); H_s = elevation difference between the pump and the sources of water after drawdown called as suction head and $V^2/2g$ is called as delivery head.

Table 8.10: Head loss due to suction in *GI* pipe

DischargeLit/sec.	Pipe diameter, centimeters										
	4.0	5.0	6.0	7.0	8.0	10.0	12.5	15.0	20.0	25.0	30.0
2.0	15.5	4.5	1.70	0.73	0.37	-	-	-	-	-	-
3.0	32.0	10.0	3.80	1.68	0.83	0.25	-	-	-	-	-
4.0	56.0	17.5	7.30	3.00	1.50	0.45	0.13	-	-	-	-
4.5	71.5	22.5	8.80	3.80	1.85	0.55	0.17	-	-	-	-
5.0	87.0	28.0	10.80	4.70	2.30	0.68	0.22	-	-	-	-
5.5	-	33.0	12.40	5.70	2.70	0.83	0.26	0.095	-	-	-
6.0	-	40.0	15.50	6.80	3.20	0.96	0.32	0.118	-	-	-
6.5	-	47.0	18.30	8.00	3.80	1.15	0.36	0.140	-	-	-
7.0	-	54.0	21.50	9.30	4.50	1.30	0.42	0.17	-	-	-
7.5	-	62.0	24.00	10.60	5.20	1.50	0.47	0.18	-	-	-
8.0	-	70.0	28.00	11.60	6.00	1.80	0.55	0.21	-	-	-
8.5	-	80.0	31.00	13.30	6.80	2.00	0.62	0.23	-	-	-
9.0	-	90.0	36.00	15.00	7.50	2.20	0.68	0.27	-	-	-
9.5	-	100.0	38.00	17.00	8.30	2.50	0.76	0.29	-	-	-
10.0	-	-	43.00	19.00	9.40	2.80	0.85	0.32	0.065	-	-
12	-	-	63.00	27.00	13.00	4.00	1.23	0.47	0.10	-	-
14	-	-	78.00	37.00	18.00	5.50	1.65	0.63	0.13	-	-
16	-	-	-	47.00	23.00	7.20	2.20	0.83	0.17	0.055	-
18	-	-	-	60.00	30.00	9.00	2.80	10.5	0.22	0.68	-
20	-	-	-	72.00	37.00	11.00	3.31	1.30	0.27	0.080	-
22	-	-	-	86.00	45.00	13.50	4.10	1.60	0.33	1.100	0.038
24	-	-	-	-	52.00	16.00	4.80	1.90	0.40	0.120	0.045
26	-	-	-	-	62.00	18.00	5.60	2.20	0.47	0.145	0.055
28	-	-	-	-	72.00	22.00	6.60	2.50	0.55	0.165	0.060
30	-	-	-	-	80.00	25.00	7.50	2.80	0.63	0.19	0.070
35	-	-	-	-	-	33.00	10.30	4.00	0.85	0.27	0.095
40	-	-	-	-	-	-	13.50	5.20	1.10	1.10	0.128
45	-	-	-	-	-	56.00	17.00	6.50	1.40	0.43	0.16
50	-	-	-	-	-	70.00	21.00	8.30	1.70	0.55	0.20

When the total design head and rate of pumping is known, the capacity of the motor to run the pump can be calculated by the formula:

$$HP = \frac{Q_{pump} \, x \, H_t}{75 \, x \, \eta} \tag{8.41}$$

where,

Q_{pump} = rate of pumping, lit/sec; H_t = total design head, m and = total efficiency which is the product of efficiency of drive, efficiency of pump and that of motor. Value of total efficiency, is assumed as 0.75.

Sample Calculation 8:

A level area (slope 0 percent) of 6 ha (200 m x 300 m) is to be irrigated by a sprinkler system. The crop in the area is groundnut and the soil type is silt loam. Field capacity (*FC)* and permanent wilting point (*PWP*) of the soil are 30 and 20 percent by weight. Bulk density of the soil is 1.5 gm/cc. The crop is to be irrigated when the moisture content reduces to 25 percent by weight. Root zone depth of the crop is assumed as 1 m. Peak evapotranspiration demand of the crop is 6.7 mm/day. Water application rate of the soil is 15 mm/hr. Application efficiency is 75 percent. Operating pressure of the sprinkler is 3 kg/cm^2. Sprinklers are spaced at 20 m along the lateral and the laterals originate perpendicularly from the main line with spacing between two laterals being 20 m. Height of the riser is 1m. Assume no rainfall and no groundwater contribution. Water is pumped from a well and fed to the main line. Pump is located at a distance of 5 m from the water level in the well and the pump is located at a distance of 20 m below the ground level having sprinkler settings. Assume the inner diameter of the sprinkler, main, suction and delivery pipe are 10, 12, 12.5 and 12.5 cm, respectively. Assume the lateral and main pipes are portable aluminum pipes and suction and delivery pipes are GI pipes. Assume any other data not given in the problems.

Solution:

Step I:

Depth of water to be applied in one irrigation, d is:

$$d = \frac{(FC - Optimum\ moisture\ content)\ xBD\ x\ D}{100}$$

$$= \frac{(30-25)\,x\,1.5\,x\,1}{100} = 0.075\,m = 75\,mm$$

Since, application efficiency is 75 percent, gross irrigation depth = $\frac{75}{0.75} = 100\,mm$

Step II: Given peak evapotranspiration, (ET_{peak}) = 6.7 mm/day

Irrigation interval = $\frac{Gross\ depth\ of irrigation}{ET_{peak}} = \frac{100}{6.7} = 15\ day$

Step III: Layout of Sprinkler Irrigation System:

Consider the layout of irrigation system as shown in figure below. There will be 15 laterals connected to the main with spacing of laterals at a distance of 20 m. Since the spacing of sprinkler along each lateral is 20 m, there will be 10 sprinklers. In addition the first and the last sprinkler in the lateral will be at a distance of 10 m from and within the field boundary. Similarly, the first and the last lateral along the main will be at a distance of 10 m from and within the field boundary as shown in figure above. The pump having suction and delivery pipe is connected to the main half way along its length i.e. there will be 150 m distance of the main either side of the central line having connection of suction and delivery pipe. For economic consideration, it is assumed that the lateral with sprinklers is arranged on wheels and is moved and fixed at the 15 locations on the main in 15 days to irrigate 6 ha area. Area irrigated each day is:

Step IV: System Capacity and Water Application Rate:

Assume that an area of 4000 m^2 is irrigated by a pump which works 12 hr per day.

(a) Main and Lateral:

A = 4000 m^2, Gross irrigation depth = 100 mm = 0.1 m and time of irrigation, t = 12 hr = 43200 sec.

Capacity of the System = Flow rate (Q) is:

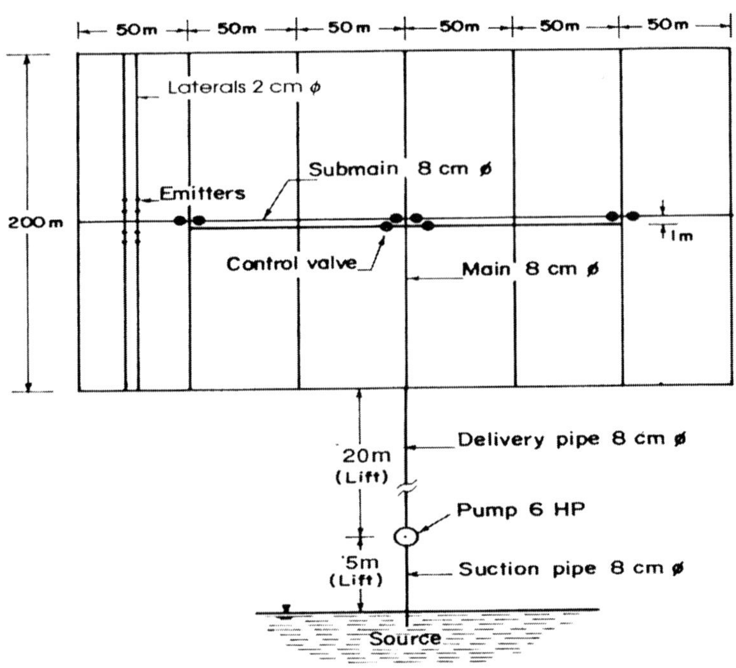

Sprinkler system layout

$$A = \frac{60000\, m^2}{15} = 4000\, m^2 = 200\, m\, x\, 20\, m$$

(b) Sprinkler Nozzle

There are 10 sprinklers with nozzles (each sprinkler having one nozzle) on each lateral.

Capacity of each sprinkler nozzle, q = 9.26/10 = 0.926 lit/sec

Area of water application by each sprinkler nozzle = 20 m x 20 m = 400 m^2.

Hence, water application rate of each sprinkler $\frac{0.926\, x\, 3600}{1000\, x\, 400} = 0.0083\, m/hr = 8.3\, mm/hr$ which is less than the given water application rate of the soil (15 mm/hr) and so the calculated application rate is acceptable.

Step V: Design of Sprinkler Nozzle:

Let the size (diameter) of the sprinkler nozzle be d and hence area of each sprinkler nozzle (a) is:

$$a = \frac{\pi\, d^2}{4}$$

The discharge of sprinkler nozzle, q = 0.926 lit/sec = 0.000926 m^3/sec

Operating pressure head of sprinkler, h = 3 kg/cm^2 = 30 m of water

Now area of sprinkler nozzle (a) can be computed as:

$$a = \frac{q}{C\sqrt{2gh}}$$

where,

C (coefficient of discharge) is assumed as 0.95.

$$a = \frac{0.000926}{0.95\, x \sqrt{2\, x\, 9.81\, x\, 30}} = 4.018\, x\, 10^{-5}\, m^2$$

Diameter of the sprinkler nozzle is = $d = \sqrt{\frac{4\, a}{\pi}}$

$$\boldsymbol{D} = \sqrt{\frac{4\, x\, 4.018\, x\, 10^{-5}}{\pi}} = 7.152\, x\, 10^{-3}\, m = 7.15\, mm$$

Step VI: Design of Laterals:

Length of the lateral, $L = 190$ m

Flow rate in lateral, $Q = 9.26$ lit/sec

Inner diameter of the lateral pipe, $D = 100$ mm

Head loss due to friction in each lateral, H_f is:

$$H_f = 4.10\ x\ 10^6 \frac{K_s\ L\ Q^{1.9}}{D^{4.9}} x\ F$$

Assume value of K_s for portable aluminum pipe as 0.4. There are 10 sprinklers on the lateral and hence from Table 8.9, value of correction factor for multiple outlet pipe flow, F is 0.363.

$$\text{So, } H_f = \frac{0.4\ x\ 190\ x\ 9.26^{1.9}\ x\ 0.363\ x\ 4.1\ x\ 10^6}{100^{4.9}} = 1.23\ m$$

Step VII. Design of Main:

Length of the main, $L = 140$ m

Flow rate in lateral, $Q = 9.26$ lit/sec

Inner diameter of the lateral pipe, $D = 120$ mm

Head loss due to friction in main, $H_{f,\ main}$ is:

$$H_f = \frac{0.4\ x\ 190\ x\ 9.26^{1.9}\ x\ 0.363\ x\ 4.1\ x\ 10^6}{100^{4.9}} = 1.23\ m\ x\ F$$

Assume value of K_s for portable aluminum pipe as 0.4. Main is not a multiple outlet pipe and so, $F = 1.0$ (Table 8.9).

$$\text{So, } H_{f,\,main} = \frac{0.4\ x\ 140\ x\ 9.26^{1.9}\ x\ 1.0\ x\ 4.1\ x\ 10^6}{120^{4.9}} = 1.02\ m$$

Step VIII. Total head required by the system, H_{Sys} is given as:

$$H_{Sys} = H_0 + H_f + H_{f,\,main} \pm H_e + H_r + H_{b,m}$$

Given, H_o = Operating pressure head, 3 kg/cm² = 30 m

H_f = Head loss due to friction in lateral = 1.23 m (calculated as above)

$H_{f,\ main}$ = Head loss due to friction in main = 1.02 m (calculated as above)

H_e = Elevation difference = 0 (since level field)

H_r = Riser height = 1 m (given)

$H_{b,m}$ = Head loss due to bend and other minor losses which is neglected.

Putting all these values, total head required by the system, H_{Sys} is calculated as:

H_{Sys} = 30 + 1.23 + 1.02 + 0 + 1 + 0 = 33.25 m

Step IX. Selection of Pump and Motor:

Total head of the pump, H_t is:

$$H_t = H_{Sys} + H_{ft} + H_j + H_s + \frac{V^2}{2g}$$

Given H_{Sys} = 33.25 m (calculated as above).

H_s and H_j are suction and delivery heads, respectively and the values are 5 and 20 m, respectively (given).

Given that suction and delivery pipe are GI pipes of 12.5 cm size. Friction loss in 12.5 cm diameter GI pipe for a flow rate of 9.26 lit/sec is read from Table 8.10 as 0.72 m/ 100 m length of pipe.

Head loss due to friction in both suction and delivery pipe of total length 25 m (H_{ft}) = (0.72/ 100) x 25 = 0.18 m

Velocity head, $$\frac{V^2}{2g} = \left(\frac{Q}{\frac{\pi d^2}{4}}\right)^2 x \frac{1}{2g}$$

$$= \left(\frac{0.00926}{\frac{\pi x 0.125^2}{4}}\right)^2 x \frac{1}{2 x 9.81} = 0.04\ m$$

Putting all these values, total head of the system (H_t) is computed as:

$H_t = 33.25 + 0.18 + 20 + 5 + 0.04 = 58.47\ m$

The discharge of the pump, Q = 9.26 lit/sec (Calculated). Assume the total efficiency of the pump and motor, η = 0.75.

Hence, *HP* of electric motor is calculated as:

$$HP = \frac{Q\ x\ H_t}{75\ x\ \eta} = \frac{9.26\ x\ 58.47}{75\ x\ 0.75} = 9.6 \approx 10$$

Connect two centrifugal pumps each driven by a 5 HP motor connected in series (as shown in figure above) for the sprinkler irrigation system.

8.6.7 Evaluation of Sprinkler Irrigation System

The basis objective of sprinkler irrigation is to supply water to the crop just like rainfall and meet the crop water requirement with maximum water utilization efficiency. This deepens upon how much water to be used at what rate to sprinkler and how best to distribute so that it will be uniform throughout the crop. The uniformity of distribution is regulated by the spacing of sprinkler, type of sprinkler nozzle and operating pressure. Low pressure result in large drops sprinkled away from the sprinkler and high pressure, on the other hand, breaks the water particles into smaller units. Wind velocity distorts the distribution pattern. Uniformity of distribution is expressed by distribution efficiency which is defined as:

$$E_d = (1 - \frac{\bar{y}}{\bar{d}})\, x\, 100 \tag{8.42}$$

where,

Ed = distribution efficiency; = average depth of water stored along the run during irrigation and = Average numerical deviation from.

In case of sprinkler irrigation uniformity is evaluated by calculating uniformity coefficient (*Cu*) as (Murty, 2004):

$$Cu = \left(1 - \frac{\{\sum |x|\}}{m\ n}\right) x\ 100 \tag{8.43}$$

where,

= absolute value of deviation of the individual observation of the depth applied from the mean value *m* and *n* = number of equally spaced observations of depth.

Data on uniformity coefficient are useful to select suitable combination of spacing, discharge, nozzle size, and operating pressure to get high irrigation efficiency.

Pattern efficiency can also be used to study the efficiency of the sprinkler irrigation system. Total depths of water from the sprinkler caught by each can is collected from which the average depth can be computed. Then the minimum depths of water collected by one-fourth of the total can used are recorded from which the average depth of water is computed. Then pattern efficiency, *Pe* is estimated as:

$$Pe = \frac{Average\ of\ lowest\ one - fourth\ of\ the\ cans}{Avearg\,e\ of\ all\ depth}\, x\ 100 \tag{8.44}$$

Sample Calculation 9:

Determine the uniformity coefficient of a sprinkler irrigation system with the following data of depth of water (cm) collected in each can in 1 hour. The field is square in size with dimensions of 24 m x 24 m. Cans are spaced at 6 m in the square field.

S	8.9	7.6	6.6	S
8.1	7.6	9.9	10.2	8.3
8.9	9.1	9.1	9.4	8.9
9.4	7.9	9.1	8.6	9.1
S	7.9	6.6	6.8	S

Note: S indicates the position of the sprinkler

Solution:

The computations are shown in the tabular form as follows:

Observation (Depth of water collected, cm)	Frequency, *f*	Depth x frequency	Absolute deviation of depth from average	Deviation x frequency
10.2	1	10.2	1.6	1.6
9.9	1	9.9	1.3	1.3
9.4	2	18.8	0.8	1.6
9.1	4	36.4	0.5	2.0
8.9	3	25.7	0.3	0.9
8.6	1	8.6	0.0	0.0
8.3	1	8.3	0.3	0.3
8.1	1	8.1	0.5	0.5
7.9	2	15.8	0.7	1.4
7.6	2	15.2	1.0	2.0
6.8	1	6.8	1.8	1.8
6.6	2	13.2	2.0	4.0
Total	21	178.0		17.7

Given total number of cans, $n = 21$

Average depth of water collected by all the 21 cans, $m = 178/21 = 8.5$

Sum of all the absolute deviation values of the individual depths from the average = 17.7

Hence uniformity coefficient, $Cu = 100 \times (1 - 17.7/178) = 90.05$ percent.

8.7 Trickle Irrigation

Trickle irrigation is the frequent, slow application of water either directly onto the land surface or into the root zone of the crop. Trickle irrigation is based on the fundamental concepts of irrigating only the root zone of the crops rather than irrigating the entire field, and maintaining the water content of the root zone at near optimum level. The system is designed to operate daily for nearly the entire day and to supply water to only the root zone of the crop.

Types of Trickle Irrigation

Trickle irrigation encompasses four methods of irrigation (James, 1988).

They are: (i) Drip irrigation (ii) Subsurface irrigation (iii) Bubbler irrigation and (iv) Spray irrigation.

8.8 Drip Irrigation

Drip irrigation is the slow, nearly continuous application of water as discrete drops. Water can be applied at single point on the land surfaces through devices called as emitter or as a line source from either closely spaced emitters or tubes with continuous or equally spaced openings that discharge water a drop at a time. Discharge rates for the point source emitters are generally less than 12 lit/hr and less than 12 lit/hr per metre length of lateral for line source emitters (James, 1988). Subsurface irrigation involves the use of point and line source emitters to apply water below the soil surface. Subsurface irrigation is different than the subirrigation, which is the method where in the root zone of the crop is irrigated through or by water table control. Discharge rate of subsurface irrigation is generally in the same range as the drip irrigation rate. In bubbler irrigation, water is applied to the land surface as a small stream. Water is delivered to the point of application in the root zone by tubes that are attached to the buried laterals. The tubes may be as large as 10 mm or more in size. The rate of discharge depends on the size of the tube and/or length. Since the size of the tube is large, bubbler system is less prone to clogging as compared to subsurface or drip irrigation. The discharge rate is also. Discharge rate is, however, less than 225 lit/sec. In spray irrigation, small sprinkler like devices (often called as micro-sprinkler) spray water as a mist over the land surface.

Micro-sprinklers can be spaced to cover the entire land surface as with conventional sprinkler systems. Discharge rate is usually less than 115 lit/hr. In this method, there is less chance to clogging compared to drip or subsurface irrigation (Rup, 2004; Dipsika, 2005). Losses due to wind and evaporation are more in this method. Among the various methods of trickle irrigation, drip irrigation is most popularly used and hence, in this book, only drip irrigation system is described.

8.8.1 Advantages of Drip Irrigation

Followings are the advantages of drip irrigation (Anonymous, 2009; Rao, 2006)).

(i) Water is directly applied to the root zone of the crop and hence deep percolation and surface runoff are reduced. This increases the application efficiency.

(ii) There is elimination of land leveling as required by the surface irrigation methods and hence cost of irrigation system is less.

(iii) Irrigation can be done in any type of soil, topography and even in steep slopes.

(iv) Irrigation can be done with poor quality water including saline water.

(v) Irrigation is frequent and low and maintains favourable moisture range always in the root zone. This saves a lot of irrigation water and increase water-use-efficiency.

(vi) Weed growth is less.

(vii) Along with irrigation water, fertilizer, herbicides and pesticides can also be applied to the crops. This increases the input use efficiency and reduces the cost of cultivation.

(viii) Intercultural operation can be done easily without any interruption.

(ix) Need for construction of irrigation and drainage channels as is required in surface irrigation is eliminated. This saves the land which otherwise would have been lost for construction of channels and this saved area can be used for cultivation.

(x) Quality of crops including fruits and vegetables are good.

(xi) Requires less operating pressure and so less energy for operation of the irrigation system as compared to sprinkler irrigation.

(xii) For widely spaced crops like frit trees, the system is less costly than sprinkler irrigation.

(xiii) It is possible to vary the supply as per the need of water by the plants at different growth stages.

8.8.2 Disadvantages of Drip Irrigation

Followings are the disadvantages:

(i) Initial cost is very high which limits for its large scale operation.

(ii) Requires high technical know-how than any other irrigation methods.

(iii) There is chance of clogging of emitters/drippers especially by chemical and biological materials. Clogging cause poor water distribution and if it continues for a long period, then crop may be damaged..

(iv) Maintenance cost is also high.

(v) A salt accumulation problem may occur in this system when only a portion of the root zone is wet and saline water is being used for irrigation. This may damage the crop.

(vi) Possibility of water leakage from pipes exists.

8.8.3 Adaptability of the System

The adaptability of the system may be described in the following ways.

(a) **Suitable to slopes:** Suitable to land slopes of 2 percent or less. Crops can be grown on contours in undulated lands.

(b) **Suitable soil:** Suitable to most of the soils. However, emitter discharge may be kept more for sandy soil as compared to clay soil.

(c) **Suitable to water:** Water should be free from suspended particles, biological and chemical pollutants.

(d) **Suitable to crops:** Suitable to all most all crops except some high water requiring crops like rice, jute etc. Better suited to fruits like apple, banana, grape, guava, lemon, mango, orange, papaya, pineapple, watermelon etc. Vegetables like brinjal, chilly, tomato, potato, onion, cabbage, cauliflower, cucumber, peas etc. flowers like jasmine, rose, cash crops like tea, coffee, rubber, sugarcane, cotton, tobacco, cocoa etc. and nuts like almond, coconut, cashew nut etc. respond well to drip irrigation and give better yields.

8.8.4 Components of Drip Irrigation System

The whole system of drip irrigation system constitutes the following sub systems:

A: Water source like open/tube well, ponds, canals, rivers, reservoirs etc.

B: Pumping system like an electric or diesel operated pump with motor assembly. Sometimes gravity flow may be used to convey water to the drip units.

C: Distribution system or pipe line system consisting of main, sub-main, lateral, emitters/drippers, micro-tubes etc.

D: Ancillary components like fittings, valves (by-pass valve, gate valve, control valve, flush valve, air release valve), filters, regulators, pressure gauge, ventury/fertilizer applicators etc.

Pumping System

If the water source is at a higher elevation than the crop field, water flows under gravity, otherwise a pump will be required to lift water from the water source and delivers it to pipe line system. Capacity and discharge pressure are the principal factors to select a pump. Pump capacity is selected on the basis of types of crops and their water requirement, area of crop coverage in a given time and system efficiency. Discharge pressure is calculated basing on the desired operating pressure, friction losses in the systems, water level in the source from the ground level and change of elevation within the field and fitting and bends used in the system. The water pressure required to be developed by the pump should be sufficient to maintain the desired pressure at the laterals. The laterals may be designed to operate under pressure as low as 0.15 to 0.20 kg/cm^2 and as high as 1.0 to 1.75 kg/cm^2. A pressure drop of 0.5 to 1.0 kg/cm2 may be anticipated in the head of the drip system including the filters. There is a further drop of pressure in the laterals and the emitters. The water coming out of the emitters is almost at atmospheric pressure. Volute centrifugal pumps, in general, and deep well turbine pumps, in particular, are used to lift water.

Fertilizer Tank

A fertilizer tank is provided at the head of the drip irrigation system for applying soluble fertilizer along with irrigation water directly at the root zone of the crops.

Filter

A filter unit which cleans the suspended impurities in the irrigation water so as to prevent blockage of holes and passages of drip nozzles is an essential part of drip system. The efficiency of drip irrigation system depends on the effective performance of the filter units. The filters used in the drip system are: (i) Hydro-cyclone filter (ii) Media sand filter (iii) Screen filter and (iv) Disc filter.

Hydro-cyclone filter also known as sand separator or centrifugal filter is used to separate heavier particles like sand, silt, stone but cannot separate algae, clay etc. It is used in the primary section where water source is pond, river, canal etc. Media or sand filter is connected to main pipe line to remove the suspended foreign particles, algae etc. from the water which will otherwise clog the drippers. It is filled with special type of coarse sand of size 0.30 to 0.54 mm in size. The sand filter is cleaned by backwash action. Almost 95 to 98 percent of impurities are filtered by sand filter. A screen filter of strainer size 20 to 200 mesh is used to filter rest of the impurities. Disc filter is another improved type of screen filter in which a series of disc are stacked on top of each other. Discs contain micro grooves equivalent to mesh size of 40 to 600 mesh which do the filter action to remove fine sand and organic matters.

Ventury

Ventury works by creating partial vacuum in the system and by suction of fertilizer solution in to the irrigation system. This vacuum is created by diverting a part of water flow the end passing it through a constriction which increases the velocity of flow. Hence, there is a pressure drop and fertilizer solution is sucked from the tank through a suction pipe into the ventury and then to the delivery system. It is made of plastic, having resistance to chemical and available in $3/4^{th}$ inch to 2 inch inlet/outlet size. Fertilizer tank, sometimes, is used instead of ventury.

Valve

By pass valve to control the water pressure, gate valve to create pressure difference for fertigation, control/ball valve to control the flow, flush valve for cleaning the main or sub-mains, air release valve to release the air at the time of start and break the vacuum during the stop are used in a drip system.

Pressure gauge

It is provided to indicate pressure at filter inlet and outlet.

Pipe Line System

The pipe line system consists mainly of a main line, sub main, laterals and emitters. The main line delivers water to the sub main and the sub main to the laterals. The emitters which are attached to the laterals distribute water to the plants. Layout of a typical drip irrigation system is presented in Fig. 8.16.

Main pipe line consist of PVC/HDPE pipes of size 25 to 75 mm. They may be buried in the soil at suitable depth so that they do not hamper in tillage operation. All filters, bypass valves, fertigation unit, pressure gauge, ventury etc. are attached to the main pipe line. Sub main is made up of PVC/HDPE pipes and convey water from the main to the laterals. They may also be buried inside the soil. Pipe diameters are selected based on total lateral length and crop water requirement. Ball valves are connected at the head end of sub main to maintain required pressure and flow in pipe and flush valves are connected at the tail end for cleaning of main and sub main pipes. They are generally available in 25 to 63 mm and commonly used are 40, 50 and 63 mm. Flexible LLDPE pipes are used to convey water from the sub main line to the emitter. They are openly spread on one/both sides of the sub mains and their spacing is decided on the basis of row to row distance of the crops. Each lateral may be provided end cap at the end so that no water from the lateral goes out. Lateral pipes are available in size of 12, 16 or 20 mm. Drippers/emitters/tricklers/drip nozzles are the most important component of the drip system. They receive water from the laterals and convey it to root zone of the crops. Fig. 8.17 presents the layout of the system in a farmer's field.

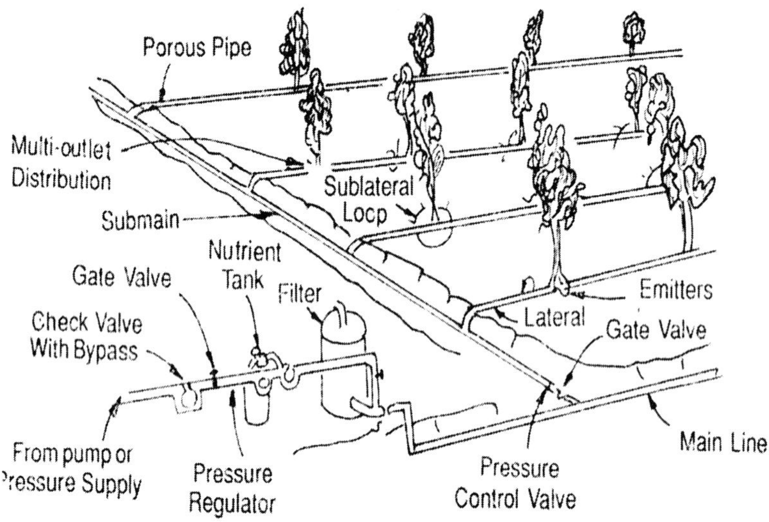

Fig. 8.16: Layout of a typical drip irrigation system

Drippers have very slow discharge varying from 2 to 14 lit/hr. The opening pressure at the emitter should be 1.5 kg/cm^2. Spacing of drippers on the laterals depends on the type of soil, crop and crop water requirement. Drippers are made up of injected hard plastics and are classified based on (i) emitting principle as orifice type, perforated pipe type, double wall type and long path type, (ii) operating pressure as low and high pressure type and (iii) discharge rate as low (2 to 4 lit/hr), medium (4 to 10 lit/hr) and high type (> 10 lit/hr). Sometimes micro tube of 1 mm inner diameter is used as emitters. Drippers can be opened and cleaned and then refixed in the laterals.

8.8.5 Design of Drip Irrigation System

Following points are to be considered for the design of drip irrigation system.

Step I. Inventory of Resources

In inventory of resources, the various information regarding location of water source (whether well or tank etc.), type of crops, topographic conditions, soil characteristics, climatic data are to be collected and analysed. Different crops require different amount of water and their spacing are different. The general layout of the system. Especially the emitters will depend on the type of the crop. It is necessary to know the general land slopes to determine the size and location of main and sub main, Various soil characteristics like infiltration rate of the soil, water holding capacity, texture, structure, and bulk density of the soil must be known while selecting emitter types, determining spacing and setting up the irrigation schedule. The climatic records like rainfall, temperature, humidity, evaporation etc. are to be gathered and analysed for determining the crop water requirement.

Fig. 8.17: Layout of drip system in farmer's field

Step II. Estimation of Depth of Irrigation, Irrigation Interval and Water Requirement of the Crop

An accurate irrigation scheduling in respect of estimating the depth of irrigation and frequency of irrigation during the period of peak water requirement of the crops are to be calculated for facilitating in design of drip irrigation. Details of estimating the depth and frequency of irrigation are discussed in Chapter 7 and estimation of water requirement of the crops is described in Chapter 5.

Step III. Calculation of System Capacity

With a proper selection of layout of the drip system, system capacity is calculated. While calculating the system capacity, it is assumed that not all the 100 percent area of the field is irrigated by drip system. Depending on the growth stages of the crops and types of crop with their spacing, percent of areas of the crop field to be served with drip vary. The system capacity comprises of computing the discharge of (i) main and sub main, (ii) lateral and (iii) emitter.

Discharge of main/sub main is computed by the formula:

$$Q = \frac{d\ x\ A}{f\ x\ t\ x\ E_a.} \tag{8.45}$$

where,

Q = discharge of main/sub main, m^3/sec; d = depth of irrigation, m; A = area to be irrigated by a main/sub main line, m^2; f = irrigation interval/frequency; t = time of operation in each irrigation interval, sec; and Ea = application efficiency in fraction.

After knowing the discharge of main/sub main, discharge of each lateral is calculated. Discharge of each lateral is calculated by dividing the discharge of main/sub main with the number of laterals arising from it. Discharge of emitter system is calculated as:

$$q = \frac{d\ x\ a}{t} \tag{8.46}$$

where,

q = discharge of emitter system, m^3/sec; d = depth of irrigation for each tree, m; a = wetted area for one tree served by irrigation water by an emitter, m^2 and t = duration of application, hr. If the discharge rate of a given emitter is specified, then the number of emitters required to be placed near each tree is calculated by dividing the discharge rate of each emitter with the discharge of emitter system calculated by Eq. (8.46) as above.

Step IV. Design of Lateral

The head loss due to friction in the lateral is estimated by the formula (Murty, 2004):

$$H_f = K\, L\, Q^{1.75}\, D^{-4.75}\, F \qquad (.8.47)$$

where,

H_f = head loss due to friction, m; K = a constant (7.85 x 10^5 for SI units for water at 20°C; L = length of pipe, m; Q = discharge of lateral, lit/sec; D = inner diameter of the pipe, mm and F = factor for multiple outlet pipe flow (Table 8.11).

Step V. Design of Main and Sub Main

Head loss due to friction in the main and sub main are calculated by the same Eq. (8.47) with respective values of *L, Q, D and F* (*F* value is taken from Table 8.11).

Step VI. Total Head Requirement of the System

Total head required by the system, H_{Sys} is summation of the (i) head required at the fartherest emitter to operate, (ii) head loss due to friction in lateral, (iii) head loss due to friction in sub main, (iv) head loss due to friction in main and (v) losses due to bend, filter and other minor losses.

Step VII. Total Head Required by the Pump

Total head required by the pump is given as:

$$H_t = H_{Sys} + H_s + H_d + H_{fs} + H_{fd} + \frac{V^2}{2g} \tag{8.48}$$

Table 8.11: Correction factor, *F* for friction loss in aluminum pipes with multiple outlets for drip system

Number of drippers	*F* values for exponent of velocity related term in friction head loss formula		
	n = 1.75	*n* = 1.80	*n* = 1.852
1	1.000	1.000	1.000
2	0.649	0.644	0.639
4	0.498	0.491	0.485
6	0.451	0.445	0.438
8	0.428	0.422	0.416
10	0.415	0.409	0.402
12	0.406	0.400	0.393
14	0.400	0.394	0.387
16	0.396	0.389	0.383
18	0.392	0.385	0.380
20	0.389	0.383	0.376
22	0.387	0.380	0.374
24	0.385	0.378	0.372
26	0.383	0.377	0.370
28	0.382	0.375	0.369
30	0.381	0.374	0.368
40	0.376	0.370	0.363
50	0.374	0.367	0.361

where,

H_t = total head required by the pump, m; H_{Sys} = total head requirement of the system, m (calculated in Step VI above); H_s *and* H_d = suction and delivery head, respectively, m; H_{fs} *and* H_{fd} = head loss in suction and delivery pipe, respectively, m and $V^2/2g$ = velocity head due to flow in suction/delivery pipe, m (V = velocity in m/sec and g = 9.81 m/sec^2). Table 8.10 represents the values of friction loss in suction and delivery pipe made up of galvanized iron (GI) for various discharges and pipe sizes.

Capacity of the motor to run the pump can be calculated by the formula:

$$HP = \frac{Q_{pump}\ x\ H_t}{75\ x\ \eta} \tag{8.49}$$

where,

Q_{pump} = rate of pumping (Discharge of the main), lit/sec; H_t = total design head, m and η =

Total efficiency which is the product of efficiency of drive, efficiency of pump and that of motor. Value of total efficiency, can be assumed as 0.75.

Sample Calculation 10:

A level area (slope 0 percent) of 6 ha (200 m x 300 m) is to be irrigated by a dripper system. The crop in the area is citrus with spacing 5 m x 5m and the soil type is silt loam. Field capacity (*FC*) and permanent wilting point (*PWP*) of the soil are 30 and 20 percent by weight. Bulk density of the soil is 1.5 gm/cc. The crop is to be irrigated when the moisture content reduces to 25 percent by weight. Root zone depth of the crop is assumed as 0.5 m. Peak evapotranspiration demand of the crop is 7 mm/day. Application efficiency is 90 percent. Operating pressure of the drip located at the farterest point is 0.5 kg/cm^2. Lateral spacing and emitter spacing are each 5 m. Assume no rainfall and no groundwater contribution. Water is pumped from a well and fed to the main line. Pump is located at a distance of 5 m from the water level in the well and the pump is located at a distance of 10 m below the ground level having emitter settings. Assume the inner diameter of the lateral, sub main and main are 20, 80, and 80 mm, respectively. Assume the lateral, main and sub main pipes are portable aluminum pipes and suction and delivery pipes are GI pipes. Suction and delivery pipes are each 80 mm in diameter. Assume any other data not given in the problems.

Solution:

Step I:

Depth of water to be applied in one irrigation, d is:

$$d = \frac{(FC - Optimum\ moisture\ content)\ xBD\ x\ D}{100}$$

$$= \frac{(30 - 25)\ x\ 1.5\ x\ 0.5}{100} = 0.0375\ m = 37.5\ mm$$

Since, application efficiency is 90 percent, gross irrigation depth = $\frac{37.5}{0.90} = 42.0\ mm$

Step II. Given peak evapotranspiration, (ET_{peak}) = 7 mm/day

Irrigation interval = $\frac{Gross\ depth\ of irrigation}{ET_{peak}} = \frac{42.0}{7} = 6\ days$

Step III. System Capacity:

Let us assume that 6 ha area is divided into 6 strips each of 1 ha area (200 m x 50 m) and assume that one of the six parts will be irrigated in 10 hours in one day. Also assume that the fraction of total area wetted by drip = 40 percent. The main line with connection of suction and delivery pipe runs at the centre of the 6 ha field and connects to two sub mains at the centre such that each sub main runs in either side of the main. Each sub main connects to the laterals on either side. Thus, there will be total 60 laterals served by each sub main. Layout of drip irrigation system is shown in figure below.

Discharge of Main

Area irrigated by the main, A = 0.40 x 1 = 0.4 ha = 4000 m^2.

Irrigation interval/frequency, f = 1;

Time of operation in each irrigation interval, t = 10 hr = 36000 sec and

Application efficiency in fraction, Ea = 0.90.

Depth of irrigation, d = 42 mm = 0.042 m.

Discharge of main and sub main is given by Eq. (8.45).

$$Q = \frac{0.0375\, x\, 4000}{1\, x\, 36000\, x\, 0.90} = 0.00467\, m^3 / \sec = 4.67\, lit / \sec$$

Discharge of Lateral

In 1 ha area, there are 20 laterals (10 in each side of the sub main having lateral length each equal to 98 m (Refer to the layout diagram). Discharge of each lateral = 4.67/20 = 0.234 lit/sec.

Discharge of Emitter

Depth of irrigation for each tree, d = 42 mm = 0.042 m (considering application efficiency of 90 percent)

Wetted area for one tree served by irrigation water by an emitter, a = 5 m x 5 m = 25 m^2. $m^{2.}$ Assume 40 percent area is wetted by the drip system.

Duration of application, t = 10 hr.

Hence discharge of the emitter system, q is given by Eq. (8.46) as:

$$q = \frac{0.042\, x\, (25\, x\, 0.4)}{10} = 0.042\, m^3 / \sec = 42\, lit / hr$$

If the discharge of each emitter available in the market is 14 lit/hr then each tree will require 3 emitters so that the crop water requirement of each tree will be fulfilled.

Step IV: Design of Lateral:

Length of each lateral, L = 98 m.

Discharge of each lateral, Q = 0.234 lit/sec.

Lateral is a multiple outlet pipe with 40 outlets and hence value of F for n = 1.75 is 0.376 (Table. 8.11).

Value of constant, $K = 7.89 \times 10^5$ for SI unit for water at 20°C.

Inner diameter of the lateral pipe, $D = 20$ mm

Hence, head loss due to friction in lateral is calculated by Eq. (8.47) as:

$$q = \frac{0.042\, x\, (25\, x\, 0.4)}{10} = 0.042\, m^3 / \sec = 42\, lit / hr \; = 1.51 \text{ m}$$

Step V. Design of Sub Main:

There are two sub mains originating from the main (one on either side of the main). Length of each sub main from the main to the fartherest lateral, $L = 148$ m.

Discharge rate of each sub main, $Q = 4.67$ lit/sec

Sub main is also a multiple outlet pipe with 20 outlets and hence value of F for $n = 1.75$ is 0.389 (Table 8.11).

Value of constant, $K = 7.89 \times 10^5$ for SI unit for water at 20°C.

Inner diameter of the sub main, $D = 80$ mm

Hence, head loss due to friction in sub main is calculated by Eq. (8.47) as:

$$H_f = 7.89\, x\, 10^5\, x\, 148\, x\, 4.67^{1.75}\, x\, 80^{-4.74}\, x\, 0.389 = 0.62\, m$$

Step VI. Design of Main:

Maximum length of the main, $L = 100$ m

Discharge rate of the main, $Q = 4.67$ lit/sec

Main is not a multiple outlet pipe and so, $F = 1.0$ (Table 8.11).

Inner diameter of the main, $D = 80$ mm

Value of constant, $K = 7.89 \times 10^5$ for SI unit for water at 20°C

Hence, head loss due to friction in main is calculated by Eq. (8.47) as:

$$H_f = 7.89\, x\, 10^5\, x\, 100\, x\, 4.67^{1.75}\, x\, 80^{-4.74}\, x\, 1.0 = 1.58\, m$$

Step VII. Total Head Requirement of the System:

Total head required by the system, H_{Sys} is summation of the (i) head required at the fartherest emitter to operate, (ii) head loss due to friction in lateral, (iii) head loss due to friction in sub main, (iv) head loss due to friction in main and (v) losses due to bend, filter and other minor losses.

It is given that the head required to operate the emitter at the fartherest end is 0.5 kg/cm^2 which is equal to 5 m of water.

Head loss due to friction in lateral, sub main and main are calculated as 1.51, 0.62 and 1.58 m, respectively.

Losses due to bend, filter and other minor losses are assumed as 1 m

Hence, $H_{Sys} = 5 + 1.51 + 0.62 + 1.58 + 1 = 9.71$ m

Step VIII. Selection of Pump:

It is given that suction head, $H_s = 5$ m

Delivery head, $H_d = 10$ m

Suction and delivery pipes are GI pipes of diameter 80 mm each. Referring to **Table 8.10** for a flow rate of 4.67 lit/sec, the friction loss in suction and delivery pipes of 80 mm G I pipes per 100 m length, are $H_{fs} = 1.90$ m *and* $H_{fd} = 1.90$ m

Head loss due to friction in both suction and delivery pipe of total length 15 m = (1.90/100) x 15 = 0.28 m

$H_{Sys} = 9.71$ m (calculated as above)

Velocity head,

$$\frac{V^2}{2g} = \left(\frac{Q}{\frac{\pi\, d^2}{4}}\right)^2 x \frac{1}{2g}$$

$$= \left(\frac{0.00467}{\frac{\pi\, x\, 0.08^2}{4}}\right)^2 x \frac{1}{2\, x\, 9.81} = 0.04\, m$$

Putting all these values, total head of the pump (H_t) is computed as (Eq 8.48):

$$H_t = 9.71 + 5 + 10 + 0.28 + 0.04 = 25.03\, m$$

The discharge of the pump, Q = 4.67 lit/sec (calculated).

Assume the total efficiency of the pump and motor, $\eta = 0.75$.

Hence, *HP* of electric motor is calculated as:

$$HP = \frac{Q\, x\, H_t}{75\, x\, \eta} = \frac{4.67\, x\, 25.03}{75\, x\, 0.75} = 2.08 \approx 2.0$$

Design charts and detail design procedure of lateral, sub main and main lines have been developed and are presented in the book "Irrigation: Theory and Practices" (Michael, 1978).

8.8.6 Emission Uniformity along Laterals

In case of drip irrigation system, head loss due to pipe friction, changes in ground surface slope as well as change in temperature of the water can cause variation in the emitter discharge along a lateral. Uniformity of emitter discharge is estimated by the formula:

$$Cu = \left(1 - \frac{\{\sum |d|\}}{N\,q}\right) x\,100 \qquad (8.50)$$

where

$|d|$ = absolute value of deviation of the individual observation of the nozzle discharge from the mean of discharges of all the nozzles along a lateral, q and N = number of equally spaced emitters along a lateral.

8.8.7 Evaluation of Drip Irrigation System

The wetting pattern in the soil profile of an individual nozzle is determined by operating a nozzle under actual field condition for various time periods. The soil moisture content at various depths and at various radial distance from the plant served by irrigation by a nozzle are measured at various time intervals. These readings of soil moisture contents are plotted to study the vertical as well as horizontal moisture distribution pattern. The effective area of coverage of the nozzle will determine the spacing of the nozzles and the number of nozzles required to wet the root spread area of the crop. While evaluating the performance of the drip irrigation on crop response various yield and yield attributing factors are to be collected and the effect of drip irrigation system on them are to be studied. Economics of the irrigation system should also be worked out with the cost of installation of the system and returns obtained from increased yield of the system. Benefit cost ratio should be more than 1.0, for the system to be economically feasible.

Questions Banks

Q1. What is the various methods of irrigation. Give a flow chart to show the application of irrigation water in various ways to crops.

Q2. What is the difference between surface and submain irrigation?

Q3. Briefly summaries the different methods of surface irrigation.

Q4. What do you mean by wild flooding? What are its merits and demerits?

Q5. Differentiate between the followings:
(i) Wild flooding and free flooding (ii) Straight and contour border (iii) Border strip and furrow irrigation (iv) Sprinkler and drip irrigation (v) Straight and corrugated furrow (vi) Straight and contour furrow.

Q6. What are the factors on which the length of the border depend on?

Q7. How does the soil and topography affect border irrigation?

Q8. What are the advantages and disadvantages of border method of irrigation>

Q9. Derive an expression for calculating the water front advance in a border strip.

Q10. What do you understand by recession flow? Draw a graph of recession and advance curve in a border method of irrigation and explain how the opportunity time is computed from it.

Q11. Describe the various design procedures of border irrigation system.

Q12. In a border strip experiment the infiltration characteristic is given by the equation $y = 0.70\, t^{0.5} + 0.6$. The irrigation applied for 1hour with constant rate of flow per unit width of the channel at the upstream side is q = 1.6 lpm/cm of border strip. The mean depth of f low over the soil surface is 6 cm. Find out the distance of water front advance.

Q13. In a border strip experiment the infiltration characteristic is given by the equation $y = 0.75\, t^{0.6} + 0.5$. The irrigation applied for 2 hour with constant rate of flow per unit width of the channel at the upstream side is q = 1.5 lpm/cm of border strip. The mean depth of f low over the soil surface is 7cm. Find out the distance of water front advance in 1 hou. Find out also the water front advance in 10, 15, 20, 30, 40 and 50 min. Draw water front advance curves.

Q14. What do you mean by check basin method of irrigation? What are its various types based on size and shape?

Q15. What are the limitations of check basin irrigation? Explain the various design factors of check basin irrigation?

Q16. Write notes on hydraulics of check basin irrigation.

Q17. What do you mean by a cutback stream? What is its importance in saving of irrigation water.

Q18. Derive an expression for calculating the application efficiency of check basin irrigation.

Q19. A check basin of size 10 m × 50 m is to be irrigated with total depth of irrigation of 7.5 cm. Accumulated infiltration time relationship is given as $d = 9.95\, t_d^{0.40}$ where d is accumulated infiltration, mm and t_d is time, min. Water front advance equation is expressed as $l = 5.6\, t_l^{0.6}$ for a stream size of 1.5 lit/sec/m width of the check basin. Find out the water application efficiency. Find out also the time required to flow into the check basin with the given stream size

Q20. A check basin of size 12 m x 12 m is to be irrigated by a stream of size 10 lit/sec. The depth of the effective root zone is 1 m. This 1 m depth of the root zone is divided into 4 profile layers and the following values of soil moisture content and other data of the basin in these four layers are recorded as below.

Depth of profile layer in (cm)	Moisture % on oven-dry basis		Apparent specific gravity, gm/cc
	Field capacity	Actual	
0 – 15	25.0	17.0	1.40
15 – 30	24.0	17.5	1.50
30 – 60	22.3	19.2	1.51
60 – 100	22.0	20.1	1.55

How long should the irrigation stream be applied to the basin to replenish the soil moisture to field capacity.

Q21. What are the adaptability factors of furrow irrigation? Write notes on characteristics and types of furrow irrigation.

Q22. Write down the various advantages of furrow irrigation.

Q23. What are the essential features of a corrugation irrigation?

Q24. Describe the design considerations of furrow irrigation with limiting considerations.

Q25. What are the factors that influence the furrow length?

Q26. A stream of size 2.2 lit/sec is initially used to irrigate a furrow of length 100 m and furrow spacing is 50 cm. When the initial furrow stream reached the tail end of the furrow, it was cut back and the cut back stream was 0.5 lit/sec. The initial furrow stream took 60 min to reach the tail end of the furrow. The cut back stream continued to irrigate the furrow for 50 min. Estimate the average depth of flow.

Q27. Narrate the procedure for evaluation of furrow irrigation in fields.

Q28. Why sprinkler irrigation is called as over head irrigation? Which types of crops can be irrigated by this irrigation?

Q29. What are the adaptability of sprinkler irrigation?

Q30. Following data were recorded from a test furrow in a sandy soil.

Stream size, lit/min	Distance, m	Advance time, min	Wetted perimeter, cm	Furrow cross sectional area at corresponding depth of flow, cm^2
120.0	20	2.00	25.0	70.0
	40	5.80	25.8	95.0
	60	11.00	26.0	105.0
	80	18.00	26.8	108.0
	100	23.50	27.0	113.0

Compute furrow infiltration.

Q31. Write down the various advantages and disadvantages of sprinkler irrigation.

Q32. What are the various classifications of sprinkler irrigation. Explain the working principle of a rotating head sprinkler.

Q33. How does a perforated type sprinkler system work?

Q34. Draw a lay out diagram of sprinkler irrigation system and describe the various components of the system.

Q35. Describe the various design procedures of sprinkler irrigation system

Q36. Determine the capacity of a sprinkler irrigation system to apply water at the arte of 1.5 cm/hr. Two 180 m long sprinkler lines are required. Sixteen sprinklers are placed at 12 m intervals on each line. The spacing between the lines is 18 m.

Q37. Explain the hydraulic design of sprinkler irrigation system?

Q38. What is the formula to compute the capacity of a pump to supply water to a sprinkler system? How can you find out the various parameters required to compute the capacity of the pump?

Q39. A levelfield of 6 ha (200 m x 300 m) is to be irrigated by a sprinkler system. The crop in the area is sunflower and the soil type is sandy loam. Field capacity (*FC*) and permanent wilting point (*PWP*) of the soil are 32 and 22 percent by weight. Bulk density of the soil is 1.5 gm/cc. The crop is to be irrigated when the moisture content reduces to 25 percent by weight. Root zone depth of the crop is assumed as

0.9 m. Peak evapotranspiration demand of the crop is 7 mm/day. Water application rate of the soil is 13 mm/hr. Application efficiency is 80 percent. Operating pressure of the sprinkler is 4 kg/cm^2. Sprinklers are spaced at 20 m along the lateral and the laterals originate perpendicularly from the main line with spacing between two laterals being 20 m. Height of the riser is 0.8 m. Water is pumped from a well and fed to the main line. Suction and delivery heads are 5 and 20 m, respectively. Assume the inner diameter of the sprinkler, main, suction and delivery pipe are 10, 12, 12.5 and 12.5 cm, respectively. Assume the lateral and main pipes are portable aluminum pipes and suction and delivery pipes are GI pipes. Assume any other data not given in the.

Q40. How does the wind velocity affect the uniform application of sprinkler irrigation?

Q41. How will you evaluate a sprinkler irrigation system?

Q42. What is pattern efficiency of sprinkler irrigation system? How can it be estimated?

Q43. Determine the uniformity coefficient of a sprinkler irrigation system with the following data of depth of water (cm) collected in each can in 2 hour. The field is square in size with dimensions of 24 m x 24 m. Cans are spaced at 6 m in the square field.

S	8.8	8.0	6.8	S
8.5	7.8	9.9	10.5	8.5
9.0	9.5	9.1	9.5	8.9
9.4	7.9	9.1	8.6	10.0
S	7.9	6.8	7.0	S

Q44. What is trickle irrigation? How does it differ from drip irrigation?

Q45. What are the various types of trickle irrigation? Describe the various types of trickle irrigation.

Q46. Write down the benefits of drip irrigation.

Q47. What are the adaptability conditions of drip irrigation?

Q48. What are the components of drip irrigation? Describe the various components.

Q49. What are the various ancillary components of drip system? Describe their functions.

Q50. Describe the functions of various filters in drip system.

Q51. Discuss the various design procedures of drip irrigation system.

Q52. A level field of 6 ha (200 m x 300 m) is to be irrigated by a dripper system. The crop in the area is papaya with spacing 4 m x 4 m and the soil type is silt loam. Field capacity (*FC*) and permanent wilting point (*PWP*) of the soil are 28 and 18 percent by weight. Bulk density of the soil is 1.5 gm/cc. The crop is to be irrigated when the moisture content reduces to 20 percent by weight. Root zone depth of the crop is assumed as 0.8 m. Peak evapotranspiration demand of the crop is 6.5 mm/day. Application efficiency is 90 percent. Operating pressure of the drip located at the farterest point is 1.0 kg/cm^2. Lateral spacing and emitter spacing are each 4 m. Assume no rainfall and no groundwater contribution. Water is pumped from a well and fed to the main line. Pump is located at a distance of 5 m from the water level in the well and the pump is located at a distance of 15 m below the ground level having emitter settings. Assume the inner diameter of the lateral, sub main and main are 20, 80, and 80 mm, respectively. Assume the lateral, main and sub main pipes are portable aluminum pipes and suction and delivery pipes are GI pipes. Assume any other data not given in the problems.

Q53. How can you compute the uniformity coefficient of drippers? What is its impact on assessing the performance of drippers?

Q54. How can the performance of drip irrigation system be evaluated? Under what condition, the drip irrigation system is said to be economically feasible.

References

Anonymous 2009. Manual on Micro-Irrigation Systems-Concepts, Design, Installation and Maintenance. Jain Irrigation Systems Ltd., Jalgaon.

Dipsika, P. 2005. Sub-surface Drip Irrigation in Sand and Gravel Tube. M. Tech Thesis, Submitted to Department of Soil & Water Cons. Engg., CAET, OUAT, Bhubaneswar.

Hall, W.A. 1960. Design of irrigation border checks. Agricultural Engineering, Vol. 41: 439-442.

Hansen, V.E., Israelsen, O.W. and Stringham, G.E. 1979. Irrigation Principles and Practices. John Wiley and Sons Inc., New York.

James, L.G. 1988. Principles of Farm Irrigation System Design. John Wiley & Sons, New York, pp. 543.

Lenka, D. 2001. Irrigation and Drainage. Kalyani Publishers, New Delhi.

Lewis, M.R and Milne, W.F. 1938. Analysis of border irrigation. Agricultural Engineering, Vol. 19: 267-272.

Michael, A.M. 1978. Irrigation Theory and Practices. Vikash Publishing House Pvt. Ltd., New Delhi.

Michael, A.M. and Pandya, A.C. 1971. Water front advance in irrigation borders. Journal of Agricultural Engineering Research, Vol. 16, No. 1: 62-71.

Murty, V.V.N. 2004. Land and Water Management Engineering. Kalyani Publishers, Cuttack.

Panigrahi, B. 2011. Irrigation Systems Enginering. New India Publishinh Agency, New Delhi.

Patnaik, P.K. and Mohanty, K.S. 2010. Training Report on Study of Micro Irrigation Systems. Submitted to Department of Soil & Water Cons. Engg., CAET, OUAT, Bhubaneswar.

Pillsbury, A.F. 1968. Sprinkler Irrigation. FAO Publication, 179 pp.

Punmia, B.C. and Lal, P.B.B. 1992. Irrigation and Water Power Engineering. Laxmi Publications Pvt. Ltd., New Delhi.

Rao, T.S. 2006. Study on Sub-surface Drip Irrigation. M. Tech Thesis, Submitted to Department of Soil & Water Cons. Engg., CAET, OUAT, Bhubaneswar.

Rup, A.R. 2004. Comparative Study on Different Micro-Irrigation Systems in Tuber Rose. M. Tech Thesis, Submitted to Department of Soil & Water Cons. Engg., CAET, OUAT, Bhubaneswar.

9

Conveyance of Irrigation Water

9.1 Introduction

In order to increase the production and productivity of crops, provision of irrigation system is highly essential. A project that makes use of weir or a barrage for providing irrigation to fields is called as direct irrigation system. On the other hand, a storage irrigation system makes use of storage reservoir by construction of dams. In order to provide irrigation by both the schemes, it is necessary to have a network of canal systems. A canal is an artificial channel, generally trapezoidal in shape constructed on the ground to carry water to the fields. The canal systems that provide water to the fields include main canals, branch canals, distributaries, minors and watercourses. It is imperative to design the canal systems properly for a certain realistic value of peak discharge that must pass through them so as to provide sufficient irrigation to the command areas. The flow irrigation that provides irrigation water to the fields may be of two types i.e. Inundation irrigation and Perennial irrigation.

9.2 Classification of Canals

(a) Classification based on type of soil over which it moves:

(i) Alluvial canal

(ii) Non-alluvial canal

Alluvial Canals

The soil which is formed by continuous silt deposition is called alluvial soil. The alluvial soil is even and is having a flat surface slope. Hard foundations are generally not available in this kind of soil. The rivers flowing through such alluvial areas have a tendency to shift their courses. The river bed consists of sand of considerable thickness, and is, therefore, permeable. When ever, an irrigation structure is to be constructed on such a river, special precautions and design method is to be adopted. The canals when excavated through such soils are called alluvial canals. Canal irrigation is generally preferred in such areas as compared to storage irrigation.

Non-Alluvial Canals

Mountainous and plateau regions may suffer from weathering. Weathering over a period of time result in the formation of a rocky plain area, called non-alluvial area. The rivers passing through such areas have no tendency to shift their courses and they do not pose many problems for designing irrigation structures on them. Canals, passing through such areas are called non-alluvial canals. Storage irrigation is preferred to canal irrigation in this type of areas.

(b) Classification based on the nature of source of supply

(i) Permanent canal

(ii) Inundation canal

Permanent Canal

A canal is said to be permanent when it is fed by a permanent source of supply. The permanent canals arc madc up of masonry works having well planned regulation and distribution systems. A permanent canal is also sometimes called as perennial canal when the source of water feeding to it is from a perennial river preferably having snow fed.

Inundation Canal

Inundation canals usually draw their water supplies from the rivers whenever there is a high stage in the river. They are not provided with any headwork for diversion of river water to the canal. They may be provided with a canal head regulator.

(c) Classification based on the function of the canals

(i) Irrigation canal

(ii) Carrier canal

(iii) Feeder canal

(iv) Navigational canal

(v) Power canal

(d) Classification based on discharge

(i) Main canal

(ii) Branch canal

(iii) Distributary

(iv) Minor

(v) Water course

(e) Classification based on alignment

(i) Watershed canal

(ii) Contour canal

(iii) Side slope canal

9.3 Canal Alignment

A canal has to be aligned in such a way that it covers the entire area proposed to be irrigated with shortest possible length and at the same time its cost must be minimum. According to alignment irrigation canals can be of three types as mentioned above i.e.

(i) Watershed canal

(ii) Contour canal

(iii) Side-slope canal

(i) Watershed Canal

Watershed canal is also called as ridge canal. The dividing line between the catchment area of two drains is called the watershed (Fig. 9.1)

For canal system in plain areas, it is advantageous to align all channels on the watersheds of the areas. The canal which is aligned along any natural watershed is called a watershed canal.

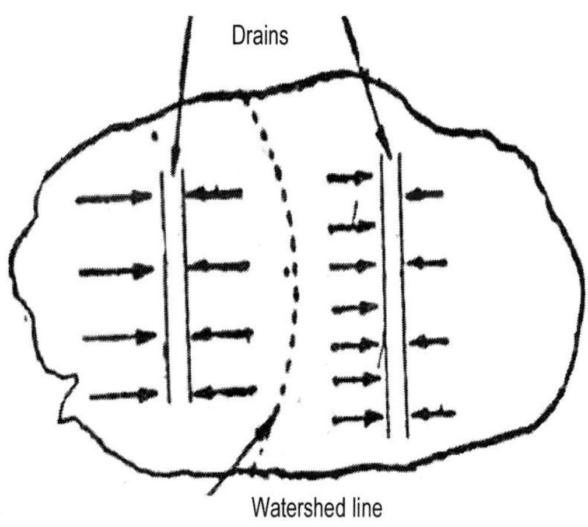

Fig.9.1: Alignment of a watershed canal along the watershed

In such a canal, water flows by gravity, on either side of the canal, directly or through small irrigation channels. More, cross-drainage works are avoided as the natural drainage will never cross a watershed because all the drainage flows away from the watershed.

(ii) Contour Canal

In hilly areas, the river flows in the valley, while the watershed or the ridge line may be hundred of metres above it. It becomes uneconomical to take the canal on top of such a ridge. The channel, in such cases, is generally aligned parallel to the area, except that the longitudinal slopes required to generate sufficient flow velocities, are given to it (Fig. 9.2).

Contour canal system requires canal falls and cross drainage works at suitable places. A contour channel irrigates only on one side, because the areas on the other side are at higher elevation.

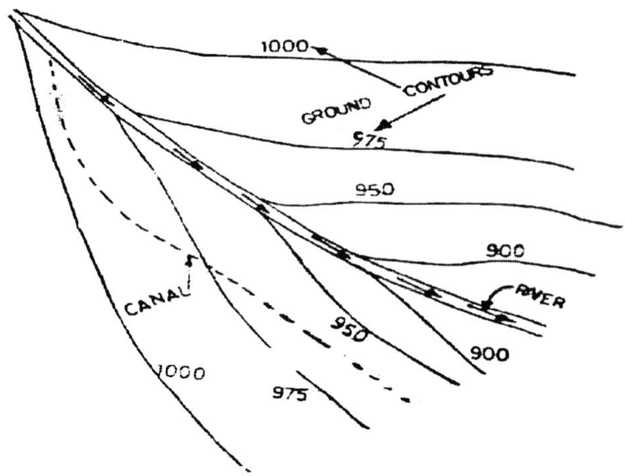

Fig. 9.2: Alignment of a contour canal

(iii.) Side Slope Canal

A side slope canal is that which is aligned at right angles to the contours, i.e. along the side slopes. Such a channel is parallel to the natural drainage flow (Fig. 9.3). It doesn't intercept cross drainage, and hence no cross drainage works are required.

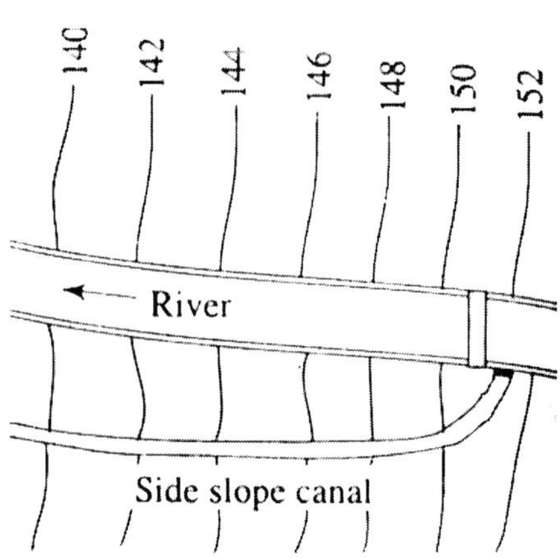

Fig. 9.3: Side slope channel

9.3.1 General Considerations for Canal Alignment

(i) The alignment of the canal should be such as to ensure the most economical way of distributing water to the field with minimum number of cross drainage works.

(ii) It should provide water to as high a command as possible.

(iii) It is preferable to align the canal on a ridge line i.e. watershed canal.

(iv) The length of the main canal from the point where it takes off from a river to a point where it mounts on a watershed should be minimum.

(v) The contour alignment should be changed in order to reduce the number of cross drainage works to a minimum.

(vi) The alignment should avoid villages, places of worship and other places of religious antiquity.

(vii) The alignment must pass through the balanced depth of cutting so that canal construction will be minimum.

(viii) The number of curves should be less. (Detail description of canal alignment in curves is discussed later on).

(ix) The alignment should not pass through rocky, brakish or cracked strata.

9.3.2 General Considerations for Alignment of a Field Channel

(i) They should be laid along the field boundaries.

(ii) They should be capable of supplying sufficient water to the tail end.

(iii) If the land is uneven, separate field channels should be provided to high and low lands.

(iv) The alignment of field channel should not pass through rocky, brakish or cracked strata.

9.4 Distribution System for Canal Irrigation

Irrigation water has to be supplied to the crop fields from the head regulator constructed over a storage reservoir or weir/barrage. The entire network of irrigation channels is called Canal System. The canal system as explained consists of:

(i) Main canal

(ii) Branch canal

(iii) Distributary

(iv) Minors

(v) Watercourses

In case of direct irrigation scheme, a weir or a barrage is constructed across the river, and water is stored on up-stream side. The arrangement is known as head works or diversion head works. Water is diverted into the main canal by means of a diversion weir. A head regulator is provided at the head of the main canal, so as to regulate the flow of water into the main canal. In storage irrigation scheme, a dam is constructed across the river, thus forming a reservoir on the upstream side of the river. The water from this reservoir is taken into the main canal through the outlet sluices. There are generally two main canals, which off-take from the reservoir, called Left Bank Canal and Right Bank Canal, as shown in Fig. 9.4 (Panigrahi, 2011).

(i) Main Canal (Head Reach)

The canal head works are generally situated in a valley, and the canal should mount the watershed in the shortest possible distance. The canal, in this reach, must be aligned very carefully and has to be generally excavated in deep cutting below natural ground surface level. Sometimes, it has to cross various drainage lines. In the portion below the head reach of the main canal, attempts are made to align the canal along the watershed and somewhat central to the commanded area. Sometimes, watershed has to be sacrificed to bypass town, villages etc. that come on the way.

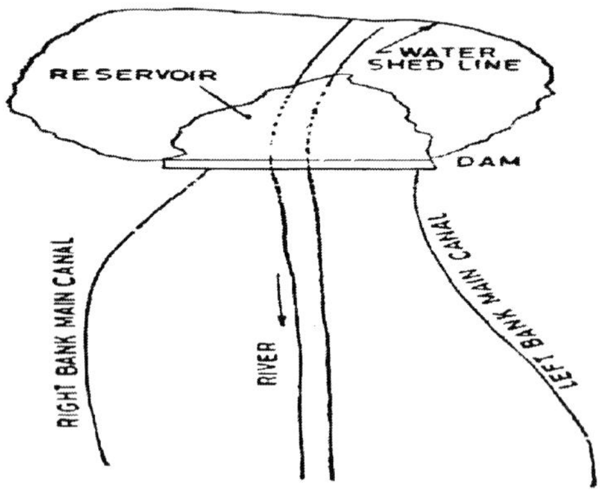

Fig. 9.4: Storage irrigation scheme

(ii) Branch Canal

The canals that originate from the main canal and take out water from it and supply to the smaller channels called as distributaries are called as branch canals.

(iii) Distributaries

Smaller channels which take off from the branch canals and distribute their supply through outlets into minors or watercourses, are called distributaries. They are aligned either as watershed channels or as side slope channels. Discharge in a distributary, is generally, less than 30 cumecs.

(iv) Minors

Sometimes, the fields in a command area are situated in far away distances such that the distance between the distributary outlet and field is very large. In such a case, small channels called minors are taken off from the distributaries, so as to supply water to the farmers at the point nearer to their fields. Discharge in a minor, is generally, less than 2.5 cumecs.

(v) Watercourses

These are small channels excavated by the farmers, so as to take water from the minors to their crop fields.

9.5 Curves in Channels

The channels should be so designed and aligned that is has less curves. But many a time the curves become inescapable. However, if curves in a channel are unavoidable, every care should be taken so that the curves have gentle slope. A curve cause disturbance of flow and results in silting on the inside (i.e convex side) and scouring on the outside (i.e concave side) (Fig. 9.5). Pitching is, therefore, sometimes proposed on the concave side so as to avoid scouring. If the discharge is more, the curve should be gentler and should, therefore, have more radius. The generally adopted minimum values of radii for different discharges are tabulated in Table 9.1 (Garg, 1987).

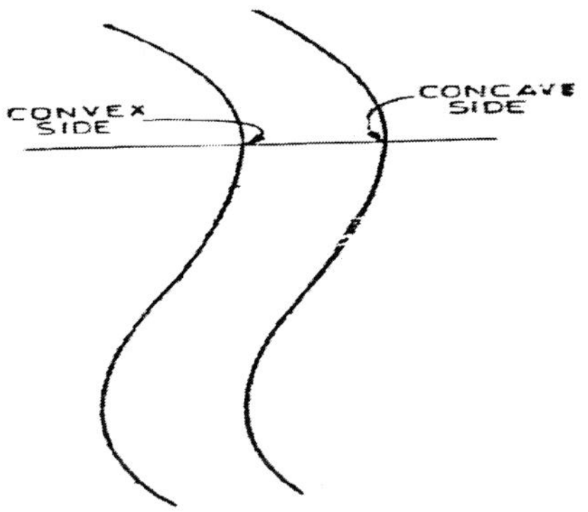

Fig. 9.5: Slope alignment in channel

Table 9.1: Minimum recommended values of curve radii of channel for different discharges

Discharge of channels, cumecs	Minimum recommended curve radius, m
< 0.5	100
0.5 – 3	150
3–15	3 00
15–30	600
30–100	900
> 100	1500

9.6 Advantages of Inundation Irrigation

The followings are the advantages of inundation irrigation.

(i) In this system, no hydraulic structure is required.

(ii) Maintenance cost is low.

(iii) Water carried by the canals contains silt which increases the fertility of the soil.

(iv) Canal runs in rainy season only and in other seasons, it remains dry. This eliminates the chances of water logging.

9.7 Disdvantages of Inundation Irrigation

(i) As there is no control over the flow of water over irrigation may spoil the crops.

(ii) Supply of water is uncertain.

(iii) Canal remains dry during the *rabi* season and hence no crop can be taken in *rabi.*

(iv) Head of canal may be damaged during time of flood.

(v) If river changes its path, the whole systems have to be abandoned.

(vi) Silting in canal is a problem.

9.8 Comparison between Inundation and Perennial Irrigation

Comparison between the two systems is mentioned as below:

	Inundation Irrigation	Perennial irrigation
1.	Irrigation is possible in rainy season only.	Irrigation is possible throughout the year.
2.	No hydraulic structure needed.	Hydraulic structures such as diversion headwork, cross-drainage works needed.
3.	Canal water carries silt that makes land fertile.	Canal water carries practically no silt and so chemical fertilizers are required
4.	Large area cannot be irrigated.	Large area can be irrigated.
5.	Silting of canal bed is a problem.	Silting is no problem.
6.	Water tax cannot be imposed.	Water tax can be imposed.
7.	Initial cost is low.	Initial cost is high.
8.	No technical knowledge is needed to operate the system.	Technical knowledge is needed to operate the system.
9.	Canal is not provided with regulator and so possibility of over irrigation is there.	Canal is provided with regulator and so there is no possibility of over irrigation.

9.9 System of Bandhara Irrigation

Bandhara irrigation is a system of minor irrigation suitable to irrigate isolated patches up to 500 ha. Bandhara is similar to weir which is constructed across a small stream to raise the water level on the up stream side to divert the water through the canal (Fig. 9.6, a). The height of the bandhara depends on the water level to be raised on the upstream side (Fig. 9.6,b).

It is constructed with brick masonry or stone machinery with R.C.C. crest. Width of the crest varies from 1 to 2 m. In order to remove the silt from the canal head, scouring sluices are provided at the bottom of the bandhara near the head reach. Surplus water is allowed to pass over the crest of the bandhara. In this system, water is directly taken from the main canal and supplied to the crop fields. The total area under the bandhara is called as *Thal* which is divided in to a number of small patches called as *Phad*. This system is also sometimes called as *Phad*

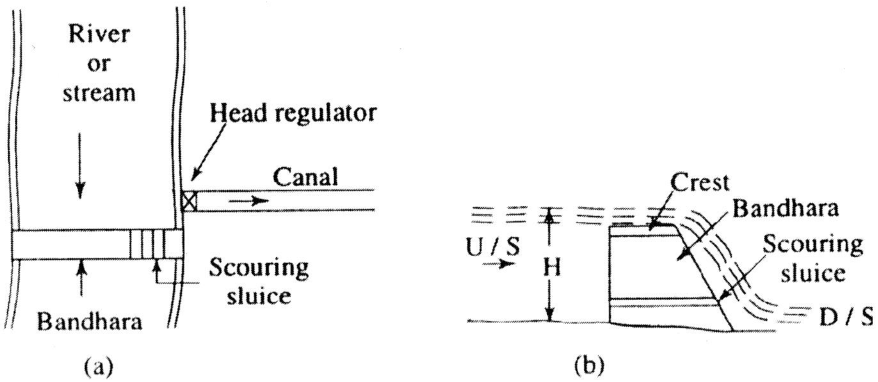

Fig. 9.6: Bandhara irrigation

irrigation system. The system is suitable for small streams. Sometimes, more than one bandhara may be constructed on the same stream at a reasonable interval to irrigate the fields (Basak, 2012). Fig. 9.7 presents a view of bandhara irrigation system.

Advantages and Disadvantages of Bandhara Irrigation

Followings are the various advantages.

(i) Water of small stream can be used for irrigation by constructing simple structures.

(ii) Cultivated area is generally close to the source. Hence, possibility of conveyance loss of water is less.

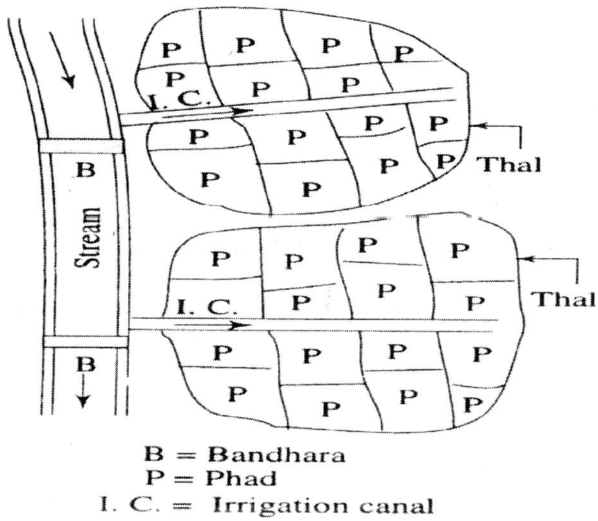

Fig. 9.7: Bandhara irrigation system

(iii) Duty of water is high as there is less loss due to conveyance.

Disadvantages are:

(i) The discharge carried by the bandhara system located adjacent to the stream is low. People residing at the tail end of the stream get less water for their uses.

(ii) Supply of water mainly depends on rainfall. So, during drought, the system is practically useless.

9.10 Canal Section

Depending on the natural ground surface and allowable bed slope of the canal, the canal section may be fully cutting or fully banking or partial cutting and partial banking. For all types of canal sections, there are several terms which need to be understood before designing and executing the canal related works. The terms are briefly mentioned below.

(a) Canal Bank

Canal bank helps to retain water in the canal to the full supply level. Section of the canal banks vary from site to site. The followings are the different forms for different site conditions.

(i) **When the canal is fully cutting type:** In this case, the banks are constructed on both sides of the canal which will work only as an inspection road. Height and top width of the bank are low and side slope is 1:1 or 2:1 according to the nature of the soil (Fig. 9.8).

(ii) **When the canal is partially cutting and banking type:** In this case, the banks are constructed on sides of the canal to retain water. The height of the bank depends on the full supply level of the canal (Fig. 9.9). Selection of canal also depends on the hydraulic gradient. The top width and the side slope of the bank should be such that the hydraulic gradient should have a minimum cover of 0.5 m from the heel of the bank.

(iii) **When the canal is full banking type:** In this case, canal and both the canal banks are constructed above the ground surface. The height of the bank required is high and its section is large due to the hydraulic gradient. A core wall of puddle clay is provided at the centre of the bank which helps the phreatic lines to cross the bank within the base width and thereby making the bank cross section minimum (Fig. 9.10).

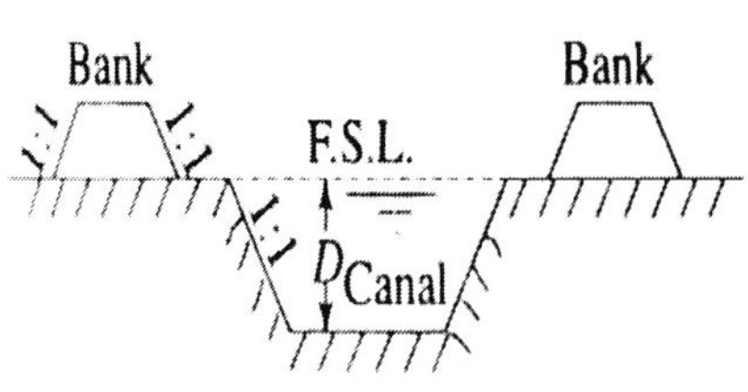

Fig. 9.8: Canal in full cutting

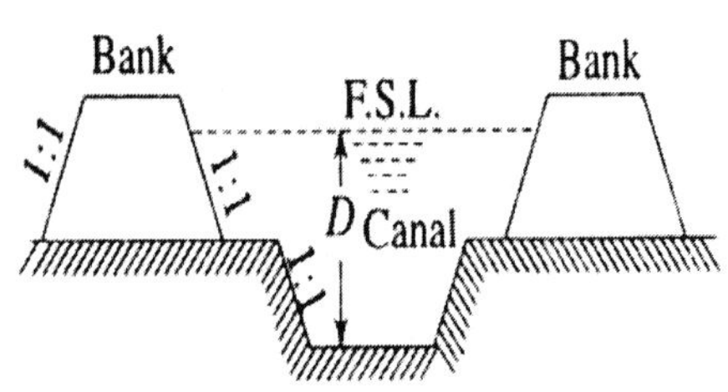

Fig. 9.9: Canal in partial cutting and partial banking

(b) Berm

The distance between the toe of the bank and the top edge of the cutting of the canal at the ground surface is called as berm. The berm is provided

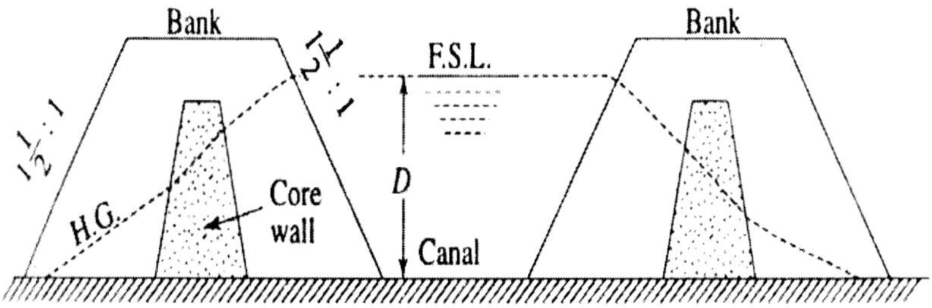

Fig. 9.10: Canal in full banking

(i) To protect the bank from erosion.

(ii) To provide a space for widening the canal section in future if necessary.

(iii) To protect the bank from sliding down towards the canal section.

(iv) If necessary, burrow pit can be excavated on the berm.

The width of the berm varies from *D to 2D*, where, *D* is the full supply depth of the canal. Minimum berm width depending on canal discharge is given in Table 9.2 (Punmia and Lal, 1992).

(c) Hydraulic Gradient

When water is retained by the canal bank, there is seepage flow inside the body of the bank and the path of the seepage line is called as phreatic line. The phreatic line slopes downward and cuts the bank at the down stream side. This sloping phreatic line is called as hydraulic gradient The soil below the line is saturated and above is dry. The pressure on the line is at atmospheric (Fig. 9.11). Hydraulic gradient depends on the soil texture. Hydraulic gradient helps in deciding the height and top width of the bank and its side slope.

Table 9.2: Minimum berm width for different discharges

Discharge, cumecs	Berm width, m
d" 4.25	0.6 +1/2 full supply depth
4.25 – 28.0	1.25 + 1/2 full supply depth
e" 28.0	1.25 + Full supply discharge/28 + ½ full supply depth

(d) Counter Berm

A counter berm is a projection which is provided on the downstream side of the bank so that the phreatic line if intersects the bank at the outer side will have a certain minimum cover (Fig. 9.12). If the phreatic line intersects the bank at the outer side, it will create instability and

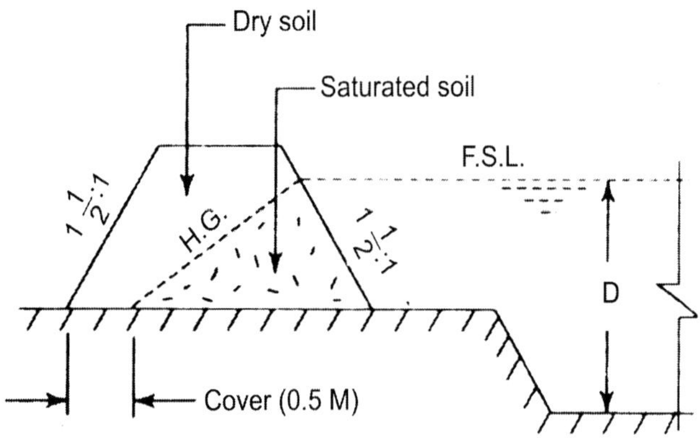

Fig. 9.11: Hydraulic gradient

consequent failure of the bank. For stability, the line must pass through the base of the bank and have at least 0.5 m cover from the outer edge of the bank (heel).

(e) Free Board

Free board is the distance between the full supply level and the top of the bank. The amount of the free board varies from 0.5 to 0.75 m (Fig. 9.13). Free board is provided to (i) keep a

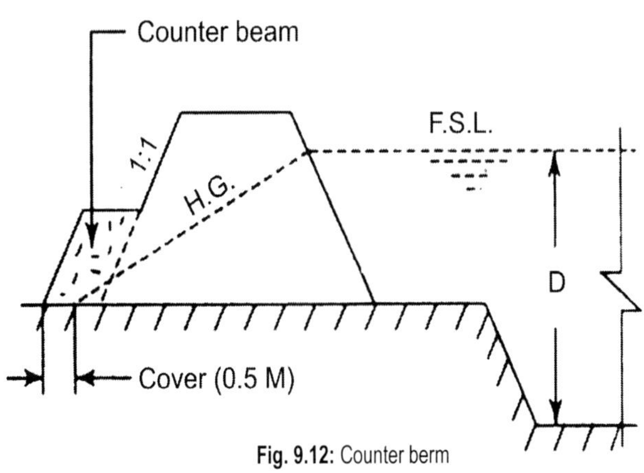

Fig. 9.12: Counter berm

sufficient margin so that the canal water does not overtop the bank in case of flood or any unprecedented reason for which the flow increases in the canal and (ii) keep the saturation gradient much below the top of the bank.

(f) Side Slope

Side slope of the bank and canal section depends on the angle of repose of the soil existing on

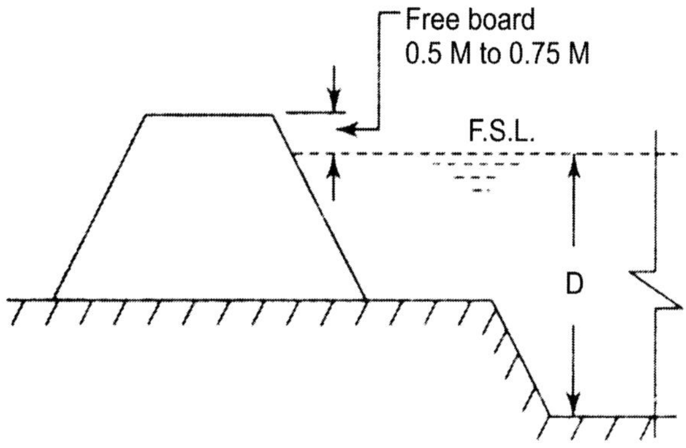

Fig. 9.13: Free board

the site. Table 9.3 presents the permissible side slope of the bank and also canal section for various types of soil.

Table 9.3: Permissible side slope for some soils (Basak, 2012).

Soil texture	Side slope in cutting	Side slope in banking
Clay	1:1	$1\frac{1}{2}:1$
Alluvial	1:1	2:1
Sandy loam	$1\frac{1}{2}:1$	2:1
Sand	2:1	3:1

(g) Service Road

Service road is also called as inspection road and is provided on the top of the bank. For main canal, it is provided on both sides whereas for branch canal, it may be provided at one side only. Width of the service road for main canal varies from 4 to 6 m where as for branch canal, it is 3 to 6 m (Fig. 9.14).

(h) Dowel or Dowla

The protective small embankment which is provided on the canal side of the service road for safety of the vehicle plying on it is called as dowel/dowla. It is provided above the full supply level (F.S.L.) with a provision of free board. The top width is generally 0.5 m and the height above the road level is about 0.5 m. The side slope is similar to the side slope of the bank (Fig. 9.14).

(i) Spoil Bank

When the canal is constructed in full cutting, the excavated earth may not be fully used in

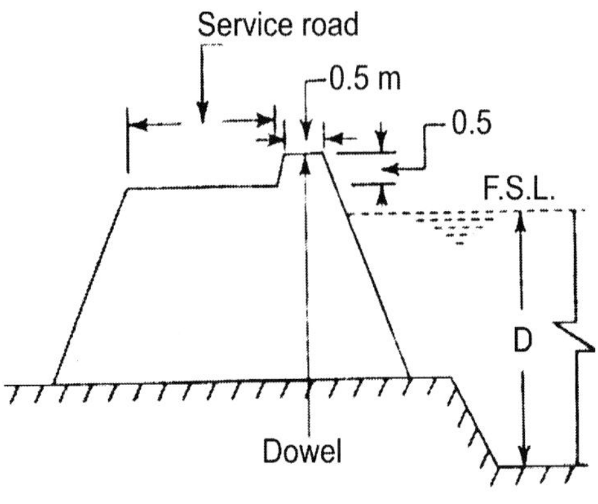

Fig. 9.14: Service road and dowel

constructing the bank. In such a case, the extra earth is deposited in the form of small banks called as spoil banks. The spoil bank runs parallel to the main bank.

(j) Borrowpit

When the canal is constructed in partial cutting and partial banking, the excavated earth may not be sufficient to construct the bank. In such a case, the extra earth is taken from some pits called as borrowpit. The borrowpits may be inside or outside the canal.

(k) Land Width

Total land width required to construct a canal must take into account the following dimensions which need to be added.

- Top width of the canal
- Twice the berm width
- Twice the bottom width of the banks
- A margin of 1 m from the heel of the bank on both sides
- Width of external borrowpit if any
- A margin of 0.5 m from the outer edge of the borrowpit on both sides if external borrowpit becomes necessary.

(l) Balancing Depth

For an economical canal section, the depth of cutting of the soils must fully meet the total soil required for construction of both the banks. The depth of cutting for such a section is known as balancing depth. This condition may not occur at all times except the case of canal constructed as partial cutting and partial banking. Following sample calculation will help to know how to calculate the balancing depth of a canal section.

Sample Calculation 1:

Find the balancing depth of a canal section (canal is fully cutting) having the following dimensions. Draw the canal section and find out the total land width required.

(a) R.L. of the top of the bank = 150 m

(b) R.L. of the bottom of the bank = 147 m

(c) Side slope of the bank = 2:1

(d) Side slope of canal section = 1:1

(e) Berm width = 1.0 m

(f) Bottom width of the canal = 10 m

(g) Top width of the bank = 3 m

Solution

R.L. of top of bank and that of bottom of bank are 150 and 147 m, respectively from which the height of the bank is calculated as (150 – 147) = 3 m

Since side slope of the bank is 2:1 and height is 3 m (calculated as above), bottom width of the bank (trapezoidal shape) = 3 + 2 x 3 x 2 = 15 m

The bank is trapezoidal in section.

Area of the bank = $2\ x\left(\frac{15+3}{2}\right)x\ 3 = 54\ m^2$

Let d be the balancing depth of the canal section which needs to be cut.

The bottom width of the canal section = 10 m (given) and side of the section = 1:1 given.

The top width of the canal section (trapezoidal in shape) = (10 + 2 d).

Area of the canal section = $\left(\frac{10+10+2d}{2}\right)x\ d = (10+d)\,x\ d$

For a balancing case, area of banking must be equal to area of cutting of canal section.

Hence, $(10 + d)\ x\ d = 54$

Solution of the above quadratic equation gives d = 3.89 m.

The balancing depth of the canal section is 3.89 m.

The total land width required is calculated as, *L = 2 x (bottom width of the bank +berm width) + top width of the canal section + 2 x margin length of 1 m (assume) from heel of the bank*

L = 2 x (15 + 1) + (10 + 2 x 3.89) + 2 x 1 = 51.78 m.

Using the given and calculated dimensions of the various parameters, the canal section with bank is drawn as given below.

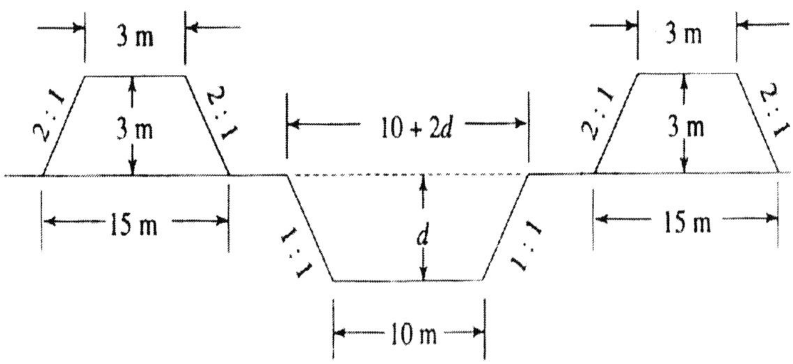

9.11 Definition of Various Terms Related to Canal Design

Silt Factor:

Design of canal section in alluvial should take into account the silt factor (*f*) because of the suspended silt in the flowing water which subsequently get deposited in the canal beds. Gerald Lacey introduced a silt factor, *f* which is a function of mean particle size of the silt (Table 9.4) and is expressed as:

$$f = 1.76\sqrt{m_r} \quad (9.1)$$

where, m_r = mean particle size of silt, mm.

Table 9.4: Values of silt factor for different particles having different sizes.

Particle	Particle size, mm	Silt factor, *f*
Very fine silt	0.05	0.40
Fine silt	0.12	0.60
Medium silt	0.23	0.85
Coarse silt	0.32	1.00

Coefficient of Rugosity (*N*):

The roughness of canal bed affects the velocity of flow. The roughness is caused due to ripples formed on the bed of the canal. R.G.Kennedy introduced a coefficient called as coefficient of rugosity/Kutter's coefficient *(N)* for calculation of mean velocity of flow. The values of *N* for earthen channels ranges from 0.0225 for very good section to 0.03 for poor channel section. If the section is good, then *N* may be taken as 0.025 (Punmia and Lal, 1992). For masonry section, *N* is 0.02 and for concrete section, it ranges from 0.013 to 0.018. Table 9.5 gives the Central Water and Power Commission (CWPC) practice of selecting the values of *N* for the canals taking off from the reservoirs carrying silt free water.

Table 9.5: CPWC practice for selecting values of *N*

Type of soil	Canal discharge, cumecs	Value of *N*
Soil other than rock	Up to 0. 14	0.03
	0.14 – 1.4	0.025
	1.4 – 14.0	0.0225
	Above 14.0	0.020
Rocky cuts	When rock portion at least 15 cm above the excavated bed level is left out in working out cross sectional area	0.035 – 0.05
	When no portion above bed level is left out	0.05 – 0.08

Value of N also depends on the nature and condition of channel surfaces and is provided in Table 9.6 (Punmia and Lal, 1992).

Table 9.6: Values of rugosity coefficient for different nature and condition of channel surfaces

Nature of channel surface	Value of *N*			
	Perfect	Good	Fair	Bad
Alluvial regime	0.017	0.020	0.0225	0.025
Loose gravel	0.0275	0.030	0.033	0.035
Rock cut, smooth and regular	0.025	0.030	0.033	0.035
Rock cut and irregular	0.035	0.040	0.045	0.050
With cobble bed and earthen banks	0.025	0.030	0.035	0.040
With earthen bed and rubble sides	0.025	0.030	0.033	0.035

9.12 Mean Velocity of Flow

Following formulae are used to calculate the mean velocity of flow in an open channel (Sharma, 1987; Sharma and Sharma, 2002).

(a) Chezy's Formula: The velocity is given by Eq. (9.2) as:

$$V = C\sqrt{R\,S} \tag{9.2}$$

where,

V = velocity, m/sec; C = chezy's coefficient; R = hydraulic mean radius, m and S = bed slope of the channel, m/m.

(b) **Manning's Formula:** The velocity by this formula is given as:

$$V = \frac{1}{N} R^{2/3} S^{1/2} \tag{9.3}$$

where,

V = velocity, m/sec; N = rugosity coefficient/Manning's coefficient; R = hydraulic mean radius, m and S = bed slope of the channel, m/m.

Selection of values of N for different channels with different conditions is discussed as earlier.

Hydraulic mean radius/hydraulic mean depth *(R)* is ratio of cross sectional area of flow (A) to the wetted perimeter of the channel (P).

In Eq. (9.2), Chezy's coefficient (C) can be computed by:

(a) Bazin's Formula

$$C = \frac{87}{1 + \frac{K}{\sqrt{R}}} \tag{9.4}$$

where,

K = Bazin's constant and R = as stated earlier.

(b) Kutter's Formula

$$C = \frac{23 + \frac{0.00155}{S} + \frac{1}{N}}{1 + \left(23 + \frac{0.00155}{S}\right) x \frac{N}{\sqrt{R}}} \tag{9.5}$$

where, N, S *and* R as stated earlier.

9.13 Discharge Capacity of the Channel

The discharge capacity or rate of flow is given as:

$$Q = A \text{ x } V \tag{9.6}$$

where,

Q = discharge, m^3/sec; A = cross sectional area, m^2 and V = mean velocity of flow, m/sec. The maximum permissible velocity of flow for bare earthen channel sections with different soil textures are given in Table 9.7 (Murty, 2004).

Table 9.7 Maximum permissible velocity of flow for bare earthen channel sections

Soil texture in bare channel section	Maximum permissible velocity, cm/sec
Sand and silt	45
Loam, sandy loam and silt loam	60
Clay loam	65
Clay	70
Gravelly soil	100

N.B. In case of lined channels and channels with vegetations, higher velocities are allowed.

9.14 Most Economical Channel Section

The most economical channel section is the one that gives the maximum rate of discharge with a minimum cross sectional area. For an economical channel section, the earth obtained from cutting can be fully utilized in forming the banks which is possible for a channel formed by partial cutting and partial banking. For a given channel section, the discharge will be maximum when the mean velocity of flow is maximum. For a given land slope, velocity will be maximum when the hydraulic mean radius is maximum i.e. wetted perimeter is minimum. Theoretically, the most efficient cross section for an open channel is semi-circle Conditions for the most economical trapezoidal channel sections are as follows:

Consider a trapezoidal channel section having bottom width, b and depth of flow d.

Let θ be the angle of repose of the section i.e. the sloping side makes an angle θ with the horizontal (Fig. 9.15). Let the side slope of the section be $n : 1$ (n horizontal to 1 vertical).

Now, $\tan \theta = 1/n$ (9.7)

Top width of the section = $(b + 2\,n\,d)$ and sloping side = $d\sqrt{(n^2+1)}$

The wetted perimeter, $P = [b + 2\,d\sqrt{(n^2+1)}]$ (9.8)

Cross sectional area, $A = (b + n\,d)\,d$ (9.9)

Hydraulic mean radius, $R = A/P = \dfrac{(b+n\,d)\,d}{\left[b+2\,d\sqrt{(n^2+1)}\right]}$ (9.10)

From Eq. (9.9), $b = \dfrac{A}{d} - n\,d$ (9.11)

Putting the value of b from Eq. (9.11) in eq. (9.8) we get,

$$P = \frac{A}{d} - n\,d + 2\,d\sqrt{n^2+1} \qquad (9.12)$$

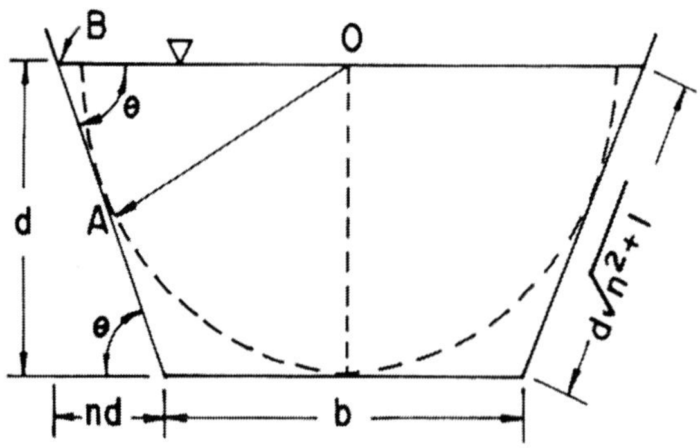

Fig. 9.15: Trapezoidal channel section

For an economical channel section, P should be minimum and $\frac{dP}{dd} = 0$

Differentiating Eq. (9.12) with respect to depth, d and equating it to zero gives

$$A = d^2 \left[2\sqrt{n^2+1} - n\right] \tag{9.13}$$

Putting the value of A from Eq. (9.13) in Eq. (9.11) and simplifying we get

$$b = d\left[2\left(\sqrt{n^2+1}\right) - n\right] - n\,d \qquad 9.14)$$

From Eq. (9.7), $n = cot\,\theta$ (9.15)

Using Eq. (9.15) and Eq. (9.14) and simplifying we get

$$b = 2\,d\,\tan\theta/2 \tag{9.16}$$

For a rectangular section, $\theta = 90^o$

For economical rectangular section, putting $\theta = 90^{o,}$ in Eq. (9.16), we get $b = 2\,d$ and $R = d/2$

Hence, for the most economical rectangular channel section, depth of flow should be half of the base width or hydraulic mean radius is half of depth

9.15 Unlined Canal Design in Non-Alluvial Soil

Canals in non-alluvial soil have stable base. The coefficient of rugosity plays a major role in the design. But the silt factor has little role. Velocity of flow is close to critical velocity. So, the mean velocity given by Chezy's or Manning's equation can be applied for the design.

Sample Calculation 2

Design an irrigation canal to carry a discharge of 24 cumecs with permissible mean velocity of

2.88 km/hr. Bed slope of the canal section is 0.02 percent and side slope is 1:1. Assume Chezy's constant (C) as 44. Assume any other data not given in the problem.

Solution

Given discharge, $Q = 24\ m^3/sec$

Mean velocity of flow, $V = 2.88$ km/hr = 0.80 m/sec

Cross sectional area, $A = Q/A = 24/0.80 = 30\ m^2$

Let base width of the canal be B and depth be D. Since side slope is 1:1, top width is $(B + 2D)$

Cross sectional area = $(B + D)\,D$

Hence, $(B + D)\,D = 30$ (9.17)

Mean velocity by Chezy's formula is

$$V = C\sqrt{R\,S}$$

Given, $V = 0.80$ m/sec, $C = 44$ and $S = 0.02$ percent = 0.0002

Hence, $0.80 = 44\sqrt{R\ x\ 0.0002}$

$R = 1.65$

Wetted perimeter, $P = B + 2\sqrt{2}\,D$

Hydraulic mean radius, R = Cross sectional area/wetted perimeter = $\dfrac{(B+D)\,D}{B+2\sqrt{2\,D}}$

Hence, $\dfrac{(B+D)\,D}{B+2\sqrt{2\,D}} = 1.65$ (9.18)

From Eq. (9.17) and (9.18), $\dfrac{30}{B+2\sqrt{2}\ D} = 1.65$ (9.19)

Simplifying Eq. (9.19), $B = 18.18 - 2.\ 83\ D$ (9.20)

Putting the value of B from Eq. (9.20) in Eq. (9.17) and simplifying,

$1.83\ D^2 - 18.18\ D + 30 = 0$

Solution of the above quadratic equation gives $D = 7.84$ m or 2.09 m.

Putting $D = 7.84$ in Eq. (9.20), $B = -\ 4.00$ (impossible) and when $D = 2.09$, B = 12.27 m (acceptable).

Hence the design dimension of the canal section is that the base width and depth of flow be 12.27 m and 2.09 m, respectively. Accordingly, the top width will be 16.45 m.

Sample Calculation 3:

Design the most economical trapezoidal canal section to carry a discharge of 20 m^3/sec with mean velocity of flow of 3.06 km/hr. Side slope of the canal is 1.5:1. Assume Bazin's constant, K = 1.30. Also calculate the allowable bed slope of the canal.

Solution

Given discharge, Q = 20 m^3/sec

Mean velocity of flow, V = 3.06 km/hr = 0.85 m/sec

Cross sectional area, $A = Q/A = 20/0.85 = 23.53$ m^2

Let base width of the canal be B and depth be D. Since side slope is 1.5:1, top width is $(B + 3 D)$

Cross sectional area $A = (B + 1.5\ D)\ D$

Hence, $(B + 1.5\ D)\ D = 23.53$ (9.21)

Wetted perimeter, P = $B + 2\left\{\sqrt{D^2 + (1.5\ D)^2}\right\} = B + 3.60\ D$

Hydraulic mean radius, $R = \dfrac{(B + 1.5\ D)\ D}{B + 3.60\ D}$ (9.22)

For an economical canal section, $R = D/2$ (9.23)

From Eq. (9.22) and (9.23),

$$\frac{(B + 1.5\ D)\ D}{B + 3.60\ D} = \frac{D}{2} \quad (9.24)$$

Simplifying Eq. (9.24), we get

$B = 0.60\ D$ (9.25)

Putting the value of B from Eq. (9.25) in Eq. (9.21) and simplifying we get D = 3.35 m, B = 2.01 m and $R = D/2$ = 1.675 m.

Using the values of Bazin's constant, K = 1.30 and R = 1.675 and putting these values in Eq. (9.4), we get Chezy's constant, C = 43.5

Velocity by Chezy's formula is $V = C\sqrt{R\ S}$

Hence, $0.85 = 43.5 \times \sqrt{1.675\ S}$

Simplifying the above equation, $S = 0.00022 = 0.022$ percent = 1:4374

Hence, base width and depth of the canal section are 2.01 and 3.35 m, respectively and bed slope is 0.022 percent.

Sample Calculation 4:

Design a canal section to carry a discharge of 40 m^3/sec with mean velocity of flow of 3.42 km/hr. Side slope of the canal is 1:1. The ratio of base width to depth of flow in the section is given as $B/D = 6.5$. Assume value of coefficient of rugosity, $N = 0.0225$.

Solution:

Given discharge, $Q = 40$ m^3/sec

Mean velocity of flow, $V = 3.42$ km/hr = 0.95 m/sec

Cross sectional area, $A = Q/A = 40/0.95 = 42.11$ m^2

Let base width of the canal be B and depth be D. Since side slope is 1:1, top width is $(B + 2D)$

Cross sectional area $A = (B + D)\,D$

Hence, $(B + D)\,D = 42.11$ (9.26)

Given $B/D = 6.5$ which gives $B = 6.5\,D$

Putting $B = 6.5\,D$ in Eq. (9.26), $D = 2.37$ m and hence, $B = 15.40$ m.

Wetted perimeter, $P = B + 2\sqrt{2}\,D$

Putting $B = 15.40$ and $D = 2.37$, $P = 22.10$ m

Hence, hydraulic mean radius, $R = 42.11/22.10 = 1.90$ m

Putting the values of $R = 1.90$, $N = 0.0225$ (given), $V = 0.95$ (given) and using Manning's formula (Eq. 9.3),

$$0.95 = \frac{1.90^{2/3}\ x\ S^{1/2}}{0.0225} \tag{9.27}$$

Simplification of Eq. (9.27) gives

Bed slope, S = 0.000194 = 0.019 percent = 1: 5155.

Hence, base width and depth of the canal section are 15.40 and 2.37 m, respectively and bed slope is 0.019 percent.

9.16 Unlined Canal Design in Alluvial Soil

Water carried by the irrigation canals contains a lot of derbies, silts and many other foreign materials. Often, the alluvial canals carry a lot of alluviums which get deposited in course of time in the channel beds and side. Water moving with a given velocity and a certain depth can carry in suspension, only a certain amount of silt of a certain nature. If water of a given velocity and depth is not fully charged with silt (that it can carry in suspension) it will scour the bed and sides of the channel, till it is fully charged with silt. Hence, if the velocity of flow in the channel is more, the bed and the banks are likely to the eroded, and similarly, if the velocity is less, the silt which was formerly carried in suspension is likely to be dropped. Silting and scouring in channels is very common and must be avoided by proper designs. Scouring lowers the full supply level and causes loss of command area. It may also cause breaching of canal banks and failure of foundations of irrigation structures. Silting interferes with the proper working of a channel, as the channel section gets reduced by silting, thereby reducing its discharging capacity. Therefore, while thinking to design a proper functioning channel, one must think to design such a channel in which neither silting nor scouring takes place. Such channels are known as stable channels or regime channels.

9.16.1 Regime Channels

When the character of the bed and bank materials of the channel are same as that of the transported materials and when the silt charge and silt grade are constant, then the channel is said to be regime. The channel will be in the state of 'regime', if the flow is such that 'sitting and scouring' need no special attention. Such a state is not easily possible in rivers, but in artificial channels, such a state can be obtained by properly designing the channel. Design of regime channels is based on mainly two approaches: (i) Kennedy's theory and (ii) Lace's theory.

9.16.1.1 Kennedy's Theory

R.G. Kennedy, an Executive Engineer of Punjab P.W.D. carried out extensive investigations on some of the canal reaches in the upper Bari Doab Canal System. He selected some straight reaches of the canal section, which had not posed any silting and scouring problems during the previous 30 years or so. From the observations, he concluded that the silt supporting power in a channel cross–section was mainly dependent upon the generation of the eddies, rising to the surface. These eddies are generated due to the friction of the flowing water with the channel surface. The vertical component of these eddies try to move the sediment up, while the weight of the sediment tries to bring it down, thus keeping the sediment in suspension. So if the velocity is sufficient to generate these eddies, so as to keep the sediment just in suspension, silting will be avoided. Based upon this concept, he defined the critical velocity (V_0) in a channel as the mean velocity which will just keep the channel free from silting or scouring and related it to the depth of flow by the equation.

$$V_0 = c\,D^n \tag{9.28}$$

where c and n are constants depending upon silt charge.

Values of c and n were found to be 0.55 and 0.64, respectively.

Therefore, Eq. (9.28) can be written as

$$V_0 = 0.55\,D^{0.64} \tag{9.29}$$

But the formula (Eq.9.29) is true only for the Upper Bari Doab Canal System and not for any other reaches. Later realizing this lacuna in the above equation, he introduced a silt grade factor depending upon the type of soil. This factor is named as critical velocity ratio *(C.V.R.)* (denoted by symbol *m*) which depends upon the silt grade. The modified Kennedy's formula is given as (Sharma, 1987)

$$V_0 = 0.55\, m\, D^{0\text{-}64} \tag{9.30}$$

V_0 = Critical velocity in the channel in m/sec and D = Depth of flow in the channel, m.

For sands coarser than the standard, the values of *m* are taken from 1.1 to 1.2, and for sands finer than the standard, *m* is taken as a value between 0.9 to 0.8. Generally a system of canal, higher *C.V.R.* is assumed in head reaches and a lower value towards lower reaches. Table 9.8 provides values of *m* for different types of silt laden water (Garg, 1987):

The values of constant *c* in Eq. (9.28) for various grades of materials may be assumed (Table 9.9) as follows (Punmia and Lal, 1992).

Table 9.8: Recommended values of critical velocity ratio, *m* for different silt carried channel

Sl. No.	Type of silt	Value of m
1.	Silt of river Indus (Pakistan)	0.70
2.	Light sandy silt in North Indian River	1.0
3.	Light sandy silt, a little coarser	1.1
4.	Sandy, loamy silt	1.2
5.	Debris of hard soil	1.3

Table 9.9: Values of constant *c*

Types of materials	Values of *c*
Light sandy silt	0.53
Coarser light silt	0.59
Sandy loam	0.65
Coarse silt	0.70

Design Procedure: Kennedy's design procedure is based on the following steps.

i. Assume a trial depth, *D* and then compute the critical velocity V_0 by the above Eq (9.29).

ii. Determine area by dividing discharge by velocity (velocity calculated as in step (i)).

iii. Then determine channel dimensions.

iv. Finally, compute the actual mean velocity (*V*) that will prevail in the channel of this cross-section by Chezy's equation (Eq. 9.2) with Chezy's constant C calculated either by Bazins formula (Eq. 9.4) or by Kutter's formula (Eq. 9.5).

v. If the two velocities V_0 and *V* works out to be the same, then the assumed depth is all right, otherwise change it and repeat the procedure. Thus several iterations are required to get the dimensions of the channel.

Sample Calculation 5:

An irrigation channel has side slope 1:1 and base width 3 m. It has flow depth of 1 m. The bed slope of the channel is 0.04 percent. Assume Manning's rugosity coefficient N = 0.028. Determine if the channel will be regime channel. Assume critical velocity ratio as 1.0.

Solution:

Given $N = 0.028$ and $S = 0.04$ percent = 1/2500

Base width = 3 m and side slope is 1. The depth of flow = 1m.

Hence top width of the channel = 5 m.

Area of the trapezoidal section = (5 +3)/2 = 4 sq. m.

Wetted perimeter = $3 + 2\sqrt{2}$ = 5.83 m.

Mean hydraulic radius, $R = A/P$ = 0.69 m.

Using the above values of *n, R and S*, velocity by Manning's formula, V = 0.56 m/s.

Using Kennedy's formula, velocity $V_0 = 0.55\ m\ D^{0\text{-}64} = 0.55 \times 1.0 \times (1.0)^{0.64} = 0.55\ m/s.$

Since the two velocities are nearly same, the channel is stable i.e. regime.

Sample Calculation 6:

Design an irrigation channel to carry a full supply discharge of 6 cumecs having bed slope of 0.2 m per kilometer length. The value of rugosity constant N is 0.0225 and assume the critical velocity ration, m as 1.0. The side slope of the channel is 1:1.

Solution:

Let us assume the full supply depth, D as 1.5 m.

If the base width is B and given side slope 1:1, the top width of the section is $(B + 3)$.

Cross sectional area is $A = (1.5\ B + 2.25)$

Wetted perimeter is $P = (B + 2\sqrt{2} \times 1.5) = (B + 4.24)$

$$\text{Hydraulic mean radius, R} = \frac{1.5\ B + 2.25}{B + 4.24}$$

Now, critical velocity, $V_0 = 0.55 \times m \times D^{0.64}$

$$= 0.55 \times 1 \times 1.5^{0.64} = 0.707 \text{ m/sec}$$

Given, discharge, Q = 6 m^3/sec.

Hence, cross sectional, A = Discharge/velocity = 6/0.707 = 8.49 m$^{2.}$

Therefore, $1.5\ B + 2.25 = 8.49$ m^2 which on simplification gives B = 4.16 m

Putting the value of B = 4.16 in computation of wetted perimeter, we get, $P = B + 4.24 = 4.16 + 4.24 = 8.40\ m$

$$\text{Hence, R} = \frac{1.5\,B + 2.25}{B + 4.24} = \frac{1.5\,x\,4.16 + 2.25}{4.16 + 4.24} = 1.0\ m$$

Putting $N = 0.0225$, $S = 0.02$ *percent* $= 0.0002$ *and* $R = 1.0$ in Kutter's formula (Eq. 9.5), the Chezy's constant (C) is computed as 44.5.

Velocity by Chezy's formula is:

$$V = C\sqrt{R\,S} = 44.5\,\sqrt{1\,x\,0.0002} = 0.629\text{ m/sec}$$

Since, critical velocity ≠ velocity by Chezy's formula, the assumed value of *D* is not correct and we have to decrease the depth (depth is decreased to decrease the critical velocity).

Let the depth be 1.20 m.

Critical velocity, $V_0 = 0.55\ x\ m\ x\ D^{0.64}$

$= 0.55 \text{ x } 1 \text{ x } 1.2^{0.64} = 0.618$ m/sec

Cross sectional, A = Discharge/velocity = 6/0.618 = 9.71m^2

With base width, *B*, depth 1.2 and side slope 1:1, we get cross sectional area, $A = 1.20\,B + 1.44$

$1.20\,B + 1.44 = 9.71$

$B = 6.89$ m

Wetted perimeter, $P = 6.89 + 2\sqrt{2} \text{ x } 1.20 = 10.28$ m

Hydraulic mean radius, $R = 9.71/10.28 = 0.94$ m

Putting $N = 0.0225$, $S = 0.02$ *percent* $= 0.0002$ *and* $R = 0.94$ in Kutter's formula (Eq. 9.5), the Chezy's constant (C) is computed as 43.97

Velocity by Chezy's formula is:

$$V = C\sqrt{R\,S} = 43.97\,\sqrt{0.94\,x\,0.0002} = 0.603\text{ m/sec}$$

Since the critical velocity ≈ velocity by Chezy's formula, the assumed value of *D* (=1.2 m) may be taken as the base width of the channel section. The base width of the channel section is 6.89 m.

Sample Calculation 7:

Design an irrigation channel to carry a full supply discharge of 10 m^3/sec. The ratio of base width to depth of flow (*B/D*) = 4 and side slope of the section is 1:1. Assume the critical velocity ration, $m = 1.0$ and Bazin's constant, $K = 1.3$.

Solution:

Let the depth of flow be D and since $B/D = 4$, base width $B = 4D$

Cross sectional area, $A = (4\,D + D)\,x\,D = 5\,D^2$

Critical velocity, $V_0 = 0.55\,x\,m\,x\,D^{0.64} = 0.55\,x\,D^{0.64}$

Cross sectional area, $A = \dfrac{10}{0.55\,x\,D^{0.64}}$

Hence, $5\,D^2 = \dfrac{10}{0.55\,x\,D^{0.64}}$

Solution of the above equation gives $D = 1.64$ m

Base width, $B = 4\,D = 6.56$ m

Hence, cross sectional area = 13.45 m^2 and wetted perimeter = 11.20 m and hydraulic mean radius = 1.20 m.

Critical velocity, $V_0 = 0.55\,x\,D^{0.64} = 0.55 \text{ x } 1.64^{0.64} = 0.749$ m/sec.

From Bazin's formula, $C = 39.7$

By Chezy's formula, velocity, V is:

$$V = 39.7 \text{ x } \sqrt{(1.20).S}$$

Since, *C.V.R. (m)* = 1.0 (given), we have critical velocity = velocity by Chezy's formula.

Hence, $0.749 = 39.7 \text{ x } \sqrt{(1.20).S}$

Simplifying, $S = 0.000297 = 0.0297$ percent ≈ 0.03 percent.

Drawbacks in Kennedy's Theory

Followings are the drawbacks in Kennedy's theory.

(i) The theory is limited to average regime channel only.

(ii) The design of channel is based on trial and error method.

(iii) The value of critical velocity ratio was fixed arbitrarily.

(iv) Silt charge and silt grade were not considered.

(v) Kennedy did not notice the importance of *B/D* ratio.

(vi) No account was taken of silt concentration and bed load.

(vii) No slope equation was given except depending on Kutter's equation.

Kennedy's equation could not be made universally applicable even after the introduction of critical velocity ration. Hence, regional formulae were developed to suit the requirements of different canal systems by various investigators. The equation of critical velocity was in the same form as Eq. (9.28) but the value of c and n are different for different canal systems as reported by Punmia and Lal (1992) (Table 9.10).

Table 9.10: Values of *c* and *n* for different canal systems

Serial No.	Canal system	*c*	*n*
1.	Godavari delta	0.391	0.55
2.	Krishna Western delta	0.530	0.52
3.	Lower Chenab canal	0.567	0.57
4.	Gyptian canals	0.283	0.73

9.16.1.2 Lacey's Theory

Dimensions, width, depth and slope of a regime channel to carry a given a discharge loaded with a given silt charge are all fixed by nature. This idea was first put forward by Lacely. According to Lacey, a channel will be in regime if it flows in coherent unlimited alluvium of the same character as that transported and silt grade and silt charge are all constant. Lacey states that the silt is kept in suspension due to the force of vertical eddy. According to him, the eddies are generated from the sides and beds of the channels, both normal to the surface of the generation. So, vertical component of the eddies generated from the sides will also support the silt. Lacey, so, assumed hydraulic mean depth (scour depth) as a variable instead of depth as proposed by Kennedy. According to Lacey's theory, there is a relationship between velocity, V and hydraulic mean radius/scour depth, R in the form $V = R^{1/2}$. As a measure of silt grade, Lacey introduced a silt factor, f and he proposed an equation (Sharma and Sharma, 2002) as:

$$V = K\sqrt{f\ R} \qquad (9.31)$$

where K is a constant. The value of K proposed is $K = \sqrt{\dfrac{2}{5}}$

Hence Eq. (9.31) in SI unit can be written as:

$$V = \sqrt{\left(\frac{2}{5} f\ R\right)} \qquad (9.32.)$$

$$\text{Hence, } f = \frac{5V^2}{2R} \qquad (9.33)$$

The value of f depends on average size of boundary materials in the channel and is given by the equation:

$$f = 1.76\ m_r^{1/2} \qquad (9.34)$$

where m_r is average particle size, mm

He also proposed the following relationship between area, A vs. velocity, V:

$$A f^2 = 140\ V^5 \tag{9.35}$$

Lacey's regime flow equation is:

$$V = 10.8\ R^{\frac{2}{3}}\ S^{\frac{1}{3}} \tag{9.36}$$

where, S is bed slope in m/m.

Eqns. (9.32, 9.35 and 9.36) are called as Lacey's fundamental equations from which other flow equations are derived which are given as below.

$$V = \left(\frac{Q f^2}{140}\right)^{\frac{1}{6}} \tag{9.37}$$

$$P = 4.75\sqrt{Q} \tag{9.38}$$

$$S = \frac{f^{3/2}}{4980\ R^{1/3}} \tag{9.39}$$

$$S = \frac{f^{5/3}}{3340\ Q^{1/6}} \tag{9.40}$$

$$Q = \left(\frac{f^{5/3}}{3340\ x\ S}\right)^6 \tag{9.41}$$

Eqns. (9.39 and 9.40) are called as Lacey's regime slope equation.

Design Procedure: Lacey's design procedure is based on the following steps.

Step I. Calculate the silt factor.

Step II. Find out velocity, $V = \left(\frac{Q f^2}{140}\right)^{\frac{1}{6}}$

Step III. Find out the cross sectional area, $A = Q/V$

Step IV. Find out the wetted perimeter, $P = 4.75\sqrt{Q}$

Step V. Find out base width, B and depth, D of the channel section since area and wetted perimeter are known.

Step VI. Calculate R = $\dfrac{5V^2}{2f}$

Step VII. Also calculate *R* as ratio of *A* and *P*. Both the values of *R* should be same which serves as a check from step I to V.

Step VIII. Find out the slope, $S = \dfrac{f^{5/3}}{3340\, Q^{1/6}}$ or as S= $S = \dfrac{f^{3/2}}{4980\, R^{1/3}}$

Sample Calculation 8:

Design a channel section to carry a full supply discharge of 30 cumecs having side slope 0.5 :1 and mean diameter of the silt particle is 0.33 mm.

Solution:

Silt factor, f = 1.76 $\sqrt{0.33}$ = 1.0

Mean velocity, $V = \left(\dfrac{Q f^2}{140}\right)^{\frac{1}{6}}$

$$= \left(\frac{30\, x\, 1^2}{140}\right)^{\frac{1}{6}} = 0.773\, m/\sec$$

Cross sectional area, $A = Q/V$ = 30/0.773 = 38.8 m^2

Wetted perimeter, $P = 4.75\sqrt{Q}$

$= 4.75\ \sqrt{30} = 26$ m.

Since side slope is 0.5:1, base width and depth are *B* and *D*. respectively

Area, $A = B\,D + \dfrac{D^2}{2}$ and wetted perimeter, *P* = $B + D\sqrt{5}$

Hence, $D = \dfrac{P - \sqrt{P^2 - 6.944\, A}}{3.472}$

$$= \frac{26 - \sqrt{26^2 - 6.944\, x\, 38.8}}{3.472} = 1.67\, m$$

Base width, B = P – D x $\sqrt{5}$ = 26 – 1.67 x $\sqrt{5}$ = 22.6 m.

Hydraulic mean radius, R = Area/wetted perimeter = 38.8/26 = 1.49 m

$$\text{Again, R} = \frac{5\,V^2}{2\,f} = \frac{5\,x\,0.773^2}{2\,x\,1} = 1.49\,m$$

Since values of R by both methods are same, calculation is checked.

$$\text{Slope,}\quad S = \frac{f^{5/3}}{3340\,Q^{1/6}} = \frac{1^{5/3}}{3340\,x\,30^{1/6}} = 1/5900 = 0.017\,percent$$

So, the base width and depth of the section are 22.6, 1.67 m, respectively and bed slope is 0.017 percent.

Sample Calculation 9:

The slope of an alluvium channel is 1:5000. Lacey's silt factor is 0.9 and side slope of the channel is ½:1. Find out the channel dimensions and peak discharge that the channel can carry.

Solution:

Given, slope, S = 1:5000 and f = 0.9

The regime slope equation is from which

$$Q = \left(\frac{f^{5/3}}{3340\,x\,S}\right)^6 = \left(\frac{0.9^{5/3}\,x\,5000}{3340}\right)^6 = 3.98\,m^3/\sec$$

The other regime slope equation is $S = \frac{f^{3/2}}{4980\,x\,R^{1/2}}$ from which

$$R = \frac{f^3}{(4980\,x\,S)^2} = \frac{0.9^3\,x\,5000\,x\,5000}{4980} = 0.736\,m$$

Wetted perimeter, $P = 4.75 \text{ x } \sqrt{Q} = 4.75 \text{ x } \sqrt{3.98} = 9.84$ m

Cross sectional area, $A = R\,x\,P$ = 0.736 x 9.84 = 6.95 m^2.

With ½:1 side slope and assuming base width B and depth D, we get

$$A = B\,D + \frac{D^2}{2} = 6.95 \text{ and}$$

$$P = B + \sqrt{5}\,D = 9.84$$

Solution of above two equations give $B = 7.54$ m and $D = 0.87$ m.

Hence the design dimensions of the channel is base width is 7.54 m, depth is 0.87 m and the channel will carry a peak discharge of 3.98 cumec.

Drawbacks in Lacey's Theory

Followings are the drawbacks in Lacey's theory.

(i) The concept of true regime is theoretical and cannot be achieved practically.

(ii) The various equations are derived considering silt factor f which is not at all constant.

(iii) Concentration of silt is not taken into account.

(iv) Silt charge and silt grade are not clearly defined.

(v) The equations are empirical and based on the available data from a particular type of channel. And may not be true for other channels.

(vi) Characteristics of regime channel may not be same for all cases.

(vii) Lacey did not take into account the silt left in the channel by water that is lost in absorption which is as much as 12 to 15 percent of the total discharge of the channel.

(viii) Lacey introduced semi ellipse as ideal shape of a regime channel which is not correct.

9.16.1.3 Comparison between Kennedy's and Lacey's Theory

Comparison between Kennedy's and Lacey's Theory is as follows:

Serial No.	Kennedy's theory	Lacey's theory
1	It states that the silt carried by the flowing water is kept in suspension by the vertical components of eddies which are generated from the beds of the channel.	It states that the silt carried by the flowing water is kept in suspension by the vertical components of eddies which are generated from the entire wetted perimeter of the channel.
2	It gives relation between velocity and depth.	It gives relation between velocity and hydraulic mean radius.
3	In this theory a factor, m called as critical velocity ratio is used to make the equation applicable to different channels with different silt grades.	In this theory a factor, f called as silt factor is used to make the equation applicable to different channels with different silt grades.
4	In this theory, Kutter's equation is used for finding out the mean velocity.	This theory gives an equation for finding out the mean velocity.
5	This theory gives no equation to find out bed slope.	This theory gives equation to find out bed slope.
6	In this theory, the design is based on trial and error method.	This theory does not involve trial and error method.

9.17 Design of Lined Canal

Lined canalss are not designed by Kennedy's or Lacey's theory as they are rigid and do not carry any silt laden water. Manning's equation can be used to design these canals. The design considerations are:

(i) The section should be economical,

(ii) Velocity should be maximum so that cross sectional area becomes minimum

(iii) The capacity would not be reduced due to silting if any.

There are generally two types of lined canal sections. They are (i) circular section and (ii) trapezoidal section.

Circular Section

The bed is circular with its centre at the full supply level and radius equal to full supply depth *D*. The sides are rectangular to the curve. However, the side slopes are generally taken as 1:1 (Fig. 9.16).

Trapezoidal Section

In this section, the horizontal bed joined to the side slope by a curve of radius equal to full supply depth *D*. The side slopes are generally taken as 1:1 (Fig 9.17). Trapezoidal sections carry more discharge than the circular section. Design parameters of circular as well as trapezoidal sections are provided in Table 9.11 (Basak, 2012).

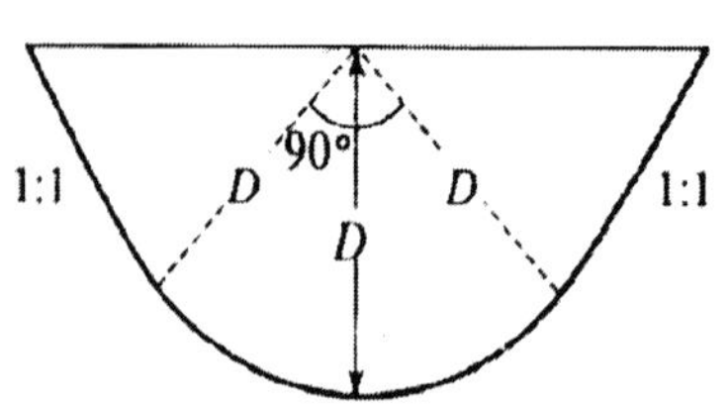

Fig. 9.16: Circular channel section

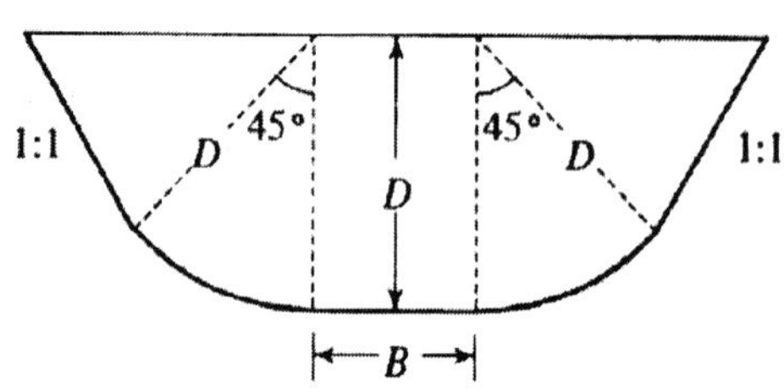

Fig. 9.17: Trapezoidal channel section

Table 9.11: Design parameters of circular and trapezoidal sections

Design parameters	Side slopes					
	Circular section			Trapezoidal section		
	1:1	1.5:1	1.25:1	1:1	1.5:1	1.25:1
Cross sectional area, *A*	$1.785\ D^2$	$2.088\ D^2$	$1.925\ D^2$	$BD + 1.785\ D^2$	$B\ D + 2.088\ D^2$	$BD + 1.925\ D^2$
Wetted perimeter, *P*	3.57 *D*	4.176 *D*	3.85 *D*	*B* + 3.570 *D*	*B* + 4.176 *D*	*B* + 3.85 *D*
Hydraulic mean radius, *R*	0.5 *D*	0.5 *D*	0.5 *D*	*A/P*	*A/P*	*A/P*
Velocity, *V*	Manning's formula	Manning's formula	Manning's formula	Manning's formula	Manning's formula	Manning's formula
Discharge, Q	A x V	A x V	A x V	A x V	A x V	A x V

Sample Calculation 10:

Design a lined circular canal section to carry a discharge of 20 cumecs. Assume bed slope of the canal as 0.02 percent, side slope 1:1 and rugosity coefficient, N as 0.0225.

Solution:

For a circular section with 1:1 side slope, the cross sectional area, $A = 1.785\ D^2$ and wetted perimeter, $P = 3.57\ D$ and hydraulic mean radius, $R = 0.5\ D$ **(**Table 9.11**).**

Velocity by Manning's formula (Eq. 9.3) is:

$$V = \frac{1}{N} R^{2/3}\ S^{1/2}$$

$$= \frac{1}{0.0225} x\ (0.5\ D)^{2/3}\ x\ 0.0002^{1/2} = 0.38\ D^{2/3}$$

Again, $Q = A\ x\ V = 1.785\ D^2\ x\ 0.38\ D^{2/3}$

$20 = 0.678\ D^{8/3}$

$D = 4.61\ m$

Area, $A = 1.785\ D^2 = 1.785 \text{ x } (4.61)^2 = 37.93$ m2.

Wetted perimeter, $P = 3.57\ D = 3.57 \text{ x } 4.61 = 16.45$ m.

Sample Calculation 11:

Design a lined trapezoidal canal section to carry a discharge of 200 cumecs. Assume bed slope of the canal as 0.02 percent, side slope 1.25:1 and rugosity coefficient, N as 0.018. Permissible velocity is 1.75 m/sec.

Solution:

Given $Q = 200$ m^3/sec and permissible velocity, $V = 1.75$ m/sec

Cross sectional area, $A = Q/V = 200/1.75 = 114.28$ m^2.

For a trapezoidal section with 1.25:1 side slope,

Cross sectional area, $A = BD + 1.925\ D^2$ and

Wetted perimeter, $P = B + 3.85\ D$

Hence, $BD + 1.925\ D^2 = 114.28$ (9.42)

Velocity by Manning's formula, $V = \frac{1}{N} R^{2/3}\ S^{1/2}$

$$1.75 = \frac{1}{0.018} R^{2/3}\ (0.0002)^{1/2}$$

$R = 3.32$ m

Wetted perimeter, $P = A/R = 114.28/3.32 = 34.42$ m

Hence, $B + 3.85\ D = 34.42$ (9.43)

From Eqns. (9.42) and (9.43), we get

$D = 13.47$ or 4.4 m.

From Eq. (9.43), when, $D = 13.47$ m, B = -17.44 m (impossible). When, $D = 4.4$ m, $B = 17.5$ m.

Hence, the base width of the canal section is 17.5 m and full supply depth is 4.4 m.

9.18 Garret's Diagram for Channel Design

Garret's diagram gives graphical solution of design of irrigation channel based on Kennedy's theory. The diagram has the discharge plotted on abscissa. The ordinates in the left hand side indicate the slope and that in the right hand side indicates critical velocity and depth of flow. Discharge lines are curves and bed width lines are shown dotted (Figs. 9.18 to 9.20). The procedure of designing the channel is presented in the following steps.

Step I. The discharge, bed slope, rugosity coefficient and value of critical velocity ratio are given. Find out the discharge for which the channel is to be designed.

Step II. Follow the discharge line and find out its intersection with slope lines. If required interpolation may be done to locate the discharge line. Mark the intersection point.

Step III. Draw a vertical line through the point of intersection. This will intersect several bed width curves. Each point of intersection or vertical line and bed width curves give a depth and critical velocity corresponding to this bed width on right hand side ordinate.

Step IV. Choose a pair of bed width, depth and critical velocity corresponding to point of intersection obtained in step III.

Step V. Calculate the area of channel section corresponding to bed width and depth obtained in Step IV.

Step VI. Calculate velocity corresponding to this area obtained in step V.

Step VII. Now find out the ratio of velocity obtained in step VI to critical velocity. This should be equal to 1.0 or to the value of *m* given for channel design.

Step VIII. Repeat the procedure with other values of bed width and depth till step VII is satisfied.

It is to be noted that the diagrams are prepared for a trapezoidal channel with side slope ½:1. It is further to be noted that from the *nomogram* provided at the top, the same curves can be used for different values of rugosity coefficient, *N*. In the *nomogram*, a vertical arrow has been shown. It represents the value of *N* for which the curves have been drawn. When the same curves are used for other values of *N* (marked on the right and left hand side of central value), the point of intersection of discharge and slope curves has to be shifted to the extent given in the *nomogram* and also in the same direction, for drawing the vertical lines.

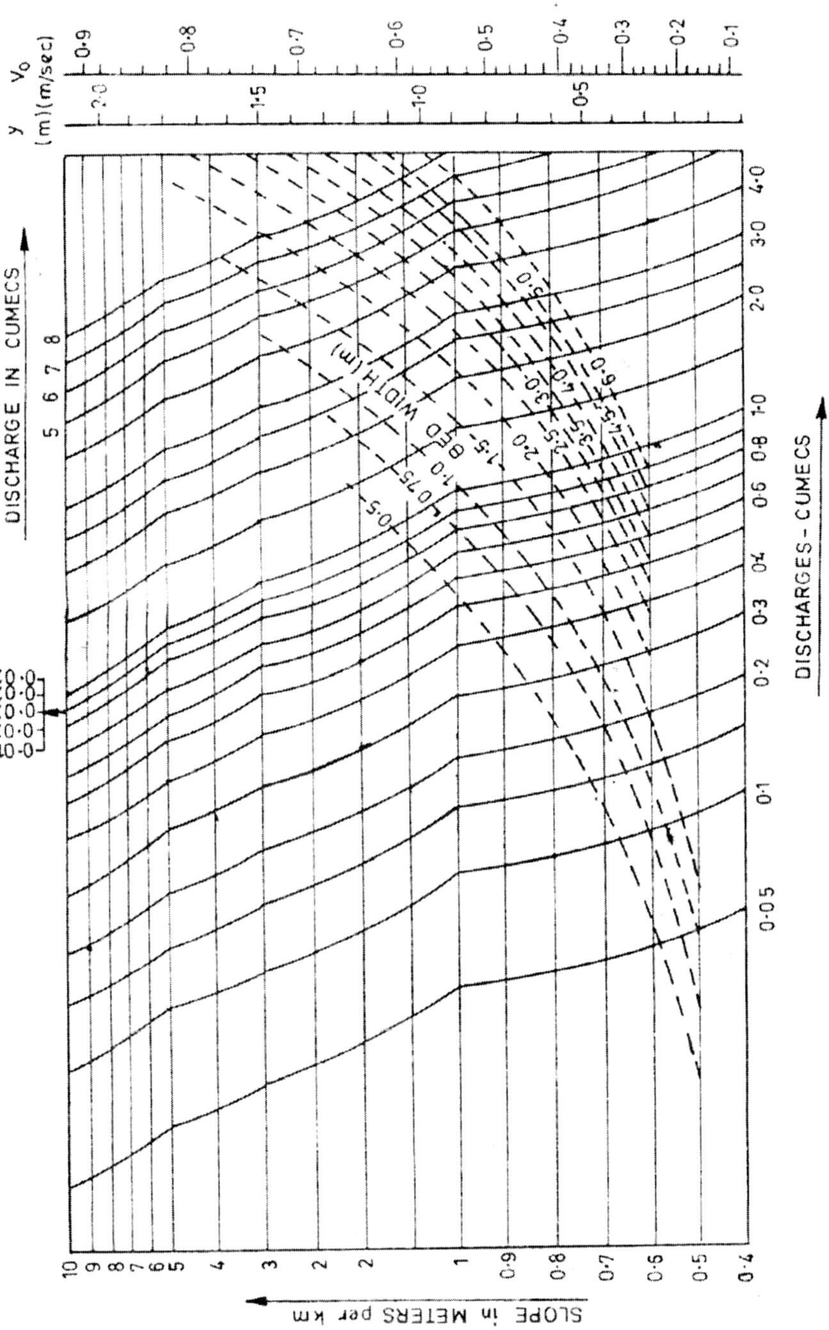

Fig. 9.18: Garret diagram

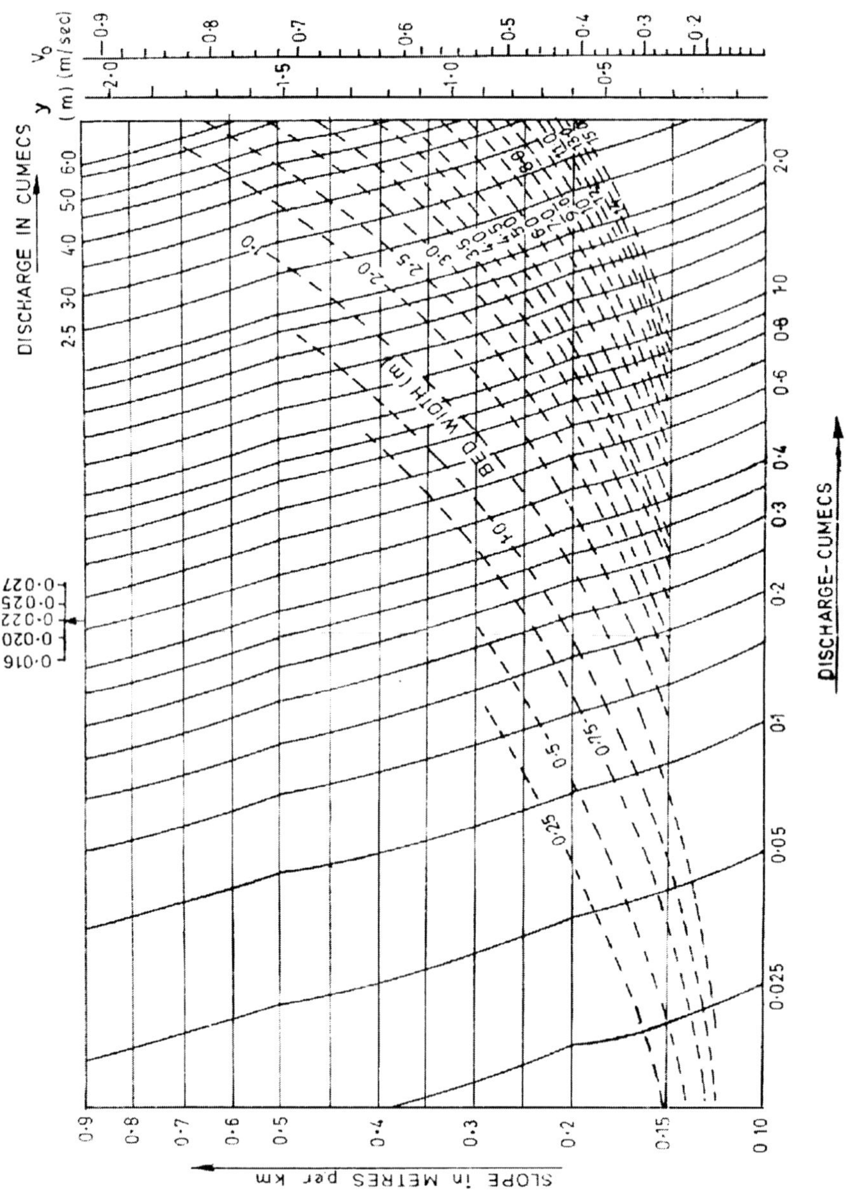

Fig. 9.19: Garret diagram

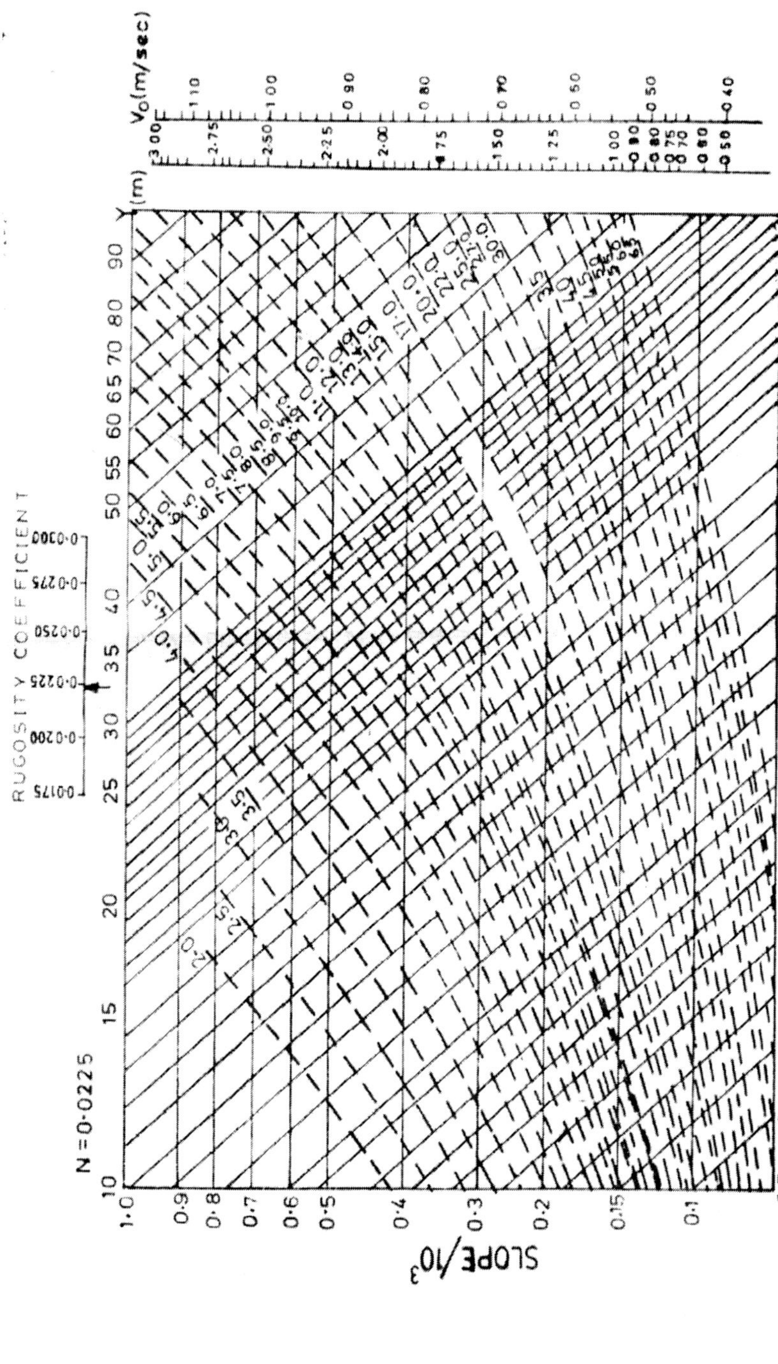

Fig. 9.20: Garret diagram

Sample Calculation 12:

Design an irrigation channel laid on bed slope of 0.02 percent to carry 30 cumecs of discharge. Critical velocity ratio is 1.1. Assume rugosity coefficient, $N = 0.0225$ and side slope ½:1.

Solution:

Given, $Q = 30$ m^3/sce

Bed slope, $S = 0.0002$

Critical velocity ratio, $m = 1.1$ and so mean velocity = 1.1 x critical velocity

Rugosity coefficient, $N = 0.0225$.

The solution of the sample calculation is mentioned as below.

Using Fig 9.20, the point of intersection of given discharge and slope line is found out and then from this intersecting point, a vertical line is drawn. Choose any point of say 12 m, as first approximation. Depth and critical velocity are read as 2.3 m and 0.95 m/sec from the ordinate curve in the right hand side corresponding to bed width of 12 m. With given side slope of ½:1, bed width of 12 m and depth of 2.3 m, the area of section was computed as 30.25 m^2.

With known values of $Q = 30$ cumecs and computed area of section of 30.25 m^2, the mean velocity is computed as 30/30.25 = 0.99 m/sec. Now critical velocity ratio, m = mean velocity/ critical velocity = 0.99/0.95 = 1.04.

Since value of m as computed is 1.04 < 1.1 (given), we have to go to second approximation and let the bed width 12.5 m. Repeating the same procedures we get, depth = 2.25 m, critical velocity = 0.92 m/sec, area of section = 30.66 m^2, mean velocity = 0.98 m/sec and critical velocity ratio = 1.07.

Since, value of m as computed is 1.07 < 1.1 (given), we have to go to third approximation and let the bed width 13.0 m. Repeating the same procedures we get, depth = 2.15 m, critical velocity = 0.90 m/sec, area of section = 30.26 m^2, mean velocity = 0.99 m/sec and critical velocity ratio = 1.1.

Hence, the bed width of the section is 13.0 m, depth is 2.15 m, and side slope is ½:1 and the channel are laid on slope of 0.02 percent.

9.19 Lacey's Regime Diagram for Channel Design

Lacey's equations for design of irrigation channel have been converted into graphical solutions. Fig.(9.21(a)) is used to determine the bed width and depth and Fig.(9.21 (b)) is used to determine the regime slope. These diagrams are prepared for the channel sections having side slope of ½:1. Procedure of the channel design using these figures is presented in following steps.

Step I. To find out the bed width and depth use Figs. 9.21 (a). Find out the point of intersection of discharge Q and silt factor, f.

Step II. Read the value of bed width from abscissa and value of depth from the ordinate.

Step III. Use Fig. (9.21 (b) to determine the regime slope. Find out the point of intersection of discharge Q and silt factor, f.

Step IV. Read the value of ordinate corresponding to this value of point of intersection and determine the bed slope, S.

Sample Calculation 13:

Design an irrigation channel in alluvial soil according to Lacey's regime diagrams with the help of the following data.

Full supply discharge = 15 cumecs.

Lacey's silt factor, f = 1.0

Side slope = ½:1

Solution:

For discharge of 15 cumecs, use Fig. 9.21 (a) to read the values of bed width and depth. Values of bed width and depth are read corresponding to the point of intersection of silt factor line f with discharge line, Q. The values of bed width and depth are read as 15.1 m and 1.38 m, respectively.

Similarly use Fig.9.21 (b) to find out the regime slope. The value of the slope is read as 1:5250.

9.20 Underground Conveyance System

Underground pipelines are now used to convey water with minimum loss from the source to the field. An underground pipeline water distribution system consists of buried pipes for conveying water to different points on the farm. The system has many advantages over the open channel irrigation system. The main advantages are mentioned below (Michael, 1978).

9.20.1 *Advantages of Underground Water Conveyance*

- Since the pipes are laid underground, cultivation can be done above the pipeline without any interference. No culverts and other structures are required. Open channels often take 2 to 4 % of the land area out of cultivation, which is saved by adopting the underground pipeline water distribution system.
- Efficient use of farm machinery can be made because fields are not cut into short runs by channels.
- When properly installed, the pipelines have long life and low maintenance costs.
- They are essentially leak proof .There placement below the ground level prevents damage to pipelines and eliminates water loss by seepage and evaporation.
- Since underground pipelines operate under pressure, they can be laid uphill or down hill, thus permitting delivery of irrigation water to areas not accessible to open channels.
- There is no chance of clogging of the pipelines by vegetations and other materials.

- They can be suited to undulating topography.
- They can irrigate small patches of land and also isolated patches of lands.
- They need little maintenance cost.

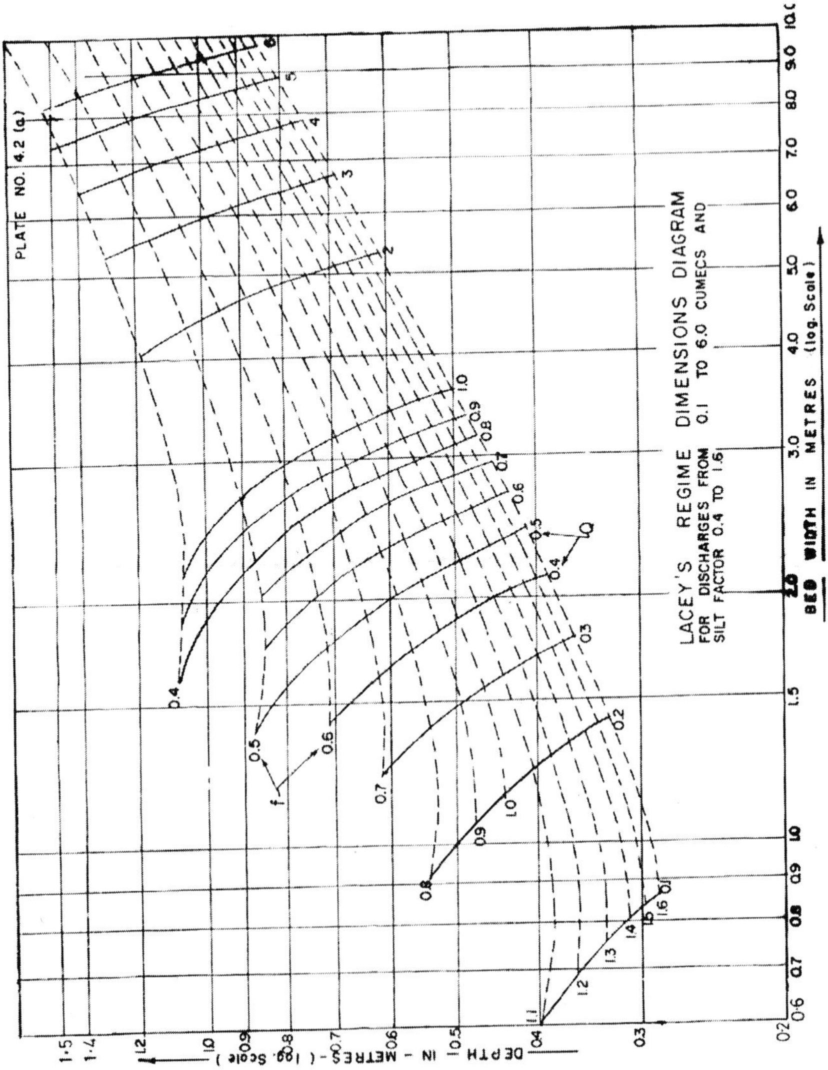

Fig. 9.21: (a) Lacey's diagram for bed width-depth relationship

9.20.2 Disadvantages of Underground Water Conveyance

Common disadvantages with underground pipe systems are (Michael, 1978)

- Underground pipeline irrigation distribution systems have a higher initial cost, as compared to open channels.
- Generally, underground pipelines are not very suitable under canal irrigation system because of insufficient head and silt laden water.

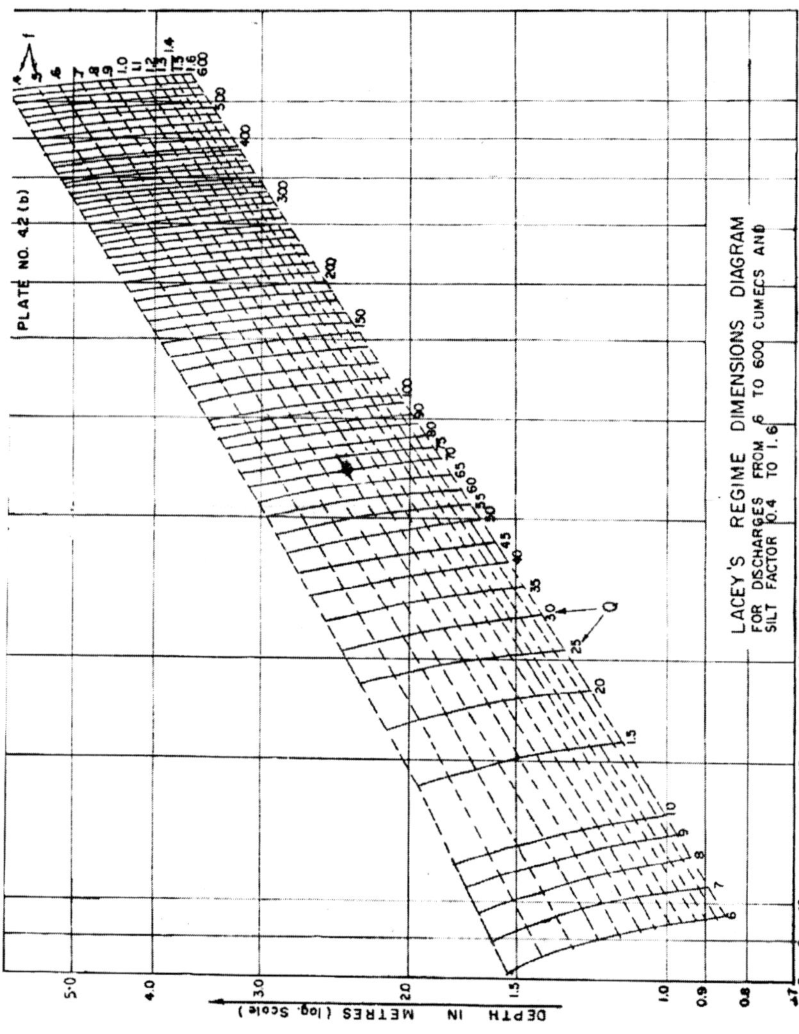

Fig. 9.21: (b) Lacey's diagram for regime slope

- Workmanship of joints between two pipes causes re-work after testing. Improper joint may cause leakage of water.
- Damage to piping from excavation or drilling.
- Damage to piping during backfill, compaction, and surface paving.
- Damage to piping from ground or surface loads or thermal changes.
- Thermal expansion / contraction cause leaks at pipe joints and sump fittings.
- Development of longitudinal cracks in the pipe, usually at the top or both at top and bottom.
- There is chance of surging or intermittent flow of water.

9.20.3 Classification of Pipeline Systems

Underground pipe lines may be classified based on working pressure as (i) low pressure system where working pressure is < 10 m, (ii) medium pressure system where working pressure is 10 - 20 m and (iii) high pressure system where working pressure is > 20 m.

Low pressure systems are used for water conveyance whereas medium pressure systems are used with drip and sprinkler systems. Low pressure system can be classified on the method of pressure control into open, semi-closed and closed systems. It can further be divided into gravity, pumped or mixed systems based on hydraulic head (Murty, 2004).

9.20.4 Materials for Pipelines

The choice of materials for pipes depends on the condition under which the pipes are to operate and the relative costs.

Underground pipeline water distribution systems are usually constructed of concrete or PVC pipes. Occasionally, however, vitrified clays pipes and cement asbestos pipes are used in small holdings. Sometimes, mild steel pipes are used but they are very costly. Concrete pipelines are economical. They are usually constructed with factory made reinforced concrete pipe. In case of small farms, where high pressure is not involved, concrete pipes may be constructed of non-reinforced home-made or factory-made concrete pipes. If sulphates are high, concrete should be avoided or the pipe should be made of with special sulphate resistant cements. Plastic or PVC pipe lines are being used in tube well command areas. These are suitable almost under any kind of field environment. Plastic pipes and pipes made of thermoplastic resins are especially suitable in the smaller diameter for high pressure underground pipelines used for sprinkler irrigation, and water distribution systems in pubic or cooperative irrigation system. Asbestos cement pipes can be easily installed but are often subject to leakage, unless carefully handled; they have a long service life and are adapted to a wide range of water pressure.

Non-Reinforced Concrete Pipes (N.R.C.C.)

Non-reinforced concrete pipes are commonly used when the pipe is not suitable to high pressures. Normally, they are suitable for operating heads (pressures) not exceeding 6 meters (0.6 kg/

cm^2). Non-reinforced concrete pipes are cheaper than reinforced pipes and are ideal for small to medium size farms. They may be hand made or machine made. Handmade pipes are cast in lengths of about one meter and may have bell joints or tongue and groove joints. Steel or wooden forms are used to cast concrete pipes.

Reinforced Concrete Pipes (R.C.C.)

R.C.C. pipes used in India are available in sizes of 15 cm, 22.5 cm, 30 cm and 60 cm. Their lengths vary from 2 to 2.5 m.. The concrete mixes used are 1: 1.5:3, 1: 2: 2, 1: 2.5:2, 1: 2: 3 and 1: 2: 4 (cement: sand: coarse aggregate). The sand used varies from fine sand to coarse sand and is to be free from organic material and other impurities. The size of the coarse aggregate usually ranges from 10 to 15 mm. The reinforcement consists of wooden steel wire conforming to standard specifications. It is so embedded that steel and concrete act together.

High Density Polyethylene Pipes

The PVC (Poly Vinyl Chloride) is a thermoplastic material produced by the polymerisation of vinyl chloride as a sole or principal monomer. The main raw materials required for manufacture of PVC pipes are acetylene and hydrogen chloride or ethylene chloride. The PVC pipes are produced by extrusion of thermoplastic in which the PVC contents are between 85-90% and the rest are stabilisers in the form of lubricants, filters impact strength modifiers and pigments.

Asbestos Cement Pipes

Asbestos cement pipes used as irrigation pipelines are of two types. First type is a pressure pipe made according to IS:1592 and the other type is a non-pressure pipe manufactured according to IS:1626. The asbestos cement pressure pipes are joined together with asbestos cement couplers, cast iron detachable joints or with a concrete collar. The collar joint is used under low heads only.

9.20.5 Testing Irrigation Pipes

There are two standard tests for irrigation pipes. They are test for measure of water absorption and strength. The water absorption in standard irrigation pipe should not exceed eight percent of the original dry weight of the pipe .To test the water absorption, samples of the pipe are boiled for about five hours and the gain in weight by moisture absorption is noted. The hydrostatic pressure requirement prescribes the minimum hydrostatic pressure the pipe line may sustain without excessive leakage through the shell of the pipe. Fig. 9.22 shows the layout for the three- edge bearing test to determine the strength of concrete pipes.

The tests are made on a standard testing machine. The load is applied parallel to and along a vertical plane passing through the axis of the pipe. The lower bearing consists of two wooden strips 2.5 to 3.8 cm in height and about 5 cm wide at the base. The strips are straight and are fastened to a rigid uniformly supported wooden block at least 15 cm wide. The upper bearing is a rigid wooden block. The upper and lower bearings extend to the full length of the pipe. The load is applied continuously and the load at which the pipe is crushed is noted. The load per linear meter of the pipe is calculated by dividing the total recorded load by the length of the pipe.

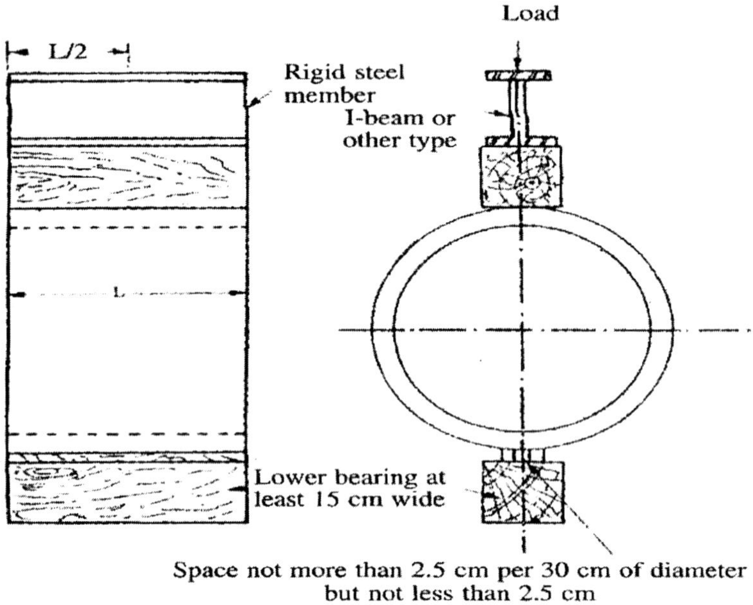

Fig. 9.22: Layout for the three-edge bearing test of concrete pipes

9.20.6 Discharge Capacity of Pipelines

The discharge through a pipeline is determined by Darcy's equation as (Michael, 1978)

$$V = \sqrt{\frac{H\,d\,g}{2\,f\,l}} \tag{9.44}$$

where,

V = velocity of flow of water through the pipe (cm/s), H = available head causing flow (difference in elevation between water level in the pump stand and the outlet point) (cm), d = diameter of pipe (cm), g = acceleration due to gravity (981 cm/sec^2), l = length of pipe (cm) and f = Darcy's roughness coefficient (Table 9.12).

9.20.7 Spacing of Pipelines

Pipelines are to be spaced such that they suit the field layout and soil type. Along the upper boundary of the field, a main line is generally drawn. When the land slope is nearly in opposite directions, the pipelines may be drawn at the middle of the field so that a common pipeline serves irrigation in both sides of the slope of the field. When the existing land slope is only in one direction only, the spacing between the pipelines may be 100 to 200 m in range. Sandy soil needs closure spacing whereas the heavy soil like clay needs wider spacing. Sometimes the pipelines are laid diagonally along the field that reduces the cost of installation of pipelines.

Table 9.12: Values of Darcy's roughness coefficient, f, for pipe flow

Diameter of pipe, cm	Velocity, cm/sec				
	30	60	90	120	150
Galvanised iron pipe					
15.0	0.0056	0.0047	0.0043	0.0042	0.0038
22.5	0.0051	0.0044	0.0041	0.0038	0.0036
30.0	0.0047	0.0042	0.0038	0.0035	0.0034
45.0	0.0044	0.0037	0.0035	0.0033	0.0032
60.0	0.0042	0.0035	0.0033	0.0032	0.0030
90.0	0.0037	0.0033	0.0031	0.0029	0.0029
120.0	0.0035	0.0032	0.0029	0.0028	0.0027
Cast iron pipes					
15.0	0.0061	0.0059	0.0058	0.0057	0.0055
22.5	0.0057	0.0055	0.0054	0.0053	0.0052
30.0	0.0054	0.0052	0.0051	0.0050	0.0048
45.0	0.0052	0.0051	0.0050	0.0048	0.0047
60.0	0.0050	0.0048	0.0045	0.0043	0.0041
90.0	0.0042	0.0040	0.0037	0.0036	0.0034
120.0	0.0036	0.0034	0.0032	0.0032	0.0031
Concrete pipe					
15.0	0.0102	0.0094	0.0090	0.0086	0.0084
22.5	0.0080	0.0074	0.0070	0.0067	0.0064
30.0	0.0067	0.0062	0.0059	0.0056	0.0054
45.0	0.0056	0.0051	0.0048	0.0046	0.0044
60.0	0.0050	0.0046	0.0044	0.0042	0.0040
90.0	0.0044	0.0040	0.0038	0.0036	0.0034
120.0	0.0039	0.0037	0.0035	0.0033	0.0030

9.20.8 Installation of Concrete Pipelines

The process of laying the pipelines involves the following steps:

(i) Selection of depth and grade of laying

(ii) Digging to proper depth and grade

(iii) Lowering the pipe and squeezing

(iv) Sealing the joints and

(v) Backfilling the trench.

Concrete pipes should be laid underground with at least 45 cm soil cover so that it is prevented from breaking. Width of trench should leave sufficient space for working and should be more than 70 cm. Grade of the pipeline should be the same as the average land slope along the lines in which pipe is to be laid. Each pipe should be laid on firm ground. Bottom of the trench should be smooth without any large stones, clods etc. The fill should be properly compacted after laying of the pipelines. The pipes are first strung along the trench and then tilted into the trench. Heavy R.C.C. pipes may be wrapped at the ends by ropes and then lowered by gradually releasing the ropes till the pipes have rested on the bottom of the trench. Pipes are laid in a straight line at the centre of the trench (James, 1988).

9.20.9 Structures for Underground Pipeline

Specialised structures are used with underground pipeline systems to control the water and protect the pipeline from damage. These include the structures at the inlet, water control and diversion structures, air release vents and end plugs. These structures are briefly described as follows.

Inlet Structure

Water may enter a pipeline by gravity from an open channel or it may be pumped from a well or any other source. An inlet structure is required to develop the full flow capacity of the pipeline, to maintain water surface elevations sufficient to distribute water to the different points on the farm, to ensure that excessive pressures do not cause damage, and keep trash from entering the pipe. Fig.9.23 shows a gravity inlet structure in which water enters into the pipelines from an open channel.

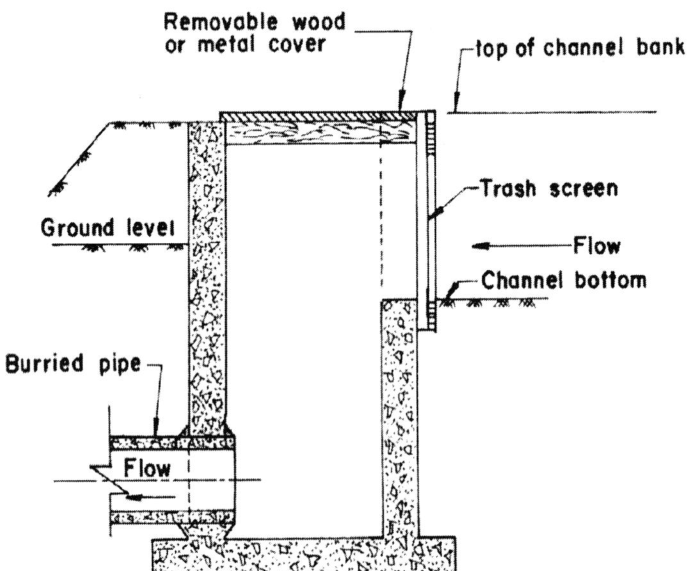

Fig. 9.23: Inlet structure

Pump Stands

A vertical pipe extending above the ground and connected to the underground pipeline system is known as pump stand (Fig.9.24). Pump stands are located at the inlet of the underground pipeline system. It is larger than the size of the pipeline. The larger capacity of the stand permits dissipation of the high velocity stream and release of the entrapped air before the water enters the pipeline. To enable the mason to make mortar joints from inside the stand, its minimum inside diameter should be about 60 cm. Any air entrained by the high velocity stream coming from the pump will have an opportunity to escape at the pump stand. Entry of air into the pipeline can cause air pockets which restrict the flow of water, surging flow conditions and development of excessive pressure. Pump stands must extend upward to a point where water will not overflow except when unusual pressure occurs. The elevation of the water surface in the stand must be sufficient to permit discharge of water through the outlets at any point on the farm. This should also include the pressure head required to overcome friction in the pipes and valves.

To install the pump stand a concrete bed is made. The thickness of bed should be about 10 cm for pump stands up to 3m high and 15 cm when the height is more than 3m.The diameter of the concrete bed should be at least 30 cm more than the outer diameter of the pipe used as pump stand. After laying the concrete and before the bed hardens the pipe is placed vertically in the centre of the bed and a concrete of 1:2:4 mix is added to a thickness of about 5 cm.

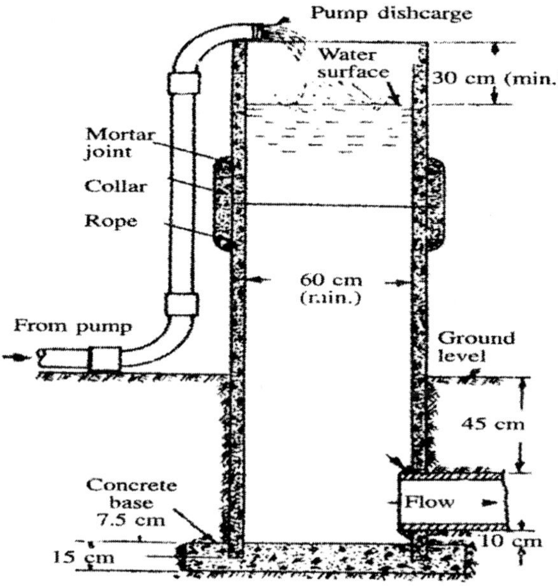

Fig. 9.24: Pump-stand for underground pipelines

Gravity Inlets

When water enters the pipeline from an open channel a gravity inlet is used. A screen is fixed to the inlet to keep trash out of the pipeline. The top of the structure is usually provided with a cover to prevent accidents and to keep trash from blowing in. If the irrigation water contains appreciable quantities of sand, a trap can be built into the inlet structure to remove most of the suspended material. From the stand functions as a sand trap also, it has an extra large diameter to ensure low velocity of water and its bottom is set at some distance (about 60 cm) below the bottom of pipeline. When the canal water contains considerable quantities of debris and weed seed a debris screen is provided at the inlet to clean the water before it enters the pipeline. Commonly, is allowed to fall through a fine screen the stream into a gravity inlet. The screen will need frequent cleaning.

Gate Stands

Gate stands are provided to control the flow laterals when pipelines branch off in different directions. They also prevent high pressure and act as surge chambers .They are made in the same way as pump stands, but are provided with slide gates or valves to control flow of laterals

Air Vents

Air vents are vertical pipe structures to release air entrapped in the pipe line and to prevent vacuum. Entrapped air must be removed to permit an even flow and avoid the danger of water hammer. They are installed near the pump stand, at all high points in the line, at sharp turns at points where there is a downward deflection of more than 10^0; directly downstream from any structure that may entrap air, and at the end of the pipeline.

End Plug

End plug is provided where the line terminates. The function of the end plug is to close a line and to absorb the pressure developed at the end of the line.

Outlets

Outlet structures are necessary to provide controlled delivery of water to the field at any desired location. These should be easy to open and close, be of proper size to provide the flow required, and be so constructed that the water release will not cause soil erosion. The most common outlet consists of a concrete riser pipe to bring the water from the buried pipeline to the ground surface and an attached valve to control the flow of water.

Riser Valves

Riser valve is a screw valve installed on the top of riser pipe .A handle and cap plate is attached to a screw type shaft that moves up or down as the handle is turned. When the valve is closed the cap plate fits the circular edge of the valve case provided with a rubber seal to make it water tight. When the plate is lifted by turning the handle, water is released from all sides of the valve .The rate of flow can be controlled by raising or lowering the valve plate. The valve stem is usually made of brass, cast iron or other rust resistant material. The seal is made of synthetic rubber cast in a suitable die. Riser valves are of two type; the valve for releasing water directly to fields or open channels and hydrant valves which are adapted for fixing a portable hydrant on it.

Hydrants

Hydrants are device placed over riser valve outlets as a means of connecting portable gated pipes to the pipe line. They are portable so that they can be moved from one valve outlet to another to serve the portion of the field which is being irrigated at a particular time. Portable hydrants are constructed of cast aluminum or cast iron. They have a double screw arrangement which is connected to the handle on the riser valve.

Gated Pipes

Gated pipes are portable lines with uniformly spaced used for releasing irrigation water to furrows, border strip or check basins. When connected to buried pipe lines through hydrants, they permit conveyance of the water in an enclosed system from the source to the head end of the irrigation strip. Seepage losses are reduced to a minimum .They also provides a convenient means of regulating the flow of water into the furrows or border strips (Hansen *et al*., 1979).

9.20.10 Design of Underground Pipeline Irrigation System

The layout of the underground drainage system should be kept as simple as possible, with the minimum number of changes in direction and gradient. Access points should be provided only if blockages could not be cleared without them. An underground pipelines irrigation system must be properly designed to handle efficiently the required flow throughout the system. If pipelines are too small in diameter, pumping cost is increased and the capacity of the system may be seriously limited. On the other hand, pipelines larger than necessary add to the cost of the system and may cause uneven flow. Pump stand must be high enough to allow sufficient operating head for the pipeline. Stands higher than necessary may permit high heads of water to build up, leading to excessive line pressure.

The pipelines size and heights of pump stands and control stand must be determined so as to obtain a balance water distribution and provide trouble-free, economical operation. The working pressures in the pipelines are kept within one-fourth the internal bursting research of the pipe. When it is necessary to design pipelines with higher head, reinforced concrete pressure pipes are used. The sizes of the outlets are selected to suit the flow required at diversion points.

In designing an underground pipeline system, the head loss due to friction in the pipeline and other factors must be taken into account. Loss of head due to friction in the pipes between two points in a pipe at a distance l apart is given by Darcy's formula as:

$$h_f = \frac{f\, l\, V^2}{2\, g\, R} \tag{9.45}$$

where,

f = Darcy's roughness coefficient (dimensionless) which is given in Table 9.12; l = length of pipes, m; V = velocity of water in pipe, m/sec; g = acceleration due to gravity (9.81 m/sec^2) and R = hydraulic mean radius, m.

Hydraulic mean radius, R is:

$$R = \frac{A}{P}$$

where,

A = area of cross-section of pipe, m^2 and P = wetted perimeter, m.

Since, $A = \frac{\pi D^2}{4}$, $P = \pi D$ (where D is diameter of pipe),

Hydraulic mean radius, R is

$$R = \frac{D}{4}$$

Therefore,

$$h_f = \frac{2 f l V^2}{g D} \tag{9.46}$$

Table 9.13 gives the value of friction loss in concrete pipe at various discharge rates.

Table 9.13: Friction loss in concrete pipes at various discharge rates

Discharge rate, lit/sec	Friction loss, m/100 m length of pipe					
	Diameter of pipe, cm					
	20	25	30	35	40	45
2.8	0.01					
5.7	0.03					
8.5	0.07					
11.3	0.12	0.04				
14.2	0.19	0.06				
17.2	0.26	0.09	0.04			
22.7	0.45	0.15	0.06			
28.3	0.68	0.23	0.09	0.04		
34.0	0.95	0.32	0.13	0.06		
39.6	1.23	0.43	0.18	0.08	0.04	
45.3	1.62	0.55	0.22	0.11	0.06	
51.0	2.02	0.68	0.28	0.13	0.07	
57.0	2.45	0.83	0.34	0.16	0.08	0.05
68.0	23.43	1.16	0.48	0.22	0.12	0.07
79.0	4.57	1.54	0.63	0.30	0.16	0.09
91.0	5.85	1.97	0.81	0.38	0.20	0.11
102.0	7.27	2.46	1.01	0.48	0.25	0.14
113.0	8.84	2.98	1.03	0.58	0.30	0.17
127.0		3.71	1.53	0.72	0.38	0.21
142.0		4.51	1.86	0.88	0.46	0.26
156.0		5.38	2.21	1.05	0.55	0.31

Entrance losses in pipelines may be calculated from the following relationship:

(i) Pipe end flush with the wall of the pump stand:

$$h_e = 0.5 \frac{V^2}{2g} \tag{9.47}$$

(ii) Pipe end protruding from wall of pump stand:

$$h_e = 1.0 \frac{V^2}{2g} \tag{9.48}$$

(iii) Pipe with bell entrance

$$h_e = 0.10 \frac{V^2}{2g} \tag{9.49}$$

where,

h_e = head loss at pipe entrance, m and V and g are the same as defined earlier.

The other losses which may be significant are the velocity head in the pipe and the head in the pipe and the head loss through riser valves. The velocity head due to flow in pipe (h_v) is obtained from the formula:

$$h_v = \frac{V^2}{2g} \tag{9.50}$$

Head loss through riser valve is given in Table 9.14.

The water surface elevation in pump stand and control stands should be sufficient to distribute water through all the risers in the field and to overcome the losses due to friction and other causes. The hydraulic gradient shows the height to which water will rise in open strand pipes or vents at any place along the line for a given flow. The hydraulic gradient line at any point in the pipeline may be obtained by deducting the total losses in the pipeline up to that point from the elevation of the water surface in the stand pipe. Changes in slope, sharp bends, and other factors will cause a shift in the hydraulic gradient and should be considered. The designer should plan the system so that the pipeline flows full and the hydraulic gradient remains above the pipeline to prevent surges and trouble due to entrapped air. In case gated pipes are used, the pressure required to overcome the friction in them is to be added to the system requirement.

Table 9.14: Head loss through riser valve

Discharge, lit/sec	Head loss in centimeters of water					
	Diameter of pipe, cm					
	15	20	25	30	35	40
10	2	1				
20	10	3	1			
30	26	7	3	1		
40	50	13	5	2	1	
50	83	22	8	4	2	1
60	126	34	12	6	3	2
80	250	64	24	11	5	3
100		105	40	18	9	5
125		165	65	29	15	7
150		250	97	44	22	12
175			136	65	30	17
200			185	83	41	22
225			245	100	55	30
250				137	69	37
275				170	83	45
300				210	100	56
350				300	145	77
400					195	103
450					250	135
500						170
550						210

Sample Calculation 14

A concrete pipe line has to carry water across a distance of 1 km from a pump which discharges 100 lit/sec. The filed has a down slope of 0.1 percent along the line the pipes are laid. It is required that a maximum pressure in the pipeline should not exceed 4 m of water. What size of a pipe is required? How high the pump stand be to operate the system.

Solution:

Given that the permissible pressure in the pipeline = 4 m

Gain in pressure head due to downfield gradient = 0.1/100 x 1000 – 1 m

The total available pressure head = 4 + 1 = 5 m over a length of 1000 m i.e. 0.5 m/100 m length.

From Table 9.13, it is observed that a 35 cm diameter pipe will deliver a discharge of 102 lit/sec and the friction loss will be 0.48 m/100 m length of the pipeline. Hence, a 35 cm diameter pipe is selected.

The water will rise to a height of 10 x 0.48 – 1.0 = 3.8 m.

Allow aw free board of 30 cm and entrance head of 50 cm.

The stand pipe will be 3.8 + 0.30 + 0.50 = 4.60 m.

Hence, the pump stand be 4.6 m to operate the system.

9.21 Structures in Conveyance System

Structures in the irrigation conveyance system are required to control the flow of water and also distribute water. Sometimes, the irrigation channels are designed on steep slopes and water moving in such step slopes attains high erosive velocity and cause severe soil erosion in earthen channels. In such a case, structures like drop and chute spillway are required to prevent soil erosion in earthen channels.

9.21.1 Drop Structures

These are used to convey water from a higher elevation to a lower one. A series of drops in the channel act as control points and reduce the slope of the channel beds. They may be two types: (i) Open drop structure and (ii) Pipe drop structures.

Open drop structures can be made up of timber, concrete, brick or stone masonry. The structure has a head wall and head wall extension, side wall, wing walls, stilling basin/apron, longitudinal and end sills and cutoff walls (**Fig. 9.25**). Water enters the structure through the head wall, falls on the apron, and then leaves the structures. The stilling basin including apron, longitudinal and end sill helps in dissipation of energy of falling water. The width of the stilling basin is kept equal to width of the channel. The length of the stilling basin should be such that the water jet from head wall should fall within the stilling basin. The crest length of the weir i.e. head wall is given by formula as:

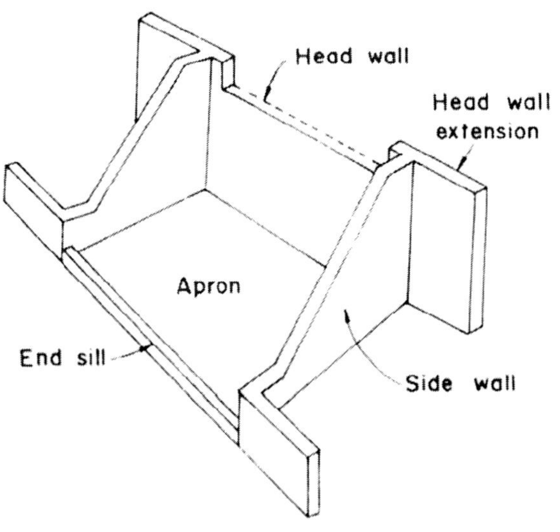

Fig. 9.25: A typical drop structure

$$Q = C\,L\,h^{3/2} \tag{9.51}$$

where,

Q = peak discharge to be conveyed by the structure, m^3/sec; L = crest length, m; h = head over crest, m and C = coefficient of flow = 1.66 when L and h are in metres and Q is in m^3/sec. A free board of 15 to 30 cm is to be added to h.

Detail design of drop and pipe structures and chute spillway are available in any standard book of Soil and Water Conservation Engineering.

Sometimes construction of an open drop structure is not possible without disturbing an existing bund or dam. In such cases, water can be safely discharged from higher to lower level by providing a pipe drop structure. This type of structure allows discharge of water through a pipeline, leaving the dam or bund undisturbed. The velocity of water in pipe drop spillways using various sizes of pipes can be calculated from Bernoulli's theorem as:

Available head = Friction loss in pipeline + Velocity head + Head loss at pipe entrance + Head loss at bend i.e.

$$H = \frac{4\,f\,l\,V^2}{2\,g\,d} + \frac{V^2}{2\,g} + K_1\,\frac{V^2}{2\,g} + K_2\,\frac{V^2}{2\,g} \tag{9.52}$$

where,

H = water level difference between upstream and downstream ends of the structure, m; V = velocity of flow in pipe, m/sec; f = friction coefficient of pipe $\approx$ 0.01; l = length of pipe, m; g = acceleration due to gravity, 9.81 m/sec^2; d = diameter of pipe, m and K_1 *and* K_2 are constants taken as 0.5 and 0.25, respectively.

Discharge of the pipeline is given as:

$$Q = A\,V \tag{9.53}$$

where,

Q = discharge, m^3/sec; A = cross sectional area of the pipe, m^2 and V = velocity of flow in pipe, m/sec.

Sample Calculation 15:

Given the water level difference between the upstream and downstream end of the structure is 1 m, diameter of the pipe 10 cm, length of pipe is 3 m, determine the discharge capacity of the pipe. Assume f = 0.012.

Solution:

Given, H = 1 m, d = 10 cm, f = 0.012 and l = 3m.

Using Eq. (9.52),

$$1 = \frac{4\,x\,0.012\,x\,3\,V^2}{2\,x\,9.81\,x\,0.1} + \frac{V^2}{2\,x\,9.81} + 0.5\,\frac{V^2}{2\,x\,9.81} + 0.25\,\frac{V^2}{2\,x\,9.81}$$

$V = 2.5$ m/sec.

Cross sectional area of pipe, $A = \frac{\pi\,x\,d^2}{4} = \frac{\pi\,x\,0.1\,x\,0.1}{4} = 0.0078\,m^2$

Hence, discharge, $Q = A \times V = 0.0078 \times 2.5$

= 0.0195 cumec = 19.5 lit/sec.

9.21.2 Chute Spillway

Where the irrigation channel has to pass through steep land slope, chute spillway can be used to convey the water from higher to lower levels. On steep slopes, chute spillway is more economical than a series of drop structures. Chute spillway consists of inlet through which water enters into the structure, conduit to allow water pass through and outlet through which finally water will go out. It is constructed of concrete, bricks laid in cement mortar or precast concrete sections. The stilling basin helps in energy dissipation of the falling water. A check gate in the inlet helps in controlling the water level in the upstream channel (Fig. 9.26). To minimize the problem of settling and undermining, chutes are constructed on solid ground or on fill that has been carefully compacted. The discharge capacity of chute spillway can be calculated by using Manning's formula. Chute sections are usually rectangular in section. Length of the stilling basin may vary from 1.0 to 1.2 m. Depth of the stilling basin is about 10 cm below the bed level of the downstream channel. When water enters from the highly sloppy section of the chute to the flat stilling basin, velocity suddenly changes and form hydraulic jumps in the downstream section. To prevent overflow of water due to this hydraulic jump, the height of the walls in the downstream section should be increased as compared to the channel section of the chute spillway.

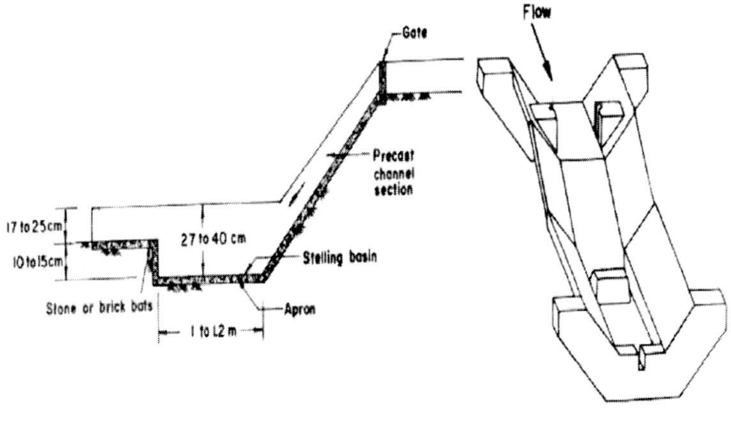

Fig. 9.26: Schematic diagram of a chute spillway

9.21.3 Water Diversion and Control Structures

Water from the farm irrigation channels is diverted to branch channels or into fields by means of junction boxes, gates or other forms of outlets. Fig. 9.27 shows the view of a typical junction box. Metallic gates are used in junction boxes to divert water. Sometimes, the gates are called as check gates. Check gates in an irrigation channel form an adjustable dam to control the elevation of water at upstream end. To apply water from channels into fields, it is needed to raise the water level. Water level in the channel should be at least 8 to 12 cm above the ground surface in order that siphon tubes or pipe turnouts may be used effectively. A check structure may consist of masonry or metallic wall built across the channel and provided with a suitable check gate. The crest of the check gate is at the same level as the bottom of the upstream channel. Specially designed rubber strips are used to obtain a watertight seal. Two types of sealing strips may be used i.e. guide strip and the sill strip. The length of the strip may vary with size of gates.

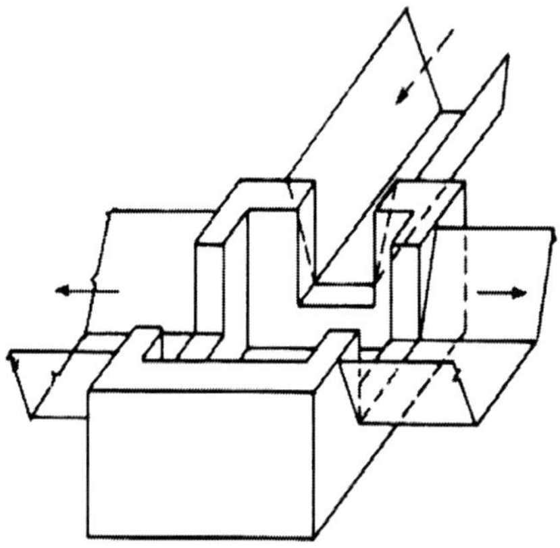

Fig. 9.27: A typical junction box

9.21.4 Circular Orifice Panel

Circular orifice panel consists of an inclined concrete panel with a circular opening of suitable diameter. Another circular concrete block exactly fits into this opening. The shape of the opening and the lid are such that the joint will be leak proof. Leakage can be controlled by applying some amount of mud. A little application of grease is helpful if the panel is used for casting the lid. Vibrating platforms will produce better quality castings. The lid after casting is subject to grinding for obtaining fine finish and ultimately a watertight joint.

9.21.5 Portable Check Dams

Portable check dams are made of canvas cloth or thin plastic sheets or sheet metal. Canvass or plastic dams are supported on a wooden or pipe cross bar with suitable loops. A loop provided at the bottom helps in anchoring the dam to the channel bed. The sheet metal dams are more durable than the canvas or plastic dams. These are driven into the channel banks to form a seal. Suitable dams can be made of 18 gauge mild steel sheets cut to suit the cross section of the unlined channel. The canvas or plastic dams are more convenient to use in both lined and unlined irrigation channels. The sheet metal dams are only suitable for unlined irrigation channels. Canvas or plastic dams should be removed immediately after irrigation and carefully washed out, dried and stored so that they may last for several seasons.

9.21.6 Diversions

In carrying out water to different farms or to different fields in the same farm, it is required for the water course or the main channel to divert the stream into proper channel. For this purpose, two, three or four way diversions made of check gates at the channel junction will serve the purpose.

9.21.7 Turnouts

Turnout is required when water is taken from a lateral to field channel or from a field channel to fields. Turnouts may be portable or built-in type. They are sometimes equipped with gates to control the flow of water. If only a part of the total flow is to be diverted through a given turnout, a more constant discharge is obtained by using an orifice type device rather than a weir type structure. One of the common turnouts is a concrete or metal pipe with a slide gate on the inlet (Fig. 9.28).

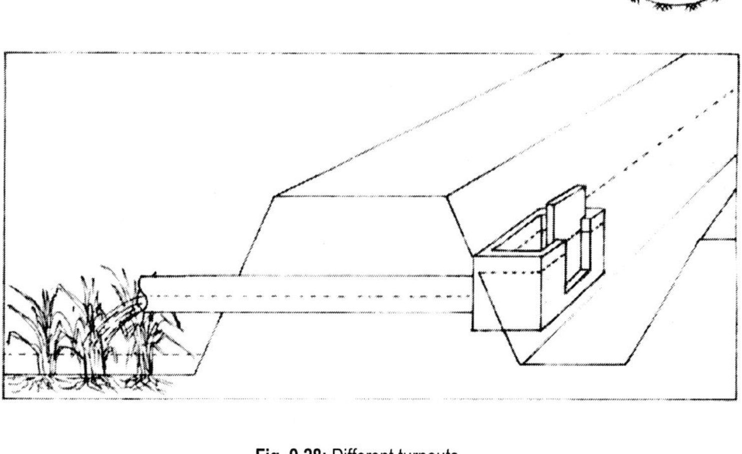

Fig. 9.28: Different turnouts

Capacity of pipe turnout can be determined by (Murty, 2004):

$$Q = C\ A\sqrt{2gh} \tag{9.54}$$

where,

Q = discharge, m^3/sec; C = coefficient usually taken as 0.8; A = cross sectional area of the pipe, m^2; h = head of flow, m and g = acceleration due to gravity, 9.81 m/sec^2.

The most common turnouts are box turnouts, spiles and siphon tubes. Prefabricated concrete turnouts are convenient to use.

9.21.8 Spiles

Spiles are made of bamboo, concrete or baked clay pipes. The size of the spiles varies from 2.5 to 10.0 cm or more depending on the flow of water desired and depth of water at the inlet and outlet of the pipe. They have no gates for water control. The length is made slightly larger than the width of the field channel. The spile should be placed at the lowest portion of the bank to reduce erosion at the outlet. Spiles are suitable to use in case of irrigation by furrows as one spile can be used for each furrow.

9.21.9 Siphon Tubes

Siphon tubes are used to convey water from the channel to the crop fields for application. They can be conveniently used in furrows also. Plastic, rubber or aluminum pipes are commonly used. Each siphon tube is about a metre or more in length. They are commonly available in sizes ranging from 2.5 to 7.5 cm. Before use, the tubes are to be primed. Siphon tube is dipped in water source and completely filled with water. Keeping one end in the water source in immersed condition, the other end is tightly closed with hand and is taken out and released at the outlet point to which water has to be applied as irrigation. Bigger size aluminum siphon pipes are equipped with a small hand operated pump for priming. Siphon tube may work under both submerged and free condition [(Figs 9.29 (a) and (b)].

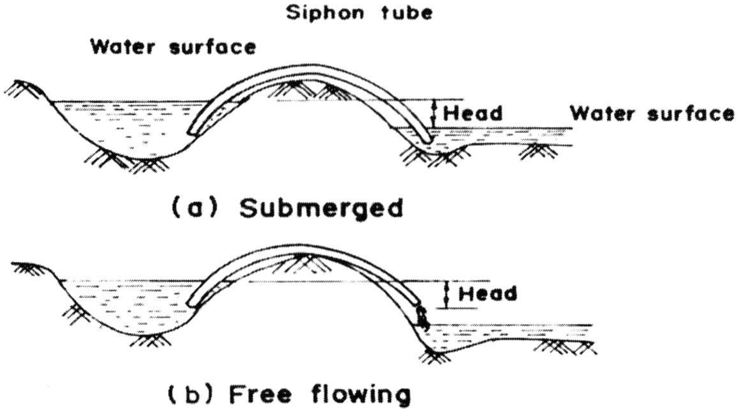

Fig. 9.29: Siphon tube used for water conveyance

The discharge of siphon tube is given by equation (Murty, 2004):

$$Q = C\ A\ \sqrt{2gh} \tag{9.55}$$

where,

Q = discharge, lit/sec; C = coefficient usually taken as 0.65 x 10^{-3}; A = cross sectional area of the pipe, cm^2; h = head of flow, cm and g = acceleration due to gravity, 981 cm/sec^2. The head is the difference of water level at the upstream and downstream end of the tube. If the outlet is not submerged, the effective head *(h)* is the vertical distance from the water level over the inlet end to the centre of the discharge end.

9.21.10 Structures at Channel Crossing

Sometimes irrigation channel has to pass over roads, hillsides or drainage ways. For most of the cases, a straight pipe or inverted siphon, depending on the conditions, can accomplish this job. When the siphon is laid over a drainage way, it is often called as flume. When it crosses a ridge, highway, or bund, it is called as culvert.

9.21.11 Flumes

Flumes are used to carry irrigation water across the stream, canal gullies, ravines, etc. They may be made of open channel or pipe which are often supported by pillars or may be fixed to bridge. The open channel type flumes are semi-circular metal pipes, or rectangular or trapezoidal wooden channels. The supporting structures may be made of timber, steel or concrete. When pipes are used as flumes, they should be placed below the water surface on the upstream end to ensure that they flow full.

9.21.12 Culverts

Culverts are commonly used at channel crossings when the road fill is sufficiently high and the channel bed lies on the field surface on either side. Culvert consists of masonry headwalls at the inlet and outlet ends connected together by a buried pipeline. The soil cover over the pipeline should be at least 30 cm, but preferably 45 cm. At the outlet of the culvert, a stilling basin is required to prevent soil erosion downstream. For ditch crossing, low cost culverts made of concrete pipes of adequate capacity may be used.

9.21.13 Inverted Siphon

When a channel has to cross wide depression or where the road surface lies close to the field surface, an inverted siphon can be used which is more economical. The structure consists of an inlet and outlet tank connected together at bottom by a pipe. A check gate is usually provided at the inlet end to control the water surface level in the upstream channel. The bottom of the tank is kept about 15 cm below the bottom of the pipe to collect silt of the flowing water and so the tanks need occasional cleaning. The size of the tank should be large to facilitate in cleaning and should be about 60 cm x 60 cm. Pipes of the inverted siphon must be protected against damage by traffics which is done by covering soil 30 cm or more above the pipe.

9.22 Cross-Drainage Structures

In an irrigation project, when the network of main canals, branch canals, distributaries etc. are provided, then these canals may have to cross the natural drainage structures like river, streams, nallahs etc. at different points within the command area of the project. Suitable structures may be constructed at the crossing point for easy flow of water in the canal and drainage channels in respective directions. These structures are called as cross-drainage works. Types of cross-drainage work depend on the relative bed levels, maximum water levels and relative discharge of canals and drainage structures.

When the bed level of the irrigation canal is well above the highest flood level of the drainage structures like river, streams etc., *aqueduct* is used. In a hydraulic structure, where the canal is taken over the drainage, but the drainage water cannot pass clearly below the canal, siphonic action is required to carry water in the canal and the hydraulic structure is called as *siphon aqueduct*. This structure is suitable when the bed level of the canal is below the highest flood level of the drainage structures like river, streams etc. When the bed level of the drainage is above the full supply level of the canal, hydraulic structure used to carry water is called as *supper passage.* When the canal bed level is low such that the drainage is taken over the irrigation canal, and canal water passes below the drainage under siphonic action, the hydraulic structure is called as *siphon super passage.* Sometimes, both the canal and drainage beds are at the same level, In such a case, the hydraulic structure required is called as level crossing. Level crossings are required for the crossing of large drainages like rivers with main canals. However, if the drainage structure and canal are small, then simple *inlet and outlet* structures are provided to allow water to flow in their respective directions. No hydraulic structures are needed in such case.

Detail designs of above mentioned various cross-drainage works are beyond the scope of this book. For details, the readers may read any standard book of Hydraulic Structures including Garg (1987).

Question Banks

Q1. Differentiate between direct irrigation system and storage irrigation system.

Q2. What is a canal? What are the differences between canal and channel?

Q3. Describe the different classifications of canals used in irrigation.

Q4. Differentiate between the followings:

(i) Alluvial canal and non-alluvial canal (ii) Inundation irrigation and perennial irrigation
(iii) Permanent canal and inundation canal

Q5. What are the various types of irrigation canals based on alignment?

Q6. Write short notes on the followings:

(i) Watershed canal (ii) Contour canal and (iii) Side-slope canal

Q7. What are the advantages of a watershed canal?

Q8. How is a side-slope canal aligned?

Q9. What are the general considerations of canal alignment?

Q10. What are the general considerations of field channel?

Q11. What do you mean by canal irrigation system? Briefly describe the various distribution systems of canal irrigation.

Q12. What are the uses of a diversion head work?

Q13. With a label sketch, describe the working of a storage irrigation scheme.

Q14. How the curves in a channel section are is designed based on alignment system?

Q15. Write down the various advantages and dis-advantages of inundation irrigation system.

Q16. What is a bandhara irrigation system? How does it function?

Q17. What is *Phad* and *Thal* in bandhara irrigation system?

Q18. With a label sketch explain the construction of bandhara irrigation system.

Q19. What are the advantages and dis-advantages of bandhara irrigation system?

Q20. Write notes on the followings with figures:

(i)Canal banks (ii) Berm (iii) Hydraulic gradient (iv) Counter berm (v) Free board (vi) Service road (vii) Dowel (viii) Spoil bank (ix) Borrow pit (ix) Balancing depth (x) Silt factor (xi) Coefficient of rugosity.

Q21. Write the formula of the following terms:

(i)Manning's formula (ii) Chezy's formula (iii) Bazin's formula (iv) Kutter's formula

Q22. Find the balancing depth of a canal section (canal is fully cutting) having the following dimensions. Draw the canal section and find out the total land width required.

R.L. of the top of the bank is 152 m, R.L. of the bottom of the bank = 148 m, side slope of the bank = 2:1, side slope of canal section = 1:1, berm width = 1.2 m, bottom width of the canal = 8 m and top width of the bank = 2.5 m.

Q23. What is a regime channel? It is possible to have a true regime channel?

Q24. What do you mean by economical channel section? Derive the relationship $b = 2d \tan \theta$ in a trapezoidal channel section.

Q25. Design an irrigation canal to carry a discharge of 20 cumecs with permissible mean velocity of 3.0 km/hr. Bed slope of the canal section is 1:5000 and side slope is 1:1. Assume Chezy's constant (C) as 44. Assume any other data not given in the problem.

Q26. Design an irrigation channel to carry a full supply discharge of 5 cumecs having bed slope of 0.02 percent. The value of rugosity constant N is 0.0220 and assumes the critical velocity ration, m as 1.0. The side slope of the channel is 1:1.

Q27. What are the two main approaches of design of regime channel?

Q28. Describe the principle of Kennedy's theory for design of unlined alluvial channel?

Q29. Write down the various design steps for unlined alluvial channel based on Kennedy's theory.

Q30. Design an irrigation channel to carry a full supply discharge of 12 m^3/sec. The ratio of base width to depth of flow (B/D) = 4 and side slope of the section is 1:1. Assume the critical velocity ration, m = 1.0 and Bazin's constant, K = 1.5.

Q31. What are the drawbacks of Kennedy's theory?

Q32. Point out the differences between Kennedy's theory and Lacey's theory.

Q33. Describe the principle of Lacey's theory for design of unlined alluvial channel.

Q34. Design a channel section to carry a full supply discharge of 200 cumecs having side slope 1:1 and mean diameter of the silt particle is 0.35 mm.

Q35. Write down the drawbacks of Lacey's theory.

Q36. What is a lined channel? What are the design considerations of lined channel?

Q37. Design a lined trapezoidal canal section to carry a discharge of 100 cumecs. Assume bed slope of the canal as 0.02 percent, side slope 1.5:1 and rugosity coefficient, N as 0.020. Permissible velocity is 1.80 m/sec.

Q38. What is the use of Garret's diagram? Write down the various steps for design of irrigation channels by Garret's diagram?

Q39. Design an irrigation channel laid on bed slope of 1:5000 to carry 25 cumecs of discharge. Critical velocity ratio is 1.1. Assume rugosity coefficient, $N = 0.0225$ and side slope ½:1.

Q40. Describe the various design steps for design of irrigation channels in alluvial soil by Lacey's regime diagram.

Q41. Design an irrigation channel in alluvial soil according to Lacey's regime diagrams with the help of the following data.

Full supply discharge = 10 cumec; Lacey's silt factor, $f = 1.0$ and side slope = ½:1

Q42. What do you mean by underground water conveyance system? Write down the various advantages and dis-advantages of underground water conveyance system.

Q43. Classify the pipeline systems for underground water conveyance system.

Q44. Describe the different materials used for pipelines in conveyance of water.

Q45. With a neat label sketch, explain how are the strengths of irrigation pipelines are tested.

Q46. How the concrete pipelines are are installed in the field to carry irrigation water?

Q47. Enumerate the different structures used in underground water conveyance system.

Q48. Explain the procedure of design of underground pipeline system in details.

Q49. A concrete pipe line has to carry water across a distance of 1.5 km from a pump which discharges 120 lit/sec. The filed has a down slope of 0.1 percent along the line the pipes are laid. It is required that a maximum pressure in the pipeline should not exceed 4 m of water. What size of a pipe is required? How high the pump stand be to operate the system.

Q50. What are the uses of drop and chute spillway in conveyance of irrigation water in the fields? Draw the figures of these structures and explain the working principles.

Q51. What is a pipe drop structure? How does it help in conveyance of water in the channels?

Q52. Narrate the working and functions of junction boxes and gates in conveyance of water in the channels.

Q53. Wrote the functions of the following structures used in conveyance of water.

(i) Portable check dam (ii) Circular orifice panel (iii) Turnouts (iv) Spiles (v) Siphon tubes (vi) Inverted siphon tubes (vii) Flumes and (viii) Culverts.

Q54. What is a cross-drainage structure? How does it help in conveyance of irrigation water in the canal commands?

Q55. Write notes on the followings:

(i) Aqueduct (ii) Siphon aqueduct (iii)Supper passage (iv) Siphon Supper passage

Q56. Water is flowing in a canal which has to pass a drainage channel. Both the canal and drainage beds are at the same level. Which type of cross-drainage structure is to be used to convey water in the canal without any obstruction? Explain the working of such a cross-drainage structure.

References

Basak, N.N. 2012. Irrigation Engineering. Tata McGraw Hill Education Pvt. Ltd., New Delhi.

Garg, S.K. 1987. Irrigation Engineering and Hydraulic Structures. Khanna Publishers, New Delhi.

Hansen, V.E., Israelsen, O.W. and Stringham, G.E. 1979. Irrigation Principles and Practices. John Wiley and Sons Inc., New York.

James, L.G. 1988. Principles of Farm Irrigation System Design. John Wiley & Sons, New York, pp. 543.

Michael, A.M. 1978. Irrigation Theory and Practices. Vikash Publishing House Pvt. Ltd., New Delhi.

Murty, V.V.N. 2004. Land and Water Management Engineering. Kalyani Publishers, Cuttack.

Panigrahi, B. 2011. Irrigation Systems Enginering. New India Publishinh Agency, New Delhi.

Punmia, B.C. and Lal, P.B.B. 1992. Irrigation and Water Power Engineering. Laxmi Publications Pvt. Ltd., New Delhi.

Sharma, R.K. and Sharma, T.K. 2002. Irrigation Engineering including Hydrology, S. Chand and Company Ltd. New Delhi.

Sharma, R.K., 1987. A Text book of Hydrology and Water Resources. Dhanpat Rai and Sons, Delhi.

10

Water Lifting Device

10.1 Introduction

Irrigation is an age old practice. It has been practiced for thousands of years. It is chiefly used to irrigate rice in wetter areas; especially in Asian countries and dry land crops including vegetables and fruits in dry land areas having low rainfall in all most all countries. Devices for irrigation water lifting range from age old indigenous water lifts to highly efficient pump. Unlike canal irrigation where water is fed to the fields through gravity system, water from the surface and sub-surface sources is lifted either mechanically by pump or manually by any indigenous lifting devices. Irrigation by lifting water from a lower level to a higher level is called as lift irrigation. Several types of indigenous water lifting devices are in vogue in India since ancient times. These are animal or/and manually operated ones, the animal operated ones being more efficient than the manually operated ones. The earliest water lifts were simple man-powered devices, many of which are still being used with modifications in different forms. With gradual improvement, principles of simple machines, like the liver, pulley, screw etc. were employed in smooth lifting of water from surface and sub-surface sources. The selection of a suitable water lifting device for a particular situation depends on the following factors:

(i) Characteristics of source of water and the lifting device

(ii) Amount of water to be lifted

(iii) Depth of water table

(iv) Type and amount of power available and

(v) Economic status of the farmers.

From a purely economic point of view, the cost of lifting water with human or animal power is relatively higher than that with the mechanically operated pumps. On the other hand, researches reveal that the cost of lifting a unit quantity of water through a unit depth from the surface and sub-surface source is lowest by using a pump operated by electric power. However, the relative cost of lifting water through engine-operated pumps is about 15-25 percent more than with electric operated pumps. The cost of lifting water through animal power is still more being 10 to 20 times more than the pumps operated by electricity. Human operated water lifting devices are the costliest, being 3 to 4 times more than the animal operated devices. In many of the rural areas, the scarce availability of mechanical and electrical powers have made the human and animal operated water lifting devices more popular although these systems are not economically advantageous. In addition, other factors that make human and animal operated water lifting devices more popular are small land holdings, lack of adequate mechanical skills amongst farmers in operating pump sets and their low economic status.

10.2 Types of Water Lifting Devices

The water lifting devices are mainly divided into two sources i.e. indigenous water lifts and pumps. Each source is again divided into several sub-divisions. Following chart presents the different water lifting devices as used in practice.

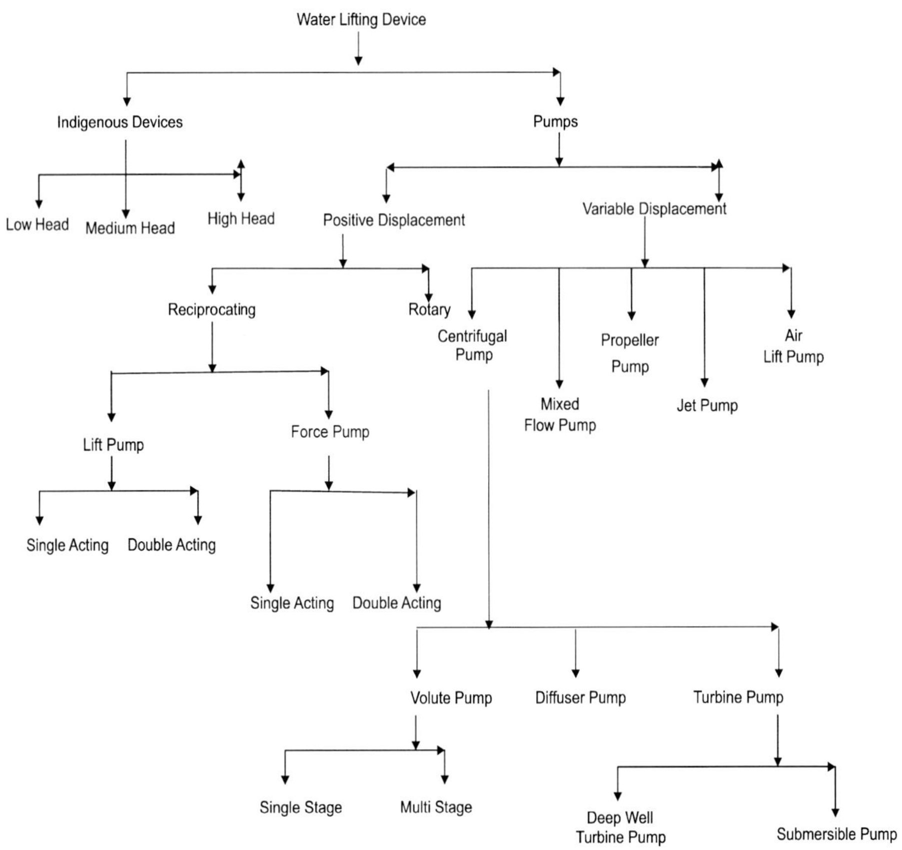

10.3 Indigenous Water Lifts

Several types of water lift devices are in vogue in India since time immemorial. They are mainly divided into human/manually operated and animal operated devices. Table 10.1 presents the various types of indigenous water lifts commonly used in India along with their performance and adaptability (Michael, 1978).

Table 10.1: Performance and adaptability of common types of indigenous water lifts

Name of device	Kind of power	Optimum lift, m	Discharge, $(x10^3)$, lph	Remarks
Swing basket	Two men	0.9 – 1.2	14-19	Mostly used in rice grown areas in eastern and southern India
Don	Single man	0.5 -1.0	9 - 13	Commonly used in rice grown areas in eastern India
Archemedian screw	Single man	0.5 – 1.2	14-19	Commonly used in deltaic region of Andhra Pradesh
Water wheel	One pair of bullocks and one man	1.0- 1.2	40-60	Commonly used in canal irrigated areas of north India
Persian wheel	One pair of bullocks buffaloes and one man	5 -10	14-18	Preferred water lifts in northern or India during first half of twentieth century
Chain pump	One pair of bullocks and one man	3- 6	15-20	Used in parts of U.P.
Rope and bucket lift	One pair of bullocks and one man	4 - 6	10-15	Commonly used in southern India, Deccan region and parts of Rajasthan
Circular two bucket lift	One bullock and one man	4 - 5	12-14	Used in parts of Tamilnadu
Counterpoise bucket lift	Single man	1.2 – 4.0	8-11	Commonly used in southern India, Deccan region and Bihar

The indigenous water lifts may be operated either manually or by animal powers. They may be sub-grouped into three main types based on their height of lifting water from the source. These three types are (i) low head, (ii) medium head and (iii) high head type.

10.3.1 Manually Operated Water Lifts

10.3.1.1 Low Head Water Lifts

Low head water lifts are suitable for lift up to 1.2 m. Followings are different types of low head water lifts used in fields.

Swing Basket (Sena)

It is the most ancient water lifting device used in may parts in the country especially in the rice belts of eastern and southern India.. It consists of a basket or shovel like scoop to which four ropes are attached (Fig. 10.1). It is operated by two persons standing face to face with each person holding two ropes in his two hands in same side of the swing basket.

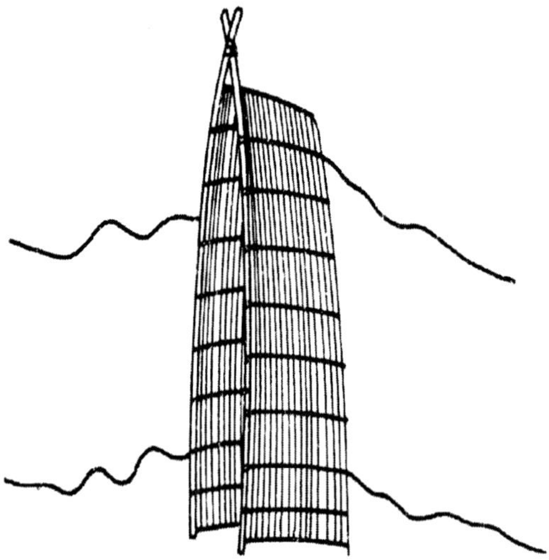

Fig. 10.1: View of the swing basket

They lower the scoops into the water ditch and when the scoop is filled with water they lift the scoop to a height where the water has to be discharged to the field or to a field channel to lead water to the fields. This is done by manipulation of their hands up and down and to and fro. Water by swing basket can be lifted to a height of 0.9 to 1.2 m. Capacity of swing basket is 8 to 12 lit/swing. The average discharge capacity is 14000 to 19000 lit/hr.

Sedua

It is a small device made of a piece of wood which is grooved inside. It is used to lift water to a small height (15 – 25 cm) and the capacity is also very less. Only one person can operate the equipment. It is mostly used to irrigate the vegetables and the discharge capacity is about 1000 lit/hr.

Don/Janta/Karim

Don is also called as janta/karim/gutter in different regions of the country. It is an oscillating trough which is boat shaped closed at one end and open at the other. The closed end is tied with a rope to a pole to work as a pivot (Fig. 10.2.). The closed end of the trough is tied with a rope to a long wooden pole which is pivoted as a liver on a post. A weight like large stone is fixed to the shorter end of the lever. The open end is hinged to the discharge point. A man stands on a platform made of bamboo or wooden plank above the water source and by applying force he pushes the close end of the don into the water source. When the don is filled with water, he lifts the don by applying jerk by hand so that the water from the don is discharged to the field through the open end. It lifts water to from a depth of 0.5 to 1.0 m and the discharge capacity varies from 9000 to 13000 lit/hr (Panigrahi, 2011). It is used in rice grown areas in eastern India. Several modifications and improved designs are available in oscillating-trough-type water lifting (don) devices. In one of the modified design, in the seesaw trough, the operator shifts his weight back and forth. An outboard flap valve facilities filling water in the don. The modified designs help in lifting more discharge than the conventional types. .

Fig. 10.2: A man with don busy in irrigation

Archemedian Screw

It is also called as Egyptian screw. It is most efficient water lifting device operated manually. It is said to have been invented by Archimedes in about 200 B.C. It is popular in Godavari delta of India and in Egypt to lift water from shallow sources where the lift is 0.5 to 1.2 m. The device consists of a wooden or metal wheel drum with interior portion having fittings just like screws or auger like spiral. The drum rotates by a handle fixed to a central spindle or called as central drive shaft (Fig 10.3). The screw may be welded steel or plastic type. The screw is encased in a sheet-metal or wooden cylinder called as drum. Wooden cylinders are made of boards of 1.25 to 5.0 cm width and about 1 cm thick. Iron bands, spaced about 30 cm apart, holds the boards of the cylinders in place. The diameter of the cylinder ranges from 40 to 60 cm and length varies from 1.5 to 2.7 m. Water moves up through the screw and discharges at the upper end of the cylinder. The device can be operated by a single man who can work up to four hours at a stretch. Sometimes, two men facing each other handle the device or they can work one after the other. This will facilitate in continuous work and in getting more discharge. The discharge capacity varies from 14000 to 19000 lit/hr.

Performance of the device depends on the angle of inclination of the screw, speed of rotation and depth of submergence. The angle of inclination of the screw ranges from 30 to 35°. Higher speed of rotation of the screw reduces the discharge. The optimum discharge is obtained at speed of about 40 rpm. Full submergence of the screw reduces the discharge. The screw was observed to give optimum result when the bottom was submerged a little more than half (about 1 cm above the centre of the shaft).

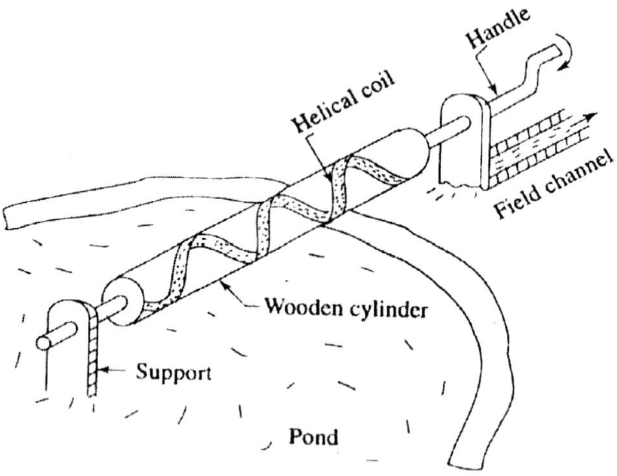

Fig. 10.3: View of Archemedian screw

Water Wheel/ Pedal Wheel

Pedal wheel is a simple pedal operated water lifting device. It is a traditional device which was in extensive use in south-East Asian countries and the Far East till the introduction of pumps. In India, it is mostly used in coastal regions and canal irrigated areas of northern India. In Kerala, it is called as Chakram and is mostly used to irrigate rice crop. It consists of small paddles/wooden blades mounted radially to a horizontal shaft, which moves in close fitting concave trough, thereby pushing water ahead of them. The shaft is driven by a bullock wheel drive similar to the drive used in a Persian wheel. The wheel, when rotated by pedaling the treadles with arms and upper parts of the body of the operator, pushes water to the field surface through the trough. The number of blades in the wheel ranges from 8 to 24 depending on the size of the wheel. The size of the wheel is small when the lifting head is small whereas for larger head of about 1.2 m, the size is more.

10.3.1.2 Medium Head Water Lifts

Medium head water lift are suitable when the lifting head is within the range of 1.2 to 10 m. Manually operated medium head water lifts used for small scale irrigation are described below.

Counterpoise Bucket Lift

It consists of a long wooden vertical post to act as a fulcrum from which a beam swings up and down like a see- saw. A bucket is tied to one end (usually the longer end) of the pole. A weight, usually large stone or a ball of dry mud is fixed to the shorter end of the pole. The weight serves as a counterpoise to a bucket suspended by a rope or a rod attached to the long arm of lever. A man at the fulcrum operates the bucket up and down attached to the beam. He pulls down the rod or rope, using his body weight and strength and until the bucket is filled with water. Then he raises the water full with bucket by his hand and counter-poise helps in raising bucket up. Then he discharges the water into the conveying channels. After the bucket is emptied, it is brought down by giving a jerk with hand and then filled with water by dipping the bucket. Then it is lifted up and water is emptied. The cycle thus continues (Fig. 10.4). Generally the working range of the counterpoise lift is within 1.2 to 4 m. The discharge capacity varies from 8000 to 11000 lit/hr (Murty, 2004).

Fig. 10.4: View of counterpoise as water lifting device

The counterpoise is adjusted to balance the weight of the full bucket. The proper weight of the counterpoise lift can be computed as follows:

Let W_1 = Weight of the bucket, kg

W_2 = Weight of water in bucket, kg

W_3 = Weight of counterpoise, kg

Z_1 = Lift or motion of load, m

Z_2 = Motion of effort, m

Let a and b be the lengths of the weight arm and force arm of the oscillating lever.

Velocity ratio is defined as ratio of motion of effort to motion of load.

$$Velocity\ ratio = \frac{Z_2}{Z_1} = \frac{a}{b}$$

Now, $(W_1 + W_2)\,b = W_3\,a$

Hence, $\dfrac{W_1 + W_2}{W_3} = \dfrac{a}{b} = \dfrac{Z_2}{Z_1}$

$$W_3 = \frac{(W_1 + W_2)\,b}{a} \qquad (10.1)$$

Tenda/Picottah/Dhonkli/Sadoop

It is one of the oldest water lifting devices used in ancient countries of the world. It is still used in many villages in India. It is known as picottah in South India. It is also used in some other countries like China, Japan and Egypt. It an adaptation of the conventional counter-poise lifts. This is built with a massive wooden framework to enable the operator to walk on the pivoted pole. In operating the device, one or two men walk back and forth on top of the pole. The pole is notched along its upper surface to provide a foothold. A hand rail is provided on the pole. Walking of the man on the pole causes a shift of body weight which causes bucket to be lowered and raised. The device can lift water from a depth ranging from 5 to 8 m. The discharge capacity is low (8000 to 11000 lit/hr).It is mainly used to irrigate vegetable crops.

Chain Pump

The chain pump is a traditional water lifting device and is dated back to the early part of the first century. They are usually human operated or animal driven. In case of animal driven chain pumps, the arrangements of the driving mechanism is same as in a Persian wheel as discussed later. In human operated chain pumps, a handle is fixed directly to the drum shaft over the chain works. It is also suitable for operation with windmill. It can work over a wide range of operating heads. For low heads, loose fitting washers are adequate. But for high heads, tight fitting plugs rather than washers may be needed to minimize back leakage Chain pump consists of an endless chain which is provided with leather discs or washers spaced at about 25 cm interval. The chain passes over a notched wheel called as winch shaft mounted on a suitable

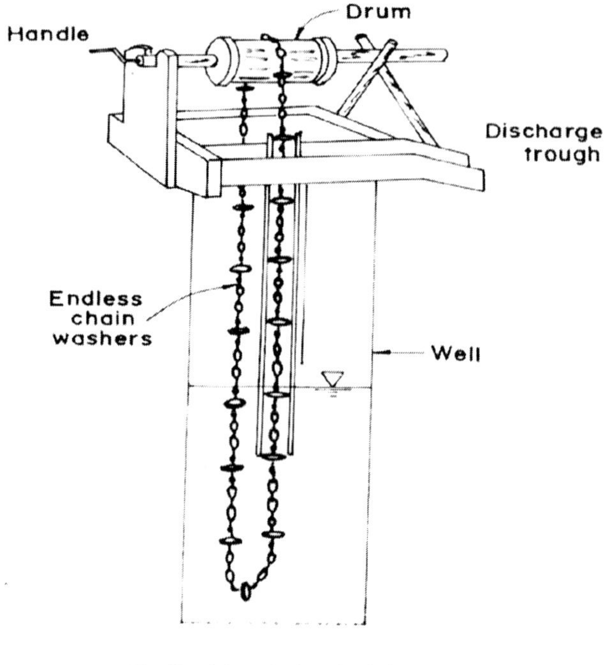

Fig. 10.5: Schematic view of a chain pump

platform fixed on top of the well. A crank handle of wood or metal is bolted or welded to the winch shaft. On one side of the chain is a pipe of about 10 cm diameter, with a flared opening at the bottom and connected to a trough at the top (Fig. 10.5). The bottom of the pipe is submerged about 60 to 90 cm below the surface of water of the source. The disc washers have the same size as the inside diameter of the pipe. When the wheel is turned, each disc lifts a certain amount of water. Chain pumps are efficient device for lifting water from shallow open wells of small size and works better when the lift is limited to 5 m. The discharge capacity varies from 15000 to 20000 lit/hr.

10.3.1.3 High Head Water Lift

Water lifting manually from deep wells is limited to domestic water supply and at times for small scale irrigation to vegetables and orchards; the discharge being very small. Rope-and-bucket lift is a type of high head water lift which is commonly used in practice.

Rope and Bucket Lift

This is also called as Mote, Charasa or Pur. It is used to lift water from deep well from a depth ranging from 10 to 30 m. It may either be operated manually or by animals. The device may be operated singly or in multiples of two or more working simultaneously, depending on the yield of the well and requirements of irrigation water. It essentially consists of a bucket and pulley arrangement. A rope is attached to the bucket which passes over a pulley set on a wooden frame work fixed on the top of the well. A man draws the rope in his hands (Fig. 10.6). The rope is first slaked and the bucket connected to the lift is lowered down and allowed to fill it with water from the well. When the bucket is full with water, the rope is drawn in downward direction and the bucket with water raises up. When the bucket comes above the ground surface or comes to the top of the well wall, the man drawing the rope brings the bucket out and tilts so that all water is discharged to another container meant to store for domestic use or the water may be directly discharged into the trough of the channel to be conveyed for irrigation purpose.

Fig. 10.6 Rope and bucket lift

10.3.2 Animal Powered Water Lifts

Use of animals in water lifting provides the equivalent power of 5 to 10 men. Animals can lift more water in a shorter time than humans. The human power saved in irrigation by lifting water by animals can be better utilized in other farming operations, thus increasing farming efficiency. The main disadvantages of animal power are that the animals require feed throughout the year while the irrigation period is not continuous throughout the year. Thus, they have to be fed fodder all through the year. It is always advantageous to utilize the same animals for tillage and other farm works, post harvest operations like threshing, milling etc. Animals commonly used to operate water lifting devices are bullocks/ox, camel, horse, donkey, buffalo and mule. The power of the draft animals varies from animals to animals. The power of bullocks ranges from 0.3 to 0.6 HP whereas for buffalo, camel, donkey, mule and horse, it ranges from 0.6 to 1.3, 0.5 to 0.9, 0.1 to 0.6, 0.3 to 0.7 and 0.6 to 1.0 HP, respectively. Commonly used animal powered water lifts are described as follows. These animal powered water lifts can be grouped into (i) low head (ii) medium head and (iii) high head water lifts (James, 1988).

10.3.2.1: Low Head Water Lifts

Circular Mot (Baldeo Balti)

This device is an adaptation of the manually operated oscillating trough for animal operation. It can facilitate to overcome the difficulty of lifting water from wells deeper than 6 to 7 m by tenda. They are very common in use in Uttar Pradesh, Bihar and Andhra Pradesh. The mot bucket is a hemispherical container with a flexible tubular tail piece of leather attached to it at its base. The leather bucket is fixed to strong iron ring with iron frame work at the top. The bucket is suspended from rope called mot rope. The length of the mot rope passing over the overhead pulleys is so arranged that when one bucket, which is filled with water, is rising up; the other is lowered down empty (Fig. 10.7).

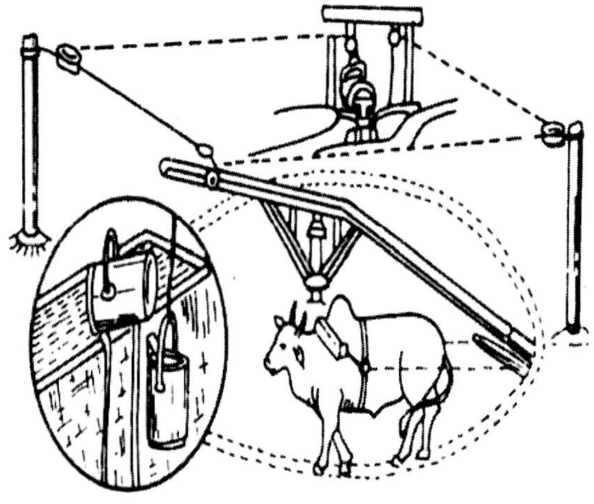

Fig. 10.7: Lifting water by circular mot

Water Wheels

The animal driven water wheels, frequently called as *jalar* in north India, is commonly used to lift water from canals up to 1.2 m lift to irrigate crops in fields at higher elevation. The device consists of small sheet metal paddles mounted radially on a horizontal shaft. The wheel works in a close fitting concave masonry trough constructed at the end of the water course. The wheel is driven by a bullock gear drive, similar to that of a Persian wheel. The wheel, when rotated, pushes water into a trough at the inlet of the field channel.

Sakia

It is a sophisticated water wheel used to lift water from a ditch to the crop fields. It is most popular in Egypt. *Sakias* are made of galvanized steel. They have diameters ranging from 2 to 3 m. The wheel of *sakia* consists of several individual compartments which curve inwards towards the centre of the wheel. Each compartment has a large inlet at the outer circumference of the wheel and an outlet at its hub. These compartments are called as *kados*. As the *sakia* turns, a *kados*.submerges and fills with water. When the wheel completes rotation, *kados* is raised out of water and empties water towards the central outlet, which is now lower than the rest of the compartment. The water discharges from the outlet at the hub of the wheel into a field channel which ultimately carries it to crop fields. *Sakias* are driven by draught animals including bullocks, buffaloes, and sometimes by donkeys. The discharge of a *sakia* depends on its diameter, width, static lift, and speed of revolution. The discharge of a 3 m diameter *sakia,* lifting water from 1 m lift, varies from 72000 to 75000 lit/hr.

10.3.2.2 Medium Head Water Lifts

Medium head water lifts are suitable to lift water from a head of 4 to 10 m. These include Persian wheel, chain pump, rope-and-bucket lift with self emptying bucket and circular two-bucket lift.

Persian Wheel

The Persian wheel or Saqiya (also called as *rahat* in Urdu), is a water lifting device made of two gear wheels and an endless chain of buckets. Traditional types of Persian wheels use earthenware pots and jute or coir rope for the endless chain. Modern Persian wheels use sheet metal buckets and hook chains made of mild steel bars. The capacity of the buckets range from 7 to 14 lit. The endless chain of buckets is mounted on an open spoked drum. The gear wheels provide the driving mechanism. Two parallel loops of chain, joined by spacing bars and having metal or earthen ware pots attached to them at intervals passes over the drum and loop into the water in the well. A horizontal shaft extends from the central axis of the drum to a small vertical cogwheel or gear. The teeth of this vertical gear mesh with the teeth of a large horizontal wheel gear. A horizontal beam is attached to the shaft to which the animals are hitched. In some models, the long horizontal shaft is eliminated and the animals go around the well instead of moving on a separate platform. When the Persian wheel is operated, the bucket drum revolves, causing the buckets at the lower end of the chain to get filled with water and carried to the top with their open mouth upward. When the buckets pass over the top of the drum, they spill water into a trough which conveys it to an irrigation channel (Fig. 10.8). The device is capable of lifting water from both shallow and deep wells. Optimum lift of the device varies from 5 to 10 m. But its efficiency gets reduced when the lift exceeds 7.5 m. It is estimated that one Persian wheel can irrigate up to one hectare of land. This device is extensively used in India and other developing countries.

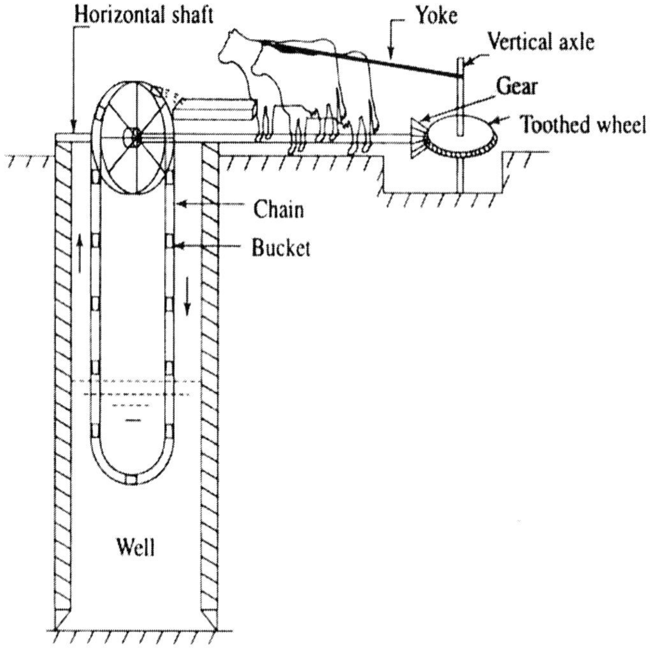

Fig. 10.8: View of Persian wheel

Chain Pump

The chain pump as described earlier in manually operated medium head water lift, can be easily adapted for operation by animals. In case of animal driven chain pumps, a drive mechanism similar to the one used in Persian wheels can be adopted. The animal operated chain pump is generally equipped with a 15 cm diameter pipe and matching disc washers. In a chain pump, the water is directly discharged into a trough at the ground surface.

Rope-and-Bucket Lift with Self Emptying Bucket

The self-emptying type rope-and-bucket lift is suitable for operation with a pair of average size draught animals. It works better when the lift is within 9 m.at which the discharge is about 8000 lit/hr. The device consists of a sheet metal or leather bucket of about 100 to 150 lit capacity. A leather tube is fixed at the bottom of the bucket. The bail of the bucket is attached to a heavy rope which passes over a pulley. Another rope is fastened to the lower end of the spout. This second rope passes over a roller fixed to the lip of the receiving trough at the ground surface. Both ropes are tied together with adjustable length and then attached to the animal yoke. When the bucket reaches top of the well, the spout follows its own rope over the roller, while the bucket continues to rise upwards towards the pulley. When the spout is extended downward, water is discharged through it into the trough. Then the animals are made to walk up the ramp to the place of starting.

Circular Two-Bucket-Lift

The device uses two buckets of about 70 lit capacities each which are alternatively raised, emptied, lowered and filled. When one bucket is filled and lifted, the other empty one is lowered into the well. A rope and pulley arrangement along with a central rotating lever permits reciprocating action while the bullocks move on a circular path. The buckets are provided with flap valves at the bottom for filling. When the bucket is lowered into the well, the valves open allowing water to get in and fill the bucket. The valves get closed when the bucket moves up because of weight of water. The up and down movement of the buckets are guided by two vertical rods fixed at the wall of the well. When the bucket reaches the top of the well, it gets tilted automatically due to the loop on the rod allowing water to get discharged into the receiving trough.

10.3.2.3 High Head Water Lift

High head water lifts can lift water from depths more than 10 m. Rope-and-bucket lift is a high head water lift operated by animal power.

Rope-and-Bucket Lift

It is used to lift water from deep well from a depth ranging from 10 to 30 m. The device may be operated singly or in multiples of two or more working simultaneously, depending on the yield of the well and requirements of irrigation water. It essentially consists of a bucket or bag of about 150 to 200 lit capacities, made of leather or galvanized iron sheet and pulley arrangement. A rope is attached to the bucket which passes over a pulley set on a wooden frame work fixed on the top of the well which may be finally fixed to the yoke of bullocks. The bullocks walk down on an earthen ramp sloped at an angle of 5 – 10 degrees, thereby obtaining some advantage of their body weight in exerting the necessary force to lift the water. The rope and bucket lift can be worked by a pair of bullocks and two men. When two pairs of bullocks are used, one pair walks up the slope while the other is drawing up the bucket. The discharge capacity varies from 6000 to 10000 lit/hr. Sometimes, instead of bullocks, two buffaloes or a camel provide power to lift the water.

10.3.3 Wind Powered Water Lifts

Wind power has been extensively used for centuries to lift water. But the disadvantages of this technique are that wind power required to lift water is not reliable; the optimum velocity of wind being uncertain at all time. The repair and maintenance cost is high. Furthermore, the technique is not cost effective with respect to the output i.e. amount of lifting of water. Wind mill is used to utilize the wind power to lift water. Wind mill consists of a large diameter vane wheel or rotor fixed on the top of a high steel or wooden tower and a reciprocating pump at the bottom. The upper end of the piston rod is fixed eccentrically to the vane wheel. The lower part of the suction pipe is always immersed in the well storing water. The strong wind rotates the vane wheels which works the pump and allowed water to be lifted up. The delivery pipe discharges water to a reservoir from where it can be later used. Fig. 10.9 shows a modern wind mill in operation.

Fig. 10.9: View of a modern wind mill

The discharge from a wind mill is intermittent depending on the availability of sufficient wind power. Its capacity depends on wind velocity and frequency. Wind mill can be suitably used to lift water from shallow and moderately deep wells. A wind velocity of over 6.5 km/hr is necessary for the satisfactory operation of the wind mills. The regions of potential areas of wind powered water lifts are the coastal areas. Even in these areas, high wind velocity is limited to summer seasons when there is hardly any crop except summer vegetables and fodders. In the main crop growing seasons i.e. *kharif* and *rabi,* the wind velocity is not adequate to provide wind energy to lift water. This restricts the wide adaptability of wind mill as water lifting device.

10.3.4 Water Powered Water Lifts

Water powered water lifting devices have been used for centuries. They are cost-effective where there is a large flowing river or there is high water drops. Hydraulic ram (hydrams) is a device used to lift water using water power.

10.3.4.1 Hydraulic Ram (Hydram)

Hydraulic ram is a device that pumps water by using hydel power from a lower elevation to a higher one. The device is suitable especially in hill areas where there is considerable slope in rivers and streams which could be harnessed to operate the machine.

Construction

Hydraulic ram consists of an inclined supply pipe starting from a supply channel and terminating

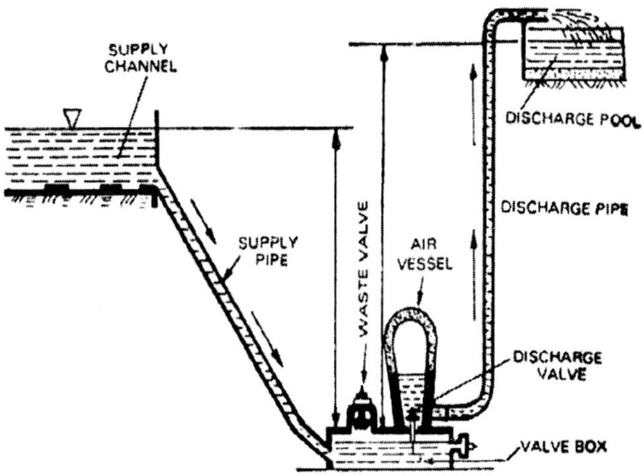

Fig. 10.10: Hydraulic ram used for lifting water

in a valve box. The valve box is fitted with a waste valve opening inwards and a discharge valve opening outwards, delivering water under pressure into an air vessel. Finally water from the air vessel is delivered through the discharge pipe to the discharge point at a higher level than the level of the supply channel (Fig. 10.10).

Function

Hydraulic ram can lift water from a head ranging from 1 to 100 m. Hydraulic ram works like an impulse pump. The impulse is developed at the expense of the dynamic inertia possessed by a running stream of water. Water flow is set up from the supply channel to the waste valve which is opened through an inclined supply pipe. The velocity of flow is increased under the influence of the supply head until the dynamic pressure on the side of the valve becomes sufficiently great to overcome its weight. The valve now closes rapidly and the supply is retarded in the supply pipe which gives rise to a rapid increase of pressure in the valve box and finally opens the delivery pipe. Water from the waste valve passes into the air vessel through the delivery valve. Air in the air vessel gets compressed and forces water to flow out from it to the discharge/delivery pipe. As the water from the air vessel goes out and there is retardation in the supply pipe, delivery valve closes and a backward motion of water is produced. This backward motion reduces pressure in the valve box below that corresponding to the static head and consequently the pressure in the valve box is reduced rapidly till a time when the waste valve reopens. The whole cycle of operation is repeated. For better working, the ram should be as close as possible to the supply channel so that the impulse gets reduced and consequently the deliver head gets reduced.

The main advantages of hydraulic ram are that (i) once installed, they need hardly and running cost and (ii) the machine can operate continuously throughout the day giving regular water supply. Expected life time is about 20 to 30 years.

Efficiency of Hydraulic Ram

The discharge decreases as the lift (effective delivery head) increases and drop (static head/ effective supply head) decreases. Effective supply head (H) and effective delivery head (h_d) are measured with respect to the level of the waste valve. Efficiency of the ram (η) is given as (Michael, 1978):

$$\eta = \frac{q\, h_d}{(Q+q)\,H} \tag{10.2}$$

where,

q = volume of water delivered by the ram and Q = water flowing from the supply channel through the supply pipe. Other terms of the equation are defined as above.

10.3.5 Other Sources of Water Lift

Other sources of water lift are solar energy and biological energy. Solar energy though abundant, is expensive with respect to the output of water lift and hence is not widely practiced for water lifting. They need more research and modification in their constructional parts to make it cost-effective. Energy can be obtained from biological sources, in particular from gases produced by the bacterial decomposition of animal and human excreta, sometimes also from the plant residues. The use of agricultural byproducts to make fuel for heat engines can be regarded as a biological or renewable energy source which can be used to run pumps to lift water for irrigation.

10.4 Pumps

Pumps are the mechanical devices used to lift water from water sources. They are used both for irrigation and drainage. Drainage water from the waterlogged areas are lifted by them to a higher elevation where they can be used for irrigation. They are also used beside canals to lift groundwater and feed them to the canal water to augment the irrigation water. They play a crucial role in modern agriculture and are responsible to increase the irrigation potential. It is needless to mention that pumping water for irrigation has been practised since many centuries in Egypt, India and China and many other countries where irrigation plays crucial role in agriculture. One of the devices used in Egypt and India in early period, known as shaduf, makes use of the principle of lever with a suspended fulcrum and a counterweight. The bucket, suspended from long end of the pole, is sometimes made of leather stiffened near the top with a wooden hoop. The operator throws his weight on the sweep, the bucket fills and the counterweight raises it to the next higher channel into which the water is poured. With the shaduf, one man can lift about 1 lit/sec water from a lift of 2 m (Hansen *et al.*, 1979).

In modern agriculture as practiced now, pumping machinery of high efficiency is used. Substantial improvements have been made in the design and operation of pumps. Many types of pumps with many makes and models are now available which can be used suiting to different situations. Most of them are cost effective and can lift water from very high lifts.

The pumps are generally divided into two categories i.e. positive displacement pump and variable displacement pump. The positive displacement pumps discharge water irrespective of their head of operation. The variable displacement pumps have variable discharges rates depending on their heads or depth from which they lift water. The discharge varies inversely with the pressure head. Centrifugal pumps are the most important variable displacement pumps used in agricultural operations.

10.4.1 Positive Displacement Pump

Positive displacement pump discharges the same volume of water irrespective of its head against which it operates. This type of pump must be powered to meet the maximum load resulting from its discharge capacity and the greatest head against which it will operate. Since the discharge capacity is low, these types of pumps are not very popular in irrigation and drainage works. They are preferred in domestic use.

10.4.1.1 Reciprocating Pump

This type of pump works with a piston movement that displaces water in a cylinder. Hence, it is also called as piston or displacement pumps. Inside a hollow cylinder, a cylindrical piston/ plunger piece reciprocated backward and forward with the help of a connecting rod. A valve controls the flow. The connecting rod may be attached to a driving wheel (Fig. 10.11). The discharge capacity of reciprocating pump depends on the size of the cylinder and length and speed of the stroke. To prevent leakage, several types of packing including cup-leathers are used (Basak, 2012). Reciprocating pumps can be classified as shallow well or deep well type reciprocating pumps.

Reciprocating pump for shallow wells are usually lift types using atmospheric pressure to raise water from the wells. The piston is in the pump body, near the handle. Handle is used to move piston up and down manually and this up and down movement of the piston displaces air from

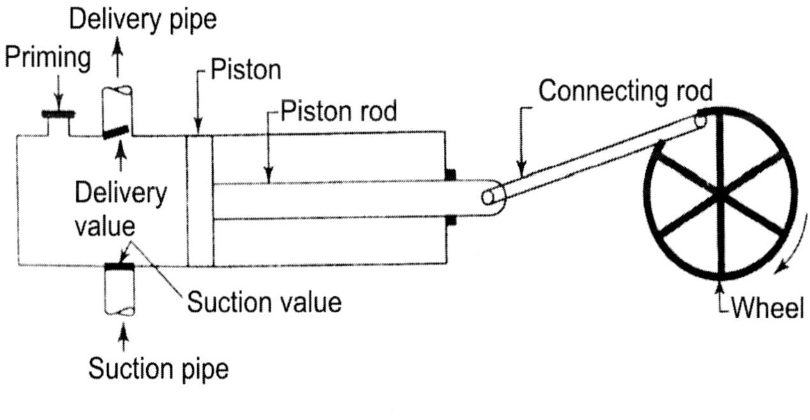

Fig. 10.11: Reciprocating pump

the pump chamber. This creates vacuum inside the pump chamber. The atmospheric pressure in the well below the pump position being more than the pressure inside the pump chamber, some amount of water from the well is forced into the pump chamber. Finally this water goes out through delivery pipes. Theoretically, it can lift water from a height of up to one atmosphere or 10.33 m. But since there are losses due to friction, the actual head against which it works is 6.5 to 7.0 m.

Reciprocating pumps can operate manually or may be operated by animal powers. An animal powered duplex reciprocating pump developed by Khepar *et al.* (1975) is very popular in lifting water from shallow tube wells. The animal operated pump is suitable to lift water from shallow tube wells and open wells when the depth of water level in the well is within 6 m. It gives discharge of about 7 lit/sec against a head of 4 m.

Deep well reciprocating pump can lift water from deep wells. In this case, the pump cylinder with its plunger and valves are introduced into the water in the well. The plunger is connected to the pump handle, or any mechanically powered crank shaft by a connecting rod. As the plunger with its upper valve moves upwards, the water on the top of the valve is forced upward through the delivery pipe. A vacuum is created inside the pump chamber which forces the lower valve to open and some water from the well is pushed in. The cycle is repeated.

Deep well reciprocating pumps may be manually operated or it may be driven by mechanical power. Manually operated deep well reciprocating pumps can lift water up to a depth of about 45 m.Mechanically driven pumps may be operated by engines or motors. They usually employ a crank shaft arrangement called as the working head, placed at the top of the well, to change the rotating motion of the power to that of reciprocating motion at the piston. The pump works at low speed. Hence, a speed reduction device is used in the power transmission system. Power is transmitted to the working head by a belt drive.

Single Acting Reciprocating Pumps

Single acting reciprocating pumps have one discharge stroke for every two strokes of the piston. Water is delivered during alternate strokes of the piston. The flow through the delivery pipe is, thus, intermittent. An air vessel fitted to the delivery side near the delivery valve makes flow continuous (Fig. 10.12). During the delivery stroke of the piston, the air in the air vessel gets compressed to a pressure greater than that corresponding to the head of water at the bottom of the delivery pipe. During the suction stroke, the pressure of the air in the air vessel maintains the flow of water through delivery pipe, thus ensuring continuous flow.

Double Acting Reciprocating Pumps

These are constructed with pistons and valves so arranged that water is pumped on both inward and outward movement of the piston (Fig. 10.13).

Duplex and Triplex Pumps

These pumps consist of two or three pistons, respectively, and are designed to pump a continuous flow of water with minimum pulsation, often against high pressure.

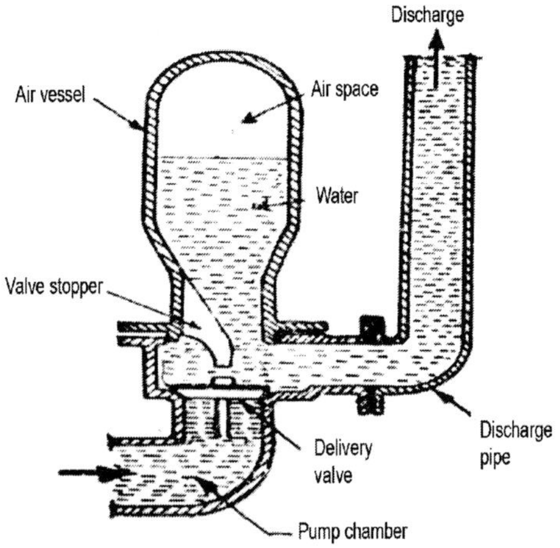

Fig. 10.12: Air vessel fitted to delivery side of single acting pump

Forced Required to Operate a Reciprocating Pump

Let a = Area of cylinder, m^2

l = Length of stroke, m

h = Total height through which water is raised, m

P = Force required to lift the piston (neglecting friction, and weight of piston and connecting rod), kg

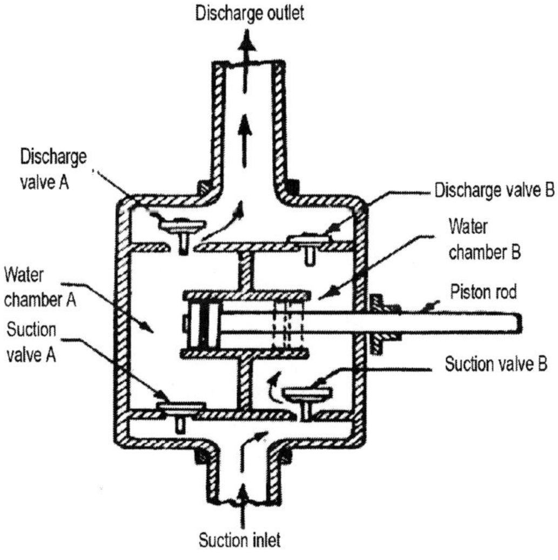

Fig. 10.13: Sketch showing double acting reciprocating pump

w = Specific weight of water, kg/m^3.

Weight of water raised in one stroke = $w \,.\, a.\, l = 1000\, a.\, l$

Work done in one stroke = $1000\, a.\, l.\, h = P.\, h$

Hence, $P = 1000\, a.l$

Thus, the pull on the pump rod is equal to the weight of a column of water whose base is equal to the area of the piston and whose height is the total head. Hence, when the friction is neglected, P is independent of the diameters of the suction and delivery pipe.

Sample Calculation 1:

A single acting reciprocating pump has piston diameter 20 cm and stroke 30 cm. The piston makes 40 double strokes per minute. Suction and delivery heads are 5 and 10 m, respectively. Fine out the discharge capacity of the pump and power requirement of the pump in H.P. Assume efficiency of suction and delivery strokes are 60 and 75 percent, respectively.

Solution:

Given, diameter of the piston, d = 20 cm and length of stroke, l = 30 cm.

Area of the piston, $a = \dfrac{\pi\, d^2}{4} = \dfrac{3.142 \, x\, 0.20 \, x\, 0.20}{4} = 0.03142\, m^2$

Volume swept by piston per stroke = $a\, l$ = 0.03142 x 0.30 = 0.009426 m^3

The pump makes 40 double strokes per minute.

Hence, discharge of the pump, Q = Volume swept by piston per stroke x Number of strokes/ min

= 0.009426 x 40 = 0.377 m^3/min = 377 lit/min = 6.28 lit/sec.

Average force in suction = *w. a. suction head / efficiency of suction stroke*

= 1000 x 0.03142 x 5 / 0.60 = 261.83 kg

Average force in delivery = *w. a. delivery head / efficiency of delivery stroke*

= 1000 x 0.03142 x 10 / 0.75 = 418.93 kg

Hence, power required by the pump is

H.P. = (Total suction and delivery force x distance covered in m/min) /4560

$= (261.83 + 418.93)\, x\, 40\, x\, 0.30/4560 =$ 1.79 H.P.

10.4. 1.2 Rotary Pumps

Rotary pumps are used as boosters in irrigation pumping. But they are commonly sued in pumping lubricating oils. The pump body is a plain housing with inlet and outlets pipes and openings for shaft which carry the gears or cams. One of the gears is the driving gear and is rotated by an outside source of power. The other one is the idler gear driven by the driving gear. The gears are fitted closely in the housing and mesh with minimum clearance. They rotate and squeeze the water between the teeth as they mesh together and force water out through the discharge opening. This creates a partial vacuum and forces some amount of water to get inside along the inlet side. The discharge remains constant regardless of pressure.

10.4.2 Variable Displacement Pump/ Rotodynamic Pumps

Variable displacement pumps have variable discharges depending on their head or depth from which they lift water. The discharge varies inversely with the pressure head. Unlike positive displacement pump, the variable displacement pumps requires the greatest input of power at a low head because of increase in discharge as the pumping head is reduced. The variable displacement pumps like centrifugal pumps, mixed flow pump and propeller pumps are predominantly used in irrigation. They range from low discharge with high heads to high discharge with low heads (Punmia and Lal, 1992).

10.4.3 Centrifugal Pumps

Centrifugal pumps are the most important variable displacement pumps used in agricultural operation. The main advantages of centrifugal pumps are

(i) They are simple in construction,

(ii) Easy to operate,

(iii) Low in initial cost

(iv) Produce a constant steady discharge

(v) Wearing parts are few and

(vi) Adapted top direct motor or engine drives without the use of costly gears.

10.4.3.1. Principles and Operation of Centrifugal Pumps

Centrifugal pumps are rotary machines consisting of two basic parts. They are (i) rotary element which is a circular impeller and (ii) a stationary element which is a tight fitting casing. The impeller wheel has 3 to 12 radiating curved vanes. The vanes are connected to the impeller wheel through a shaft. The impeller is set inside the casing. The casing has an inlet pipe which acts as suction pipe and an outlet pipe which acts as delivery pipe. The casings may be of two types i.e. volute type or diffuser type. Accordingly the centrifugal pump may be volute type (Fig. 10.14) or diffuser type (Fig. 10.15).

In a diffuser type centrifugal pump, a diffuser consisting of a series of guide vanes or blades surrounds the impeller. The impeller rotates inside the casing with the help of motor assembly and draws water into the casing. Water enters into the casing through the eye of the impeller. The casing then, by virtue of centrifugal force, throws water from the casing to deliver end. Rotation of the vanes produces high velocity which is then transformed into pressure head. By changing the form of the vanes, different characteristics are obtained. By enlarging the diameter

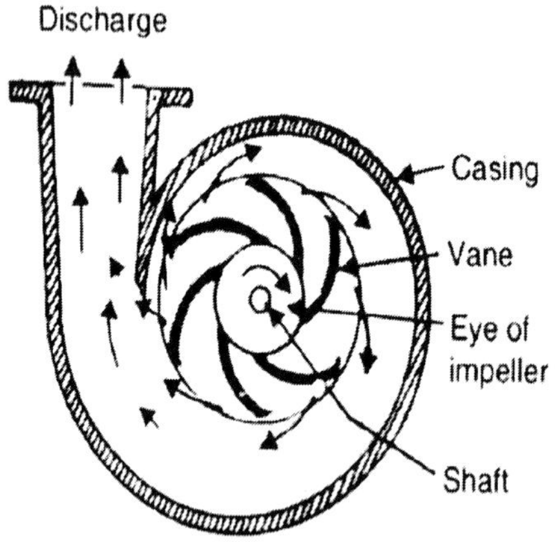

Fig. 10.14: Volute type centrifugal pump

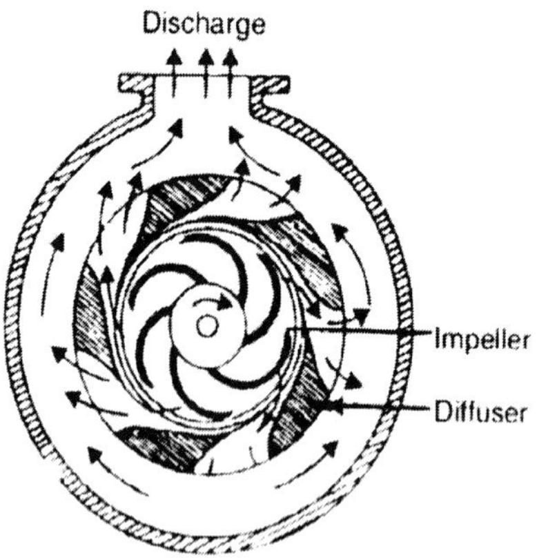

Fig. 10.15: Diffuser type centrifugal pump

of the inlet eye and width of the impeller, the amount of water that the pump delivers against a given head is increased. For balancing the flow of water into the casing of the pump, suction pipe should be connected at both sides of the impeller. In case of multi-stage pumps, balancing is done by using back to back mounting of impellers in pairs. A bed plate is provided at the base of the pump body for mounting the pumps and driving motor or engine. This provides a solid foundation on which they can be installed as a unit (Kumar *et al.*, 2012).

Centrifugal pumps need priming before it starts operating. For priming a foot valve is required. The foot valve is attached to the end of the suction pipe which deeps in water. Foot valve has the following functions:

(i) It helps in building up pressure to start the pump.

(ii) It helps in filtering out the foreign materials which would have otherwise passed through the suction pipe and clogged the pump casing and vanes. This is done by fitting a strainer to the foot valve.

(iii) It also holds water in the suction pipe.

Before priming is done the suction pipe and casing are fully filled with water and all the air bubbles are expelled out. The foot valve should be dipped in depth up to at least 60 cm of water in the source. After priming the impeller is rotated and water is drawn inside the casing. Centrifugal action then causes water to move out from casing to the delivery end. For priming, sometimes an auxiliary piston pump is used to fill the pump casing and suction pipe with water.

A centrifugal pump works better when the depth of water in a source is up to 7 m below the pump line and any depth above 8 m, decreases the discharge. The discharge of the pump is inversely proportional to the head. The discharge capacity of the pump varies from 40,000 to 64,000 lit/hr. High pressure centrifugal pumps are designed for use in applications requiring high discharge pressures and flows. Typically these pumps produce heads in excess of 240 feet. The pump has a 2 inch suction port and up to three discharge ports of varying size for added versatility. The impellers used in this case are a closed design type. Diffuser used is more compact to generate high pressures (Garg, 1987).

10.4.3.2 Classifications of Centrifugal Pumps

Centrifugal pumps can be classified on the following basis.

(a) Based on type of casing:
(i) Volute type and (ii) Diffuser type/turbine pump

(b) Based on number of stages:
(i) Single stage and (ii) Multi-stage

(c) Based on impeller types:
(i) Open, (ii) Semi-open (iii) Closed and (iv) Non-clog

(d) Based on type of suction inlet:
(i) Single suction and (ii) Double suction

(e) Based on construction of casing:
(i) Horizontally split and (ii) Vertically split

(f) Based on axis of rotation:
(i) Horizontal pump and (ii) Vertical pump

(g) Based on methods of drive:
(i) Direct coupled (ii) Belt-driven and (iii) Gear-driven.

(h) Based on working head:
(i) Low head (ii) Medium head and (iii) High head

Volute Centrifugal Pump

This type of pump has a casing made in the form of a spiral or volute curve. The volute casing starts with a small cross-sectional area near the impeller periphery and then gradually enlarges towards the delivery end of the pump. The casing is proportioned to reduce gradually the velocity of water as it flows from the impeller to the deliver end of the pump. This change in velocity head changes the pressure head and affects the flow. The most common type of pump used in irrigation in India is horizontal axis volute type centrifugal pump.

Diffuser or Turbine Pump

As mentioned earlier, in a turbine pump, the impeller is surrounded by diffuser vanes. The diffuser vanes have small openings near the impeller and get enlarged gradually to their outlet diameters where the liquid flows unto the chamber and around to the pump discharge. A major part of the conversion of velocity into pressure takes place between the diffuser vanes. Turbine pumps are usually used for high head conditions whereas volute type pumps are used for low head large capacity conditions. Turbine pumps are preferred for use in deep tube wells.

Single Stage and Multi-Stage Pump

In a single stage pump, total head is developed by a single impeller. In case of a multi-stage pump, two or more impellers are used on common shaft acting in series in a single casing. The liquid is conducted from the discharge of the preceding impeller or stage to the suction of the following impeller causing a pressure or head increase as it passes through each stage. As the number of stages (impellers) increase, the head and power requirement of the pumps also increase. Centrifugal type impellers used for irrigation may be open, semi-open, closed or non-clog type. A semi-open type impeller is used in case the pumping water carries some amount of silt. The impellers may be made of bronze or cast iron if ordinary water is used for pumping. However, to pump, brakish or salt water, impellers made of gun metal are preferred.

Single Suction and Double Suction Pumps

In a single suction pump, liquid enters the impeller from one side of the suction pipe whereas in a double suction pump, it enters from both sides *i.e* a double suction pump has two impellers. The double suction impellers are similar to two single suction impellers cast back to back. They support balancing condition to the pump and hence the use of thrust bearing for balancing is avoided.

Horizontal Centrifugal Pumps

A horizontal centrifugal pump is less costly and mostly used for irrigation. It is easier to install and is more accessible for inspection and maintenance. It has a vertical impeller mounted on a horizontal shaft. The pump should be installed above the water surface, but as close to it as possible. The suction lift should not exceed 4.5 to 6.0 m for satisfactory performance (Lal, 1969).

Vertical Centrifugal Pumps

Vertical centrifugal pump has a horizontal impeller mounted on a vertical shaft. It can be lowered to any depth required to pump water and the vertical shaft is extended to the surface where the power is applied. It can either be exposed or submerged type. In a submerged type, the impeller and the suction entrance remains submerged below the water level and so no priming is required. Volute type vertical centrifugal pumps are usually restricted to pumping heads of about 15 m.

10.4.3.3 Installation of Centrifugal Pumps

Centrifugal pumps must be properly installed to ensure its prolonged trouble free service. The pump should be installed as close to the water surface as possible. It should be installed in a clean, dry and well ventilated place. The place of installation should be so chosen that there is the shortest and most direct suction and delivery pipe which will ensure high discharges. The pump should be installed over a rigid foundation to absorb all vibrations. For pump foundation, cement concrete base, steel girders and wooden beams are used. Direct connected pumping sets are always mounted in a horizontal position on a level foundation. Concrete foundations with foundation bolts imbedded in the concrete, form the best type foundation. Bolts should be placed in pipe sleeves and anchored in concrete foundation. Washers are fitted to the bolts so that they do not slip through the pipe sleeves. Top of the foundation may be kept roughened, cleaned and dampened before placing the pump set in position. Wedges may be inserted between the top of foundation and pump unit for proper leveling of the pump set. Suction and delivery pipe connections are then lightly tightened up. Then the space between the base plate and foundation surface is thoroughly grouted. Finally concrete is allowed to cure for 2 days after which final tightening of the bolts are done.

Before starting the pump, alignment of the pump and driver must be checked. A straight edge placed across the coupling halves help in checking the parallel alignment. Angular alignment may be checked with a feeler gauge. On belt driven units, the pump and driver shafts must be parallel. The pulleys should also be aligned. The pulleys on both the pump and driver unit should be mounted as close to the bearing housing as possible to minimize overhang. On flat belt and V- belt drives, the tight side of the belt should be at the bottom. Normally, the belt speed should not exceed 1500 m/min. Suction and delivery pipes should be as short as possible with minimum bends. They should not contain a lot of fittings including elbows, tees etc. so that frictional losses will be minimum. Sharp angle bends must be avoided. The leakage in the suction pipes and its joints must be avoided. The size of the suction pipe should be equal to or larger than the suction opening of the pump. The size is so selected that the velocity of water in the suction pipe does not exceed 3 m/sec so that cavitation is avoided.

Foot valve fitted to the suction pipe at its bottom should remain submerged in water by at least 60 cm below the top water surface. The strainer fitted at the bottom of the foot valve should be placed at least 1 m above the bottom of the well so that there is no clogging and pump does not get chocked up by mud and other foreign materials. A non-return valve/reflux may be fitted to the delivery pipe as close as to pump as possible to carry water to distant places or under high head conditions. Sluice valve fitted in the delivery pipe just after the reflux valve helps in obtaining smooth and steady flow.

10.4.3.4 Electrical Connection

A pump will discharge more efficiently if a right type of motor is used in the pump along with its proper installation. Use of suitable controls for the operation and protection of the motor will also increase the efficiency of the pump. Motor and control wiring and grounding should be perfect as per the manufacturer's recommendations. Wiring should be of sufficient capacity, based on a maximum drop of 2 percent of the line voltage at full load current. Alternating current, 50 cycles 400 to 440 volts, 3-phase electric power supply is usually available in the farm. Four wire 3-phase supplies are taken from the local sub-station. Out of these four wires, three are phase wires and the fourth one is the neutral.

The 3-phase system is usually used for operating 3-phase electric motors. In the 3-phase system, the neutral wire of the service entrance to all buildings must be grounded for safety reasons. A 3-phase squirrel cage induction motor is used for pumping. They are low in initial cost, have low repair and maintenance cost and give smooth running. If maintained properly, they have long life span. Under low head and low discharge conditions, they may be run by single phase induction motors. A 3-phase motor has the followings electrical accessories. They are (i) energy meter, (ii) volt meter, (iii) ammeter, (iv) indicator lamps, (v) main switch and (vi) starter. *Direct-*

on-line or push button starters are used commonly for motors up to 5 H.P. and star-delta starters are used for motors having more than 5 H.P. Starters are usually equipped with overload releases to protect the motor from 'single phasing' and overload. The overload releases may be of the magnetic type or thermal type and could be adjusted to 'trip off' and disconnect the motor in case of power failure in any of the phases.

10.4.3.5 Operation, Maintenance and Troubling Shooting of Centrifugal Pumps

Pump Operation

Before starting the pump for the first time, following points should be considered (Michael, 1978).

(i) Pump should be properly aligned.

(ii) Ensure that engine or motor will drive the pump in the direction indicated on the body of the pump.

(iii) Gland should be tightly and evenly adjusted and the pump shaft revolves freely when turned by hand.

(iv) There should not be any air passage in the suction pipe.

(v) The foot valve should not have any leakage.

(vi) Before operation, priming should be done properly.

(vii) Proper lubrication must be done as and when required.

(viii) Before water from the source becomes dry, the pump must be stopped.

(ix) At the time of starting, the sluice valves are kept closed. When the pumps reaches full speed, the sluice valve is opened gradually, till the desired amount of water is being delivered.

(x) Before stopping the pump, the discharge valve is closed so that water hammer is avoided.

Pump Maintenance

Following points are to be considered as regard to maintenance:

(i) In every month, the bearing temperature should be checked. Bearings get hot due to either shortage of lubricants or excess of it.

(ii) Lubricants in ring oil bearing should be drained out and oil wells and bearings should be washed with kerosene once in every three months. Check the wear in the bearings and replace, if excessive.

(iii) Gland packing is to be replaced once in every six months. Alignment of the pump and driver are also to be checked once in every six months.

(iv) The unit should be thoroughly checked once in a year. Remove the bearings, clean them and remove the packing and examine the war in the shaft sleeve. Disconnect the coupling valves and check the alignment. Inspect the foot valve and check them.

Pump Troubles

The common pump troubles and their causes are mentioned as below (Michael, 1978):

(A) No water delivered

(i) Priming not done properly

(ii) Motor speed too low

(iii) Either suction head or discharge head too high

(iv) Impeller or suction pipe completely plugged

(v) Rotation of motor in wrong direction

(B) Sufficient water not delivered

(i) There may be air leakage in suction line

(ii) Suction lift too high

(iii) Motor speed too low

(iv) Delivery head is too high

(v) Impeller or suction pipe partially plugged

(vi) Foot valve size is too small

(vii) Motor rotating in wrong direction

(viii) Suction inlet is not fully submerged

(ix) Wearing rings are worn

(C) Enough pressure not developed

(i) Motor speed too low

(ii) Liquid pumped contains excess amount of air or gas

(iii) Motor direction wrong

(iv) Viscosity of liquid too high

(v) Wearing rings worn

(vi) Impeller diameter too small

(D) Pump works for sometime and then loses prime

(i) Suction line has air leakage

(ii) Liquid pumped contains excess amount of air or gas

(iii) Suction line has air pocket

(iv) Water seal tube clogged

(v) Suction lift too high

(vi) Suction inlet is not fully submerged

(E) Pump needs excess power

(i)Motor speed too high

(ii) Pump discharge more than anticipated

(iii) Viscosity of liquid too high

(iv) Alignment not well

(vi) Shaft is bent

(vii) Wearings ring worn

(viii) Motor runs in opposite direction

(F) Pump noisy or vibrates

(i) Impeller or suction pipe partially pluged

(ii) Suction lift too high

(iii) Alignment not well

(iv) Shaft bent

(v) Bearing worn

(vi) Foundation not rigid

(vii) Shortage of lubrication

10.4.3.6 Cavitation

The term cavitation means formation of cavities filled with liquid vapour due to a local pressure drop and their collapse as soon as the vapour bubbles reach regions of high pressure. In a centrifugal pump, as water flows through the suction line and enters the eye of the impeller, velocity increases and pressure decreases. If the pressure falls below the vapour pressure, water gets vapourized and some amount of vapour along with water gets inside the suction line. Water moving beyond the impeller reaches regions of high pressure and the cavities of vapour collapse. The collapse of vapour pockets causes the noise incident of cavitation.

Cavitation occurs due to improper design, installation and operation of pumps. Caviatation may be mild or severe. A severe cavitation may cause excessive noise and vibration and may destroy the impeller and other parts of pump. Cavitations may be controlled by (i) suspending the pump operation at heads and capacities much lower than the design values, (ii) avoiding high suction lifts and high velocities of water in suction line and (iii) limiting pump speed to manufacture's recommendation.

10.4.3.7 Water Hammer

Water hammer or hydraulic shock occurs when the water flowing through a pipe undergoes a sudden change in velocity such that the kinetic energy of the flowing liquid is converted to a dynamic pressure wave which may produce a substantial impact in rebounding back and forth in the pipe lines. Water hammer is caused at starting and stopping of a pump, at an abrupt change of pump speed, in case of power failure or rapid closure of a valve in the piping system. Reflux valves and surge pipes are installed in long pipe lines to reduce the effect of water hammer.

10.4.4 Vertical Turbine Pumps

A vertical turbine pump or called as deep well turbine pump, is a vertical axis centrifugal or mixed flow type pump. It consists of stages which accommodates rotating impellers and stationary bowls possessing guide vanes. These pumps are used where the pumping water level is below the limits of volute centrifugal pump. They are used for high lifts and have high efficiencies under optimum operating conditions. However, they have higher initial cost and are more difficult to install and repair. The pressure head developed depends on the diameter of the impeller and the speed at which it is rotated. Since the head from which it lifts water is high, a series of impellers are required in this case. The total head developed is proportional to the number of impellers.

The pump has three main parts. They are (i) pump element, (ii) discharge column and (iii) discharge head. The pump element is made up of one or more bowls or stages. Each bowl assembly consists of an impeller, a diffuser and a bearing. It has a strainer fitted at the bottom to screen out debris and coarser materials getting entered into the pump. The discharge column connects the bowl assembly and the discharge head assembly and conducts water from the bowl to the discharge head. The discharge head assembly consists of the base from which the discharge column, bowl assembly and shaft assembly are suspended (Fig. 10.16). Based on the method of lubrication of the pump, the turbine pumps may be oil lubricated or water lubricated type. The selection of oil or water lubricated pump depends on the conditions of operation. When the pumped water contains fine soil particles, oil lubricated type is preferred. For drinking purpose, water lubricated type pump is preferred. In situations where the pump remains suspended without any use for frequent periods, oil lubricated type pump is to be used. Water lubricated type is not used since the rubber bearings become dry and get out of shape when the pump remains idle for considerable period. However, the cost of water lubricated pump is little less than the oil lubricated type pump.

10.4.5 Submersible Pumps

A deep well turbine pump close-coupled to a small diameter submersible electrical motor is termed as a submersible pump. The motor is fixed directly below the water level in the well or water level in the underground surface. Thus, the pump elements and motor operate under submersed condition. This type of assembly is used in a deep well where a long shaft would not be practical. It is also suitable for flood prone areas. It is also preferably used in public grounds where an above-ground pump house would be inconvenient to build.

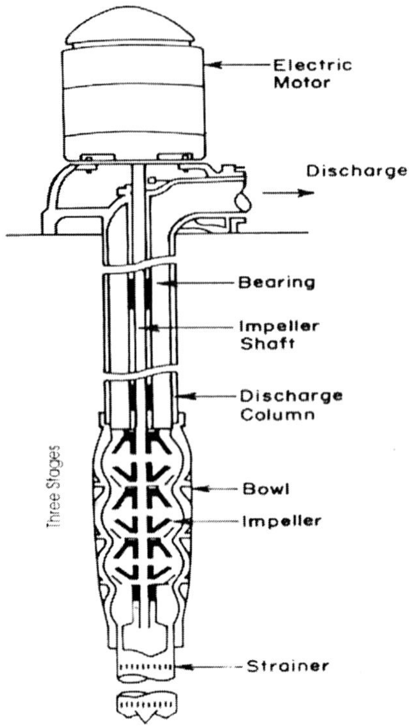

Fig. 10.16: Turbine pump

The submersible pump consists of (i) a pump and motor assembly (ii) a discharge column (iii) a discharge head assembly and (iv) a water proof cable to conduct the electric current to the submerged motors (Fig. 10.17). The construction of the pump element is similar to vertical turbine pump. Impellers made up of bronze are mounted by stainless steel propelling shaft. Impellers may be closed or semi-open type. The closed impellers are used where it is necessary for the pump to develop high pressure. Water enters the pump through a screen located between motor and the pump. The electric motor used should have same diameter as the size of the pump bowl. Its length is more than the type of motor used in other pumps. It is a squirrel cage induction motor which may be of the dry or wet type.

Submersible pumps are easy to install. The pump motor assembly is lowered first into the well adding pipe lengths as required. Care must be taken not to damage the outside sheathing of the water proof cable with pipe tools while lowering the pump and piping. A pipe clamp is used at the top of the well to support the pump and the delivery pipe. A sluice valve is fitted at the discharge end of the pump. The control board consisting of the switch, starter, and meter should be enclosed in a water-proof box. The pump is started with the sluice valve closed or slightly open. During the initial run, the physical quality of water is checked. If water being pumped contains sand and other impurities, then it should not be worried since most of the newly drilled wells when pumped contain silt laden water. The pump should never be stopped. The sluice valve is kept partially closed and pumping continues till clear water comes up.

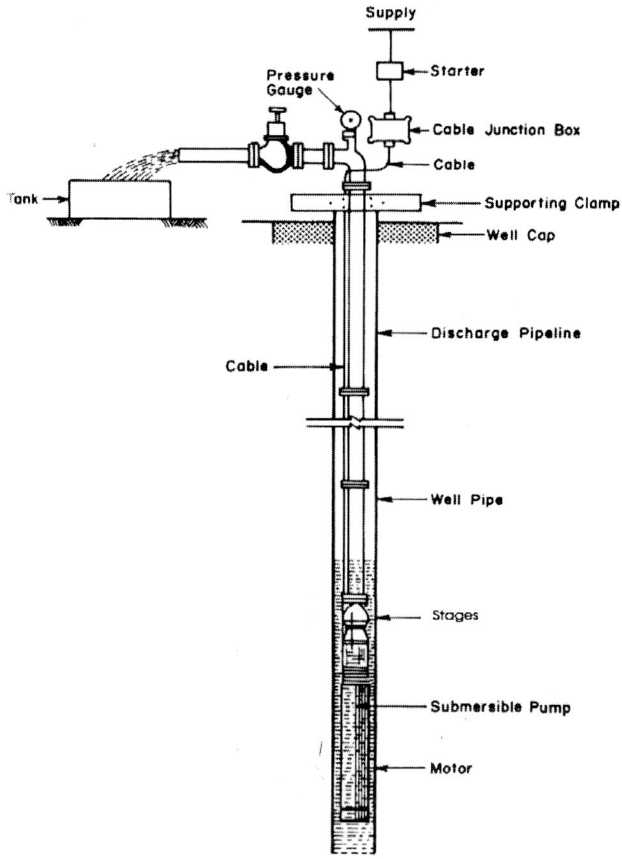

Fig. 10.17: Submersible pump

When the water is reasonably clear, the sluice valve is opened a bit more and again the quality of water is checked. The valve is so regulated that finally clear water comes up. Submersible pumps need little care and maintenance when it is correctly installed and operated. After continuous 2 years of service life (6000 hours of operation), the pump can be brought out of the well/bore hole and overhauled as per the instruction of the manufacturer.

10.4.6 Propeller and Mixed Flow Pump

The flow through the impeller of a propeller pump is parallel to the axis of the drive shaft rather than radial and so the propeller pumps are called as axial-flow pumps. Propeller and mixed flow pumps are used for low to medium heads and high discharge operations. By varying the vane pitch and curvature of the impeller, a wide range of head-capacity requirements can be met without changing the impeller diameter. Multi-stage approaches can be employed to obtain a higher head. To minimize the cavitations, impeller of the pump should be submerged deep enough in the water surface. Both of these pumps are extensively used for drainage. They are also suitable to pump water for irrigation from canals, rivers etc. to adjacent fields. The principal parts of propeller and mixed flow pumps are similar to deep well turbine pumps in that they have a head, an impeller and a discharge column. A shaft extends from the head, down the centre of the column to drive the impeller. They may be oil or water lubricated.

Propeller pumps are preferred to handle large volume of water at comparatively low heads and at speeds much higher than those of ordinary centrifugal pumps. They are not easily clogged. They usually pump water from a head of 1.0 to 2.5 m and are available in sizes varying from 20 to 120 cm. A mixed flow pump combines features of both the vertical turbine pump and propeller pump. The head is developed partly by centrifugal force and partly by the lift of the impeller vanes on the liquid as in a propeller pump. It operates under medium head and has high discharge. The head usually varies from 5 to 10 m.

10.4.7 Jet Pump

The jet pump consists of a combination of a centrifugal pump and a jet mechanism. The centrifugal pump is placed at the ground surface and furnishes the driving head and capacity for the jet placed in the well below the water surface. The main parts of a jet pump constitute a nozzle and a venturi. The flow passes through the nozzle having reduced area so that flow passes with high velocity. This creates a low pressure area around the venture which draws more water from the well. In general, they are low efficient pumps having highest capacity of about 35 percent. The jet pumps are generally used for low capacity deep well pumping where it is necessary to locate the moving parts of the pump and the primer at the ground surface. Advantages of jet pumps are:

(i) It can be installed inside very small size wells as small as 5 cm inside diameter.

(ii) Suction lift is high. Suction lift usually obtained with low to medium size jet pumps vary from 12 to 18 m.

(iii) It can be installed with the moving parts offset from the well.

(iv) Simple in installation, and have relatively low cost and maintenance.

(v) All moving parts of the pump are located at the ground surface and so are easily accessible.

10.4.8 Air Lift Pump

It consists of a casing pipe in which an educator pipe is introduced. An air pipe is also introduced into the casing pipe. An air diffuser is connected to the bottom end of the air pipe. The air diffuser is introduced into the educator pipe in upward direction. Compressed air is forced into the air pipe which produces a mixture of air and water that rises up in the form of bubbles. This mixture has low specific gravity than water in the casing pipe. This creates pressure differential between the water in the educator pipe and that in the casing forcing some water to move up in the educator pipe. Finally this raised column of water discharges out through the outlet (Fig. 10.18). The efficiency of air lift pump depends on the submergence depth of the air pipe. The depth of submergence should be about two-third of the length of air pipe (Basak, 2012).

10.4.9 Low Lift Hand Pump

This is a cheap pump and used by many villagers for lifting water from a small depth. It can be operated by hand by a single person. It has a wooden handle connected at one end of the pump bowel and a suction pipe at the other end. The suction pipe dips in the water body and draws water. The delivery pipe is short and connected to the other end of the suction pipe very near to the handle (Fig. 10.19). The delivery pipe cannot throw water to a great distance but delivers water to the conveying channel. When the handle is lifted up, suction is created in the pump chamber/bowel and some water rushes inside the chamber. Then when the handle is lowered down, the water from the pump chamber discharges out. The pump can operate for a low head up to 3 m and has discharge capacity 4000 to 5000 lit/hr. It is ideal to irrigate the crops with low water requirement such as vegetables, flowers, horticultural crops and in kitchen garden. Its weight is not very high being limited up to 40 kg. Therefore it is portable and so can be easily handled by a single person (Panigrahi, 2011).

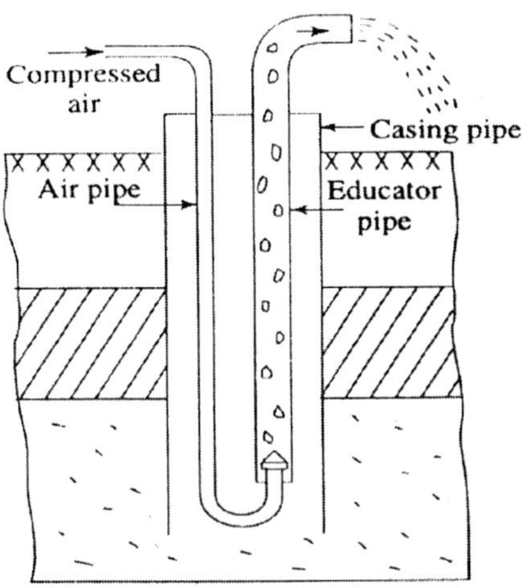

Fig. 10.18: Air lift pump

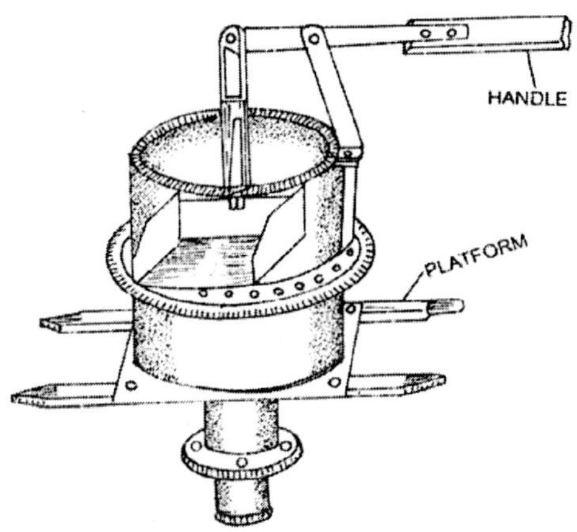

Fig. 10.19: A schematic view of low lift hand pump

10.4.10 Specific Speed of Pump

Operating characteristics of a pump is characterized often by specific speed. Specific speed gives relationship between discharge, head and speed. It is defined as the speed of a geometrically similar pump when delivering one cumec of discharge against a total head of 1 m (Church and Lal, 1973).

It can be expressed mathematically as:

$$n_s = \frac{n\sqrt{Q}}{H^{\frac{3}{4}}} \tag{10.3}$$

where,

n_s = specific speed, rpm; n = pump speed, rpm; Q = discharge of the pump, cumce and H = total head, m.

Sample Calculation 2:

A centrifugal pump discharges 0.05 cumec of water against a total head of 30 m. The speed of the pump 1500 rpm. What is the specific speed of the pump?

Solution:

Given pump speed, n = 1500 rpm

Discharge, Q = 0.05 cumec and

Total head, H = 30 m

Specific speed, ns as per Eqn. (10.3) is:

$$n_s = \frac{1500\ x\ \sqrt{0.05}}{30^{\frac{3}{4}}} = 26.2 \text{ rpm.}$$

Specific speed ranges from 10 to 300 rpm in single suction impeller. For a multi-stage pump, it is computed on the basis of the head per stage. The specific speed of a double suction impeller is divided by $\sqrt{2}$ to get the specific speed of a single suction impeller.

Generally, lower the specific speed, the higher the head that can be developed per stage by the pump. From Eqn. (10.3), it is seen that for a given head-discharge value, higher specific speed pumps require will run at higher speed and will be of smaller dimensions. Many other hypotheses can be concluded by analysis of Eqn. (10.3). One such important hypothesis is that at same speed and head condition, a higher specific speed pump will deliver higher discharge and vice-versa.

10.4.11 Some Terminologies Used in Pump Characteristics

Capacity: It is the volume of water pumped per unit time and generally expressed in lit/sec.

Static Suction Lift: It is the vertical distance from the static water level (determined when the pump is not operating) to the centre line of the pump.

Total Suction Lift: The summation of static suction lift and losses due to friction as well as entrance is called total suction lift.

Static Suction Head: It is the vertical distance from the centre line of the pump to the free level of water to be pumped.

Total Suction Head: It is the difference between the static suction head and losses due to friction in suction pipes and fittings and any pressure head existing on the suction supply. Total discharge head is also estimated as the reading of a pressure gauge at the discharge of the pump, expressed in metres of water, and corrected to pump centre line, plus the velocity head at the point of gauge attachment.

Static Discharge Head: It is the vertical distance from the centre line of the pump to the discharge water level.

Total Discharge Head: It is the sum of the static discharge head, friction and exit losses in the delivery line plus the velocity head and pressure head at the point of discharge. Total discharge head is also estimated as the reading of a pressure gauge at the discharge of the pump, expressed in metres of water, and corrected to pump centre line, plus the velocity head at the point of gauge attachment.

Total Static Head: It is the vertical distance from the suction water level to the discharge water level or sum of the static suction lift and static discharge head. .

Friction Head: It is the head expressed in metres of water required to overcome friction caused due to flow of water in pipes and in fittings.

Total Head/System Head: It is total static head plus losses due to friction in pipes, fittings, entrance loss and bend loss *etc.*, velocity head and operating head if any.

Net Positive Suction Head (NPSH): It is the total suction head minus the vapour pressure of water at the pumping temperature, both expressed in same units i.e. metre of water. Vapor pressure of water at different temperatures is given in standard physical tables. The NPSH is a characteristic of the pump and is usually given in pump characteristic curves. For operation of a centrifugal pump without cavitation, the atmospheric pressure should be more than the sum of the suction lift, all friction losses in pipes, fittings, elbows, bends, valves *etc.* and factor of safety considering to altitude of the site. The Eq. (10.4) gives the maximum practical suction lift of pumps (Michael, 1978):

$$H_p = H_a - H_f - H_s - F_s \tag{10.4}$$

where,

H_p = maximum practical suction lift, m; H_a = atmospheric pressure at water surface, m which is 10.3 m at sea level); H_f = friction losses in pipes, fittings, elbows. bends, valves etc. on the suction line, m; H_s = total suction head (which is sum of net positive suction head and saturated vapour pressure of water), m and F_s = factor of safety which is a correction of H_a for altitude and is a reduction value of 0.36 m for every 300 m rise of altitude.

For satisfactory performance of the centrifugal pump, friction losses and suction lift should be kept as minimum as possible. For this reason, the suction pipe is taken larger than the discharge pipe and the pump is placed as close as possible to the water surface of the source.

10.4.12 Power Requirement of Pump

Water Horse Power (WHP)

It is the theoretical horse power required for pumping. The power requirement of the pump motor is expressed by the term horse power (H.P.) The power of the pump motor relates head with discharge. The water horse power is given as:

$$W.H.P. = \frac{\omega . Q . H}{75} \tag{10.5}$$

where, ω = unit weight of water, 1000 kg/m^3; Q = discharge to be delivered, cumecs; H = total head against which the pump has to work, m and η = efficiency of the pump set generally taken as 65 percent i.e. 0.65.

The total head is the sum of total suction and delivery head, velocity head and friction losses in pipe including losses at the bends and strainer openings etc. Velocity head is expressed as:

$$Velocity\ head = \frac{V^2}{2g} \tag{10.6}$$

where, velocity head is given, m; V = velocity of water flow in the pipe, m/sec and g = acceleration due to gravity, 9.81 m/ sec^2.

Friction loss is expressed as:

$$h_f = \frac{f^{/}\ l . V^2}{2gd} \tag{10.7}$$

where, h_f is the frictional loss head, m; l = total length of pipe including vertical and horizontal portions, m; V = velocity of water flow in pipe, m/sec; g = acceleration due to gravity, 9.81 m/ sq. sec and d = diameter of pipe in which water flows, m and $f^{/}$ = coefficient of friction which is generally taken as 0.024. .

Friction loss in strainer of a pump is given as:

$$h_f = \frac{K_s . V^2}{2g} \tag{10.8}$$

Friction loss in foot valve is:

$$h_f = \frac{K_f . V^2}{2g} \tag{10.9}$$

where, h_f is the frictional head loss, m; K_s and K_f are constants having values 0.95 and 0.80, respectively. Other terms are defined earlier.

Shaft Horse Power (S.H.P.)

Considering the efficiency of the pump as η, the shaft horse power ($S.H.P.$) is given as:

$$S.H.P. = \frac{\omega .Q .H}{75 .\eta} \tag{10.10}$$

Break Horse Power (B.H.P.)

It is the actual horse power required to be supplied by the engine or electric motor for driving the pump. With direct driven pump (drive efficiency 100 percent) $B.H.P. = S.H.P.$ With pumps driven by belts or other indirect devices, $B.H.P.$ is given as:

$$B.H.P. = \frac{S.H.P.}{Drive\ efficiency} = \frac{W.H.P.}{Pump\ efficiency\ x\ Drive\ efficiency} \tag{10.11}$$

Horse power input to electric motor ($H.P._{motor}$) is given as:

$$H.P._{motor} = \frac{W.H.P.}{Pump\ efficiency\ x\ Drive\ efficiency\ x\ Motor\ efficiency} \tag{10.12}$$

Sample Calculation 3:

A pump has to irrigate 2 ha wheat crops and the depth of irrigation is 5 cm. The pump runs for 10 hours to irrigate the crop. If the total head against which the pump has to run is 8 m, determine the power requirement of the pump. Assume the pump has efficiency 65 percent. Assume driving efficiency as 100 percent. If the motor has an efficiency of 70 percent, find out also the horse power of the motor to drive the pump.

Solution:

Total quantity of water to be supplied for irrigation = 2 x 5 = 10 ha cm which is equivalent to 1000 m^3.

Time for which the pump runs = 10 hr

Hence, discharge of the pump = 1000 /10 x 3600 = 0.0278 cumec.

Total head = 8 m and the pump efficiency = 0.65

Hence the pump horse power = (0.0278 x 8 x 1000)/ (75 x 0.65) = 4.56 H.P. = 5 H.P.

Hence a 5 H.P. pump can be selected for irrigation.

If the motor has an efficiency of 70 percent, then the H.P. of the motor required = 4.56/0.70 = 6.51 = 7 H.P.

Sample Calculation 4:

A direct centrifugal pump has to lift water at the rate of 150 lit/sec. Calculate the horse power of the engine for a suction head of 7 m, delivery head of 3 m, coefficient of friction is 0.024, diameter of the suction and delivery pipes are 15 and 10 cm, respectively. Neglect minor losses. Assume a pump has efficiency as 80 percent.

Solution:

Discharge of the pump, Q = 150 lit/sec = 0.150 m^3/sec

Suction head = 7 m and delivery head = 3 m

Diameter of suction pipe, d_s = 15 cm and that of delivery pipe, d_d = 10 cm

Velocity of water in suction pipe, $V_s = \dfrac{4\,x\,Q}{\pi\,x\,d^2_s} = \dfrac{4\,x\,0.150}{\pi\,x\,0.15^2} = 8.48\,m/\sec$

Velocity of water in delivery pipe, $V_d = \dfrac{4\,x\,Q}{\pi\,x\,d_d^{\,2}} = \dfrac{4\,x\,0.150}{\pi\,x\,0.10^2} = 19.10\,m/\sec$

Velocity head in suction pipe = $\dfrac{V_s^{\,2}}{2g} = \dfrac{8.48^2}{2\,x\,9.81} = 3.67\,m$

Velocity head in delivery pipe = $\dfrac{V_d^{\,2}}{2g} = \dfrac{19.10^2}{2\,x\,9.81} = 18.60\,m$

Friction loss in suction pipe = $h_f = \dfrac{f'\,l\,.V^2}{2gd} = \dfrac{0.024\,x7\,x\,8.48^2}{2\,x\,9.81\,x\,0.15} = 4.10\,m$

Friction loss in delivery pipe $h_f = \dfrac{f'\,l\,.V^2}{2gd} = \dfrac{0.024\,x3\,x\,19.10^2}{2\,x\,9.81\,x\,0.10} = 13.38\,m$

Total head, H = Suction head + delivery head + friction loss in suction and delivery pipe + velocity head in suction and delivery pipe = 7.0 +3.0 + 4.10 + 13.38 + 3.67 + 18.60 = 49.8 m.

As the pump is direct drive, shaft horse power is equal to break horse power.

Hence, $B.H.P. = \dfrac{\omega\,.Q\,.H}{75\,.\eta} = \dfrac{1000\,x\,0.150\,x\,49.75}{75\,x\,0.80} = 124\,HP$

10.4.13 Pump Characteristics

In order to use modern pumps most efficiently, it is essential to select pumps well adapted to the particular conditions of operation and to obtain relatively low expenses on pumping irrigation water. Efficiency of the pumping system decreases, when the head is excessively high or when the pumps discharges low amount of water than the manufacture's recommended value at a

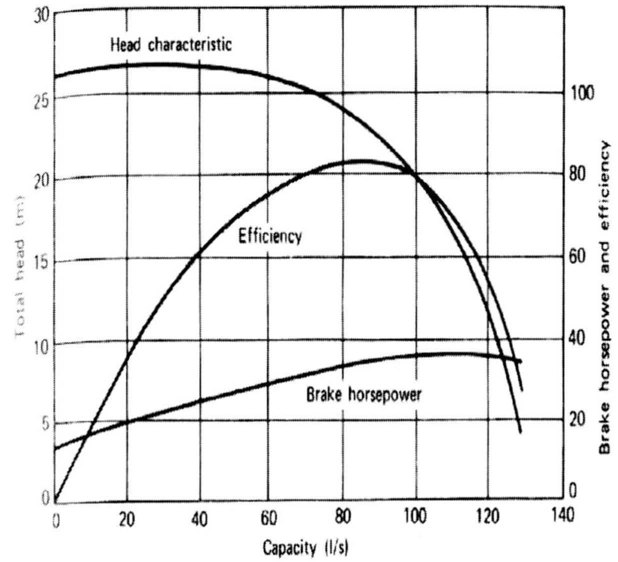

Fig. 10.20: Pump characteristic curve of a single pump

head lower than the normal and cause the efficiency to be low. The interrelations between head, discharge, efficiency and power of a pump are usually represented by curves which are designated as "characteristics" curves (Fig. 10.20, Hansen *et al.,* 1979).

Pump characteristic curves are also called as performance curves of pumps. Knowledge of pump characteristics enables one to select a pump that fits operating conditions and thus attain a relatively high efficiency with low operating cost. In a pump characteristic curve, the head, power input and efficiency are plotted in ordinates and capacity id represented in abscissa at a constant pump speed. The net positive suction head may also be plotted as ordinate. About 6 to 12 values are taken during the pump test to plot the points. Smooth curves are then drawn joining the points.

The head-capacity of the pump characteristics curves shows how much water a given pump will deliver with a given head. This curve signifies that as the head increases, the capacity of the pump gradually decreases and vice-versa. The efficiency curve represents that the efficiency decreases from zero with nil discharge to a maximum value of 82 percent when the discharge is about 86 lit/sec. The head corresponding to this highest value of the efficiency is about 23 m. At higher head of more than 23 m and discharge more than 86 lit/sec, the efficiency gradually decreases at becomes zero at zero head. The break horse power curve indicates that the power of the pump increases over most of the range as the discharge increases, reaching a peak value at a somewhat higher rate of discharge than that which produces maximum efficiency.

A composite characteristic curve may also be drawn with different impeller diameters. In a composite characteristic curve, several curves, representing different pump speeds with different impeller diameters are drawn on the same graph. This type of graph shows a number of head-capacity curves for one impeller diameter and different speeds or head-capacity curves for different impeller diameters and one speed.

Effect of Speed and Impeller Diameter on Pump Performance

The performance of a centrifugal pump can be changed by changing the impeller diameter and/or speed.

Effect of Change of Speed: When the speed of a centrifugal pump changes, the operation of the pump also changes as follows:

$$Q = Q_1\left(\frac{n}{n_1}\right)$$

$$H = H_1\left(\frac{n}{n_1}\right)^2$$

$$P = P_1\left(\frac{n}{n_1}\right)^3$$

$$\text{Or,}\quad \frac{n}{n_1} = \frac{Q}{Q_1} = \sqrt{\frac{H}{H_1}} = \sqrt[3]{\frac{P}{P_1}} \qquad (10.13)$$

where,

n = New desired speed, rpm; Q = Capacity of the pump at the desired speed n, lit/sec; H = Head at the desired speed n for capacity Q, m; P = BHP at the desired speed n at H and Q; n_1 = Speed at which the characteristics are known/original speed, rpm; Q_1 = Capacity at speed n_1, lit/sec; H_1 = Head at capacity Q_1 and speed n_1, m and P_1 = BHP at speed n_1 at H_1 and Q_1.

Effect of Change of Impeller Diameter: When the impeller diameter of a centrifugal pump changes, the operation of the pump also changes as follows:

$$Q = Q_1\left(\frac{D}{D_1}\right)$$

$$H = H_1\left(\frac{D}{D_1}\right)^2$$

$$P = P_1\left(\frac{D}{D_1}\right)^3$$

$$\text{Or,}\quad \frac{D}{D_1} = \frac{Q}{Q_1} = \sqrt{\frac{H}{H_1}} = \sqrt[3]{\frac{P}{P_1}} \qquad (10.14)$$

where,

D = changed diameter of the impeller, mm and D_1 = original diameter of the impeller, mm. Other terms of the Eq. (10.14) are defined earlier.

Sample Calculation 5:

The original speed of a pump is 1000 rpm. The values of discharge, head and power of the pump at this original speed of 1000 rpm are found 150 lit/sec, 25 m and 35 HP, respectively. Find out the discharge, head and power at a speed of 2000 rpm.

Solution:

Given, $n_1 = 1000$ rpm

$Q_1 = 150$ lit/sec

$H_1 = 25$ m

$P_1 = 35$ HP and

$n = 2000$ rpm

The new discharge is $Q = Q_1\left(\frac{n}{n_1}\right) = 150\left(\frac{2000}{1000}\right) = 300$ lit/sec

New head is $H = H_1\left(\frac{n}{n_1}\right)^2 = 25\left(\frac{2000}{1000}\right)^2 = 100$ m

New power is $P = P_1\left(\frac{n}{n_1}\right)^3 = 35\left(\frac{2000}{1000}\right)^3 = 280$ HP.

10.4.14 Selection of Pumps

Selection of a suitable water lifting device depends on type of sources of water, amount of water to be lifted, depth of pumping water level, type and amount of power available and finally economic status of the farmer. Every pump installation has different operating head and discharge conditions. Hence, it is necessary to select the pump such that it operates under maximum efficiency with the given head and the discharge. This is done by plotting the characteristic curves both or the well and for the pump. Normally, the most suitable pump is selected based on economics. Sometimes, manufacturer's catalogue s are consulted to identify pumps of the proper type that are capable of supplying the discharge and head requirements of the irrigation system.

Criteria and Procedure for Selection of Pumps

The main factors influencing the selection of the pumping sets are:

(i) Requirement of irrigation water by the crops under irrigation

(ii) Yield of the sources of water like well, streams, rivers etc.

(iii) Availabity and cost of the type of pump and kind of energy.

The pump discharge must meet the maximum water demand of the crop and/or cropping pattern per day in peak periods. The rate of pumping depends on the area under different crops, water requirements of the crops, rotation period i.e. interval between two consecutive irrigations of a crop and duration of pumping each day. It may be estimated by the formula:

$$Q = \frac{1000\ x A\ x\ y}{36\ x\ R\ x T} = \frac{27.78 \times A \times y}{R \times T} \tag{10.15}$$

where,

Q = pump discharge, lit/sec; A = area of land under crop, ha; y = depth of irrigation water, cm; R = rotation period, days and T = duration of pumping, hr/day.

Sample Calculation 6:

A farmer wishes to have his own pump set for the following cropping pattern in his farm of 6 ha area. What right size of the centrifugal pump should he select?

Crop	Area to be irrigated, ha	Depth of irrigation, cm	Rotation period, days	Period of work, hrs/day
Wheat	2.0	7.5	12	10
Maize	0.4	7.5	20	10
Vegetables	0.4	7.5	10	10
Mustard	2.2	5.0	40	10
Sunflower	1.0	5.0	20	10

Discharge rate of the pump,

$$= 27.78\ \frac{A\ x\ y}{R\ x\ T}$$

$$= 27.78 \left(\frac{2\ x\ 7.5}{12\ x\ 10} + \frac{0.4\ x\ 7.5}{20\ x\ 10} + \frac{0.4\ x\ 7.5}{10\ x\ 10} + \frac{2.2\ x\ 5.0}{40\ \ 10} + \frac{1.0\ x\ 5.0}{20\ x\ 10} \right)$$

= 6.18 lit/sec.

10.4.15 Irrigation Pumping Costs

In order to estimate the costs of irrigation water obtained by pumping and to compare these costs with costs of water from gravity system, it is essential to compute all pumping costs in terms of volume of water delivered annually to the irrigated areas. The cost of pumping includes (i) fixed cost and (ii) operating cost.

Fixed Cost

Fixed cost includes interest cost, depreciation and taxes if any. The interest is computed as:

$$\text{Annual Interest} = \frac{(\text{Cost of installation - Salvage value}) \times \text{Interest rate}}{2} \quad (10.16)$$

Depreciation is computed since there is loss in the value of the pumping plant due to operation or age and is estimated as:

$$\textit{Annual Depreciation} = \frac{(\text{Original cost - Salvage value})}{\text{Life span of system in years}} \quad (10.17)$$

The estimated life span of centrifugal pump is 16 years whereas that of electric motor and diesel engine is 25 and 14 years, respectively (Michael, 1978).

Operating Cost

Operating cost includes the followings:

Fuel/energy consumption: Energy consumption of electric motor is given as:

$$\textit{Energy consumption in kilowatt-hours} = \frac{B.H.P. \times 0.746}{\textit{Motor efficiency}} \quad (10.18)$$

Motor efficiency varies from 80 to 90 percent.

Energy consumption by diesel engine which is commonly used in irrigation pumping is 0.23 litres per break horse power hour. Cost per hour of operation is estimated as:

Cost per hour of operation= B.H.P. x Fuel consumed in lit/hr x Cost of fuel per lit.

Lubricating oil and grease: This is negligible for electric driven engines. For diesel and petrol engines, the requirement is 4.4 lit/1000 HP hours.

Repair and maintenance: Yearly repair and maintenance cost of centrifugal pumps is calculated by dividing the estimated life of the pump with total cost.

In addition to the above costs, the operator wages as per the prevailing rate in the area of operation is to be considered while computing the total operating costs.

Sample Calculation 7:

A centrifugal pump with 5 HP water horse power is installed in a well and operates by a direct coupled 3-phase electric motor. Efficiencies of the pump and electric motors are 70 and 75 percent, respectively. The pump operates for a total of 2000 hours in 200 days a year. Compute the annual cost of pumping. The cost of pump and motor are Rs. 10,000 and 15,000, respectively. Total cost of suction and delivery pipes and other accessories are Rs. 5,000. Costs of electrical fittings are Rs. 5000. Cost of electric energy is Rs. 4/unit. Assume the interest rate as 7 percent. Salvage value of pump and motor may be assumed as Rs. 1000 and Rs. 1500, respectively and that of other items may be neglected. Operator spends daily 1 hour in running the pump and his wage is Rs. 20 per hour daily.

Solution:

Total cost of installation = Cost of pump + cost of motor + cost of pipes including fittings + cost of electrical fittings = Rs. 10,000 + Rs. 15,000 + Rs. 5,000 + Rs. 5,000 = Rs. 35,000.

Salvage value of pump and motor = Rs. 1,000 + Rs. 1,500 = Rs. 2,500.

Given interest rate is 7 per cent.

Computation of fixed cost: Fixed cost is sum of annual interest and depreciation cost.

$$Annual\ Interest = \frac{(Cost\ of\ installation - Salvage\ value)\ x\ Interest\ rate}{2}$$

$$= \frac{(35{,}000 - 2{,}500)\ x\ 7}{2\ x\ 100} = \text{Rs.}1137.50$$

Depreciation:

(i) Pump $= \dfrac{(10{,}000 - 1{,}000)}{16} = Rs.562.50$

(ii) Motor $= \dfrac{(15{,}000 - 1{,}500)}{25} = Rs.\ 540.00$

(iii) Pipe and fittings: $\dfrac{(5{,}000 - 0)}{25} = Rs.\ 200.00$

(iv) Electrical accessories $= \dfrac{(5{,}000 - 0)}{25} = Rs.\ 200.00$

Total depreciation = Rs. 1502.50

Hence, fixed cost = Rs. 1137.50 + Rs. 1502.50 = Rs. 2640.00

Operating Cost: Operating cost is sum of cost of electrical energy, repair and maintenance cost of pump and wages of the operator.

Annual electrical energy consumption $= \dfrac{5\ x.746\ 2000}{0.79\ x\ 0.75}$

= 14209.50 kilowatt hours = 14209.50 units

Cost of electrical energy: Rs. 2.00 x 14209.50 = Rs. 28419.00

Repair and maintenance cost of pump = Rs. 10,000 / 16 = Rs. Rs. 625.00

Operator's wages = Rs. 20 x 200 = Rs. 4000.00

Total operating cost = Rs. 28419.00 + Rs. 625.00 + Rs. 4000.00 = Rs. 33,044.00

Therefore, total annual cost of pumping = Fixed cost + Operating cost

= Rs. 2640.00 + Rs. 33,044.00 = Rs. 35, 684.00

Question Banks

Q1. What is lift irrigation? What are the differences between flow irrigation and lift irrigation?

Q2. What are the applications of water lifting device in irrigation of agricultural lands?

Q3. What do you mean by indigenous water lifting device? What are its uses?

Q4. What are the different types of indigenous water lifting devices?

Q5. Although the manual and animal operated water lifting devices are costly, still they are adopted for irrigation purpose? Describe why?

Q6. What are various criteria for selection of a suitable water lifting device?

Q7. What are the differences among the low, medium and high head water lifting devices?

Q8. Write notes on the following low head manual water lifting devices.

(i) Swing basket (ii) Sedua (iii) Don (iv) Archemedian Screw (v) Water wheel

Q9. How does a Persian wheel work?

Q10. With a diagram, explain the construction and working of Archemedian screw.

Q11. With a well label diagram, describe the construction and function of a counter poise bucket lift. Derive an expression for estimation of weight of counter poise lift.

Q12. How does a tenda lift water from a water body?

Q13. Describe the mechanism of working of a chain pump.

Q14. How the water from a well is get lifted by a rope-and-bucket lift?

Q15. What are the advantages and dis-advantages of animal operated water lifting devices?

Q16. Briefly narrate the working principles of following animal operated water lifting devices.

(i) Circular mot (ii) Water wheels (iii) Sakia (iv) Saqiya (v) Circular two-bucket-lift (vi) Chain pump

Q17. What are the advantages and dis-advantages of wind powered water lifting device?

Q18. How does a wind mill help in lifting water from a well?

Q19. With a diagram, write down the construction and functions of hydraulic ram. How can the efficiency of a hydraulic ram be increased?

Q20. Can bio-diessel be used as a source to run pump and lift water? Write the merits and demerits of using bio-fuel as an energy source to lift water.

Q21. Briefly explain the role of pumps in increasing the irrigation potential and agricultural productions?

Q22. What is shaduf? How does it help in lifting water?

Q23. What are the various classifications of pumps? Of all these pumps, which one is commonly used in irrigation?

Q24. What are the main differences between positive displacement pump and variable displacement pump?

Q25. Reciprocating pump is a positive displacement pump. Justify why? With a neat sketh, explain the principle of operation of reciprocating pump.

Q26. Differentiate between the followings.

(i) Single acting and double acting reciprocating pumps
(ii) Duplex and triplex pump

Q27. Derive an expression for force required to operate a pump.

Q28. A single acting reciprocating pump has piston diameter 15 cm and stroke 20 cm. The piston makes 50 double strokes per minute. Suction and delivery heads are 5 and 8 m, respectively. Fine out the discharge capacity of the pump and power requirement of the pump in H.P. Assume efficiency of suction and delivery strokes are 60 and 75 percent, respectively.

Q29. A double acting reciprocating pump has a piston diameter 20 cm and piston rod diameter of 5 cm connected to one side of piston only. Length of the piston stroke is 30 cm. Speed of the crank moving the piston is 50 rpm. Suction and discharge heads are 5.0 and 15.0 m, respectively. Determine the power required to operate the pump.

Q30. How does a rotary pump work?

Q31. What are the characteristics of variable displacement pumps?

Q32. What are the chief advantages of centrifugal pumps?

Q33. With diagram, write the principles and operations of centrifugal pumps.

Q34. Classify the different basis of centrifugal pumps?

Q35. Differentiate between the followings.

(i) Volute type and diffuser type centrifugal pump

(ii) Single stage and multi stage centrifugal pump

(iii) Horizontal and vertical centrifugal pumps

(iv) Single suction and double suction pumps

Q36. Describe the procedures for installation of centrifugal pumps for lifting water for irrigation.

Q37. Write the different electrical accessories required for working of a pump. How are the electrical connections done in setting up of a three phase electrical motor to operate a pump?

Q38. Before starting the pump for the first time, what are the points to be considered?

Q39. What are the points to be considered for maintenance of pumps?

Q40. What are the different common pump troubles? Mention their causes and remedies.

Q41. A pump runs but lifts less amount of water. What are the causes and how can you rectify it.

Q42. What is cavitation? How it occurs and how can it be controlled?

Q43. What do you mean by water hammering? What are its ill effects? How can it be checked?

Q44. What is a vertical turbine pump? How is it constructed? Write down the principles of working of this pump.

Q45. Discuss the construction and working principles of the following pumps used in lifting irrigation water.

(i)Submersible pump (ii) Air lift pump (iii) Propeller and mixed flow pump (iv) Jet pumps

(v) Low lift hand pump

Q46. Define specific speed of pump. What is the usefulness of specific speed in operation of pump? Give its mathematical expression.

Q47. A centrifugal pump discharges 0.08umec of water against a total head of 20 m. The speed of the pump 1000 rpm. What is the specific speed of the pump?

Q48. Define the followings:

(i) Static suction lift (ii) Static suction head (iii) Static discharge head (iv) Total discharge head (v) Total head (vi) Net positive suction head (vii) Water horse power (viii) Shaft horse power (ix) Break horse power

Q49. A pump has to irrigate 5 ha crops and the depth of irrigation is 5 cm. The pump runs for 8 hours to irrigate the crop. If the total head against which the pump has to run is 10 m, determine the power requirement of the pump. Assume the pump has efficiency 65 percent. Assume driving efficiency as 80 percent. If the motor has an efficiency of 70 percent, find out also the horse power of the motor to drive the pump.

Q50. A direct centrifugal pump has to lift water at the rate of 100 lit/sec. Calculate the horse power of the engine for a suction head of 6 m, delivery head of 2 m, coefficient of friction is 0.024, diameter of the suction and delivery pipes are 10 and 6 cm, respectively. Neglect minor losses. Assume a pump has efficiency as 80 percent.

Q51. Draw a pump characteristic curve and explain its use in the filed of irrigation.

Q52. Describe the effect of speed and impeller diameter on the performance of the pump.

Q53. The original speed of a pump is 1200 rpm. The values of discharge, head and power of the pump at this original speed of 1200 rpm are found 100 lit/sec, 20 m and 30 HP, respectively. Find out the discharge, head and power at a speed of 2000 rpm.

Q54. Mention the main factors influencing the selection of the pumping sets.

Q55. A farmer wishes to install a pump for the following cropping pattern in his farm of 5 ha area. What right size of the centrifugal pump should he select?

Crop	Area to be irrigated, ha	Depth of irrigation, cm	Rotation period, days	Period of work, hrs/day
Sesamum	1.0	5.0	10	10
Maize	0.4	6.0	15	10
Vegetables	0.4	7.5	10	10
Mustard	2.0	5.0	30	10
Sunflower	1.2	5.0	20	10

Q56. A centrifugal pump with 5 HP water horse power is installed in a well and operates by a direct coupled 3-phase electric motor. Efficiencies of the pump and electric motors are 65 and 70 percent, respectively. The pump operates for a total of 2000 hours in 180 days a year. Compute the annual cost of pumping. The cost of pump and motor are Rs. 12,000 and 18,000, respectively. Total cost of suction and delivery pipes and other accessories are Rs. 6,000. Costs of electrical fittings are Rs. 4000. Cost of electric energy is Rs. 3.50/unit. Assume the interest rate as 7 percent. Salvage value of pump and motor may be assumed as Rs. 1000 and Rs. 1500, respectively and that of other items may be neglected. Operator spends daily 1 hour in running the pump and his wage is Rs. 25 per hour daily.

References

Basak, N.N. 2012. Irrigation Engineering. Tata McGraw Hill Education Pvt. Ltd., New Delhi.

Church, A.H. and Lal, J. 1973. Centrifugal Pumps and Blowers. Metropolitan Book Co. Pvt. Ltd. , Delhi

Garg, S.K. 1987. Irrigation Engineering and Hydraulic Structures. Khanna Publishers, New Delhi.

Hansen, V.E., Israelsen, O.W. and Stringham, G.E. 1979. Irrigation Principles and Practices. John Wiley and Sons Inc., New York.

James, L.G. 1988. Principles of Farm Irrigation System Design. John Wiley & Sons, New York, pp. 543.

Khepar, S.D; Kaushal, M.P. and Katyal, A.K. 1975. Development of an animal drawn lift irrigation pump. Journal of Agricultural Engineering, Vol, XII, No. 3 & 4.

Kumar, S; Kumar, V and Sahu, R.K. 2012. Fundamentals of Agricultural Engineering. Kalyani Publishers, New Delhi, pp. 565.

Lal, J. 1969. Hydraulic Machines. Metropolitan Book Co. Pvt. Ltd. , Delhi, pp. 631.

Michael, A.M. 1978. Irrigation Theory and Practices. Vikash Publishing House Pvt. Ltd., New Delhi.

Murty, V.V.N. 2004. Land and Water Management Engineering. Kalyani Publishers, Cuttack.

Panigrahi, B. 2011. Irrigation Systems Enginering. New India Publishinh Agency, New Delhi.

Punmia, B.C. and Lal, P.B.B. 1992. Irrigation and Water Power Engineering. Laxmi Publications Pvt. Ltd., New Delhi.

11

Losses of Water in Canal and Control

11.1 Introduction

The entire amount of water that is diverted at the canal head regulator does not reach at the field for use of irrigation. Major amount of it is lost on way during its movement in the canal systems. During the passage of water from main canal to outlet at the head of the water course, water may be lost either by evaporation from the surface or by seepage through the peripheries of the channel. These losses are sometimes very high, and may go up to 50 % of the total water diverted into the main canal. Evaporation losses are low and may go to a maximum of about 7 percent.of total losses. The major loss of water in the conveyance system is the seepage loss. Seepage loss depends on the channel geometry and evaporation loss is proportional to the area of free surface. In determining the designed channel capacity, a provision of these water losses must be made. The provision for the water lost in the watercourses and in the field however, already made in the outlet discharge factor, and hence no extra provision is made on that account. However, excluding the above two losses, there is operational loss which when added to the evaporation and seepage losses, the total losses goes up the maximum 50 percent level. The operational loss includes spillage, over-toping, leakage through rodent holes, dead storage, initial infiltration and evapotranspiration from the vegetation along the watercourse. Evaporation and seepage losses of the canals are discussed below.

11.2 Evaporation Loss

Evaporation is the process of conversion of water from liquid state to gaseous state and its transfer into the atmosphere in this form. It is a natural process and continues because of falling of incident rays of solar radiation on water bodies. It is a surface phenomenon. Rate of evaporation depends upon the climatic factor like temperature, wind velocity, relative humidity of air as well as nature and size of water body. In summer season, these losses may be more but seldom exceed about 7% of the total water diverted into main canal. The evaporation losses depending on water spread area of the canal is a function of side slope of canal and depth of flow. If the side slope of the canal is flatter ,then water spread area will be more and hence evaporation loss will be more and vice versa .For a given side slope, water spread area depend upon the depth of water in the canal. If the depth of water in canal is more, the losses will be less.

Rate of evaporation can be determined by (i) Analytical method (ii) Empirical method (iii) Energy budget method and (iv) Pan evaporation method. There are two analytical methods to compute evaporation losses. They are water budget method and mass transfer method. Detail discussions of various above mentioned methods for determining the evaporation rate are already presented in Chapter 3 entitled "Storage of Irrigation Water". Different measures for reduction of evaporation loss are also mentioned in Chapter 3 entitled "Storage of Irrigation Water".

11.3 Seepage Loss

Seepage is the most dominant process by which water is lost in the canal. Thus, for the effective operational planning and management of an irrigation system, a dependable forecasting of the seepage is very important. It is necessary to prevent seepage loss to save the costly irrigation water .Seepage loss varies spatially due to different soil types and hydraulic conditions. The bed material of the canal is a dominant factor deciding seepage loss. If the bed material comprises heavy soil, then seepage loss will be less. Another factor that influences seepage loss is the ground water table position. If the water table is farther from ground surface ,then the hydraulic gradient causing downward flux of seepage water will be more .For a newly constructed earthen channel, seepage loss is very high. With progress of time, siltation reduces seepage loss. Seepage loss also depends upon wetted surface area which in turn depends on side slope and depth of flow.

There may be two different conditions of seepage. They are (i) Percolation and (ii) Absorption.

Percolation:

In percolation, there exists a zone of saturation which continuously extends from the water surface level of the channel to the underlying ground water table. Direct flow of water continues in this saturated zone. The loss of water in the channel depends on the vertical distance between the tip of water level in the channel and the position of ground water table (H) as shown in Fig. 11.1.

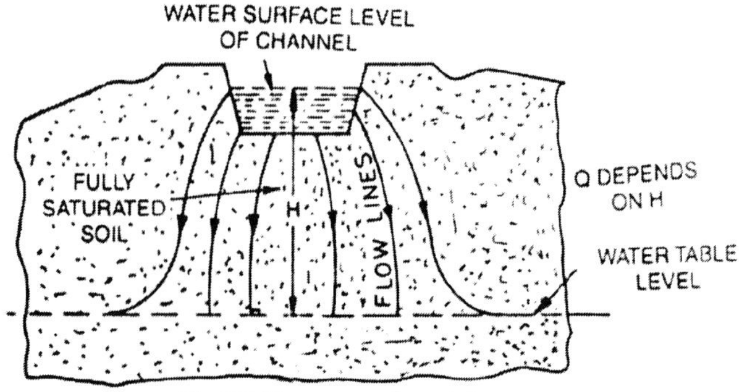

Fig. 11.1 Phenomenon of percolation

Absorption

In absorption, there is a small saturated soil zone that extends from the channel section carrying water. This small saturated zone is surrounded by a zone of decreasing saturation. The zone below the unsaturated zone having decreasing saturation up to the ground water table is again saturated. But this saturation is due to the capillary rise of ground water. Thus the intermediate unsaturated zone is surrounded by two saturated zone (Fig. 11.2).

In this case, the rate of loss of water from the channel is independent of seepage head, H but depends on the water head, h (i.e. distance between water surface level of channel section and bottom of the first saturated zone) plus the capillary head, h_c as shown in Fig. 11.2 (Garg, 2012).

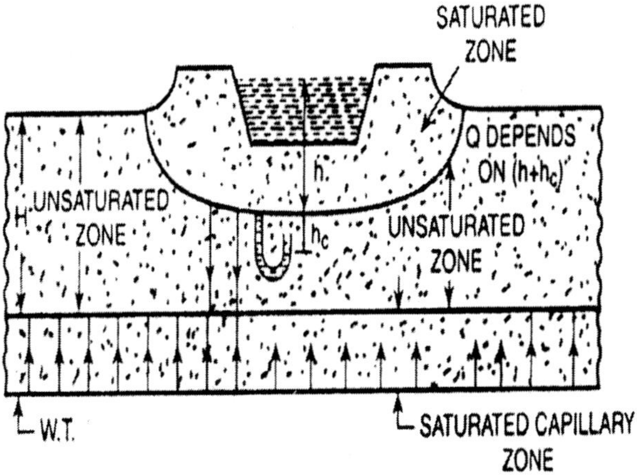

Fig. 11.2: Phenomenon of absorption

11.3.1 Factors Affecting Seepage Loss

Seepage losses depend upon the following factors:

- Type of seepage *i.e.* whether percolation or absorption
- Soil permeability
- The condition of canal. The seepage through a silted canal is less than that from a new canal.
- Amount of silt carried by the canal. The more the silt, lesser are the losses.
- Velocity of canal water. The more the velocity lesser will be losses.
- Cross section of the canal and its wetted perimeter

11.3.2 Computation of Seepage Loss

Analytical methods are available for computation of seepage loss in canal system by water balance method. Seepage loss can also be measured by (i) ponding method (ii) inflow-outflow method (iii) seepage meter (iv) wells (v) laboratory test of permeability of soils and (vi) special methods including electrical resistance and tracing of natural and radioactive salts. The method best suited to a canal will depend on the depth and velocity of flow, ability to drain the canal, materials in the bed and rate of seepage. The three most commonly used methods are described below (Hansen *et al.,* 1979).

Inflow-Outflow Method

In this method, a canal section is chosen which must be as far as possible straight and uniform. The flow entering into and going out of the section is measured. Difference of the two readings gives the seepage losses in the section. Allowances must be given to consider the rainfall and evaporation occurring during seepage measurement. The average wetted area of the section is computed and then seepage rate per unit area is estimated.

Ponding Method

In this method, a selected canal section is blocked with earthen dam/dyke to create a ponding area. Information of the canal geometry and ponding sites is collected. Water into the ponding area is fed. Water level readings using pre-installed staff gauges or pressure transducers for several consecutive days are taken. By knowing the difference of water level that recedes during two consecutive sampling period and cross sectional area of the water in the channel section, the total volume of water lost is calculated. Subtracting the evaporation loss and rainfall if any from the total loss of water, seepage loss is computed. Then, the unit seepage rate over the wetted area is calculated. This method is the most accurate method to measure the seepage loss provided there is no leakage of water through the dyke.

Seepage Meter

Seepage meters are used to obtain measurement of seepage from relatively smaller areas of canal section. Measurements can be taken without disturbing the flow in the canal section. It consists of a cylinder with a dome or cone on the top to allow for trapped air to be removed from the system through a valve attached to the top of the cone. The cylinder is pressed gently into the soil and the unit is filled with water. To obtain results indicative of actual seepage, several readings are required with representative readings taken on the sides as well as on the bottom of the canal.

11.3.3 Empirical Formulae for Channel Losses

The channel losses can be determined by using certain empirical formulae as mentioned below (Garg, 2012).

(a) $$\Delta Q = \frac{(B+D)^{2/3}}{200} \tag{11.1}$$

where,

ΔQ = channel loss in cumecs per km length of channel; B = bed width of channel, m and D = depth of water in channel, m.

This formula is used in Uttar Pradesh.

(b) $\Delta Q = 1.9\, Q^{1/6}$ (11.2)

where,

ΔQ = channel loss in cumecs per million sq. m of wetted perimeter; and Q = discharge in cumecs.

The above formula is used in Punjab.

11.3.4 Advantages of Lining

Possible advantages of lining a canal include:

Water Conservation

Water suitable for human consumption, livestock and irrigation is limited in many parts of the world. Since precipitation is the only source of all fresh water and since it varies from place to place, from year to year and from season to season, water must be collected, stored and transported, if it is to be available when and where it is required. The loss of irrigation water in a canal system occurs during its conveyance through canal, sub-canal, distributary, minor, water course and finally during application in the field. An important reason for lining a canal can be the reduction in water losses and thereby conserving water. Lining a canal will not completely eliminate these losses, but roughly 60 to 80% of the water that is lost in unlined irrigation canals can be saved by proper lining.

Seepage Control

If canal banks are highly permeable, the seepage of water will cause very wet or waterlogged conditions, or even standing water on adjacent fields or roads. Lining of such a canal can solve this problem, since the permeability of a lined canal bank is far less than that of an unlined bank, or may even be zero, depending on the lining material. It is obvious that a lined canal costs about 2 to 2.5 times as much as an unlined canal, but where the seepage is heavy, the saving of costly irrigation water may itself be sufficient to fully justify the expenses on lining. Fig.11.3 shows the photographic view of an unlined channel in Hirakud canal command which shows that there is heavy seepage loss consequently resulting in low discharge in the canal. On the other hand Fig.11.4 shows the photographic view of a concrete lined channel which represents that the seepage losses in the channel is very less which consequently increases the channel carrying capacity.

The seepage loss at various sections of the canal distribution system is not the same. Lining the different sections would reduce the seepage loss of varying amount. In a study made by the Central Water and Power Commission, the losses during the various phases of water conveyance was found to be as shown in Table 11.1.

Fig. 11.3: Photographic view of an unlined channel in Hirakud canal command

Fig. 11.4: Photographic view of a concrete channel of Hirakud canal system

Table 11.1; Losses of water at various sections of canal distribution system for lining of different section
Lining of various sectionsLosses at various sections of canal distribution system (percent)

	Main canal	Distributary	Water course	Field loss	Total loss
Entire system lined	15	7	22	27	71 (29)*
Only main canal lined	4	7	25	30	66 (34)*
Both canal and distributary lined	4	2	26	32	64 (36)*
Whole system lined	4	2	6	42	54 (46)*

*Values in parenthesis indicate percentage of net utilization.

Prevention of Water-logging

Seepage losses in unlined canal sections move downward and add to the underlying groundwater table. This process ultimately raises the groundwater table such that plant roots get submerged with it. This phenomenon of submersing the roots of the plants by the groundwater table is called as water-logging. Water-logging causes decreased bacterial activity and hence reduced crop yield. It also causes salinity and alkalinity problem and renders vast areas to be unproductive. Thus, the canal which is originally built up as a boon to provide irrigation to crops soon becomes a curse as vast areas of irrigated lands become water-logged and become uncultivable.

Reduced Canal Dimensions

The roughness resistance to flow of a lined canal is less than that of an unlined canal, and thus the flow velocity will be higher in the lined canal when the canal bed slope is the same. Moreover, the hard surface of the lining material allows a higher velocity compared to an earthen canal surface as it is not so easily eroded. Further, a lined canal section has less weed growth compared to an unlined section. The velocity of flow in a weed infested channel will be less than that of a lined section. Canal discharge is the product of the cross-section of a canal and the velocity of the flow. Therefore, with the higher velocity allowable and obtainable in lined canals, the canal cross-section for a lined canal can be smaller than that of an unlined canal for a given discharge. Typically a lined channel will have 40% of the unlined surface area for a given discharge. Therefore even at the same loss rate per unit area there will be a saving in water. Thus, a lined section increases channel capacity and continuously reduces the required channel section. This reduces the area required for construction of channel and hence reduces the project cost.

Reduced Maintenance

A surface lining, such as concrete, brick or plastic, on the canal prevents the growth of plants and discourages hole-making by rats or termites, and so the maintenance of a lined canal can be easier and quicker than that of an unlined canal. Moreover, the higher velocity that can safely be allowed in the lined canal prevents the small particles of soil carried in the water from settling out, accumulating and causing siltation. The bed and sides of lined canals are more stable than those of unlined canals and are thus less susceptible to erosion. Periodical plugging of holes burrowed by rats, insects etc. is constantly required in an unlined channel. If these are not repaired then breaches in channel section may occur. However, the provision of adequate lining reduces the danger of breaches and thus stabilizing the section.

Increase in Command Area

A lined canal is designed not only in smaller in cross section but also shorter in length. Since higher velocity of flow in lined canal is allowed, steeper gradient of canal section can be provided which will allow water to move at a faster rate and reach at the tail end of the command. Moreover, water can spread to larger areas and thus, command area under irrigation is increased.

Elimination of Flood Dangers

An unlined channel section is susceptible to breaches which will cause not only loss of water in the channel but also danger of flood to adjoining areas. Water escalating thorough the small breaches of channel section, washes away a large section and through this large breach, large amount of water rushes to the surrounding areas causing flood and havoc. Lining of the section makes the section stable and chances of flood are eliminated.

11.3.5 Disadvantages of Canal Lining

Disadvantages of canal ling are:

- Initial cost of canal lining is very high. It makes the project expensive with respect to output.
- It involves much difficulty for repairing the damaged section of lining.
- It increases the time period to complete the project.
- If the outlets are required to be shifted or new outlets are required to be provided, it becomes difficult because of dismantling the lined section of the canal.

11.3.6 Justification of Lining

Irrigation water is a very precious commodity. But huge amount of water is lost on transit system from the canal head regulator to the outlet of water courses. These losses are mainly attributed to seepage loss. However, lining of the canal system though can reduce appreciable amount of seepage losses but their huge investment makes it difficult to line the entire system. The huge capital outlay incurred on lining of the canal system must be justified in terms of monetary benefit obtained from increased yield of the crops due to additional area covered under irrigation and at the same time increased yield of the prevailing crops in the existing areas. To justify lining of canal, the following points must be kept in mind (Basak, 2012).

Capacity of Reservoir

Reservoirs are constructed to store water. If for some points of view, it is not possible to increase the reservoir capacity but additional areas are required to be brought under irrigation, then seepage losses in conveyance should be checked. Seepage loss can be checked only by lining the canal systems.

Reduction of Dam Height

In unlined canal system, there is huge losses of water due to seepage. This loss causes less amount of water to be available at the tail end of the canal system. Further, the losses decreases the irrigation requirements of crops and decreases crop yield. To satisfy the full irrigation requirement of crops and allow water to reach at the tail end in sufficient quantity, reservoir capacity has to be increased. Reservoir capacity can be increased either by increasing the reservoir area or increasing the dam height. If the reservoir area cannot be increased, then there are no other options than increasing the dam height. But increasing dam height, needs huge investment and at the same time it creates environmental and socio-economic problems. Hence under such juncture, the canal has to be lined to minimize the seepage loss so that dam height need not be increased.

Location of Command Area

If the command area is located far away from the reservoir site, then length of the conveyance system increases which makes more loss of water due to seepage. It is true in some cases that not a drop of water reaches at the tail end of the canal because of heavy losses of water in the canal sections. Under such conditions, it is essential to line the canals.

Nature of Land

If the command area of the project has fertile land, then cares should be taken such that minimum area is lost for construction of canal. In such a case, it is essential to line the canals to reduce seepage loss so that smaller cross section of the canal will carry greater discharge.

Nature of Topography

If the canal is constructed on a high fill or a canal is constructed partly on a rocky surface and partly on permeable strata, the seepage loss will be high. In that case lining is needed to check excess seepage loss.

Economics of Study

The economics of lining of the canal should be studied in terms of net present value, benefit-cost ratio, internal rate of return and pay-back-period and then justification of linings must be decided. Economics of different linings have to be studied and compared and then the low cost technology must be proposed.

11.3.7 Economics of Canal Lining

Annual Benefits

Irrigation water can be sold to other farmers for use in their crop fields at certain rate. If the rate is assumed as Rs. R_1 per unit cubic meter of water, and let there be savings of m cubic metre of water, then total revenue obtained from selling of water is Rs. $m.R_1$. Lining will also reduce repair and maintenance cost. Let the average cost of annual uptake of unlined channel be Rs. R_2. If p is the percentage of savings achieved in maintenance cost by lining the canal, then total amount saved is Rs. $p.R_2$. If other benefits are Rs. R_3, then the total benefits are computed as (Garg, 2012):

$$Total\ annual\ benefits = m\,R_1 + p\,R_2 + R_3 \tag{11.3}$$

The other indirect benefits are reduction of water-logged areas, controlling breaches in canals, reducing cost of drainage in adjoining lands *etc.* Actual benefit evaluation of these items are difficult and may be approximated.

Annual Costs

If the capital expenditure required to line the canal is Rs. *C* and the lining has a life span of say *Y* years, then annual depreciation charge will be Rs. *C/Y.* If the rate of annual simple interest is *r*, then a locked up capital of Rs. *C* would earn annual interest value of Rs. *C (r/100).* Since, the capital value of the asset decreases from *C* to zero in Y years, the average annual interest value may be taken as Rs. $\frac{C}{2}\left(\frac{r}{100}\right)$.

The total annual cost of lining is given as below (Garg, 2012):

$$\text{Total annual cost of lining} = \frac{C}{Y} + \frac{C}{2}\left(\frac{r}{100}\right) \tag{11.4}$$

The economic indicators are calculated as follows:

$$\text{NPV} = \left(m\,R_1 + p\,R_2 + R_3\right) - \left[\frac{C}{Y} + \frac{C}{2}\left(\frac{r}{100}\right)\right] \tag{11.5}$$

where,

NPV = net present value

$$\text{B.C.R.} = \frac{m\,R_1 + p\,R_2 + R_3}{\left[\frac{C}{Y} + \frac{C}{2}\left(\frac{r}{100}\right)\right]} \tag{11.6}$$

where,

B.C.R. = benefit-cost-ratio

The value of p in the Eqns. (11.5) and (11.6) is assumed as 0.4.

The project is justified when the net present value is positive and benefit cost ratio is more than 1.0.

Sample Calculation 1:

An unlined canal has seepage loss of 3.0 cumecs per million sq. m of wetted area and has to be lined with 10 cm thick cement concrete lining that costs Rs. 300/10 sq. m. Using the following data, work out the economics of lining.

Annual revenue per cumec of water from all crops = Rs6.00 lakhs

Discharge in the channel = 90.0 cumecs

Area of the channel = 45 sq. m

Wetted perimeter of the channel = 20 m

Wetted perimeter of the lining = 18 m

Annual maintenance cost of unlined channel per 10 sq. m = Re. 4.00

Assume any other data not provided.

Solution:

Let us consider 1 km reach of canal section.

Hence, wetted surface per 1 km length of reach = 20 x 1000 = 20,000 sq. m

Annual Benefit

Seepage

Seepage loss in unlined canal @ 3.0 cumecs per million sq. m of wetted perimeter

= (3.0 x 20,000)/10^6 = 0.06 cumecs/km

Assume seepage loss in lined canal as 0.01 cumecs per million sq. m of wetted perimeter.

Hence, seepage loss in lined canal = (0.01 x 20,000)/ 10^6 = 0.0002 cumecs/km.

Net saving of water due to lining = 0.06 - 0.0002 = 0.0598 cumecs/km.

Annual revenue saved per km of channel = Rs. (0.0598 x 6.0) lakhs = Rs. 35,880.00

Saving in maintenance

Annual maintenance cost of unlined channel for 10 sq. m = Rs. 4.00

Total wetted perimeter per 1 km length = 20,000 sq. m

Hence annual maintenance charge of unlined channel per km length = Rs. 8,000.00

Assuming 40 percent of the above maintenance cost is saved by lining the channel, the annual maintenance charge of lined channel becomes = Rs. 3,200.00

Total annual benefit per km = Rs. 35,880.00 +Rs. 3,200.00 = Rs.39,080.00

Annual cost

Area of lining per km of channel = 18 x 1000 = 18,000 sq. m

Cost of lining per km of channel @ Rs. 300/10 sq. m = Rs. (18,000 x 300)/10 = Rs. 5,40,000.00

Assume life of the lining = 40 years.

Depreciation cost per year = Rs. 5,40,000.00/40 = Rs. 13,500.00

Assuming 7 percent interest rate,

Average annual interest = $\frac{C}{2}\left(\frac{r}{100}\right) = Rs.\frac{5,40,000.00}{2} x \frac{7}{100} = Rs. 18,900.00$

Total annual cost = Rs. 13,500.00 + Rs. 18,900.00 = Rs.32,400.00

Hence, net present value = Rs.39,080.00 - Rs. 32,400.00 = Rs. 6,680.00

Benefit cost ratio = Rs. 39,080.00 / Rs. 32,400.00 = 1.21

Since, benefit cost ratio is more than 1.0 and there is positive net present value, the lining is justified.

Sample Calculation 2:

A canal of length 10 km and of discharge capacity 3 cumecs is proposed to be lined with boulder. The total cost of lining is computed as 5 lakhs. The life of the lining may be considered 40 years. Estimate the economics of lining with the following data,

Rate of interest = 7 percent

Saving in seepage loss by providing lining = 2 percent

Annual revenue for irrigation water = Rs. 60.00 per ha m

Base period of the crop = 150 days

Maintenance cost per km of unlined canal = Rs. 3000.00

Maintenance cost per km of lined canal = Rs. 1500.00

Additional benefit per km = Rs. 1000.00

Solution:

Annual benefit/km length of canal

Discharge of the canal = 3 cumecs

Base period = 150 days

Volume of water flowing in the canal in 150 days = 150 x 3 = 450 cumec day which is equal to = 450 x 8.64 = 3888 ha-m

Saving in seepage loss by providing lining = 0.02 x 3888 = 77.76 ha-m.

Amount of saving = 77.76 x 60 = Rs. 4,666.00

Saving in maintenance cost/km = Rs. 3000 –Rs. 1500 = Rs. 1,500.00

Additional benefit/km = Rs. 1000.00

Total benefit = Rs. 4,666 + Rs.1,500 + Rs.1,000 = Rs. 7166.00

Annual cost/km of canal

Annual depreciation charge/km = $\frac{5,00,000}{40\,x\,10} = Rs.\,1250.00$

Average annual interest /km = $\frac{5,00,000}{2\,x\,10}\,x\,0.07 = Rs.\,1750.00$

Total annual cost/km = Rs. 1250.00 + Rs. 1750.00 = Rs. 3,000.00

Hence, annual savings/ km or net present value/km = Rs.7,166 – Rs.3000 = Rs. 4,166.00

Benefit cost ratio = Rs.7,166.00/Rs3,000.00 = 2.39

Since, benefit cost ratio is more than 1.0 and there is positive net present value, the lining is justified.

11.3.8 Methods to Reduce Seepage Losses

Since, irrigation water is very costly, efforts should be made to reduce the seepage loss. Seepage loss can be reduced broadly by two mechanisms. They are (i) reducing wetted surface area and (ii) using a cost effective sealant.

11.3.9 Reducing Wetted Surface Area

Seepage increases with the wetted area and so special considerations during design must be given to minimize the wetted area per unit of discharge. This may be achieved by increasing the side slope i.e. steepening the sides of the canal as well as increasing the depth of water in canal section. Cluff (1977) has suggested compartmentalized canal section in which the canal section is divided into two or more sections and as one section is emptied, the water is discharged out from another section. By doing so, the wetted area and top water surface area of the canal section can be considerably reduced and so seepage and evaporation loss of water can be reduced to a great extent. But design of compartmental canal section is difficult and hardly it is practiced in seepage control in canal water distribution.

11.3.10 Use of Sealants

A newly constructed canal has self sealing property by silting. The discharge carried by the canals carries some silt and clay which gets deposited in the side and bottom of the soil in the canal and reduces the pore spaces of the soil. This, consequently results in the reduction of the seepage loss of the canal. In some cases, puddling caused due to trampling by animals causes the compaction of pore spaces of the soil which has a consequential affect in checking seepage losses.

A number of sealants/ lining materials are now available to reduce the seepage loss. Lining of the canal though costly can reduce the seepage loss and improve the effective flow capacity of the canals. Different lining materials used to reduce the seepage losses are plastic film, soil cement lining, bitumen lining, clay lining, cow dung lining, brick cement lining *etc.* The permanent and most effective and costly is brick masonry or cement mortar lining. Descriptions of a few lining materials used for seepage control in canals are presented below.

11.3.10.1 Clay Lining

Natural clay can be used for lining with varying degree of efficiency, especially when lower cost is desirable. Clay lining can be applied by two methods; (i) by placing a blanket of relatively impervious clay 15-30 cm thick over or within the permeable bed and sides of the canal and (ii) by dispersing clay in the water and having it filtered out and seal off the pores in the permeable lands. Shally (1965) observed that in some parts of Bengal and Madras, the seepage losses by heavy clay soil are so effective that lining is hardly necessary or justified. Water losses through clay is much less than bricks and is as or more pervious than ordinary cement concrete. Kamra and Dhruvanarayana (1987) reported that alkali soils having poor infiltration rates can be preferred for seepage control if available in the vicinity of the canal.. Biswas and Mallick (1997) observed that the burnt clay tiles used as a lining material can reduce the seepage losses and the saving of water due to seepage by it is about 98.8% more than the earthen materials. The average seepage rate by the burnt clay tiles was observed to be 1.54 cm^3/cm^2/day. The drawback of using clay as lining material is that it is not weather resistant and its life is comparatively short and the repair and maintenance costs are relatively higher. It also does not permit higher velocity of flow even that in an unlined channel.

11.3.10.2 Cement Concrete Lining

Structures lined with cement concrete are stronger and more stable than lined with any other material. It gives long service. Though the initial investment for lining is more, later on the repair and maintenance costs are less. It has a longer life than any other type of lining. Concrete linings have been extensively used in the developed countries. India, being a developing nation, finds it difficult to afford their high initial cost s and as such, their use is restricted only to important projects. It is alos not recommended to line the entire canal system with this type of lining except main canal and branch canal at head regulators and at the bends in canal systems. Concrete mixture usually recommended for lining is 1:3:4 (cement:sand:gravel) and thickness for lining is 4-5 cm (Biswas, 2008). The sides and bottom of the structure should be compacted at suitable moisture content. When concrete hardens, it shrinks and cracking may result. Joint must be provided at a distance not more than 2 m in order to localize and control the cracks. Cement concrete lining can withstand higher velocity more than 2.5 m/sec because of its greater resistance to erosion and is therefore preferable than any other type of lining. It can completely eliminate the growth of weeds in the canals. This type of lining is preferred in canals in full banking.

Preparation of Sub-grade

The success of cement concrete lining depends to a great extent upon a stable foundation. Canal foundations should be thoroughly compacted and then over it the lining is to be laid. A firm foundation eliminates the chances of cracking due to settlement of the sub-grade. For cement concrete lining, any rock or earth or predominantly sandy soil is suited as sub-grade material. For expansive clays, proper moisture and density control of the soil should be ensured during placing of the lining. If the sub-grade becomes too dry, it should be thoroughly wetted several hours before the lining is placed. However, black cotton soils needs laying of a layer of cohesive non-swelling soil like *murum* in depth of about 0.3 to 1.0 m over the given soil surface before laying the lining. In areas where there is groundwater table very near the ground surface, there may be fear of uplift hydrostatic pressure, especially during period of no flow in canal, which can be eliminated by providing a drain underneath or along the sides of the canal.

Thickness

The thickness of the subgrade may vary from 5 to 15 cm of cement concrete. Thickness of the subgrade depends on the depth of flow in the canal and discharge carried by it. Table 11.2 gives suggested thickness of subgrade for lining the canals (Garg, 2012).

Table 11.2: Suggested thickness of subgrade for canal lining

Sl. No.	Discharge in canals, cumecs	Depth of water, m	Thickness of subgrade, cm
1.	0 – 5	0 – 1	5.0 – 6.0
2.	5 – 50	1 – 2.5	6.0 – 7.5
3.	50 – 200	2.5 – 4.5	7.5 – 10.0
4.	200 – 300	4.5 – 6.5	9.0 - 10.0
5.	300 - 700	6.5 – 9.0	12.0 – 15.0

The minimum thickness of subgrade used in India is 7.5 cm. Subgrades should be laid on stable side slopes of 1.5:1 or 1.25:1.

Laying of Concrete

The subgrade soil surface should be moistened thoroughly before laying cement concrete, so that moisture is not withdrawn by the subgrade from the cement concrete. For smaller canal sections, cement concrete materials of grade M_{15} may be placed by hand where as for large canal section cement concrete lining is laid mechanically. Cement concrete materials are dumped and spread on the bed and side slope of the canal, usually in panels of not more than 3 m side, with suitable joints in between. This construction joints also serve the purpose of contraction joints that are basically needed for shrinkage of concrete and to take care of temperature stresses caused due to change in atmospheric temperature. After laying the subgrade materials, they should be properly finished so that there will be no impurities, lumps of mortar or grout *etc.* Subsequent to laying of concrete lining and after a period of 24 to 36 hours, the lining should be cured for at least 28 days. Surface drainage provisions must be provided to prevent surface and subgrade erosion and consequent damage to lining (Michael, 1978).

Joints in Cement Concrete Lining

Cracks in cement concrete lining usually occur due to warping stresses and tensile stresses. Cracks caused due to the above stresses are controlled by either by providing reinforcement or by providing joints in concrete lining. There are three types of joints used in cement concrete lining. They are:

- Expansion joints
- Construction joints and
- Contraction joints

Expansion joints are usually not provided in cement concrete lining except where the structures intersect the canal. At the intersecting structure, an expansion joint of 25 mm width, filled with sealing compound, is provided. Construction joints serve the purpose of contraction joints. They are specially required to take care of shrinkage and temperature stresses.

Construction joints are again divided into the following three types.

- Lap joint
- Lap joint with separate fillet and
- Tongue and groove joint

Advantages of Cement Concrete Lining

Followings are the advantages of cement concrete lining.

(i) Cement concrete lining has longer life than any other type of lining.
(ii) It has least permeability of all types.
(iii) It requires low recurring maintenance cost.
(iv) It permits fast construction by mechanical means.
(v) It is most resistant to erosion.

Disadvantages of Cement Concrete Lining

Followings are the disadvantages.

(i) It has high initial cost.
(ii) There is greater possibility of temperature cracking.
(iii) It requires skilled supervision for its construction.
(iv) It is less flexible and easily affected by adverse subgrade conditions.

11.3.10.3 Pre-cast Concrete Lining

This type of lining is recommended for canals in full banking. It consists of pre-cast concrete slab of size 60 x 60 x 5 cm which are set along the canal bank and bed. Rebates are provided on all the four sides of the slabs to form proper joint when the slabs are kept side by side. The joints are finished with cement mortar 1:3 (Fig. 11. 5). Expansion joints are provided at a suitable interval. The procedure of laying the slabs on the canal sides and beds are as follows (Basak, 2012).

(a) The subgrade is provided by properly ramming the soil with a layer of sand. The bed is leveled so that the slabs can be placed easily.

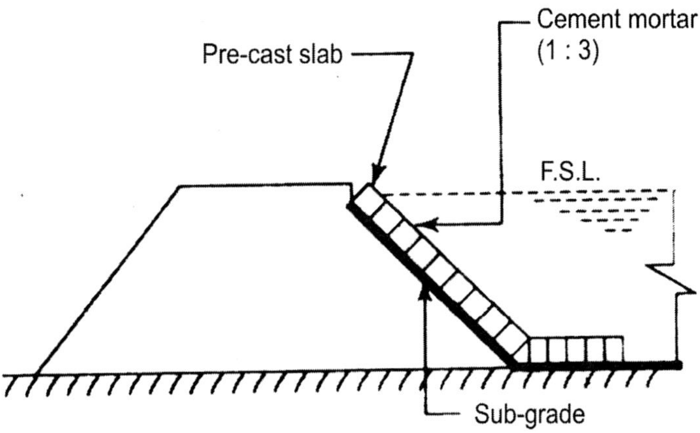

Fig. 11.5 Pre-cast concrete lining

(b) The slabs are stacked as per estimate along the course of the canal. The slabs are placed with cement mortar by setting the rebates properly.

(c) The curing is done for a week.

11.3.10.4 Cement Mortar Lining

This type of lining is recommended for canal in full cutting where hard soil or clay soil is available. The thickness of the cement mortar (1:4) is generally 2.5 cm. The subgrade is prepared by ramming the soil after cutting. Then, over the compacted subgrade, cement mortar is laid uniformly and then the surface is finished with neat cement polish (Fig. 11.6). This lining is impervious but not long lasting.

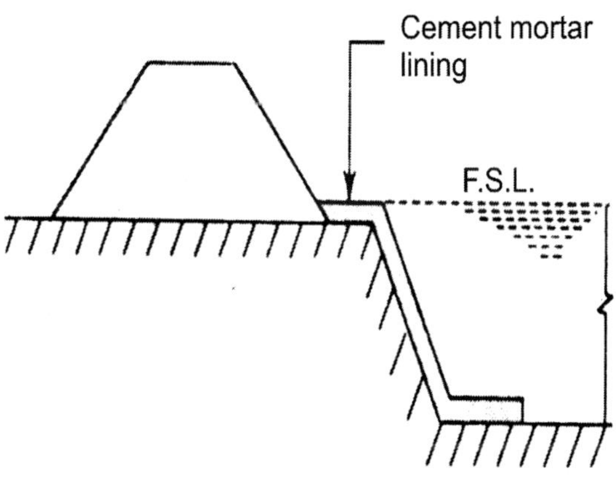

Fig. 11.6: Cement mortar lining

11.3.10.5 Lime Concrete Lining

If the hydraulic lime, surki and brick ballast are available in plenty at the site of the canal, then lining of the canal may be made by the lime concrete in proportion of 1:1:6. The procedure of laying this concrete is same as that of cement concrete lining. In this case, the thickness of the concrete varies from 15 to 22.5 cm. It needs long curing. Though its lasting is less than that of cement concrete lining, yet because of its economics, it is recommended.

11.3.10.6 Shotcrete Lining

Shotcrete is a technical term used to designate cement mortar applied under pressure through nozzle of a gun on the surface of the channel. It consists of a mixture of cement and sand in the ratio of 1:4 and is directly applied on the subgrade by a cement gun. The mortar is termed as shotcrete and the lining is called as shotcrete lining. The process is also called as guniting as a gun is used for laying the mortar. Sometimes it is also called as gunited lining. It is alos designated as plaster lining. Wire mesh reinforcement is generally used in shotcrete lining (Basak, 2012). The lining is done in two ways. They are:

(i) Dry mix method: In this method, a mixture of cement and moist sand is prepared and loaded in the cement gun. Then it is forced through the nozzle of the gun with the help of compressed air. The mortar spreads over the subgrade to a thickness which varies from 2.5 to 5.0 cm.

(ii) Wet mix method: In this process, the mixture of cement, sand and water is prepared according to the approved consistency. The mixture is loaded in the gun and forced on the subgrade.

Shotcrete lining is useful for repair works and in rehabilitation of old canals. Excavation, compaction, curing *etc.* for shotcrete lining is same as those required for cement concrete lining.

11.3.10.7 Soil Cement Lining

Hegade *et al.* (1981) reported that the seepage losses using soil cement plaster (5:1) is ideal from the combined considerations of the cost and efficiency. This lining is made up of mixture of cement and natural soil. For good results the mixture should be mixed well, laid and compacted. It should be cured for seven days with moist soil cover.

11.3.10.8 Asphalt Lining

Asphalt is a petroleum product which when mixed with sand and gravel is used as a lining material in the same way as cement. Asphalt concrete lining is cheaper than cement concrete liming. According to Chowla (1992), the life of asphalt lining varies from 10-20 years. Locally available bituminous and other low cost indigenous materials were used for lining under field conditions in Bangladesh by Khair *et al.* (1984). The following two types of prefabricated asphalt melts have shown promising results in seepage control in Bangladesh. They are (i) Gunny (coarse sack cloth made of jute) reinforced asphalt melt and (ii) Synthetic cloth reinforced asphalt melt. The results have indicated that the former lining material reduced the seepage losses considerably. This material is used as lining material at very few places. It is still under the stage of evolution. It provides a fairly cheap lining, especially where the asphalt is available within the country. It is flexible and readily confirms to the sub-grade. The disadvantages or the limitations of this type are as follows:

- It does not decrease the rugosity coefficient of the channel.
- It permits certain type of weed growth.

11.3.10.9 Boulder Lining

Boulder lining also called as dry stone lining or stone pitching, consists of lining the side slopes of an earthen canal by proper placement and packing of stones, either after laying a filter layer over the soil surface, or without any such filter, depending on the site requirement. The joints of the boulders are grouted with cement mortar (1:6). The surface is finished with cement mortar (C.M.) 1:3 (Fig. 11.7). Such linings do not control the seepage but gives stability to the side slopes of the canal. It maintains the canal sides and thus reduces the repair and maintenance cost. Since transporting cost of boulders is very high, it cannot be recommended for all cases. In hilly areas where the boulders are available in plenty, this type of lining is generally recommended.

11.3.10.10 Brick Lining

This type of lining is very popular in India because of certain inherent advantages in their use. These advantages are mentioned as follows:

Advantages

- It is easier to handle and requires less technical knowledge. Brick or concrete tiles can be laid by ordinary masons and no skilled labour is required.
- Rigid quality control is not required.
- These types of lining involve low consumption of cement and can be laid without use of any machinery.
- It requires less investment.

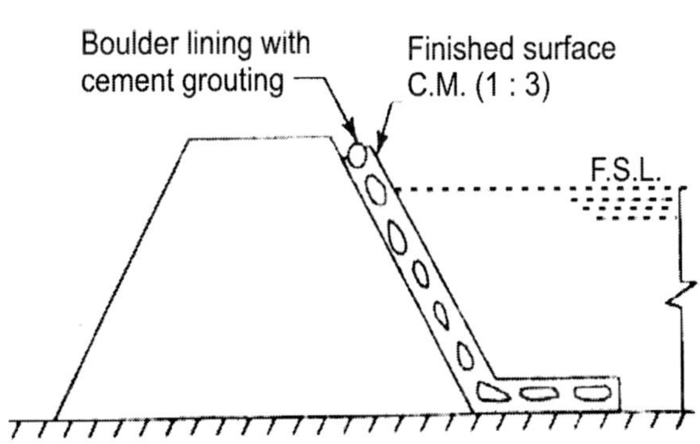

Fig. 11.7: Boulder lining

- No expansion joints are required.
- Rounded sections can be easily laid without using any formwork.
- Work can be done very quickly.
- Repair works can be done easily.

The specifications may provide either a single or a double layer of brick masonry or tiles laid in mortar. Sometimes a layer of tiles is laid over a layer of brick masonry. The top layer of tiles is generally laid in 1:3 cement mortar over 15 mm thick layer of plaster in 1:3 cement plaster (Fig. 11.8). The curing must be done perfectly.

Disadvantages

- Brick lining is not completely impervious. So, it is not that effective in seepage control as compared to cement concrete lining.
- It has low resistance against erosion.
- It is not so much durable.
- It is not suitable for curves.
- It is too light for hydel channels.

11.3.10.11 Bentonite Lining

Seepage losses in a well graded coarse soil may be reduced by mixing bentonite with clay thoroughly and forming a paste which is spread over the subgrade in the surface soil of the canal. This forms an impervious membrane which controls seepage of the canal beds and sides. Actually bentonite can absorbe several times more water and swell 8-15 times its volume. The main advantage of bentonite lining is low cost and simplicity of installation while its disadvantage is its short life. Bentonite loses its effectiveness and durability due to fluctuation resulting from wetting and drying of the canal and removal of surface layer by erosion and seepage. It cannot check the weed growth in the channels. This lining is generally recommended for small channels (Panigrahi, 2011).

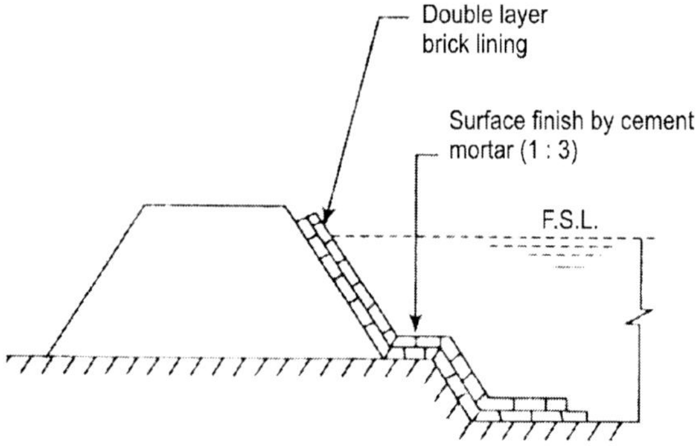

Fig. 11.8: Brick lining

11.3.10.12 Soil Deflocculents

Soil deflocculents like sodium carbonate, sodium chloride, tetrasodium polyphosphate and sodium tripolyphosphate can be used for reducing permeability of canal surfaces. Sodium chloride and sodium polyphosphate works for 6 and 8 months, respectively, but sodium carbonate is more durable. It is suggested that for successful sodium carbonate treatment at least 3 cm of soil should be put over the pervious materials and the clay content of soil should be at least 15% and cation exchange capacity should be greater than 15 mil-equivalents per gram (Reginato *et al.*, 1973).

11.3.10.13 Gleization

Gleization is a new method used for reducing seepage. It involves the use of organic matter like straw and manure. Under the anaerobic condition, the metabolic activities of certain anaerobes affect a chemical reduction process in soil which results in increased plasticity and dispersity and a decrease in permeability (Nicholaichuk, 1978).

11.3.10.14 Chemical Sealants

U.S. Bureau of Reclamation (1963) studied many chemicals including resins, silicones, linings but none was found suitable in seepage control. Even cationic asphalt emulsion, petroleum emulsion and resinous polymers were tested and found to be short livid and affected by wetting and drying and erosion.

11.3.10.15 Enzymes

Some enzymes have also been tested for seepage control. Bayer and Cluff (1971) mixed organic enzyme with lubricants as aids in compaction of soil in canal surface and found reduction in seepage rate.

11.3.10.16 Polythene Lining

Polythne films have been widely used for lining purpose in irrigation channels by many researchers. These lining can reduce the seepage losses to the minimum even zero if perfectly laid. The percentage of seepage loss by it is maximum than any other lining material. Though the initial investment for lining the structure is more, yet compared to the enormous amount of saving of costly irrigation water, the benefit cost ratio is found to be maximum. While lining the canal with polythene, cares should be taken so that there is no puncture or any gap between the cut pieces of the sheets. The sides of the canal should be constructed with steps and over the steps, the polythene sheet is laid. Then the soil is covered so that the sheet remains buried in soil. Then it is compacted and the sides of the canal is made sloppy as in the original section. By doing so, the wears and tears of the polythene sheet is reduced and its longevity increase. If properly laid, the life span of a polythene lined canal can go for 10-15 years.

It is important to note that before using any lining materials for seepage control in the canals, the economic analysis relating to life and cost of lining materials must be taken into account. At the same time, the amount of saving of irrigation water by reduction of seepage and its use in increasing the crop productivity, total crop production and other associated benefits must also be considered. Finally with present worth analysis the different economic indicators like net profit, benefit-cost ratio, internal rate of returns and pay back periods are to be worked out. The lining material that gives more net profit, benefit-cost ratio, and internal rate of returns and less pay back period is the best from seepage control point of view. Economics of lining and its justification has been discussed earlier along with solution of a sample calculation.

11.3.11 Selection of Type of Lining

The selection of type of lining should take into consideration the following points:

(i) **Imperviousness:** When the canal passes through sandy soil, there is heavy seepage loss. In that case, cement concrete lining, pre-cast concrete ling or boulder lining are recommended.

(ii) **Smoothness:** The canal surface must be made smooth so that the velocity of flow will be more and duty will be more. The lining like cement concrete or pre-cast concrete gives more smoothness to the canal surface.

(iii) **Durability:** The ultimate benefit of a project depends on the durability of the canal. To make the durability more, the canal should be lined with most strong and erosion resistant material.

(iv) **Economy:** The economics of various types of linings should be studied and the one that gives maximum benefit with minimum investment should be selected as the lining material.

(v) **Site Condition:** Lining material should be chosen as per the soil and site condition.

(vi) **Life of the Project:** The type of lining should be chosen considering the life of the project.

(vii) **Availability of Construction Materials:** Lining materials should be locally available. If it is not available at the site of the project, then transportation cost will be more and as a result, the project cost will be more. This will make the project economically unviable.

In addition to the above mentioned factors, other factors that help to select lining are:

- Structural stability
- Repairability
- Hydraulic efficiency
- Resistance to erosion

Question Banks

Q1. What are the losses of water in canal systems? Which of them are the biggest losses?

Q2. What are the ways in which operational losses occur in the canal water distribution system?

Q3. Define evaporation and how does it occur?

Q4. How does the rate of evaporation loss vary with the climatic data? Discuss the phenomenon.

Q5. What will happen to the evaporation loss if the side slope of a canal increases?

Q6. In which of the following two cases, the evaporation loss will be more?
(i) A canal having more depth of flow (ii) A canal having less depth of flow.

Q7. Write the different ways by which, rate of evaporation can be determined?

Q8. For the effective operational planning and management of an irrigation system, a dependable forecasting of the seepage is very important. Justify the statement.

Q9. How does the bed material of the canal influence seepage loss?

Q10. Does the water table position in the ground affect seepage loss? Discuss the phenomenon in field situation.

Q11. What are the two ways in which seepage occurs in canals? Discuss the percolation and absorption process in which seepage loss occurs in canals.

Q12. What are the various factors that affect seepage loss?

Q13. Explain the ponding method for measurement of seepage in a canal section.

Q14. Write down the empirical formulae used to estimate seepage loss.

Q15. What are the advantages of lining of the canals?

Q16. How does the seepage add to problems of water-logging? How can it be controlled?

Q17. Lining of canal helps in water conservation. Explain how is this statement true.

Q18. Hoe does the lining of canal help in reduction of flood hazards?

Q19. What are the different disadvantages of canal lining?

Q20. What are the points to be considered in justification of canal lining?

Q21. Why are the repair and maintenance of irrigation channel required?

Q22. Discuss the construction practices of different types of lining.

Q23. Hoe does the process of seepage control in canals help in reducing the dam height of a reservoir?

Q24. Why the study of economics in canal lining is must before selecting a particular lining material. How is the study carried out?

Q25. How are the net present value and benefit-cost ration of a project computed? What will happen to a project if the benefit-cost ratio is less than unity?

Q26. An unlined canal has seepage loss of 3.5 cumecs per million sq. m of wetted area and has to be lined with 10 cm thick cement concrete lining that costs Rs. 350/10 sq. m. Using the following data, work out the economics of lining.
Annual revenue per cumec of water from all crops = Rs 7.0 lakhs
Discharge in the channel = 90.0 cumecs
Area of the channel = 40 sq. m
Wetted perimeter of the channel = 22 m
Wetted perimeter of the lining = 20 m
Annual maintenance cost of unlined channel per 10 sq. m = Re. 6.00
Assume any other data not provided.

Q27. A canal of length 6 km and of discharge capacity 3 cumecs is proposed to be lined with boulder. The total cost of lining is computed as 5 lakhs. The life of the lining may be considered 50 years. Estimate the economics of lining with the following data,

Rate of interest = 8 percent
Saving in seepage loss by providing lining = 4 percent
Annual revenue for irrigation water = Rs. 75.00 per ha m
Base period of the crop = 145 days
Maintenance cost per km of unlined canal = Rs. 2800.00
Maintenance cost per km of lined canal = Rs. 1500.00
Additional benefit per km = Rs. 1200.00

Q28. What are the means to reduce seepage losses?

Q29. What are the roles of sealant in controlling seepage losses? Write down the different types of sealants that are used as lining materials in controlling seepage loss.

Q30. Describe in details about the various types of sealants used in canal lining with merits and demerits of each.

Q31. How is the cement-concrete lining done in the canals? Write the advantages and disadvantages of cement-concrete lining.

Q32. How is the pre-cast concrete lining adopted in field situations?

Q33. Write short notes on the following types of sealants.

(i) Cement mortar lining (ii) Boulder lining (iii) Polythene lining

(iv) Shotcrete lining (v) Brick lining (vi) Asphalt lining

Q34. Describe the various advantages and disadvantages of brick lining.

Q35. What are the advantages and disadvantages of lining the canals with polythene sheets? Describe the process of lining the canals by polythene sheets.

Q36. Narrate the different points that are to be considered for selecting a particular type of lining.

References

Basak, N.N. 2012. Irrigation Engineering. Tata McGraw Hill Education Pvt. Ltd., New Delhi.

Bayer , D.G. and Cluff, C.B. 1971. Report of water loss investigation of West Lake, EI, Dorado Country Club. Tucso, Arizona Water Resources Research Centre, Univeristy of Arizona, USA.

Biswas, R.K. 2008. A Few Low Cost Lining Materials. Technical Bulletin, Bidhan Chandra Krishi Viswavidayala, West Bengal, pp. 32.

Biswas, R.K. and Mallick, S. 1997. Performance of locally made burnt clay tiles in controlling seepage losses. J. of Agril. Engg., Vol. 34(4): 47-54.

Chawla, A.S. 1992. Canal lining and maintenance. Seminar on Irrigation Water Management, Water Management Forum, New Delhi.

Cluff, C.S (1977). The compartmented reservoir, a method of efficient water storage. Ph.D. dissertation, Colorado State University, Fort Collins.

Garg, S.K. 2012. Irrigation Engineering and Hydraulic Structures. Khanna Publishers, New Delhi.

Hansen, V.E., Israelsen, O.W. and Stringham, G.E. 1979. Irrigation Principles and Practices. John Wiley and Sons Inc., New York.

Hegade, Channappa, T.C. and Anand Ram, B.K. 1981. Water harvesting recycling in the red soils of Karnataka. Ind. J. of Soil Conservation, Vol. 9(2-3): 107-111.

Kamra, S.K. and Dhruvanarayana, B. 1987. Chemical sealant for seepage controlling farm pond. The role of Agricultural Engineering in dry land agriculture. Proceedings of 23rd Convention of ISAE, Yabilpur, India.

Khair, A., Ahmed, M and Data, S.C. 1984. Development of low cost indigenous technology to minimize water losses due to seepage in irrigation canal. AMA, Vol. XV: 77-81.

Michael, A.M. 1978. Irrigation Theory and Practices. Vikash Publishing House Pvt. Ltd., New Delhi.

Nicholaichuk, W. 1978. Seepage control in exacavated earthen reservoirs. Can. Soc. Agril. Engr., Vol. 20(2): 97-101.

Panigrahi, B. 2011. Irrigation Systems Enginering. New India Publishinh Agency, New Delhi.

Reginato, R.J., E.S. Nakayam and J.B. Miller (1973). Reducing seepage for stock tank with compacted sodium treated soils. J. Soil Cons., 28: 214-215.

Shally, H.L. 1965. Lining of Earthen Channel. Asia Publishing House, Bombay.

U.S. Bureau of Reclamation (1963). Lining for irrigation canal Department of Interior, Washington, 149 pp.

12

Salt Problem in Irrigated Agriculture

12.1 Introduction

Rainfall is the major source of water that contributes lion's share for irrigation to crops. Before it reaches the earth, it is to a great extent pure in nature. By the time it falls on the ground surface, it comes in contact with various types of gases and suspended materials of the atmosphere and also after it flows as runoff on the surface it gets dissolved with many types of salts, impurities *etc.* that render the water to be impure. Many a time the industrial waste, agricultural wastes, pesticides, fungicides, chemical fertilizers etc. get mixed with water and make it unsuitable for irrigation use. Water being a universal solvent, it dissolves all most all organic impurities, salts *etc.* which makes it objectionable for both domestic and agricultural uses.

Irrigated agriculture is dependent on an adequate water supply of usable quality. In the past, water quality concerns were neglected because good quality water supplies were plentiful and readily available. This situation is now changing in many areas. Especially, in the arid and semi-arid zones, water quality is fast deteriorating which is posing threat to irrigated agriculture.

Conceptually, water quality refers to the characteristics of a water supply that will influence its suitability for a specific use. On the other hand, it refers to the situation of how well the quality meets the needs of the user. There are three quality indicators. They are physical, chemical and biological indicators. Even a personal preference such as taste is a simple evaluation of acceptability. For example, if two drinking waters of equally good quality are available, people may express a preference for one supply rather than the other; the better tasting water becomes the preferred supply. In irrigated agriculture, emphasis is placed on the chemical and physical indicators of the water and rarely the biological indicator is considered important.

Quality needs are different for different specific uses and one water supply is considered more acceptable (of better quality) if it produces better results or causes fewer problems than an alternative water supply. For example, good quality river water which can be used successfully for irrigation may, because of its sediment load, be unacceptable for municipal use without treatment to remove the sediment. Similarly, snowmelt water of excellent quality for municipal use may be too corrosive for industrial use without treatment to reduce its corrosion potential. A number of different water quality guidelines related to irrigated agriculture are available. Each has been useful but none has been entirely satisfactory because of the wide variability in field conditions.

12.2 Some Basic Terminology

Alkali or Sodic Soil: Soil which has an excessive degree of saturation with exchangeable sodium, either with or without appreciable amount of soluble salts.

Alkaline: A chemical term referring to basic soil reaction where the pH is above 7.0.

Cation Exchange: Interchange of a cation in solution with another cation on a surface active material.

Cation Exchange Capacity (*CEC*): The total quantity of cations which a soil can absorb by cation exchange, usually expressed as milliequivalents per 100 grams.

Electrical Conductivity *(EC)*: Electrical conductivity is the reciprocal of electrical resitivity. Electrical resitivity is the resistance, in ohms, of a conductor which is 1 cm long and has a cross-sectional area of 1 sq. cm. Hence electrical conductivity is expressed as the reciprocal of ohm-cm or mhos per cm. The term electrical conductivity and specific electrical conductance have identical meanings. Small units of electrical conductivity are mmhos/cm (millimhos/cm) and μmhos/cm (micromhos/cm). One mmhos/cm is equal to 10^{-3} mhos/cm and one μmhos/cm is equal to 10^{-6} mhos/cm. One mmhos/cm is equal to one decisiemen/metre *(dS/m)*

Equivalent Weight: It is the combining capacity of an element or radical with hydrogen. It is the weight in gram of an ion or compound that combines with or replaces one gram of hydrogen.

It is given by the formula as:

$$Equivalent\ weight = \frac{Atomic\ weight}{Valency} \tag{12.1}$$

For example equivalent weight of sodium (Na^+) = $\frac{23}{1} = 23$ and that of sulphate $(SO_4^{2-}) = \frac{96}{2} = 48$

Small unit of equivalent weight is milliequivalent weight. One milliequivalent weight is equal to one thousandth of an equivalent weight.

Exchangeable Cation: A cation that is adsorbed on the exchange complex and which is capable of being exchanged with other cations.

Exchangeable sodium percentage *(ESP)*: It is the degree of saturation of the soil-exchange complex with sodium defined as follows:

$$ESP = \frac{Exchangeable\ sodium\ (milliequivalents\ per\ 100\ gm\ soil)}{Cation - exchange\ capacity\ (milliequivalent\ per\ 100\ gm\ soil)} x\,100 \qquad (12.2)$$

Leaching: The process of removal of soluble materials by the flow of water in soil.

Leaching Requirement: The fractions of water entering the soil that must pass through the root zone in order to prevent the soil salinity from exceeding a specific value.

Non-Saline Alkali Soil: Soil for which the exchangeable sodium percentage is greater than 15 and the conductivity of the saturation extract is less than 4 millimhos/cm (at 25 °C). The *pH* value of these soils generally varies between 8.5 and 10.

Saline-Alkali Soil: A soil for which the conductivity of the saturation extracts is greater than 4 millimhos/cm (at 25° C) and the exchangeable sodium percentage is greater than 15. The *pH* value of these soils may somewhat exceed 8.5.

Saline Soil: A soil which shows electrical conductivity of the saturation extract of the saturation extract more than 4 millimhos/cm (at 25° C), exchangeable sodium percentage is less than 15 and and *pH* value of the saturation paste is usually less than 8.5.

Saturation Extract: The solution extracted from a soil at its saturation percentage.

Sodium Adsorption Ratio (*SAR*): A ratio for the soil extracts and irrigation water used to express the relative activity of sodium ions in exchange reactions with soil. It is expressed as:

$$SAR = \frac{Na^{+}}{\sqrt{\left(Ca^{2+} + Mg^{2+}\right)/2}} \qquad 12.3)$$

In which the ionic concentrations are expressed in milliequivalents per litre.

pH: The *pH* value of a soil or natural water is a measure of its alkalinity or acidity. It is better to express it as a measure of the hydrogen ion concentration in water. It is found out by the logarithm, of the base 10, of the reciprocal of the hydrogen –ion concentration of pure water. *pH* of pure water is 7.0 . *pH* value of a neutral solution (neither alkaline nor acid) is 7.0. A *pH* value of more than 7.0 indicates a solution basic and less than 7.0 acidic in nature.

The important properties to diagnosis a salt affected soil are *pH*, presence of soluble salts, exchangeable sodium, cation exchange capacity and gypsum requirement. Following relationships may be employed for the evaluation of salt concentration (Michael, 1978).

(i) Salt concentration, mg/l or ppm. = 640 x *EC*, millimhos/cm (12.4)

(ii) Total cation concentration, meq/l = 10 x *EC*, millimhos/cm (12.5)

when *EC* is measured upto the range of 5 millimhos/cm at 25° C.

(iii) Osmotic pressure, atmospheres = 0.36 × EC, millimhos/cm (12.6)

(iv) Ppm/Eq. wt. = meq/l (12.7)

(v) Milliequivalents per litre to ppm = sum of the product of the milliequivalent of each ion times its equivalent weight. (12.8)

Sample Calculation 1:

(i) Express 3200 ppm salt concentration in micromhos, millimhos and mhos/cm.

(ii) Convert 3 millimhos/cm EC into osmotic pressure in atmospheres.

(iii) Express 2340 ppm calcium chloride salt concentration in meq/l.

Solution:

From Eq. (12.4), EC = ppm/640

= 3200/640 = 5 millimhos/cm

= 5 x 1000 = 5000 micromhos/cm

From Eq. (12.6), osmotic pressure, atmospheres = 0.36 x EC, millimhos/cm

= 0.36 x 3 = 1.08 atmospheres.

Equivalent weight of calcium chloride is 55.50.

From Eq. (12.7), ppm/Eq. wt. = meq/l

= 2340/55.50 = 42.16 meq/l.

12.3 Water Quality Problems

Water may be used for irrigation from a number of sources including rivers, ponds, groundwater reservoirs, lakes etc. Water used for irrigation from all these sources can vary greatly in quality depending upon type and quantity of dissolved salts. Salts are present in irrigation water in relatively small but significant amounts. They originate from dissolution or weathering of the rocks and soil, including dissolution of lime, gypsum and other slowly dissolved soil minerals. These salts are carried with the water to wherever it is used. In the case of irrigation, the salts are applied with the water and remain behind in the soil as water evaporates or is used by the crop. In the absence of drainage, the heavy application of irrigation water adds some 3 to 25 tons or more of salts each year to a hectare of land. This accumulation of salts each year make the soil unfit for agricultural uses after some years. Some of the main causes for the development of salinity and alkalinity in soils are (i) use of saline irrigation water, (ii) deposition of salts on soil surface from the high-sub soil water table, (iii) seepage from the canals, (iv) arid climate, (v) absence of good drainage and (vi) sea water intrusion in coastal areas.

The suitability of water for irrigation is determined not only by the total amount of salt present but also by the kind of salt. Various soil and cropping problems develop as the total salt content increases, and special management practices may be required to maintain acceptable crop yields. Water quality or suitability for use is judged on the potential severity of problems that can be expected to develop during long-term use. The soil problems most commonly encountered and used as a basis to evaluate water quality are those related to salinity, water infiltration rate, toxicity and a group of other miscellaneous problems.

12.4 Causes of Formation of Saline and Alkaline Soils

Some of the basic causes for formation of saline and alkaline soils are mentioned below.

Climatic Factors

Climate has a great impact on imparting pureness of water. The various climatic factors especially rainfall and temperature plays a dominant role in making water saline and alkaline. In the humid region, there is heavy rainfall which washes away the dissolved salts and soluble minerals substances with runoff and discharges them in the rivers and streams. Moreover, the percolation caused due to rainfall carries some amount of salts and mineral constituents of soil and adds them to the groundwater. Thus, the salinity of water in the humid region is less. However, in the arid and semi-arid regions, there is less distribution of rainfall and on the other hand there is more evaporation caused due to high temperature. The high evaporation rate of water decreases the water content of the soil leaving behind a high percentage of salts in the soil. Since there is low rainfall, chances of the washing away of the salts of the soil are less. Decreased rainfall causes low percolation which also hinders in carrying higher amount of salts and adding to the groundwater reservoir. This makes the arid and semi-arid soil more saline. Thus the basic cause of saline and alkaline soil is insufficient application of water which may be either from the rainfall or from irrigation.

Geological Factors

Geological factors relate to the origin of the soils. Some soils are formed from basic parent material and therefore they are saline in nature. Some soils contain salt layers in sub-soil. When such soils are deep cultivated, these salts come up to the surface soil. Some saline soils are also believed to have been formed by deposition of salts carried down by rivers from hill rocks.

Hydrological Factors

Water is the chief transporting agent for salts and its evaporation gives rise to accumulation of salts in the soil. At flood plains, deltas, coastal areas, lakes and regions of high water table, surface movement of water is negligible. Water, therefore, evaporates leaving the salts on the surface. This process also brings up salts from the lower horizons if the moisture regime is connected with ground water. In fact, this is the most important cause of salinization.

Depth to Watertable

If the depth of water-table below the soil surface is less than 1.2 m, chances of salinization of surface soil become more important. Rising water table dissolves salts falling in the way and leaves them on the soil surface after evaporation. Movement of water from lower soil layers to surface layers takes place through capillary action. The arid and semi-arid areas of the country have often poor quality ground water. The use of saline ground water for irrigation has accentuated the problem of salt accumulation in several areas. Many places of Punjab, Haryana, Rajasthan and Gujarat have shallow depth of utilizable saline groundwater which has already created problems of soil salinity. The problem is further aggravated where the ground water has high sodium content. Excessive irrigation in the canal commands with faulty water and crop management practice has caused raising groundwater table in many parts of the country which has aggravated the salinisation problems.

12.5 Soil Salinity in Coastal India

India has a vast length of coastal line amounting about 7515 km. It has vast areas suffering from soil salinity. The soil salinity areas of different places of coastal India are presented in Table 12.1. In these areas, salts accumulate in the soil profile due to higher groundwater table, sea water intrusion, tidal inundation, wind sprays *etc.* It is observed that in the coastal belts, where there is ample rainfall, severity of salinity is less. This is because, in high rainfall areas, there is dilution and washing away of salts. Runoff is high in these areas which carries away greater amount of salt from the soil profile and thereby decreasing the severity of salinity hazards.

Table 12.1: Salt affected areas of coastal India (Bandyopadhyaya and Abrol,1986)

State	Districts affected	Area (x 10^3), ha	Rainfall, mm	
			Kharif	*Rabi*
Andhra Pradesh	Visakhapatanam, West Godavari, Krishna, Guntur, Nellore	40	570	330
Odisha	Balasore, Bhadrak, Puri, Jajpur, Jagatsinghpur, Kendrapara, Ganjam	400	1140	180
Tamil Nadu	Chennai, Chingalpet, Thanjavur, Ramnathpuram	4	310	480
West Bengal	24 Praganas, North and South Midnapore, East Howrah	850	1325	177
Bihar	North East Alluvial Plain including Gopalganj	4	1280	80
Gujarat	Surat, Kaira, Ahmedabad, Surendranagar, Kutch	720	930	30
Kerala	Ernakulum, Alleppey, Canannore	16	2010	550
Maharastra	Kolaba, Thana	40	2700	130
Andaman and Nicobar Islands		15	2900	450

12.6 Effect of Irrigation, Soil and Crop Management on Soil Salinity and Alkalinity

The nature and extent of salt accumulation and the degree of soil alkalinity depend on the quality of irrigation water, soil type and its permeability, salt tolerance characteristics of the plant and climatic conditions. Soils which are light textured like sandy soil are less salinised than the medium and heavy textured soils. Alkalinity of the soil increases with the development of a hard pan layer of lime or clay. The situation becomes more complex when the depth of the water table is high and the quality of irrigation water is poor. Under such situation, the final salt balance near the root zone is governed by the combined effect of the quality of irrigation water, irrigation management, climate and water transmission properties of the soil during the cropping period.

The salt concentration of water is influenced by the relief of the land. For example, in flood plain areas, coastal belts etc., the salt concentration is usually high. In areas of high water table, salts move up and are deposited near the surface that makes the soil saline. The rate of salinization by capillary action decreases when the depth of water table is low and the soil moisture content is also low. In places where excess irrigation is practiced, there is more percolation which adds to the underlying groundwater reservoir and consequently raising the water table. Under such situation, salts present in the water move up by capillarity and get deposited near the surface. On the other hand, application of too less irrigation water than the crop needs and leaching requirements, salinity and alkalinity problems develop. At times, in spite of good quality irrigation water and good irrigation management practices, soil salinity develops because of the poor soil physical conditions which impede drainage.

12.7 Nature and Extent of Salt Problems in Irrigated Areas of India

The salt affected areas of India can be classified into three categories. They are (i) arid and semi arid regions with low rainfall and irrigated with poor quality ground water, (ii) regions of high water table and areas influenced by temporary water logging and seepage from canals and other sources, and (iii) coastal regions influenced by sea water intrusion.

Salt problems in arid and semi arid regions having annual rainfall below 450 mm are spread over the states of Rajasthan, Gujarat, some parts of Uttar Pradesh, Haryana and Punjab. In these regions, the soil salinity is mainly due to use of poor quality irrigation water. High temperature with low rainfall increases the severity of the problem in these areas.

Combined effects of poor quality irrigation water, high water table and inadequate drainage aggravate the salt problems in some of the areas of Punjab, Haryana, Uttar Pradesh and Delhi. Salt problems of some parts of Haryana like Rohtak and Gurgaon are primarily due to low permeability of the soil rather than high water table or poor quality water. The salt affected areas in the central and southern regions are mostly confined to black cotton soil regions in the semi arid regions comprising parts of Maharastra, Karnataka and Andhra Pradesh. Nearly all the deep black soils and soils in the low lying areas are potentially saline or alkanine depending on the depth and nature of alkalinisation

Along the coastal line of India, thousands of hectares of land are salt affected. The salinity in these areas is attributed to sea water intrusion. Large areas of land along the mouths of rivers and creeks in West Bengal, Odisha, Andhra Pradesh, Tamil Nadu, Maharastra and Gujrat face tidal affects and are inundated with sea water. Large areas in these regions are now barren being devoid of all types of vegetations (Michael, 1978).

12.8 Chemistry of Salty Soils

The main process occurring in soils while irrigating with poor quality water are the followings.

- Ion exchange between cations in irrigation water and those present on the soil exchange process
- Dissolution and precipitation of $CaCO_3$
- Weathering of minerals
- Hydration and dehydration of the soil as a result of fluctuation of soil moisture
- Leaching down of ions

- Upward movement of ions through capillary activity and
- Mineral nutritional characteristics of the crop growth.

Amongst these processes, cation exchange is the most important process which governs the accumulation of excessive sodium during irrigation with saline water and reclamation of alkali soils during treatment with gypsum.

A clay particle with sodium and calcium ions attached tends to hydrolyse. When a sodium ion is exchanged for a hydrogen ion, and the sodium ion combines with a molecule of water, sodium hydroxide (*NaOH*) is formed.

$$Ca\ \overset{Na}{\underset{Na}{\boxed{\text{Clay particle}}}}\quad Na + H_2O \leftrightarrow NaOH + H\ \overset{Na}{\underset{Na}{\boxed{\text{Clay}}}}\ Ca \qquad (12.9)$$

When carbon dioxide (CO_2) is present in soil air, it readily reacts with the water to form hydrogen carbonate (H_2CO_3). However, this *NaOH* reacts readily with H_2CO_3 to form sodium carbonate (Na_2CO_3).

Sodium carbonate is gradually removed with extensive leaching and the soil is left with hydrogen ions having replaced sodium ions. This increase of hydrogen ions is reflected in a lower *pH*. Where soil contains calcium carbonate ($CaCO_3$) or gypsum, calcium is dissolved into the soil solution. This available calcium is exchanged for sodium during the leaching process to obtain a normal soil.

$$Ca\ \overset{Na}{\underset{Na}{\boxed{\text{Clay particle}}}}\ \overset{Na}{\underset{Na}{}}\quad + CaCO_3 \leftrightarrow Na_2CO_3 + Ca\ \overset{Na}{\underset{Na}{\boxed{\text{Clay particle}}}}\ Ca \qquad (12.10)$$

When chemicals are added to the soil to accomplish the exchange of sodium for other ions, the products like gypsum ($CaSO_4.2H_2O$), soil sulfur (*S*), sulphuric acid (H_2SO_4), concentrated lime sulfur solution and dry lime sulfur can be used (Hansen *et al.,* 1979). .

12.9 Effect of Salinity on Crop

The primary objective of irrigation is to provide a crop with sufficient amounts of water in due time, thus avoiding water stress and consequently yield loss caused by extended periods of water stress during critical crop growth stages. However, frequent applications of irrigation water cause salt accumulation in the root zone of the crop that reduces water availability to the crop and hastening the onset of a water shortage. Salt affected soils do not allow satisfactory plant growth due to many reasons as mentioned below. Fig.12.1 represents the effect of soil salinity on growth of plant.

- Concentration of salts outside the roots goes so high that roots cannot absorb water and, therefore, plant will dry.

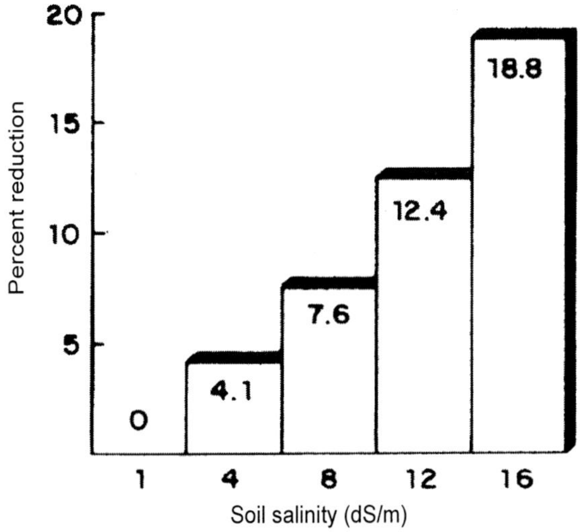

Fig. 12.1: Soil salinity affecting the plant growth

- In salt affected soils plants suffer misbalance of nutrients. Some essential nutrients become unavailable, whereas some other nutrients are available in excess which affect the plant growth. For example, in alkali soil, sodium becomes toxic to the plant due to its excessive solubility or availability.
- Excessive sodium in soil results .in dispersal of soil particles. Soil structure is destroyed and it does not provide suitable physical condition for plant growth.
- A hard pan develops in alkaline sub-soil which hampers the extension of plant roots and downward movement of water.
- Amount of organic matter and population of soil organisms go down to an unfavourable level. Soil loses its fertility and plant growth is restricted in such soils.

12.10 Tolerance of Crops to Salinity

All crops do not respond to salinity in a similar manner. Some crops can produce acceptable yields at higher level of soil salinity than others Fig. 12.2 shows how the relative crop yield is affected by soil salinity with categories from tolerant to sensitive. The extent of ability of the crop to tolerate the salinity is extremely useful in selecting the appropriate crop for the soil in the soil *in-situ*. In areas where a build-up of soil salinity cannot be controlled at an acceptable concentration for the crop being grown, an alternative crop can be selected that is both more tolerant of the expected soil salinity and can produce economical yields. Tolerance of the crops to salinity depends on many factors such as type and variety of crops, type of salt present in the soil and its concentration, *etc.* Table 12.2 presents the tolerance of different crops and yield potential affected by varying degree of soil salinity and irrigation water salinity.

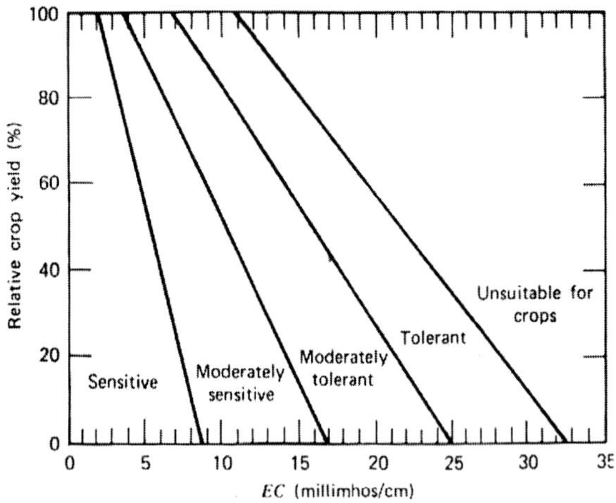

Fig. 12.2: Relative crop yield as affected by salinity

Table 12.2: Crop tolerance and yield potential of crops affected by irrigation water salinity (EC_w) or soil salinity (EC_e) (Maas and Hoffman (1977) and Maas (1984)

Crop tolerance/Yield potetial	100%		90%		75%		50%		0% (Maximum)	
	EC_e	EC_w	EC_e	EC_w	EC_e	EC_w	EC_e	EC_w	EC_e	EC_w
Field Crops										
Barley (Hordeum vulgare)	8.0	5.3	10	6.7	13	8.7	18	12	28	19
Cotton (Gossypium hirsutum)	7.7	5.1	9.6	6.4	13	8.4	17	12	27	18
Sugarbeet (Beta vulgaris)	7.0	4.7	8.7	5.8	11	7.5	15	10	24	16
Sorghum (Sorghum bicolor)	6.8	4.5	7.4	5.0	8.4	5.6	9.9	6.7	13	8.7
Wheat *(Triticum aestivum)*	6.0	4.0	7.4	4.9	9.5	6.3	13	8.7	20	13
Wheat, durum *(Triticum turgidum)*	5.7	3.8	7.6	5.0	10	6.9	15	10	24	16
Soybean *(Glycine max)*	5.0	3.3	5.5	3.7	6.3	4.2	7.5	5.0	10	6.7
Cowpea *(Vigna unguiculata)*	4.9	3.3	5.7	3.8	7.0	4.7	9.1	6.0	13	8.8
Groundnut (Peanut) *(Arachis hypogaea)*	3.2	2.1	3.5	2.4	4.1	2.7	4.9	3.3	6.6	4.4
Rice (paddy) *(Oriza sativa)*	3.0	2.0	3.8	2.6	5.1	3.4	7.2	4.8	11	7.6
Sugarcane *(Saccharum officinarum)*	1.7	1.1	3.4	2.3	5.9	4.0	10	6.8	19	12
Corn (maize) *(Zea mays)*	1.7	1.1	2.5	1.7	3.8	2.5	5.9	3.9	10	6.7
Flax *(Linum usitatissimum)*	1.7	1.1	2.5	1.7	3.8	2.5	5.9	3.9	10	6.7
Broadbean *(Vicia faba)*	1.5	1.1	2.6	1.8	4.2	2.0	6.8	4.5	12	8.0
Bean *(Phaseolus vulgaris)*	1.0	0.7	1.5	1.0	2.3	1.5	3.6	2.4	6.3	4.2
Vegetables										
Squash, zucchini (courgette) (Cucurbita pepo melopepo)	4.7	3.1	5.8	3.8	7.4	4.9	10	6.7	15	10
Beet, red (Beta vulgaris)	4.0	2.7	5.1	3.4	6.8	4.5	9.6	6.4	15	10
Squash, scallop (Cucurbita pepo melopepo)	3.2	2.1	3.8	2.6	4.8	3.2	6.3	4.2	9.4	6.3
Broccoli (Brassica oleracea botrytis)	2.8	1.9	3.9	2.6	5.5	3.7	8.2	5.5	14	9.1
Tomato (Lycopersicon esculentum)	2.5	1.7	3.5	2.3	5.0	3.4	7.6	5.0	13	8.4

(Contd.)

Crop tolerance/Yield potetial	100%		90%		75%		50%		0% (Maximum)	
	EC_e	EC_w	EC_e	EC_w	EC_e	EC_w	EC_e	EC_w	EC_e	EC_w
Cucumber (Cucumis sativus)	2.5	1.7	3.3	2.2	4.4	2.9	6.3	4.2	10	6.8
Spinach (Spinacia oleracea)	2.0	1.3	3.3	2.2	5.3	3.5	8.6	5.7	15	10
Celery (Apium graveolens)	1.8	1.2	3.4	2.3	5.8	3.9	9.9	6.6	18	12
Cabbage (Brassica oleracea capitata)	1.8	1.2	2.8	1.9	4.4	2.9	7.0	4.6	12	8.1
Potato (Solanum tuberosum)	1.7	1.1	2.5	1.7	3.8	2.5	5.9	3.9	10	6.7
Corn, sweet (maize) (Zea mays)	1.7	1.1	2.5	1.7	3.8	2.5	5.9	3.9	10	6.7
Sweet potato (Ipomoea batatas)	1.5	1.0	2.4	1.6	3.8	2.5	6.0	4.0	11	7.1
Pepper (Capsicum annuum)	1.5	1.0	2.2	1.5	3.3	2.2	5.1	3.4	8.6	5.8
Lettuce (Lactuca sativa)	1.3	0.9	2.1	1.4	3.2	2.1	5.1	3.4	9.0	6.0
Radish (Raphanus sativus)	1.2	0.8	2.0	1.3	3.1	2.1	5.0	3.4	8.9	5.9
Onion (Allium cepa)	1.2	0.8	1.8	1.2	2.8	1.8	4.3	2.9	7.4	5.0
Carrot (Daucus carota)	1.0	0.7	1.7	1.1	2.8	1.9	4.6	3.0	8.1	5.4
Bean (Phaseolus vulgaris)	1.0	0.7	1.5	1.0	2.3	1.5	3.6	2.4	6.3	4.2
Turnip (Brassica rapa)	0.9	0.6	2.0	1.3	3.7	2.5	6.5	4.3	12	8.0
Wheatgrass, tall (Agropyron elongatum)	7.5	5.0	9.9	6.6	13	9.0	19	13	31	21
Wheatgrass, fairway crested (Agropyron cristatum)	7.5	5.0	9.0	6.0	11	7.4	15	9.8	22	15
Bermuda grass (Cynodon dactylon)	6.9	4.6	8.5	5.6	11	7.2	15	9.8	23	15
Barley (forage) (Hordeum vulgare)	6.0	4.0	7.4	4.9	9.5	6.4	13	8.7	20	13
Ryegrass, perennial (Lolium perenne)	5.6	3.7	6.9	4.6	8.9	5.9	12	8.1	19	13
Trefoil, narrowleaf birdsfoot[8]	5.0	3.3	6.0	4.0	7.5	5.0	10	6.7	15	10
Harding grass (Phalaris tuberosa)	4.6	3.1	5.9	3.9	7.9	5.3	11	7.4	18	12
Fescue, tall (Festuca elatior)	3.9	2.6	5.5	3.6	7.8	5.2	12	7.8	20	13
Wheatgrass, standard crested (Agropyron sibiricum)	3.5	2.3	6.0	4.0	9.8	6.5	16	11	28	19
Vetch, common (Vicia angustifolia)	3.0	2.0	3.9	2.6	5.3	3.5	7.6	5.0	12	8.1
Sudan grass (Sorghum sudanense)	2.8	1.9	5.1	3.4	8.6	5.7	14	9.6	26	17
Wildrye, beardless (Elymus triticoides)	2.7	1.8	4.4	2.9	6.9	4.6	11	7.4	19	13
Cowpea (forage) (Vigna unguiculata)	2.5	1.7	3.4	2.3	4.8	3.2	7.1	4.8	12	7.8
Trefoil, big (Lotus uliginosus)	2.3	1.5	2.8	1.9	3.6	2.4	4.9	3.3	7.6	5.0
Sesbania (Sesbania exaltata)	2.3	1.5	3.7	2.5	5.9	3.9	9.4	6.3	17	11
Sphaerophysa (Sphaerophysa salsula)	2.2	1.5	3.6	2.4	5.8	3.8	9.3	6.2	16	11
Alfalfa (Medicago sativa)	2.0	1.3	3.4	2.2	5.4	3.6	8.8	5.9	16	10
Lovegrass (Eragrostis sp.)	2.0	1.3	3.2	2.1	5.0	3.3	8.0	5.3	14	9.3
Corn (forage) (maize) (Zea mays)	1.8	1.2	3.2	2.1	5.2	3.5	8.6	5.7	15	10
Clover, berseem (Trifolium alexandrinum)	1.5	1.0	3.2	2.2	5.9	3.9	10	6.8	19	13
Orchard grass (Dactylis glomerata)	1.5	1.0	3.1	2.1	5.5	3.7	9.6	6.4	18	12
Foxtail, meadow (Alopecurus pratensis)	1.5	1.0	2.5	1.7	4.1	2.7	6.7	4.5	12	7.9
Clover, red (Trifolium pratense)	1.5	1.0	2.3	1.6	3.6	2.4	5.7	3.8	9.8	6.6
Clover, alsike (Trifolium hybridum)	1.5	1.0	2.3	1.6	3.6	2.4	5.7	3.8	9.8	6.6
Clover, ladino (Trifolium repens)	1.5	1.0	2.3	1.6	3.6	2.4	5.7	3.8	9.8	6.6
Clover, strawberry (Trifolium fragiferum)	1.5	1.0	2.3	1.6	3.6	2.4	5.7	3.8	9.8	6.6

(*Contd.*)

Crop tolerance/Yield potetial	100%		90%		75%		50%		0% (Maximum)	
	EC_e	EC_w	EC_e	EC_w	EC_e	EC_w	EC_e	EC_w	EC_e	EC_w
Fruit Crops										
Date palm (phoenix dactylifera)	4.0	2.7	6.8	4.5	11	7.3	18	12	32	21
Grapefruit (Citrus paradisi)	1.8	1.2	2.4	1.6	3.4	2.2	4.9	3.3	8.0	5.4
Orange (Citrus sinensis)	1.7	1.1	2.3	1.6	3.3	2.2	4.8	3.2	8.0	5.3
Peach (Prunus persica)	1.7	1.1	2.2	1.5	2.9	1.9	4.1	2.7	6.5	4.3
Apricot (Prunus armeniaca)	1.6	1.1	2.0	1.3	2.6	1.8	3.7	2.5	5.8	3.8
Grape (Vitus sp.)	1.5	1.0	2.5	1.7	4.1	2.7	6.7	4.5	12	7.9
Almond (Prunus dulcis)	1.5	1.0	2.0	1.4	2.8	1.9	4.1	2.8	6.8	4.5
Plum, prune (Prunus domestica)	1.5	1.0	2.1	1.4	2.9	1.9	4.3	2.9	7.1	4.7
Blackberry (Rubus sp.)	1.5	1.0	2.0	1.3	2.6	1.8	3.8	2.5	6.0	4.0
Boysenberry (Rubus ursinus)	1.5	1.0	2.0	1.3	2.6	1.8	3.8	2.5	6.0	4.0
Strawberry (Fragaria sp.)	1.0	0.7	1.3	0.9	1.8	1.2	2.5	1.7	4	2.7

Tolerance level of a few crops and verities are mentioned below.

A. Tolerance of Field Crops

- **High salt tolerant:** Sesbania, rice, sugarcane, oats, sugarbeet, berseem, lucern, senji, fenugreek, barley.
- **Medium salt tolerant:** Castor, cotton, sorghum, pearl millet, maize, mustard, wheat, bajra.
- **Low salt tolerant:** Pulses, pigeon pea, sunnhemp, gram, peas, linseed, sesamum.

B. Varietal Tolerance

- Paddy: coarse varieties like Kranti and Saket 4, IR-8, Jaya, PR-I06, P-.2-21, Satha, Jhona.
- Jowar: M-47-3, GadraRoad, Pokaran, Texas, Hybrid-610, IS- 541, J-718, PJ-4-R, Sweet Sudan, M-34-1, T-l.
- Bajra: Khatani, Pal Manihari, Agra,JetStal, T-55,Bikaner, T-5.
- Wheat: HD-2009(Arjun), WH-711, HD-1982, HD-1553.
- Sugarcane: Co 205, Co-286, Co-280, pa5 -2878, Co-210, Co-331,Co-312,Co-321,Co-513,Co-290,Co421,CoS- . 109, Bo-10,Bo-24,Bo-3.
- Cotton: Short staple varieties

C. Tolerance of Vegetables

Salt tolerant: Garden beet, asparagus, spinach, cabbage, tomato and cole crops.

Salt sensitive: Radish, celery, green beans and tuber vegetables.

D. Tolerance of Grasses

High: Chloris guyana (rhodes grass), Cynodon dactylon (doob grass) and Paspalum notatum.

Medium: Brachiaris mutica (paragrass),Cynodon plectostachys (giant Star)

Low: Panicum antidotale (blue panic), Cenchrus ciliaris, Dichanthium annulatum, Leptochloa fusea.

E. Forest Trees and Shrubs

Acacia nilotica (babul), Butea monosperma (dhak), Prosopis spicigera (khejri), Acacia catechu (khair), Azadirachta indica (neem), Cordia lalifalia (lasora), Dalbergia sissoo (sisum), Albizzia lebbec (siris), Termindlia belleria (bahera), Emblica officinalis (aonla).

F. Fruit Trees

With a few exceptions most of the fruit trees are sensitive to salinity and prone to leaf injury due to accumulation of sodium and chloride. Stone fruit trees, citrus, strawberry avocado and papaya are very sensitive to salinity. Date palm is highly salt tolerant while grape is medium tolerant.

12.11 Reclamation of Salt Affected Soils

Reclamation of saline and alkaline soil involves techniques of removal of excess salts from root zone such that the crop yield is not appreciably hampered. Reclamation is said to be successful when the excess salts are completely removed and prevention of the reclaimed land from reversion to its previous state is achieved. Harmful salts being soluble can be removed by leaching with water which is the chief means of their transport. In alkaline soils an acid forming amendment is necessarily required to be applied before leaching, in order to flush out sodium from the soil. Temporary control of salts on irrigated land is sometimes practiced by (i) ploughing the salt surface crusts deeply into the soil, (ii) removing surface accumulation from the soil and (iii) neutralizing the effects of certain salts by use of other salts or acids.

Permanent reclamation of saline and alkali soil requires the followings steps to be adopted:

12.11.1 Bunding and Land Leveling

Strong earthen bunds of 35-40 cm height should be erected around the field boundary well before the monsoon rain begins. Total field area should be divided into small plots and should be surrounded by earhen bunds. Bunding prevents inflow or outflow of water from the fields and thus causes ponding of water in the field. Field should be completely leveled to avoid any localization of salts. A gentle slope of about 0.1 percent towards the drainage channel should be provided in draining out excess water. A trench should be dug across the slope of the plot for draining out water. The trench can be 60-75 cm deep and width may vary according to the size of the plot and intensity of leaching with water.

12.11.2 Provision of Irrigation Water

In some areas, the canal water is not fit for irrigation. However, it is often observed that ground water of those areas is of good quality. Tube wells should be installed in these areas to tap the good quality ground water which should be blended with poor quality canal water and can be used for irrigation safely.

12.11.3 Lowering of Water Table

Waterlogging is the main cause for rising of ground water. A land suffering from waterlogging should be reclaimed immediately before it is left abandoned due to salinisation and alkalinisation for permanent cultivation. The water table should be lowered to such a level that the effective crop root zone is free from inundation. This will cause proper air and water circulation in the root zone; the micro biological properties are not hampered and the growth of the crop is not affected. The first source to lower the water table is to identify the source causing to raise the water table. For a small patch of land owned by an individual farmer or a small group of farmers, it may be easy to identify the source and take remedial measures by constructing an interceptor drain around the patch of land so that excess water is drained out and ground water does not rise. But for a large patch of land like a big irrigated command areas, where large scale canal irrigation is practiced, it becomes difficult to check the rising water table by any individual farmer or small group of farmers. In this case, community approach is required. There are many ways by which ground water table can be lowered. Some of the main means are:

- Provision of surface and sub surface drainage (details of different types of surface and sub-surface drainage methods to reduce the problems of waterlogging and lowering the groundwater table is discussed in subsequent chapter)
- Provision of bio-drainage in which some deep rooted trees are grown in the high water table areas. These trees extract huge amount of water from the ground reservoir for their transpiration ultimately decreasing the water table.
- Pumping the groundwater reservoir by tube well irrigation
- Advocating crop diversification in canal commands instead of growing only rice-rice
- Adopting proper crop and water management practice
- Using water saving irrigation techniques in crop fields
- Irrigating the crops through field channels in stead of plot to plot method
- Adopting volumetric measures for irrigation water measurement
- Using correct method of irrigation scheduling
- Adopting rotational irrigation system

12.11.4 Satisfactory Intake Rates

The rate of water infiltration in to the soil depends on soil texture, structure, degree of soil dispersion, drainage facility provided and also on the depth to groundwater table. Alkali soils disperse during leaching and often become impermeable as the soluble salts are removed. Chemical amendments are then required to bring about the improvement of the soil by replacing the exchangeable sodium with calcium. Gypsum and in some cases sulfur are used as amendment materials.

The cheapest method of obtaining the satisfactory intake rates is to minimize the surface tillage. Conservation agriculture plays a vital role in increasing the intake rate of water into the soil. Leaving the soil surface rough with crop residues and formation of clods on the soil surface during ploughing time makes the surface uneven. This hinders smooth movement of water (runoff) on the soil surface and thereby enhancing infiltration. Summer ploughing and good tillage practice also helps the soil to remain shallow which helps in greater infiltration of water into the soil. In addition, growing green manures including *dhaincha (Sesbania aculeata)* makes soil soft and porous which helps in increasing the intake rate of water.

Saline water, containing less amount of sodium, can also improve the intake rate. Quality of water greatly influences the intake rates. Moreover, growing of deep rooted crops like clover and alfalfa can influence the intake rate. The soils should not be left fallow and to be covered by crops so that runoff will be less and more water will percolate into the soil.

12.11.5 Application of Soil Amendments

Gypsum is by far the most commonly used chemical amendment. Its quantity should be determined through proper soil analysis. It is, therefore, advisable to obtain representative soil samples after the fields have been leveled and properly bunded so as to work out the required amount of an amendment. For highly alkaline soils lying barren, 15 tonnes/ha gypsum is required. Gypsum in powered form should be applied 10-15 days before transplanting rice crop. It can be spread evenly in the field and mixed with top 10-15 cm soil by 2-3 shallow ploughings. The amount of the gypsum to be applied greatly depends upon the soil texture and type of the crop to be sown. Fine textured clay soil needs more gypsum than light textured soil. Deep rooted crops require more gypsum than shallow rooted ones.

When acidifiers are added in the soil, they dissolve in water and form strong acids. These acids solubilise native calcium which then replaces sodium from soil particles. Most of the soils in India are poor in calcium. Therefore, the use of gypsum is widely adopted.

For saline-alkaline soils which contain, along with sodium, a large amount of calcium, use of acid forming amendments like sulphur, iron sulphate, pyrite and sulphuric acid should be made. Amounts of other chemical amendments which can replace one ton gypsum have been shown in Table 12.3.

Table.12.3: Gypsum equivalents of some chemical amendments used for reclaiming alkali soils

Sl. No.	Amendment	Weight in tones, equivalent to one ton of gypsum
1	Gypsum	1.00
2	Sulphuric acid	0.57
3	Sulphur	0.19
4	Iron sulphate	1.62
5	Aluminium sulphate	1.29
6	Lime sulphur	0.76
7	Pyrite	0.63

12.11.6 Leaching

In this process, salts are made to dissolve in the standing water of the field and then allowed to leach down in the soil. Thus, upper 20-25 cm soil is free from salts. Amendment used in alkali soil liberates sodium from soil particles and soluble sodium salts come out in soil water. This sodium rich water must be leached down in the soil. This is done by impounding water in the field and allowing higher percolation. A flood irrigation should be given just after the application of gypsum and water should be made to stand in the field for about 4-5 days. During this period a slow downward movement of water along with salts dissolved in it takes place. If water is available in plenty, drainage of sodium mixed water should be carried out and the drained water should be disposed off safely.

Soil texture plays a crucial role in in the leaching process. Coarse textured soils of open structure as a rule have sufficiently high permeability to make leaching of alkali salts an easy task after the water table has been sufficiently lowered. Fine textured compact soils of low permeability predominate in the low-lying waterlogged areas. Thus, soil permeability is an important factor to decide the rate of leaching in the waterlogged areas. Excessive leaching, however, is detrimental for plant growth. Due to excessive leaching, plant nutrients, especially the nitrate fertilizers, may be removed from the soil which hampers the growth of plants. Excess irrigation required for leaching may cause rise of groundwater table which again creates waterlogging and salinisation problems. Water application should be just sufficient to maintain a favourable salt balance in the soil such that it accelerates the leaching and at the same time, it does not cause waterlogging and drainage problems.

Leaching Requirements

The depth of irrigation water per unit depth of soil, required to produce any specified increase in soil salinity for any given conductivity of irrigation water is calculated from the following equation:

$$\frac{D_i}{D_s} = \frac{d_s}{d_w} \cdot \frac{SP}{100} \cdot \frac{\Delta EC_e}{EC_i} \tag{12.11}$$

where,

D_i = depth of irrigation water; D_s = depth of soil; d_s = bulk density of soil; d_w = density of irrigation water; SP = saturation percentage of soil; ΔEC_e = Increase in electrical conductivity of saturation extract of the soil and EC_i = electrical conductivity of irrigation water.

Under high water table condition, the increase in soil salinity, by evaporation of groundwater can be computed by the formula:

$$\Delta EC_e = \frac{D_g}{D_s} \cdot \frac{EC_g}{SP} \cdot \frac{d_w}{d_s} . 100 \tag{12.12}$$

where,

D_g = depth of groundwater evaporated; EC_g = electrical conductivity of groundwater and other terms of the above equation are defined earlier.

It is to be noted that Eqn. (12.12) is obtained from Eqn. (12.11) after rearranging Eqn. (12.11) and substituting the depth of groundwater evaporated and its EC for the depth of irrigation water and its quality.

Leaching requirement (*LR*) also called as leaching fraction (*LF*) is defined as the fraction of irrigation water that must be leached through the root zone of the crop to keep the salinity of the soil below a specified limit. Mathematically it is written as:

$$LR = \frac{D_d}{D_i} = \frac{EC_i}{EC_d} \tag{12.13}$$

where,

LR = leaching requirement expressed as a ratio or in percent; EC_i = electrical conductivity of irrigation water, mmhos/cm; EC_d = electrical conductivity of drainage water, mmhos/cm, D_d = depth of water moving down or leached down the root zone (also called as depth of drainage water) and other terms of the above equation are defined earlier.

Water applied for irrigation to a crop is used to meet mainly its consumptive use and some part of it is used as drainage. If the depth of irrigation water is D_i, consumptive use is D_c and drainage is D_d then,

$$D_i = D_c + D_d \tag{12.14}$$

Using Eqn. (12.13) and Eqn. (12.14) we get,

$$D_i = \frac{D_c}{1 - LR} \tag{12.15}$$

Using Eqn. (12.13) and Eqn. (12.15), we get,

$$D_i = \left(\frac{EC_d}{EC_d - EC_i} \right) D_c \tag{12.16}$$

Eqn. (12.16) gives relationship in terms of the depth of water lost as consumptive use and the salt concentration of irrigation and drainage water. From this equation, one can access what should be the depth of irrigation water required to satisfy the consumptive demand of the crop and the leaching requirements. Salt concentration of irrigation water can be determined by laboratory test of irrigation water. Salt concentration of drainage water is taken on the basis of permissible salt tolerance limits of the crop. Generally salt tolerance limits are expressed in terms of *EC* of the saturation extract of the soil solution (EC_e). In most cases, EC_d *is* assumed to be approximately equal to twice as high as the corresponding value of EC_e (Michael, 1978).

Sample Calculation 2:

Determine the change in salinity level of the soil due to evaporation of groundwater over a period of 5 months when the depth of ground water evaporated is 11 cm and electrical conductivity is 12 millimhos/cm. The depth of soil influenced by salt accumulation is 40 cm. Bulk density of the soil is 1.5 gm/cm^3 and saturation percent of the soil is 60 percent. Assume density of water as 1 gm/cm^3.

Solution:

Given D_g = Depth of groundwater evaporated = 11 cm

D_s = Depth of the soil layer = 40 cm

EC_g = Electrical conductivity of groundwater = 12 millimhos/cm

SP = Saturation percentage of soil = 60

d_w = Density of irrigation water = 1 gm/cm^3

d_s = Bulk density of soil = 1.5 gm/cm^3

Change in salinity level of the soil is given as:

$$\Delta EC_e = \frac{D_g}{D_s} \cdot \frac{EC_g}{SP} \cdot \frac{d_w}{d_s} \cdot 100$$

$$= \frac{11}{40} x. \frac{12}{60} .x \frac{1}{1.5} .x\ 100 = 3.67\ mill\ mhos / cm$$

Sample Calculation 3:

Estimate the leaching requirement when the *EC* of saturation extract of the soil is 15 millimhos/cm at 25 percent reduction of yield of the crops. The EC of irrigation water is 1.5 millimhos/cm.

Solution:

Given EC_i = *EC* of irrigation water = 1.5 millimhos/cm

EC_e = *EC* of saturation extract of the soil = 15 millimhos/cm

EC_d = *EC* of drainage water = 2 x 15 = 30 millimhos/cm

$$LR = \frac{EC_i}{EC_d} = \frac{1.5}{15} = 0.10 = 10\ percent$$

Thus, in order to prevent further rise in soil salinity above 15 millimhos/cm in this particular case, it will be necessary to apply more than 10 percent irrigation water that what the crop needs as evapotarnspiration/consumptive demand.

12.11.7 Cultivation Practice

Complete reclamation of an alkali soil may take several years. It is, therefore, advisable to start cultivation soon after the application of gypsum in the soil. Rice-wheat rotation has been found to be the most effective for alkali soils under reclamation process.

Cultivation on such soils should start with *dhaincha (Sesbania aculeata)* as green manure for subsequent paddy crop. About 40- 45 days old *dhaincha* should be turned down and allowed to decompose for about one week. Then paddy seedlings should be transplanted in the field. High yielding varieties of paddy such as IR-8, Jaya and PR-I06 have been found suitable during initial stages of reclamation. The best time of transplanting is between 15 June and 15 July. Three to four seedlings per hill should be transplanted with hill to hill distance of 15 cm. There is no need to puddle the soil in first year. The crop will require about 150 kg nitrogen and 25 kg zinc sulphate per hectare. There is no need of applying phosphatic or potassic fertilizers in first year. If green manuring has been practiced, only 100 kg N/ha may be enough. After transplanting, 4.5 cm standing water should be kept for a few days. After 10-15 days of transplanting water should be drained out and the field is exposed to fresh air for 2-3 days and then again flood irrigated. Field should be kept submerged until the crop is near maturity.

Paddy should be followed by wheat crop. Wheat varieties such as HD-2009, WH- 711 and HD-1982 are recommended for alkali soils under reclamation. Sowing is done during first fortnight of November as paddy is harvested during first week of October itself and there is enough time left for field preparation. Nitrogen should be given @ 120-140kg/ha. If possible, potash @ 50 kg/ha may also be given. Spraying of urea solution on 45 days old crop has also been found beneficial, especially in water scarcity conditions. Crop should be properly irrigated. First irrigation is done 25-28 days after sowing. It should be noted that light but frequent irrigations are more beneficial.

Growing of salt and flood tolerant varieties during *kharif* and salt tolerant crops during *rabi* and summer is important in salt affected soils in coastal regions. Salt tolerant varieties of paddy acceptable for various submergence situations are to be selected for the area. Chilli, cotton, barley, linseed, sugarbeet and water melons are suitable for *rabi* and summer seasons.

12.12 Dissolved Constituents and Their Function

The constituents of water can be broadly classified into two categories i.e. (i) major constituents and (ii) minor constituents. The cations like sodium, potassium, calcium and magnesium and anions like bicarbonates, carbonate, sulphate, nitrate and chloride are the important major constituents. Boron, silicon, fluoride, nitrate, sulphide, iron, aluminum, phosphate, *pH* and organic matter are some of the important minor constituents.

12.12.1 Major Constituents

Calcium: Calcium is an essential element for plant growth. It is an important constituent of rocks and minerals. Limestone ($CaCO_3$), gypsum ($CaSO_4$), and calcium phosphate ($CaPO_4$) are very useful in agricultural fields. Calcium is found in almost all natural water, soils, plant tissues and animal bones. The chlorides and nitrates of calcium are more water soluble than the carbonates and phosphates. A soil reach in calcium is friable and is structurally good. It permits more water intake into the soil. Water rich in calcium is desirable.

Magnesium: Magnesium is profoundly found in the nature. Dolomite is a carbonate of calcium and magnesium. High concentration of magnesium is found in sea water. Magnesium is also found in the natural water. Hard water is characterized by high concentration of calcium or magnesium or both. Hard water makes the soil soft, porous and allows more water to infiltrate into the soil. Magnesium is an important constituent of green plants.

Sodium: Sodium is an alkali metal. It is a constituent of igneous rock and rock salt. Sea water contains high concentration of sodium of about 10,000 ppm. All the salts of sodium are soluble in water. It is an unique cation of all other cations. When present in exchangeable form in the soil, it creats adverse physical and chemical conditions for plant growth. In moist soil, the exchangeable sodium makes the soil impermeable. When the soil gets dried, it becomes so hard that it is difficult to work. The adverse affect of the cation increases as the soil becomes more dry. Increasing amounts of sodium on the exchange complex makes soil alkaline. Sodium makes water soft.

The adverse effect of sodium in the soil is closely related to the ratio of sodium to total cations in the irrigation water (called as percent sodium) than to the absolute concentration of sodium. Percent sodium is given by the formula (Lenka, 2001) as:

$$Percent\ sodium = \frac{Na}{Ca^{++} + Mg^{++} + K^{+} + Na^{+}} x\ 100\ meq/l \tag{12.17}$$

Increased concentration of sodium in irrigation water increases its concentration in soil solution and makes the soil sodic.

Potassium: Potassium is found in many rocks and is found in all oils. Most of the potassium salts are readily soluble in water except when it in silicate mineral forms. It is found in low concentration in water. Reaction of potassium with soils is similar to that of sodium. But its effect with soil is not much harmful. It is an essential element for plant growth. Potassium toxicity can be decreased by high calcium concentration. High concentration of potassium may induce magnesium deficiency and iron chlorides.

Carbonate: Carbonate is present in the form of lime stone and dolomite. Sodium carbonate is often present in mineral springs but in traces in most of the natural water. Sodium and potassium carbonates are water soluble but carbonates of calcium and magnesium are fairly insoluble in water but soluble in acids. If soluble carbonates are applied in irrigation water to soil, in absence of calcium and magnesium in soil, soil becomes alkaline. Sodium carbonate is not good for irrigation water and is toxic to plants.

Bicarbonates: It occurs in solution in natural water. Calcium bicarbonate is more soluble than the normal carbonate and exists in solution. Calcium and magnesium bicarbonates are present in solution in most of the irrigation water. Bicarbonates are vital for irrigation purpose and tend to maintain an excess of calcium in the soil. Carbonate, bicarbonate, carbon dioxide and cation equilibrium determine the *pH* of water

Sulphate: Sulphate occurs in nature as calcium salt and gypsum. Sodium and magnesium sulphates are readily soluble in water. Calcium sulphate is not readily soluble in water. Sulphate ion increases the soil salinity. Plants readily get sulphur in the form of sulphate.

Chloride: All natural water contains chloride. Chloride content in stream or river water receiving industrial waste is high and may reach toxic level. Sodium chloride is the common salt used in most domestic purposes. Sea water contains high concentration of sodium chloride. Some trees and vine crops show specific toxicity to chloride ion. High percentage of sodium chloride makes soil saline which affects the plant growth.

Nitrate: Principal source of nitrate is in Chile where the material is recovered as impure sodium nitrate. It is not present in igneous rocks. Some small amounts are derived from lightening discharged in the atmosphere where the nitrogen is combined into nitric oxides which when dissolved in rain water forms nitrous and nitric acid. All nitrates are soluble in water and can easily be leached. Oxidation of nitrogenous organic matter and symbiotic fixation of atmospheric nitrogen are the chief sources of nitrate in the soil. Nitrate reach water is good for crop. But high concentration of nitrate in water makes it polluted.

12.12.2 Minor Constituents

Boron: Boron principally occurs as borax, calcium borate or as boric acid.in hot mineral springs. Boron content of natural water may range from less than 0.01 ppm to ore than 30 ppm. In acid water, it occurs as boric acid and in alkaline water as boric acid and tetraborate . Boron is toxic for plant growth. Its concentration in water is low and hence, its affect on plant growth is not much noticed. Its concentration in lower quantity is essential for plant growth and at the same time the high concentration is highly toxic for plants. Irrigation water containing more than 1 ppm of boron may cause injury to more sensitive plants. Not all crops are equally sensitive to a particular level of boron concentration. Table 12.4 represents the relative boron tolerance of different agricultural crops. Toxicity symptoms, however, can appear on almost any crop if concentrations are high enough. Toxicity often accompanies or complicates a salinity or infiltration problem although it may appear even when salinity is low.

Nitrite: It is not present in normal irrigation water. However, under unaerobic conditions, it is found in closed drains and stagnate wells. Nitrite occurs as an intermediate step in the oxidation of organic nitrogen to nitrate. Since it is easily oxidized to nitrate, its harmful effects are not common.

Sulphide: It occurs as pyrite, iron sulphide, galena, and cinnabar. In natural water, it occurs as hydrogen sulphide. Deep wells may contain substantial amount of hydrogen sulphide. In aerobic condition, it is oxidized to sulphate easily and as such has no harmful affect on soil and plants. When associated with sodium and boron, it may be harmful.

Fluoride: It is an essential element for all living beings. But excess beyond certain levels may prove toxic to plants. Excess fluorine in drinking and irrigation water produces dental and skeletal disorders. Fluorine content in excess of 10 ppm in irrigation water is harmful. In arid and semi arid regions of Rajasthan, fluorine content in irrigation water is high and proves to be toxic to plants. Fluorine adsorption is more on kaolinite than on montmorillonite soils and may become toxic in light soils.

Table 12.4: Relative boron tolerance of agricultural crops

Very Sensitive (<0.5 mg/l)	Sensitive (0.5 – 0.75 mg/l)	Sensitive (0.75 – 1.0 mg/l)	Moderately Sensitive (1.0 – 2.0 mg/l)
Lemon	Avocado	Garlic	Pepper, red
Blackberry	Grapefruit	Sweet potato	Pea
	Orange	Wheat	Carrot
Moderately Tolerant (2.0 – 4.0 mg/l)	Apricot	Barley	Radish
Lettuce	Peach	Sunflower	Potato
Cabbage	Cherry	Bean, mung	Cucumber
Celery	Plum	Sesame	
Turnip	Persimmon	Lupine	
Bluegrass, Kentucky	Fig, kadota	Strawberry	
Oats	Grape	Artichoke, Jerusalem	
Maize	Walnut	Bean, kidney	
Artichoke	Pecan	Bean, lima	
Tobacco	Cowpea	Groundnut/Peanut	
Mustard	Onion		
Squash	Tolerant(4.0–6.0 mg/l)	Very Tolerant (6.0 – 15.0 mg/l)	
Muskmelon	Sorghum	Cotton	
	Tomato	Asparagus	
	Alfalfa		
	Vetch, purple		
	Parsley		
	Beet, red		
	Sugarbeet		

Source: Maas (1984).

12.13 Water Quality Standards

To classify irrigation water into various grades with respect to their suitability for irrigation, the important criteria suggested are as follows:

- Sediment concentration in water
- Concentration of potentially toxic elements present in water
- Total soluble salts
- Relative proportion of sodium to other cations
- Residual sodium carbonate content
- Bicarbonate content
- Bacterial contamination

12.13.1 Sediment Concentration

The effects of sediment present in irrigation water depend on the type of irrigated land. Fine sediment carried by the water when added to the sandy soil, it increases its fertility and water intake rate. On the other hand, sediments from eroded soil when gets deposited on the soil, it decreases its fertility, decreases its permeability and water intake rate. Canals get silted up if the sediment laden water moves in the canals. Groundwater, in general, carries less sediment compared to the canal water.

12.13.2 Concentration of Potentially Toxic Elements Present in Water

A large number of elements such as boron, selenium, molybdenum and fluorine if present in excess quantity, may prove toxic to living beings. Boron is good for plant. But its concentration above 0.3 ppm is not good for plant. Concentration above 0.5 ppm is harmful to nuts, citrus fruits, and deciduous fruits. Table 12.4 presents the relative boron tolerance of various agricultural crops Boron is generally present in soap. So, the waste water containing soap should be carefully used in irrigation.

12.13.3 Total Soluble Salts

Whatever may be the source of irrigation water, some soluble salts are always dissolved in it. However, the nature and quantity of dissolved salts depend on the sources of irrigation water, whether river, pond, wells *etc.* and its course before use. The main soluble constituents in water are calcium, magnesium, sodium and sometimes potassium as cations and chloride, sulphate, bicarbonate and sometimes carbonate as anions. Ions of some other elements like lithium, silicon, bromine, iodine, copper, nickel, cobalt, fluorine, boron *etc.* are present in minor quantities. These elements usually do not affect the quality of irrigation water as far as the total salt concentration is concerned. However, some elements like selenium, molybdenum and fluorine, if present in excess quantity, may prove toxic to living beings. Among the soluble salts, calcium, magnesium, sodium, boron, chloride, sulphate and bicarbonate are of prime importance in determining the quality of irrigation water.

Total concentration of soluble salts is better expressed as electrical conductivity. The term electrical conductivity has already been defined earlier. As per the criteria suggested by the United States Salinity Laboratory, salinity of irrigation water may be grouped into four types. They are (i) C_1 (*EC* varying from 0 – 250 micromhos/cm), (ii) C_2 (*EC* varying from 250 – 750 micromhos/cm), (iii) C_3 (*EC* varying from 750 - 2250 micromhos/cm), and (iv) C_4 (*EC* varying from 2250 - 5000 micromhos/cm). Table 12.5 presents the quality of irrigation water based on salinity classes decided as per the *EC* values.

Table 12.5 Salinity classes for irrigation water

Class of irrigation water	$EC \times 10^6$ /cm at 25°C	Salt concentration (gm/l)
C_1 (Low salinity)	0 – 250	< 0.16
C_2 (Medium salinity)	250 – 500	0.16 – 0.50
C_3 (High salinity)	500 – 2250	0.50 – 1.5
C_4 (Very high salinity)	2250 – 5000	1.5 – 3.0

The class C_1 *is* safe for use in irrigation for all most all crops and for almost all kinds of soils If this water is added to soil, there is little chance of development of soil salinity. C_2 needs moderate leaching. Normal salt tolerant crops can be grown without much salinity control. C_3 and C_4 are not suitable for irrigation. However, with adequate drainage facility they may be used to some extent. High salt tolerant crops can be grown with these types of water. Kanwar (1961) has added the fifth class C_5 which has *EC* value 5,000 – 20,000 micromhos/cm.

12.13.4 Relative Proportion of Sodium to Other Cations

Relative proportion of sodium to other cations can be studied by sodium adsorption ratio *(SAR)* and Exchangeable sodium percentage *(ESP).* Both the terms *SAR* and *ESP* have been defined earlier. Eqn. (12.2) and Eqn. (12.3) give the formula to estimate the *ESP* and *SAR*. The relationship between *ESP* and *SAR* is presented as:

$$ESP = \frac{100\left(-0.0126 + 0.01475\, SAR\right)}{1 + \left(-0.0126 + 0.01475\, SAR\right)} \tag{12.18}$$

Subsequently on the basis of cation exchange equilibria for *Na-Ca-Mg* system of five different soils in India, *ESP* is calculated (Lenka, 2001) as:

$$ESP = \frac{100\left(0.23 + 0.0042\, SAR\right)}{1 + \left(0.23 + 0.0042\, SAR\right)} \tag{12.19}$$

Salt concentration in irrigation water is less. But when it is applied to the soil, it dilutes the salt concentration of the soil. The soil solution may have 2 to 10 times more salt concentration than the irrigation water. The *SAR* value can be reduced by adding gypsum to the irrigation water or to the soil.

Table 12.6 helps in evaluating the type of irrigation water based on *SAR* (Garg, 2012).

Table 12.6: Types of irrigation water based on *SAR* and their use in irrigation

Types of water	Values of *SAR*	Use in irrigation
S_1 (Low sodium water)	0 – 10	Can be used for irrigation in all most all soils and all crops except those which are highly sensitive to sodium.
S_2 (Medium sodium water)	10 – 18	Appreciable hazardous in fine textured soils, which may require gypsum etc. but may be used on coarse textured or organic soil with good permeability.
S_3 (High-medium odium water)	18 – 26	May prove harmful to all crops in all soils and do require good drainage, high leaching, gypsum addition etc. for proper irrigation.
S_4 (Very high sodium water)	> 26	Generally not suitable for irrigation.

Depending on the *EC*, *ESP* and *pH* value, the soils are classified as saline, alkaline or saline-alkaline as shown in Table 12.7.

Table 12.7: Classification of saline and alkaline soils

Sl No.	Classification of soil	*EC*, micromhos/cm	*ESP*	*pH*
1.	Saline soil/white alkali	> 4000	<15	$\leq$ 8.5
2.	Alkaline soil/sodic soil/black alkali	< 4000	> 15	8.5 - 10
3.	Saline-alkali soil	> 4000	> 15	< 8.5

12.13.5 Residual Sodium Carbonate Content

Presence of carbonate and bicarbonate in irrigation water results in the precipitation of calcium and magnesium present in the soil and thus increases the extent of sodium hazard. To access the sodium hazard due to carbonate and bicarbonate of irrigation water, it is generally assumed that all the calcium and magnesium would precipitate as carbonates. The residual sodium carbonate (RSC) is expressed (Murty, 2004) as:

$$RSC = \left(CO_3^{\ 2-} + HCO_3^{\ -}\right) - \left(Ca^{2+} + Mg^{2+}\right) \tag{12.20}$$

where the concentrations of ions are expressed in *meq/l*. Water with *RSC* value > 2.5 *meq/l* is not suitable for irrigation. Water containing *RSC* value within 1.25 to 2.5 *meq/l* is marginal, and that containing < 1.25 *meq/l*are suitable for irrigation.

12.13.6 Bicarbonate Content

High concentration of bicarbonate ions may result in precipitation of calcium and magnesium bicarbonates from the soil solution, increasing the relative proportion of sodium ions and causing sodium hazards.

12.13.7 Bacterial Contamination

Bacterial contamination of irrigation water is not a serious problem, unless the crops irrigated with highly contaminated water are directly eaten without being properly cooked. Cash crops like cotton which are processed after harvesting, can, hence use contaminated waste waters, without any trouble.

Several other problems related to irrigation water quality occur with sufficient frequency for them to be specifically noted. These include high nitrogen concentrations in the water which supplies nitrogen to the crop and may cause excessive vegetative growth, lodging, and delayed crop maturity; unsightly deposits on fruit or leaves due to overhead sprinkler irrigation with high bicarbonate water, water containing gypsum, or water high in iron; and various abnormalities often associated with an unusual pH of the water. A special problem faced by some farmers practicing irrigation is deterioration of equipment due to water-induced corrosion or encrustation. This problem is most serious for wells and pumps, but in some areas, poor quality water may also damage irrigation equipment and canals.

In areas where there is a potential risk from diseases such as malaria, schistosomiasis and lymphatic filariasis, disease vector problems must be considered along with other water quality-related problems. Vector problems (mosquitoes) often originate as a secondary trouble related to a low water infiltration rate, to the use of wastewater for irrigation, or to poor drainage.

Suspended organic as well as inorganic sediments cause problems in irrigation systems through clogging of gates, sprinkler heads and drippers. They can cause damage to pumps if screens are not used to exclude them. More commonly, sediments tend to fill canals and ditches and cause costly dredging and maintenance problems. Sediment also tends to reduce further the water infiltration rate of an already slowly permeable soil.

Relationship between *EC* and *SAR* can be studied by Fig. 12.3 as given by the U.S. Salinity Laboratory which helps in classifying the irrigation water. In this figure, the horizontal axis represents the values of *EC* and vertical axis shows the value of *SAR*. From this figure we can find out the class of irrigation water as S_2 if *SAR* is 15 and *EC* is 200. With same value of *SAR* as 15 but *EC* is 500, the water becomes S_3 water. Increased *EC* with same value of *SAR* makes water more sodium hazardous.

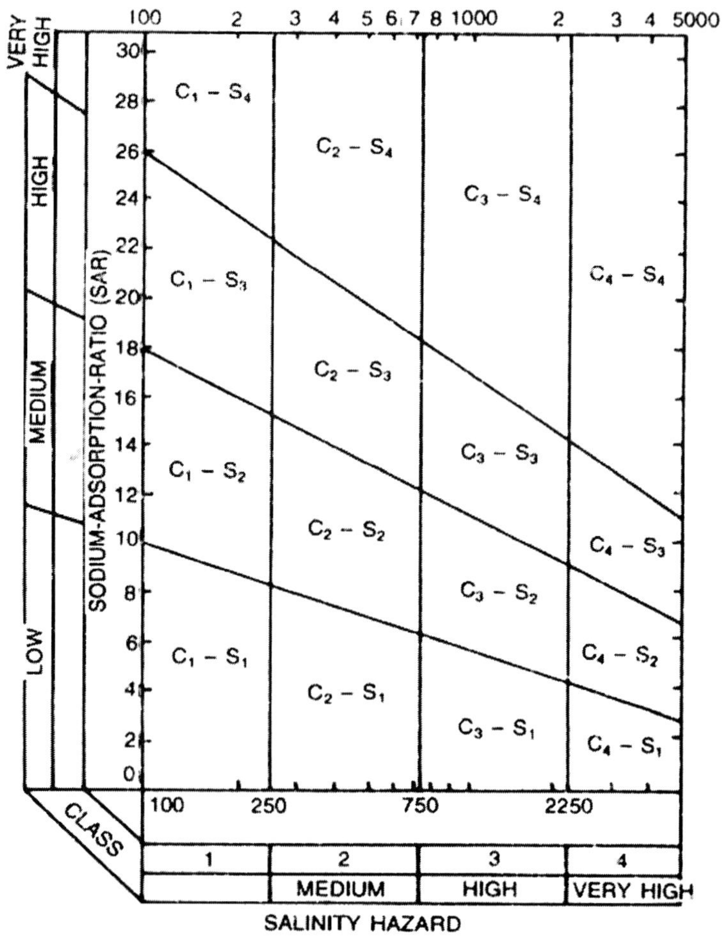

Fig. 12.3: Relationship between *EC* and *SAR*

Sample Calculation 4:

A given sample of irrigation water when tested gives the following properties. Classify the quality of the irrigation water. Concentration of sodium, calcium and magnesium are 25, 3 and 2 milliequivalent/l, respectively. The value of electrical conductivity is 200 micromhos/cm at 25^{0}C. If this water is used for irrigation in fine textured soil, what problem is expected to arise and how can it may be rectified?

Solution:

Given concentration of sodium (Na^+), calcium (Ca^{++}) and magnesium (Mg^{++}) are given as 25, 3 and 2 milliequivalent/l, respectively.

The value of *SAR* is calculated as:

$$SAR = \frac{Na^+}{\sqrt{\left(Ca^{++} + Mg^{++}\right)/2}}$$

$$= \frac{25}{\sqrt{\frac{3+2}{2}}} = 15.81$$

The value of *SAR* lies between 10 and 26 and, so, the irrigation water is medium sodium water (S_2). The value of *EC* given is 200 micromhos/cm and, hence, the water is low saline (C_1) irrigation water. Hence, the given irrigation water is C_1-S_2 type.

In fine textured soils, application of medium sodium water (S_2) may creat the following problems:

(i) Decreases the permeability of the soil.

(ii) Forms crust on soil surface when it becomes dry.

(iii) Soil becomes plastic and sticky when wet.

Gypsum may be added as soil amendment to rectify the problem.

Sample Calculation 5:

Analysis of given water sample gave the following data:

Constituents	*Ca*	*Mg*	*Na*	HCO_3	CO_3	SO_4	*Cl*
Concentration, mg/l	52.1	38.9	340.4	518.6	18.0	205.7	273.8

Value of *EC* of water = 1800 micromhos/cm. What is the quality of irrigation water? Can this water be used for irrigation and if yes, how?

Solution:

The values of the concentrations of various constituents can be converted from *mg/l* to *meq/l* by dividing the concentration of each constituent element in *mg/l* with its equivalent weight. Equivalent weight of each constituent element is calculated as the ratio of its atomic weight and its valency. Data of the concentrations of various elements were converted from *mg/l* to *meq/l* as discussed above and are presented in tabular form as below.

Constituents		*Cations*				*Anions*	
	Ca	*Mg*	*Na*	HCO_3	CO_3	SO_4	*Cl*
Concentration, *mg/l*	52.1	38.9	340.4	518.6	18.0	205.7	273.8
Concentration, *meq/l*	2.6	3.2	14.8	8.5	0.6	5.7	5.8

Value of *SAR* is given as:

$$SAR = \frac{Na^+}{\sqrt{\left(Ca^{2+} + Mg^{2+}\right)/2}}$$

$$= \frac{14.8}{\sqrt{(2.6 + 3.2)/2}} = 8.69$$

Value of residual sodium carbonate, *RSC* is given as:

$$RSC = \left(CO_3^{2-} + HCO_3^{-}\right) - \left(Ca^{2+} + Mg^{2+}\right)$$

$$= (0.6 + 8.5) - (2.6 + 3.2) = 3.3\ meq/l$$

Value of *EC* of water = 1800 micromhos/cm.

Hence, the irrigation water can be classified as C_3 (highly saline) and since SAR <10, it is low sodium water (S_2). Again the value of *RSC* is 3.3 meq/*l* which is very high and so the water cannot be safely used for irrigation.

With application of gypsum and provision of adequate and good drainage facility, the water can be used for irrigation for semi tolerant crops.

12.14 Water Quality Guidelines

Before application of irrigation water, the various water parameters need to be determined in the laboratory. Accordingly the necessary management criteria are to be adopted for use of the water. Table 12.8 gives the lists of some important water parameters that are to be tested in the laboratory along with the usual range.

Table 12.8: Laboratory determinations needed to evaluate common irrigation water quality

Water parameter	Symbol	Unit	Usual range in irrigation water
Salinity/ Salt content/Electrical conductivity	EC_w	*dS/m*	0–3
Total dissolved solids	*TDS*	*mg/l*	0–2000
Cations and Anions			
Calcium	Ca^{++}	*me/l*	0–20
Magnesium	Mg^{++}	*me/l*	0–5
Sodium	Na^{+}	*me/l*	0–40
Carbonate	CO^{--}_3	*me/l*	0– 1
Bicarbonate	HCO_3^-	*me/l*	0–10
Chloride	Cl^-	*me/l*	0–30
Sulphate	SO_4^{--}	*me/l*	0–20
Nutrients			
Nitrate-Nitrogen	NO_3-N	*mg/l*	0–10
Ammonium-Nitrogen	NH_4-N	*mg/l*	0–5
Phosphate-Phosphorus	PO_4-P	mg/l	0–2
Potassium	K^+	mg/l	0–2
Miscellaneous			
Boron	B	mg/l	0–2
Acid/Basicity	*pH*	1–14	6.0–8.5
Sodium adsorption ratio	*SAR*	(me/l)	0–15

Guidelines for evaluation of water quality for irrigation are given in Table 12.9. They emphasize the long-term influence of water quality on crop production, soil conditions and farm management, Water quality guidelines were prepared in cooperation with staff of the United States Salinity Laboratory. The guidelines are practical and have been used successfully in general irrigated agriculture for evaluation of the common constituents in surface water, groundwater, drainage water, sewage effluent and wastewater.

Table 12.9: Guidelines for interpretations of water quality for irrigation

Potential irrigation problem	Units		Degree of restriction on use		
			None	Slight to moderate	Severe
Salinity(affects crop water availability)					
EC_w	*dS/m*		< 0.7	0.7–3.0	> 3.0
TDS	*mg/l*		<450	450–2000	> 2000
Infiltration (affects infiltration rate of water into soil. Evaluate using EC_w & *SAR* together)					
SAR = 0 – 3	and EC_w	=	> 0.7	0.7 – 0.2	< 0.2
= 3 – 6		=	> 1.2	1.2 – 0.3	< 0.3
= 6 – 12		=	> 1.9	1.9 – 0.5	< 0.5
= 12 – 20		=	> 2.9	2.9 – 1.3	< 1.3
= 20 – 40		=	> 5.0	5.0 – 2.9	< 2.9
Specific ion toxicity (affects sensitive crops)					
Sodium *(Na)*					
Surface irrigation		SAR	< 3	3 – 9	> 9

(*Contd.*)

Sprinkler irrigation	me/l	< 3	> 3	
Chloride (*Cl*)				
Surface irrigation	me/l	< 4	4 – 10	> 10
Sprinkler irrigation	me/l	< 3	> 3	
Boron (*B*)	mg/l	< 0.7	0.7 – 3.0	> 3.0
Miscellaneous effects (affects susceptible crops)				
Nitrogen (NO_3 - *N*)	mg/l	< 5	5 – 30	> 30
Bicarbonate (HCO_3)				
(overhead sprinkling only)	me/l	< 1.5	1.5 – 8.5	> 8.5
	pH	Normal range 6.5 – 8.4		

Source: University of California Committee of Consultants (1974).

12.15 Trace Elements in Irrigation Water

In addition to the above mentioned elements, some trace elements may also be present in irrigation water. Their recommended maximum concentration levels in irrigation water are mentioned in Table 12.10.

Table 12.10: Recommended maximum concentrations of trace elements in irrigation water

Element	Recommended maximum concentration (mg/l)	Remarks
Al (aluminum)	5.0	Can cause non-productivity in acid soils (pH < 5.5), but more alkaline soils at pH > 7.0 will precipitate the ion and eliminate any toxicity.
As (arsenic)	0.10	Toxicity to plants varies widely, ranging from 12 mg/l for Sudan grass to less than 0.05 mg/l for rice.
Be (beryllium)	0.10	Toxicity to plants varies widely, ranging from 5 mg/l for kale to 0.5 mg/l for bush beans.
Cd (cadmium)	0.01	Toxic to beans, beets and turnips at concentrations as low as 0.1 mg/l in nutrient solutions. Conservative limits recommended due to its potential for accumulation in plants and soils to concentrations that may be harmful to humans.
Co (cobalt)	0.05	Toxic to tomato plants at 0.1 mg/l in nutrient solution. Tends to be inactivated by neutral and alkaline soils.
Cr (chromium)	0.10	Not generally recognized as an essential growth element. Con-servative limits recommended due to lack of knowledge on its toxicity to plants.
Cu (copper)	0.20	Toxic to a number of plants at 0.1 to 1.0 mg/l in nutrient solutions.
F (fluoride)	1.0	Inactivated by neutral and alkaline soils.
Fe (iron)	5.0	Not toxic to plants in aerated soils, but can contribute to soil acidification and loss of availability of essential phosphorus and molybdenum. Overhead sprinkling may result in unsightly deposits on plants, equipment and buildings.
Li (lithium)	2.5	Tolerated by most crops up to 5 mg/l; mobile in soil. Toxic to citrus at low concentrations (<0.075 mg/l). Acts similarly to boron.
Mn (manganese)	0.20	Toxic to a number of crops at a few-tenths to a few mg/l, but usually only in acid soils.

(*Contd.*)

Element	Recommended maximum concentration (mg/l)	Remarks
Mo (molybdenum)	0.01	Not toxic to plants at normal concentrations in soil and water. Can be toxic to livestock if forage is grown in soils with high concentrations of available molybdenum.
Ni (nickel)	0.20	Toxic to a number of plants at 0.5 mg/l to 1.0 mg/l; reduced toxicity at neutral or alkaline pH.
Pd (lead)	5.0	Can inhibit plant cell growth at very high concentrations.
Se (selenium)	0.02	Toxic to plants at concentrations as low as 0.025 mg/l and toxic to livestock if forage is grown in soils with relatively high levels of added selenium. An essential element to animals but in very low concentrations.
Ti (titanium)	—	Effectively excluded by plants; specific tolerance unknown.
V (vanadium)	0.10	Toxic to many plants at relatively low concentrations.
Zn (zinc)	2.0	Toxic to many plants at widely varying concentrations; reduced toxicity at pH > 6.0 and in fine textured or organic soils.

Source: National Academy of Sciences (1972)

12.16 Irrigation with Poor Quality Water

In areas where irrigation water quality is not good but there is no other choice, then one has to manage with the water. But there will be some reduction in yields. However, the yield potential in such areas can be increased by adopting the following management practices:

- Adopting the proper reclamation procedures for saline and alkaline soil including soil amendments which has been discussed earlier.
- Selection of salt tolerant crops and their varieties. This procedure has also been discussed earlier.
- Optimum fertilizers and manuring
- Proper irrigation management
- Proper soil management
- Proper seed placement

12.16.1 Optimum Fertilizers and Manuring

Soils which are irrigated with poor quality water are generally poor in their fertility status, especially with reference to nitrogen and sometimes phosphorus. Better crops can be grown by raising their fertility status. Nitrogen applied along with manures shows good response on the growth of plants. It has been observed that for wheat, barley, bajra and maize, the usual doses of fertilizers, as applied on normal soils, can be applied upto an EC value of 6.5 millimhos/cm and an exchangeable sodium percentage of about 30. However, excessive fertilization, on a highly saline alkali soil, is of no value.

Among the fertilizers, nitrogen is better than phosphorus followed by potash. Application of urea should be avoided due to its higher rate of nitrogen loss in the gaseous form under saline-alkali conditions. The efficiency of fertilizers on soils irrigated with poor quality irrigation water depends largely on the interactive effects of salts, exchangeable sodium and fertilizers. The most important determining factors are the nature and degree of salinity and sodicity, chemical reactivity and solubility of added fertilizers, nutrient requirements of crops-their availability in the soil and salt tolerance capacity of the crop.

12.16.2 Proper Irrigation Management

While irrigating with poor quality water, accumulation of salts increases with the fineness of soil texture. It is important to follow the irrigate practices such that the salinity at the root zone is kept to the minimum. For this purpose, the quantity of water and frequency of irrigation are so kept that they could meet the leaching requirement of the soil and consumptive use of the crop grown. During the periods of no crop grown, salt accumulates in the top soil layer. Under such condition, a heavy pre-sowing irrigation may help to leach the salts from the surface layer which will help the germination of the crops and improves early growth.

12.16.3 Proper Soil Management

Before using irrigation with poor quality water for a particular site, it is essential to have a detailed laboratory test of the soil with respect to its physio-chemical properties. Soil analysis should include the texture and structure study of the soil, its permeability, *pH,* lime and gypsum content, total soluble salts, their ionic compositions and exchangeable cations. Information on water transmitting properties of the soil and depth of groundwater table has to be collected. Climatic data including the rainfall and evaporation *etc.* are to be collected as they help in assessing the salt build up of an area.

Success of irrigation with saline water lies in the degree of accuracy in the predicted values of soil salinity and sodicity expected to be developed during the cropping period. Saline water can be more safely used in coarse textured soil than the fine ones. Before irrigation is done, the field affected by soil salinity should be properly leveled so that there is uniform spreading of water in the field. Uniform spreading of water will help for uniform leaching of salts from the soil otherwise it will accumulate in the places of high elevation.

Medium textured soil with *kanka*r ayers pose a problem of sodicity. Such soils can be managed well by deep ploughing or by cultivating green crops like dhaincha which will increase the water intake rate into the soil. Application of gypsum under the conditions of low water table may increase the land productivity.

12.16.4 Proper Seed Placement

Low yields are often the result of obtaining poor crop stands during the germination or early seedling stage of growth. Placement of seed is important so that there will be uniform crop stand in the entire fields so that no area is left bare which is likely to be salinized. If salinity is a problem, planting seeds in the centre of a single row raised bed will place the seed exactly in the area where salts concentrate. A double row raised planting bed by comparison offers an advantage over single row raised planting bed. The two rows are placed so that each is near a shoulder of the raised bed, thus placing the seed away from the area of greatest salt accumulation.

By this method higher soil and water salinities can be than with the single single-row planting because water moves the salts through the seed area to the centre of the ridge (Fig. 12.4). A better salinity control can be achieved by using sloping beds with seeds planted on the sloping sides and the seed row is placed just above the water line (Fig. 12.5).

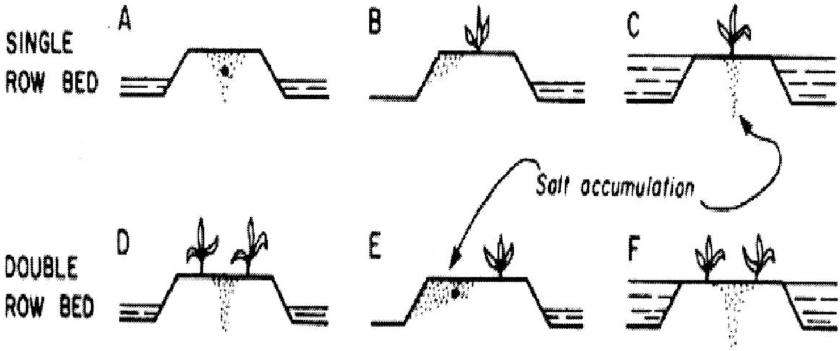

Fig.12.4: Flat top beds and irrigation practice (Bernstein *et al.*, 1955)

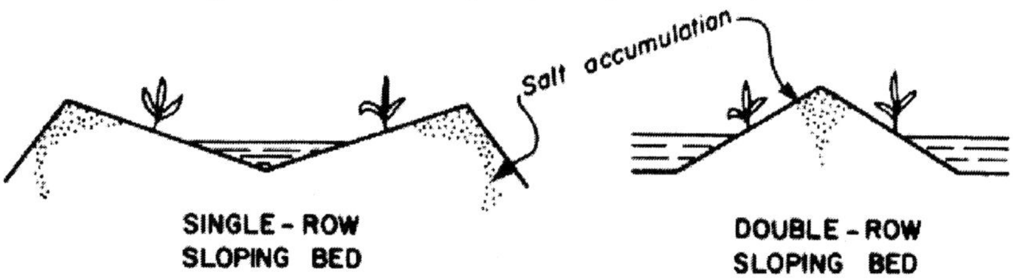

Fig. 12.5: Salinity control with sloping beds (Bernstein and Fireman 1957)

12.17 Quality of Groundwater

Well water forms an important source for irrigation. Its quality depends on the climatological and hydrological conditions. Well water in the arid regions is more saline than the water in the humid region. As recommended by Paliwal (1972), quality of ground water sources can be classified into three categories. They are (i) water quality of arid and semi-arid regions having annual rainfall below 450 mm, consisting of major parts of Rajasthan, Gujarat, Agra and Mathura district of Uttar Pradesh, some parts of Punjab and Haryana, (ii) water quality as influenced by hydrological conditions as high water table, consisting of some areas of Punjab, Haryana, Delhi, Uttar Pradesh and Rajasthan in the canal irrigated areas and river basins, and (iii) water quality of wells in some areas of the coastal regions of the country which are affected by sea water intrusion and waves of sea water *etc.* Of all the factors, aridity is the most important single factor responsible for creation of very high degree salinity of well water.

Continuous use of poor quality ground water may cause calcium deficiency. It also creates poor soil physical conditions, as both sodium and magnesium, damage the soil texture by their highly dispersive action. Groundwater of the arid regions contain more bicarbonate ions. Extensive data on the quality of groundwater of Rajasthan, Gujarat, Uttar Pradesh, Punjab, Haryana, Delhi and other parts of the country show that water of high *EC* value are predominant with sodium and chloride ions, while less saline water have relatively more calcium, magnesium and bicarbonate ions. Potassium and carbonate ions, if present, are mostly confined upto a range of 5 percent of the total salt concentration. Area with high rainfall more than 750 annual rainfall and having good drainage conditions, have better quality groundwater than the arid regions and regions with poor drainage conditions.

Salinity of groundwater is considerable influenced by the climatic conditions. It increases during summer and decreases during rainy season due to dilution of salts. Degree of salination and its ionic composition varies with the depth of groundwater table, infiltration capacity of the soil, and the rainfall characteristics of the area concerned.

Quality of groundwater varies with depth of the aquifers. Quality of water varies from place to place and also at the same place with varying depths of soil. In areas, where the quality of groundwater at the shallow aquifers is bad but at the deeper aquifers good, deep tube wells may be dug to tap the good quality water for use in irrigation.

Question Banks

Q1. What are the sources that make rainwater to be impure?

Q2. Define the following terms:
(i) Electrical conductivity (ii) Alkali/sodic soil (iii) Cation exchange capacity
(iv) Exchangeable sodium percentage (v) Saline-alkali soil (vi) Saline soil
(vii) Sodium adsorption ratio

Q3. The indicator *pH* stands for what? How can you find out the *pH* of a soil and water sample? What is the value of *pH* for neutral water?

Q4. Convert 3500 ppm salt concentration into mhos/cm, millimhos/cm, micromhos/cm and *dS/m.*

Q5. A given soil sample has 3000 ppm calcium chloride salt concentration. What will be its value of the salt concentration in *meq/l.*

Q6. What are the causes of saline and alkaline soil? Discuss each cause in details.

Q7. How does the depth to groundwater table affect the salinity of soil?

Q8. Irrigation is both bone and curse in creation of salinity? Justify the statement with suitable examples.

Q9. How does the soil salinity in the coastal regions differ from that in arid regions? What are the main causes for development of soil salinity in coastal areas? How can the soil salinity be checked in these areas?

Q10. Explain the various effects of irrigation, soil and crop management on soil salinity and alkalinity.

Q11. What are the different categories into which the salt affected areas of India can be classified?

Q12. Discuss the different categories of salt affected areas with their spreads in different parts of the country.

Q13. What are the main processes that occur in soils while irrigating with poor quality irrigation water?

Q14. Discuss how the soil salinity affects the growth of plants.

Q15. Discuss the process of chemistry of salt affected soils.

Q16. Do all crops respond in similar manner to soil salinity? Discuss how the crop yield is affected by different levels of soil salinity.

Q17. Mention the names and varieties of some field crops, vegetables, fruits and forest trees which are tolerant to soil salinity at varying degrees.

Q18. Describe the different techniques for reclamations of salt affected areas.

Q19. What are the various ways by which groundwater table of an area can be lowered down? How does the lowering of groundwater table affect the soil salinisation?

Q20. What is the chemical formula of gypsum? How does it help in reclamation of salty soils?

Q21. Write down the different chemical amendments used for reclamations of salty soils other than gypsum. Write their dozes equivalent to 1 ton of gypsum.

Q22. What does the term leaching stand for? Derive an expression for showing the relationship in terms of depth of water lost as consumptive use and salt concentration of both irrigation and drainage water.

Q23. Determine the change in salinity level of the soil due to evaporation of groundwater over a period of 4 months when the depth of ground water evaporated is 10 cm and electrical conductivity is 11 millimhos/cm. The depth of soil influenced by salt accumulation is 45 cm. Bulk density of the soil is 1.5 gm/cm^3 and saturation percent of the soil is 65 percent. Assume density of water as 1 gm/cm^3.

Q24. Estimate the leaching requirement when the *EC* of saturation extract of the soil is 12 millimhos/cm at 25 percent reduction of yield of the crops. The *EC* of irrigation water is 1.4 millimhos/cm.

Q25. Name some major and minor constituents present in irrigation water and mention their sources of origination and their utility in the field of irrigation to crops.

Q26. What is the role of boron on growth of plants? Name some crops which are very sensitive, sensitive, moderately tolerant, and tolerant to boron.

Q27. What are the main criteria by which irrigation water is classified into various grades?

Q28. Name some potentially toxic elements present in irrigation water.

Q29. Why is the water mixed with detergent not good for use in agriculture?

Q30. What are the various categories into which salinity and sodium hazards of irrigation water are classified. What is the basis of their classifications?

Q31. What is the relationship between *SAR* and *ESP*?

Q32. Write down the values of *EC, ESP* and *pH* for the following soils:

(i) Saline soil (ii) Alkaline soil and (iii) Saline-alkali soil.

Q33. How is the residual sodium carbonate *(RSC)* content determined? Write the values of *RSC* of water suitable for irrigation as well unsuitable for irrigation.

Q34. A given sample of irrigation water when tested gives the following properties. Classify the quality of the irrigation water.
Concentration of sodium, calcium and magnesium are 28, 2.5 and 2 milliequivalent/l, respectively. The value of electrical conductivity is 220 micromhos/cm at 25^{0}C. If this water is used for irrigation in fine textured soil, what problem is expected to arise and how can it may be rectified?

Q35. Analysis of given water sample gave the following data:

Constituents	Ca	Mg	Na	HCO_3	CO_3	SO_4	Cl
Concentration, mg/l	55.1	40.9	350.4	518.6	18.0	205.7	275.8

Value of *EC* of water = 1850 micromhos/cm. What is the quality of irrigation water? Can this water be used for irrigation and if yes, how?

Q36. Mention the important water quality parameters with their usual range present in irrigation water.

Q37. Mention the values of *EC, SAR* and total dissolved salt of irrigation water having no restriction on their uses.

Q38. Name some trace elements present in irrigation water with their recommended values of maximum concentration.

Q39. How can the yield potential of areas irrigated by poor quality water be increased?

Q40. Write a brief note on quality of underground water. How does it vary from surface water?

References

Bandyopadhyaya, B. and Abrol, I.P. 1986. In. Progress in Rainfed Lowland Rice. IRRI, Philippines.

Bernstein, L. and Fireman M. 1957. Laboratory studies on salt distribution in furrow irrigated soil with special reference to the pre-emergence period. Soil Science 83:249–263.

Bernstein, L., Fireman, M. and Reeve, R.C. 1955. Control of Salinity in the Imperial Valley. US Department of Agriculture, Agricultural Research Service Bulletin 41(4). 15 p.

Garg, S.K. 2012. Irrigation Engineering and Hydraulic Structures. Khanna Publishers, New Delhi.

Hansen, V.E., Israelsen, O.W. and Stringham, G.E. 1979. Irrigation Principles and Practices. John Wiley and Sons Inc., New York.

Kanwar, J.S. 1961. Quality of irrigation water as an index of suitability for irrigation purpose. Potash Rev., 13: 1-13.

Lenka, D. 2001. Irrigation and Drainage. Kalyani Publishers, New Delhi.

Maas, E.V. 1984. Salt tolerance of plants. In: The Handbook of Plant Science in Agriculture. B.R. Christie (ed). CRC Press, Boca Raton, Florida.

Maas, E.V. and Hoffman, G.J. 1977. Crop salt tolerance - Current assessment. J. Irrigation and Drainage Division, ASCE 103(IRZ): 115–134. Proceeding Paper 12993

Michael, A.M. 1978. Irrigation Theory and Practices. Vikash Publishing House Pvt. Ltd., New Delhi.

Murty, V.V.N. 2004. Land and Water Management Engineering. Kalyani Publishers, Cuttack.

National Academy of Sciences. 1972. Water quality criteria. United States Environmental Protection Agency, Washington DC. Report No. EPA-R373-033. 592 p.

Paliwal, K.V. 1972. Irrigation with saline water. I.A.R.I., Monograph- 2, Water Technology Centre, IARI, New Delhi.

University of California Committee of Consultants. 1974 Guidelines for interpretation of water quality for agriculture. University of California, Davis. 13 p.

13

Waterlogging and Drainage

13.1 Introduction

In agricultural lands, when the pores of the soil within the root zone of the crop get saturated with the subsoil water due to infiltration of high rainfall, excess irrigation or heavy seepage loss in the unlined canal systems, the air circulation within the soil pores get totally stopped. This phenomenon is called as waterlogging. In canal commands where there is continuous flow of water in the canals and canals are mostly unlined, there is heavy loss of seepage which raises the ground water table. This raising water table sometimes comes up to the ground surface which impedes surface drainage. By capillary action, under ground water rises to the crop root zone and submerges the roots of the crops for most of the time. This situation is called as waterlogging which creates salinisation and alkalinisation problems affecting the plant growth.

The height to which soil water rises above the water table by capillary action is called as capillary fringe. Generally, the height of the capillary fringe varies from 1 to 1.5 m. When the water table comes to 1.5 m below the surface of the soil, the land is said to be waterlogged. At times, due to poor drainage conditions, surface water cannot be freely drained out. The stagnate water in these low lying areas also causes waterlogging which enhances the weed growth consequently affecting the crop yield.

13.2 Waterlogging in Some of the Canal Commands

Irrigation has been considered as a boon in enhancing agricultural production. The assured supply of irrigation water in sufficient quantity and at the time of crop water demand has boosted the production, in both quantity and quality in all most all the irrigations commands. At the time of initiation of the project, not much thought was focused on drainage problems. This overlooked sights in respect of drainage created the problems of waterlogging and drainage gradually in all most all the commands. The irrigation projects which were once considered as boon soon became curse since vast areas became waterlogged and were devoided from crop cultivation. In India, waterlogging in canal command areas is increasing at an alarming rate. In the Tungabhadra project of Karnataka which was commissioned in 1953, more than 33,000 ha irrigated areas have become waterlogged and suffering from salinity. The rate of increase of the waterlogged areas in the state is more than 600 ha per annum. Out of 33,000 ha, about 20,000 ha areas are so badly affected that no crop is cultivated. Rising water table in Mahi-Kadana in Gujarat is responsible for substantial increasing of areas of waterlogging and salinity. Similarly, the Bhakra canal in Haryana has caused serious waterlogging problems in the state. Following the extension of the Sarda Sahayak canal in Uttar Pradesh which brought about additional 4 lakh ha area under irrigation, about 5 lakh ha area of the irrigated command has become waterlogged (Lenka, 2001).

Eastern India is also suffering from worse conditions due to waterlogging and associated problems. Vast areas of Odisha, West Bengal, Bihar, Assam, eastern Uttar Pradesh etc. are suffering from waterlogging. The Hirakud command of Odisha has brought about 20 percent of the irrigated areas under waterlogging. The groundwater table in the command has risen by more than 10 m since inception of the project. Waterlogging and drainage problems have received alarming dimensions in the deltaic regions of the Mahanadi delta. of Odisha. Out of 3 lakh ha CCA, 42,000 ha is unsuitable for cultivation in *kharif*, and 55,000 ha is unsuitable for cultivation in both *kharif* and *rabi* due to ill drainage condition (Anonymous, 2005)..

Waterlogging in the head reaches in the canal command is more serious than the tail reaches. In the head reaches, there is heavy seepage loss in the canals due to high discharge, intense irrigation, rice-rice cultivation and faulty water management practice associated with medium to deep submergence in the rice fields. All these problems have caused seriously to waterlogging. The problem of rising water table is becoming a national problem and a time will come in near future, when the yield of most of the irrigated commands will be so less that most of the nations will face problems of acute food shortage. Hence, though late, it is now to be thought how to tackle this situation and while initiating new irrigation projects, the dimensions of drainage should be thought of.

13.3 Causes of Waterlogging

Followings are the main causes of waterlogging.

- **Excess Irrigation**: Irrigation is essential for growth of crops. Correct irrigation scheduling should be followed while irrigation is provided to the crops. Excess than the actual demand, is harmful for crop. High frequency coupled with over dose of irrigation causes waterlogging. The inundation irrigation system as well as perennial system hastens up the waterlogging problems. Surface irrigation, including flood irrigation, border and furrow methods causes more waterlogging situations than the sprinkler or drip methods. This is because of higher percolation losses caused due to the former methods of irrigation than the later ones.

- **Seepage from Canals**: Most of the canals in irrigated commands are unlined. There is heavy seepage losses in the unlined canals through the beds and sides which move downward, adds to groundwater and raises the water table. The seepage loss is more in canals in full banking than in full cutting.

- **Inadequate Drainage Conditions**: Drainage is the removal of excess water from both the surface of the soil as well as from sub-soil layers. If the drainage facilities are not good and adequate, then the excess water caused either due to heavy rainfall or intense irrigation provided to the field cannot be drained out easily and quickly and thereby, it accumulates on the soil. Accumulation of water over a long period in a site causes waterlogging. In some cases, there is a hard pan development at a lower depth just below the ground surface. In that case, the movement of subsoil water is obstructed which ultimately causes waterlogging.
- **Obstruction in Natural Water Course:** If there is some obstructions in the natural flow path of water, then, water cannot be freely moved. Water stagnation condition occurs ahead of the obstruction and causes waterlogging. Most of the deltaic regions of the river basins face this problem. The high deposit of silts carried by the river water in the delta creates hindrance in safe removal of excess water due to flood water and heavy rainfall which adds to waterlogging problems.
- **Nature of Soil:** The soil having low permeability, like black cotton soil, does not allow more water to percolate through it. So, in case of over irrigation or flood, the water retains in this type of land and waterlogging occurs.
- **Incorrect Method of Cultivation:** If the agricultural land is not properly leveled and there is more undulations, then water cannot be quickly flown out. Water remains stagnant over the soil surface for a longer period leading to waterlogging.
- **Seepage from Reservoir:** Seepage occurs both from the canals as well as from the reservoirs. If the reservoir basins consist of permeable zones, cracks and fissures which were not detected at the time of construction of dam, then these may cause more seepage loss from the reservoir and add to waterlogging.
- **Topography of Land:** If the agricultural land is flat having no country slope and has a lot of depressions and undulations, then water will stand on the soil surface for a longer time which causes waterloging.
- **Absence of Crop Diversification:** If in an area, rice-rice cultivation is practiced for a number of years, then water table in that area will increase. This is because, rice is a more water requiring crop. More amount of irrigation water added to the field raises the groundwater table. Instead of rice-rice cropping system, crop diversification should be followed so that the second rice crop in a year in an area can be substituted by a low water requiring crop like pulses or oil seeds. The non rice crop requires less water and hence eliminates the danger of waterlogging.
- **Occasional Flood:** If the surface drainage gets congested and there is heavy rainfall and occasional flood, then the excess water will not be drained out quickly which creates waterlogging.
- **Faulty Irrigation Management:** Faulty irrigation management may be due to the following reasons.

(i) Incorrect irrigation scheduling

(ii) Continuous flow of water in the canals

(iii) Adoption of plot to plot irrigation instead of irrigation through field channel

(iv) Absence of volumetric method of irrigation measurement

(v) Water application by surface irrigation methods including flooding than by micro irrigation methods

(vi) Maintaining medium to deep ponding in rice field.

13.4 Effects of Waterlogging

The ill effects of waterlogging can be better studied under three categories as mentioned below i.e. (i) Effects on soil, (ii) Effects on crops and (iii) Effects on environment.

13.4.1 Effects on Soil

Waterlogging affects the soil physical, chemical as well as microbial activities in various ways. Some of them are:

- **Reduced soil temperature:** Increased soil moisture content due to waterlogging reduces soil temperature due to increased thermal conductivity and decreased diffusivity but beyond a critical value, diffusivity is less affected. A well drained soil gets warmer faster than a wet soil due to high specific heat of the latter. Increased moisture content also increases soil albedo. Reduced soil moisture due to waterlogging causes poor seed germination, poor seedling growth, decreased growth of roots of the plants, depresses biotic activity, and affects the microbes in the root zones to work satisfactorily. All these affect the yield of crops.
- **Poor soil structure:** Adverse effect of waterlogging on soil structure is due to reduced microbial activities on account of reduction in oxygen concentration and death of microbes. Soils tilling at high moisture content increases shearing stress and decreases soil structure. Under inadequate drainage condition, leaching is hampered and accumulation of salts in the surface as well as sub-soil layer occurs. Increased concentration of sodium disperses soil, results in single grained structure and increases drainage problems.
- **Lack of aeration:** Roots of the plants get submerged due to waterlogging. There are some microbes (*rhizobium bacteria*) present in the roots of the plants, especially in nodules of pulses and legumes. These microbes are beneficial to plants. They help in nitrogen fixation and supply nitrogenous fertilizers to plants. Because of waterlogging, there is poor aeration in the root zones of the plants. Poor aeration affects the activities of the microbes and most of them die because of shortage of oxygen concentration. Oxygen concentration in the soil atmosphere is about 17 to 20 percent. If soil surface is covered by water, or air in the pores of the soil is displaced by water, diffusion of oxygen reduces to a negligible slow rate. Oxygen concentration below 10 percent inhibits plant growth.

Aeration is the most important factor helping in increasing the rooting depth. Poor aeration due to waterlogging decreases the plant growth and affects the yield.

- **Creation of Salinization:** Salinisation and alkalization are the twin problems and occur primarily due to waterlogging. Because of waterlogging, groundwater table rises and sometimes come upto the ground level. Moreover, there is upward rise of groundwater due to capillary action. This brings up the salts into the ground surface. Because of evaporation, water gets evaporated whereas the salt gets deposited on the soil surface. Gradual build up of the salt makes soil salty.

- **Denitrifixation:** Denitrifixation occurs because of the competition for nitrogen by the soil micro-organisms that thrive in saturated soil and reduction in numbers of nitrifying organisms due to lack of aeration. Flooding or waterlogging results in fully or partially anaerobic conditions because the soil micro-organisms use up the soil oxygen faster than the atmospheric oxygen can diffuse in. The supply of oxygen is cutoff due to lack of aeration resulting in the death beneficial bacteria which affects the plant growth and yield.

- **Retards cultivation:** In waterlogged soil, there is high soil moisture content because of which it remains marshy. It is difficult to operate the farming operation and carry on the normal cultivation in marshy/wet soil. Delayed farming operation including the tilling of soil delays sowing of seeds and planting seedling in time which affects the plant growth and affects yield.

13.4.2 Effects on Crops

- **Growth of weeds:** Marshy land, created due to waterlogging, has massive growth of weeds and aquatic plants. The weeds and unwanted aquatic plants compete with the cultivated crops and retard their growth. This process ultimately affects the yield. Weed removal from the marshy land is a costly affair. It needs some extra investment to be incurred otherwise one has to compromise with the yield.

- **Diseased crops:** Waterlogging decreases soil temperature and degrades soil structure. It causes physiological disease to crops. Decay of roots, external symptoms of the foliage and fruits are some of the common problems of diseased crops.

- **Loss of cash crops:** Cash crops require a well drained soil. But waterlogging causes soil to remain moist most of the time and sometimes creates water stagnation condition on the soil surface. Ponded water on the soil surface causes decay of roots and frequent disease to crops. Thus, waterlogging prohibits cash crops to be cultivated. It also affects crop diversification to be adopted in an irrigation command. Wet soil surface does not allow other crops to be successfully cultivated other than rice.

- **Oxygen depletion:** In waterlogged soils, plant roots are submerged with the raising groundwater table which hinders the air circulation in the root zones of the crop. If soil surface is covered by water, or air in the pores of the soil is displaced by water, diffusion of oxygen reduces to a negligible slow rate. Oxygen concentration below 10 percent inhibits plant growth.

13.4.3 Effects on Environment

Waterlogging puts adverse affect on the environment. Marshy land causes mosquitoes and flies to breed which creates deadly diseases like malaria, filarial, sleeping sickness *etc.* affecting all the living beings. It impairs sanitary conditions and affects the environment.

Groundwater table rises because of waterlogging. Soil remains saturated and all the pores are filled with sub-soil water. Because of this situation, when there is high rainfall, less amount of water infiltrates into the soil which causes flood which adversely affects the environment.

13.5 Depth of Water Table

Steps are to be taken up to ameliorate the cause of waterlogging in order to in crease the crop productivity. The excess water present in and above the soil profile need to be drained out beyond a safe depth so that the waterlogging problem is avoided. Except rice which is a water loving crop, other crops cannot withstand standing water. Depth of water table below the root zone which affects the crop varies from crop to crop. Depending on the growth pattern, soil characteristics, growth of roots in the soil, desirable depth of water table varies with the crop. In general, if the water table remains 2 m below the soil surface, its ill effects are less experienced by the crops and soil productivity is not affected. On the basis of water table and its fluctuation, arable lands can be classified into the following four types. They are presented in Table 13.1 (Lenka, 2001).

Table 13.1: Classification of arable lands based on groundwater table

Classification	Description
Good	Water table lies 2 m below ground surface; only for one month, it rises up to 1.5 m below the soil surface.
Medium	Water table lies 2 m below ground surface; only for one month, it rises up to 1.2 m below the soil surface.
Poor	Water table fluctuates between 1.2 to 2 m below ground surface; ground water up to 1.2 m creates salinity problem.
Bad	Water table lies 1.2 m below ground surface, groundwater creates serious drainage problems.

13.6 Identification of Drainage Problems

There are several indicators to know whether a place is having waterlogging and drainage problem or not. Some of them are

- Accumulation of water in depressions here and there for long periods in year
- Barren patches due to accumulation of salts
- High concentration of salts in surface layers and white crustations here and there and
- Shallow water table
- Marshy and swampy land with flies and mosquitoes flying over it

Stunted crop growth, shallow root depth, yellowing of leaves, increased infestation of diseases and pests and delayed crop maturity are a few major symptoms of waterlogging too. Waterlogging problem exists in a place where the value of the hydraulic conductivity is less than 2.5 cm/hr. Sometimes the soil stratum contains impervious layers and exchangeable sodium is more than 15 percent.

13.7 Measures to Control Waterlogging

Anti-waterlogging measures can be preventive as well as curative type.

13.7.1 Preventive Measures

The preventive measures encompass reduction of inflow to the sub-soil water table by the following measures:

- Percolation from rainfall
- Percolation from canals
- Percolation from irrigation in fields
- Sub-soil water flow in groundwater reservoir

These include economical use of irrigation waters, construction of water course, restricted supply of canal waters, cropping patterns, lining of canals and water courses to minimize seepage, improved water management practice, proper irrigation scheduling, scientific water management techniques including water saving irrigation techniques and training to the farmers for optimum use of irrigation water to crops.

13.7.2 Curative Measures

Curative measures constitute extraction of excess water from root zone of the crops to restore the conditions conducive for normal growth of crops, cutting down seepage from canal system *etc.* Brief description of various anti-waterlogging measures is as given below:

- **Arresting percolation from rainfall**: When rain falls on the ground surface, some amount of it moves down as percolation. This percolation component adds to the underground water and raises it causing waterlogging. Rising of groundwater table can be checked if percolation of rainfall is decreased. Provision of surface drainage can drain out excess water from the soil surface and there by percolation component is reduced. Provision of sub-surface drainage will drain out the sub-soil water and rising groundwater and thereby the water table is lowered down. Detail descriptions and designs of some of the surface and sub-surface drainage systems are mentioned later on. in this chapter.
- **Arresting percolation from canals**: Most of the canal sections are unlined and there is heavy seepage loss from the beds and sides of the canal. Seepage losses in unlined canal sections move downward and add to the underlying groundwater table. This process ultimately raises the groundwater table. Lining of the canals and water courses will reduce seepage loss and thereby the rise of the ground water table is checked. Saving in discharge to a tune of 88 percent can be achieved by lining the whole canal systems. The

seepage loss at various sections of the canal distribution system is not the same. Lining the different sections would reduce the seepage loss of varying amount. Losses during the various phases of water conveyance are found to be as shown in Table 13.2 (Sharma, 1987).

Table 13.2: Lining of various sectionsLosses at various sections of canal distribution system (percent)

	Main canal	Distributary	Water course	Field loss	Total loss
Entire system lined	15	7	22	27	71
Only main canal lined	4	7	25	30	66
Canal+ distributary lined	4	2	26	32	64
Whole system lined	4	2	6	42	54
Whole system lined + sprinkler irrigation in field	4	2	6	6	18

Lining of canals is especially important for canals in full banking. If canal runs on high pervious sandy soil strata, then lining is most.

Seepage loss from canals can also be reduced by decreasing the wetted surface area of the canal which is obtained for reduced depth of flow in the canal. Another measure to reduce the canal seepage is to carry water to the crop field by under ground conveyance system. RCC, cement concrete or plastic pipes can be used as under ground water conveyance systems to carry water over long distance with minimum or even without any seepage loss. Although the high initial investment of laying under ground water system is a major disadvantage of the system, yet considering the saving of irrigation water, reduction of waterlogging and salinisation problem and saving of costly irrigated area, lining is recommended.

- **Provision of intercepting seepage drains:** Parallel intercepting seepage drains, at least 15 to 30 m away from the canal banks are effective to lower the water table. The intercepting seepage drains carry away the seepage water from the sub-soil layers and thereby lower the water table.

- **Arresting percolation from irrigation in fields:** Various measures coming under this category are as under:

(i) **Lining of water courses:** Most of the losses in unlined canal systems are due to unlined system of water courses. A sizeable amount of costly irrigation water can be saved by lining the water course and this saved water can be used to increase the irrigated area in a canal command and at the same time increase the crop yield. Moreover, lining of the water courses decreases seepage loss and there by acts anti-waterlogging measures.

(ii) **Controlling intensity of irrigation:** Intensive irrigation at frequent intervals increases the percolation loss of water in the crop fields which increases the water table. Cultivation of rice-rice crop requires more amount of irrigation and hence causes more losses due to percolation in fields. If crop diversification technique is allowed and instead of rice-

rice, rice and a non0rice crop is cultivated in the command, then less amount of irrigation will be required. Less application of irrigation means less increase of the ground water table. If possible, tubewell irrigation may be operated instead of canal irrigation in *rabi* season to deplete the water table.

(iii) **Economical use of water:** Crops should be irrigated with water saving irrigation techniques. This technique will save the costly irrigation water and at the same time minimize the percolation loss and thereby lower the water table. Various water saving irrigation techniques including application of high-tech irrigation method like drip and sprinkler, irrigating the rice fields by intermittent irrigation method, application of irrigation to fields through field channels, adoption of warabandi system of irrigation, proper leveling and grading of crop fields, supplying water to the fields by volumetric measurements etc. are some of the methods by which economical use of water can be done.

(iv) **Pumping from sub-soil:** Construction of shallow and deep tubewells should be done inn the high water table areas and should be pumped at regular intervals so that the groundwater table will be lowered down. In canal irrigated areas, tubewells may be installed beside the canals and should be pumped out to lower down the water table.

(v) **Seepage drains:** Most economical arrangement to lower high sub-soil water table, as anti-waterlogging measure, is the construction of 1.0 to 1.5 m deep seepage drains in the beds of the surface drains. These seepage drains are constructed for storm water disposal and are provided with suitable outfalls.

(vi) **Natural drainage rivers:** During monsoon, rising rivers charge water table along countryside but act as drainage in winter when rivers run low with minimum flow.

(vii) **Seepage-cum-pumping:** In the areas which are not suitable for shallow tubewell pumping due to soil conditions *etc.,* deep drains are dug to lower the water table. In case suitable gravity outfall is not available for such drains, water is pumped to higher level and better put into canals for irrigation.

(viii) **Local reservoirs:** Small local reservoirs like farm pond, tank *etc.* may be dug in the lowest part of the farm having waterlogged patches. These local reservoirs will collect surface runoff from the fields which may be reused for irrigation. At the same time, they collect seepage water from the underground low groundwater reservoir and there by reduce waterlogging problems.

(ix) **Cropping patterns:** Crops with high rate of evapotranspiration may be cultivated in the waterlogged areas which will extract more water from the groundwater reservoir by capillary action and deduce the water table. Another important point to remember is that if the area is not waterlogged but there is chance to have it in near future, then cropping pattern in that area has to be changed. Instead of rice-rice crop, diversified cropping systems should be advocated.

(x) **Bio-drainage:** Deep rooted trees should be planted in the waterlogged patch which will extract more underground water and decrease the water level. Trees like Cineraria draw a good quality groundwater and reduce the uprising groundwater level. Eucalyptus tree has roots extending up to 3 m but transpire water at a high rate and thus works as a biological pump. It is as such suitable as anti-waterlogging measures to take care of seepage along the canals. Bio-drainage is effective if there is adequate aeration in the root zone and the roots extend up to groundwater reservoir and draw water from the capillary zone.

(xi) **Artificial groundwater recharge:** Sometimes, there may be a barrier in the sub-soil regions which does not allow water to move down. This creates perched water table and localized waterlogging problems in an area. The hard pan of the sub-soil layer should be punctured to facilitate the percolation and push the groundwater level deep down and there by localized waterlogging problems would be avoided.

(xii) **Cleaning the drainage congestions:** In deltaic region, there is silt deposit in the river beds which retards the drainage flow from the natural drains including the rivers to the seas. Sometimes culverts, roads, buildings etc. are constructed along the path of natural drainage which hinders in smooth movement of water to the drains. These congestions should be cleared so that there is quick disposal of water in the drains.

(xiii) **Land leveling and grading:** The land should be leveled well and be provided with some grade so that there will be no obstruction to impounding water. If the field is having some undulations, then water will remain ponding in the filed here and there which will percolate down and raise the groundwater reservoir.

13.8 Land Drainage

Surface irrigation has both blessings and curse on agricultural lands. It is a blessing only if it is practiced with great care. Only optimum amount of water is to be applied to the crop in accordance with proper irrigation scheduling. Excess water that the root zones of the crop fail to retain moves down and add to the underground water and raises the water table. Sometimes the percolated water cannot move down due to an impermivious stratum like a hard pan layer in the sub-soil zone and stands over it forming a perched aquifer. This perched aquifer creates problems in drainage. Optimum agricultural production depends on provision of proper drainage system of most of the soils having problems like high water table, high salt content *etc.* There are some areas in the irrigation commands where irrigation is not required during monsoon season but there is need for drainage.

In canal irrigated areas, implementation of drainage schemes need to be taken up simultaneously with irrigation schemes so as to avoid the problems of waterlogging and development of salinisation and alkalinisation problems later on. Thus, drainage as a means of disposal of excess water is necessitated due to various factors such as (Michael, 1978):

- Water accumulation on the land surface is usually the result of heavy precipitation and/or over bank spill combined with deficiency of drainage system.
- Excess water on the land surface may be the result of waterlogging or stagnation in depressions and low areas.

- Rise of sub-surface water due to excessive irrigation and percolation from other sources.
- Water accumulation on land can also result due to seepage from canals and ponds and sub-soil flow from lands situated at higher places.
- Water stagnation due to obstruction in the natural flow of drainage as a result of construction of roads, rail lines, buildings *etc.*

13.8.1 Objectives of Drainage

Objectives of drainage are followings:

- Drainage tackles the situation of waterlogging as an anti-waterlogging measure.
- Keep water table at sufficient lower depth to prevent upward flow and evaporation of soil water which results in concentration of soluble salts in the root zone of soil.
- Prevents accumulation of excess soluble salts on the surface of irrigated lands and in the root zone of soil.
- Provide proper aeration in the root zones of the crops,
- Facilitate growth of roots so that the plants can extract more soil water from deeper zones.
- Maintains soil temperature and structure in optimum conditions for better growth of crops in soils.
- Increases leaching and drain out the harmful salts from the crop root zones.
- Reduce flood hazards.
- Improve quality and market value of land.
- Helps in smooth farming operation.
- Increases microbial activities and in nitrification.
- Improve sanitary and environmental conditions.

13.8.2 Essential Requirements of Drains

Essential requirements to be satisfied by a drain are as under:

- Admit all the flood discharge from the catchment.
- Allow water to flow unobstructed at a higher rate towards the drain.
- Have a proper capacity drain to carry away the received water to the outfall.
- Outfall conditions should be ideal.
- The land where drains are to be laid should have less weed infestation problems.
- Initial as well as maintenance cost should be low.

- Soil where drains are to be laid should be stable, no-silting and capable of avoiding sloughing of side slopes.

13.8.3 Points to be Considered for Planning a Drainage System

In understanding the drainage problems of an area and planning for drainage systems, following points need to be considered.

(i) Period and amount of water standing on the soil surface during rainy season.
(ii) Number of days for which soil is workable for agricultural operations.
(iii) Salinity of top soil layer at the time of sowing/planting.
(iv) Seasonal salinity of the root zone.
(v) Average depth of water table during the cropping season and number of days it is at critical level of the crops.
(vi) Average depth of water table during non-cropping season.

In addition to the above mentioned points other factors to be considered are:

(i) Intensity and frequency of rainfall of different durations
(ii) Infiltration capacity of the soil
(iii) Permissible depth and duration of submergence which varies from crop to crop and
(iv) Economics of drainage system.

13.9 Classification of Drains

Drains may be classified into different categories as given in chart below:

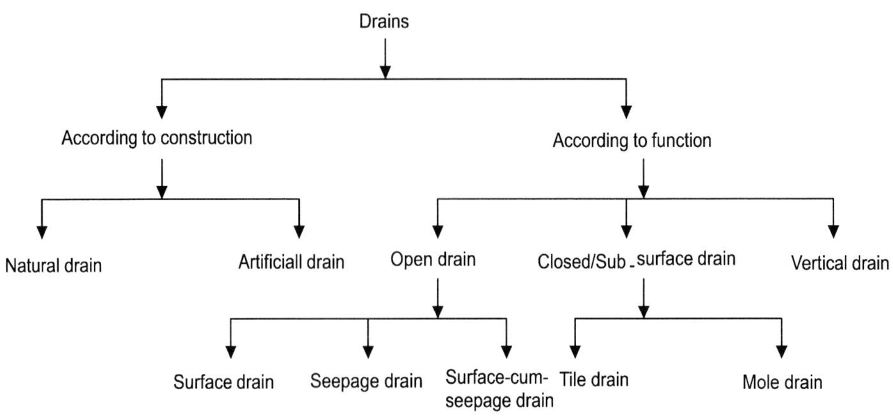

Natural Drains: These are the lowest valley line between two ridges. They occur naturally and are not man made. The drainage lines, rivers, nallahs *etc.* serve as natural drains.

Artificial Drains: These are man made structures. They are constructed on alignment along the drainage line. They are sometimes constructed across the valley to reduce length of the drain or to have proper outfall conditions.

Open Drains: Open drains are generally 1.0 to 1.5 m deep to cater for storm water, for lowering water table and reducing sloughing of the sides. Open drains are suitable and advantageous under the following conditions:

- For removing large quantities of surface water and also sub-surface water where there is low water table.
- Where slope is too flat for proper installation of tile drains.
- For lands having low value
- Where there are no satisfactory outfalls for tile drains.
- Where plant roots tend to clog the tile drains.

Surface Drains: These drains are also called as storm water drains and are used to dispose of surplus rain water so that there is no standing water on the cropped lands for more than minimum period as is not detrimental to the crops. They are also used to dispose of excess irrigation water form the field. They remove water before it has entered into the soil. Whether constructed for the purpose or not, deep surface drains with bed level below the water table also drain out seepage water from the sub-soil.

Seepage Drains: Seepage drains are constructed along the canal sides and are used to drain out the seepage water from the sub-soil layer and thereby reduce rising of water table. These drains are deep enough such that they are below the water level in the ground and has gradient of water flow towards the drain. Hence, ground water form the surrounding aquifers enter the drains and move in downward direction towards a natural outfall. Seepage drains are smaller in cross section as compared to the surface drains. They are constructed about 15 to 30 m from the toe of the canal banks.

Surface-cum-Seepage Drains: These drains serve the dual purposes of seepage and storm water drain and have bed level below the water table. During rainy season, they carry storm water and in non monsoon season, they carry seepage water. A cunnette is usually provided to cater for the small seepage water.

13.10 Surface Drainage

Safe disposal of excess water from the land surface through land shaping and grading and improved or constructed channels is called as surface drainage. Surface drainage problems occur mainly in flat or nearly flat areas, uneven surfaces with undulations, ridges creating hindrance in safe disposal of surface water and in areas without having suitable outfall. Soils with low infiltration rate and heavy soils are susceptible to surface drainage problems. Sub-surface drainage is not well feasible in deep heavy soils having low infiltration capacity and also for sub-soil layers having low hydraulic conductivity. In that case, surface drainage system is feasible.

13.10.1 Surface Drainage Systems

In surface drainage systems, water flows with a hydraulic gradient because of gravity. The system can be considered in three functional parts i.e. (i) collection system, (ii) conveyance or disposal system and (iii) outlet. Water from the individual fields is collected through the collections system and moves through the disposal system to the outlet.

In irrigated areas, the collection system consists of field drains and the conveyance system consists of intermediate and main drains. The location of the elevation of the outlet decides whether the disposal of collected water should be by surface drainage condition or by pumping (Fig. 13.1).

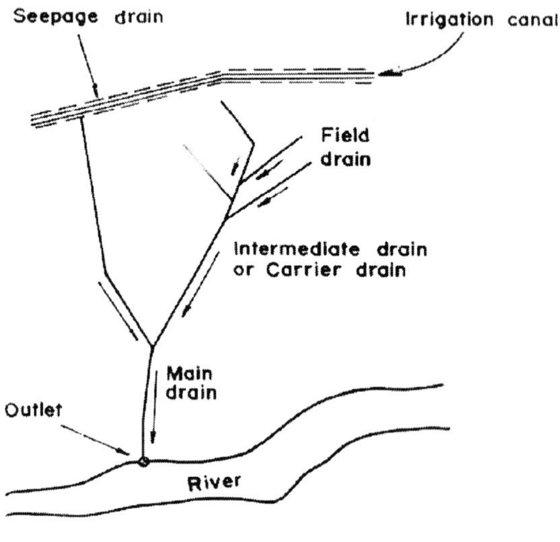

Fig. 13.1: Collection of water by surface drainage system

For quick and safe disposal of surface water, the land should be leveled without any undulations. If the filed is uneven and have depressions here and there, then it should be graded and leveled by using farm implements. Surface drains work better if the land has level shape with a small slope towards the outlet.

13.10.2 Types of Surface Drainage System

There are four types of drainage systems used in flat areas (slope less than 2 percent). They are (i) random drain system, (ii) parallel field drain system, (iii) parallel open ditch system and (iv) bedding system.

13.10.2.1 Random Drain System

This system is used where small scattered depressions or low spots are present over a field and they contain water. Shallow field ditches are constructed in the fields to collect water from these depressions and convey the same to lateral ditches. The laterals connect to the main outlet ditch perpendicularly as shown in Fig. (13.2). The drains connecting the depressions may be surface drains or underground pipe lines or sub-surface drains.

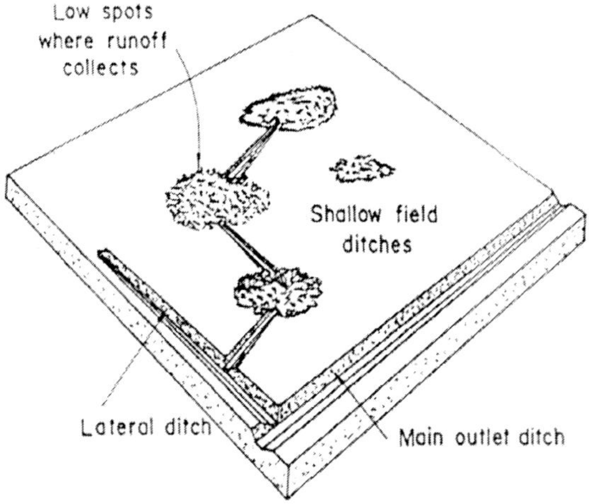

Fig. 13.2: Random drain system

13.10.2.2 Parallel Field Drain System

It (Fig. 13.3) is the most effective method of surface drainage system and is well suited for both rainfed and irrigated areas. In this system, individual field is properly graded such that it discharges water into the field drain/ditch. The field drain/ditch is also called as collector drains. The collector drains collect water from each field and disposes it to the lateral drains/field laterals which are constructed at right angle to the collector drain. The lateral drain, in turn, disposes the water to the main drain which is also constructed perpendicularly to the lateral drains. Laterals and main drains are bigger in cross section than the collector drain and have depth more than the collector drains.

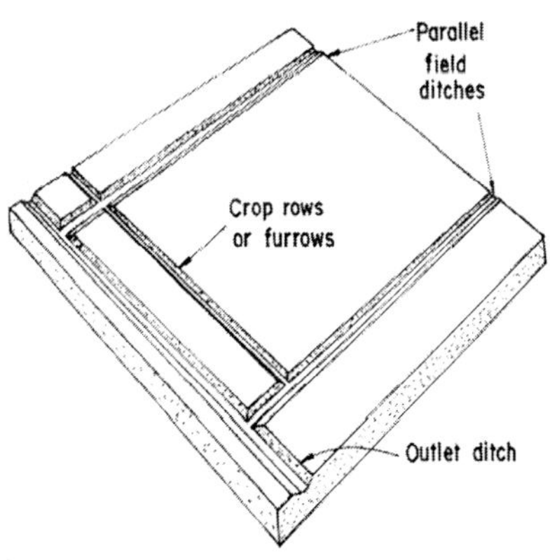

Fig. 13.3: Parallel field drain system

13.10.2.3 Parallel Open Ditch System

This system is applicable in areas which need both surface and sub-surface drainages. It is similar to parallel field drain system, except that in this system, the drains are replaced by open ditches which are generally deeper and have steeper side slopes than the field drains (Fig. 13.4). The advantage of replacing drains by ditches is that they can collect seepage water along with the surface water from the field and dispose them of. The groundwater table is lowered down and crop root zone is relieved from the problem of waterlogging. But the disadvantage is that they create hindrance in movement of farm machineries and animals. They also require large land area for their construction which is a waste for cultivation. The spacing of the ditches depends on the soil and water table conditions and may vary from 60 to 200 m. This system is also called as diversion ditch system.

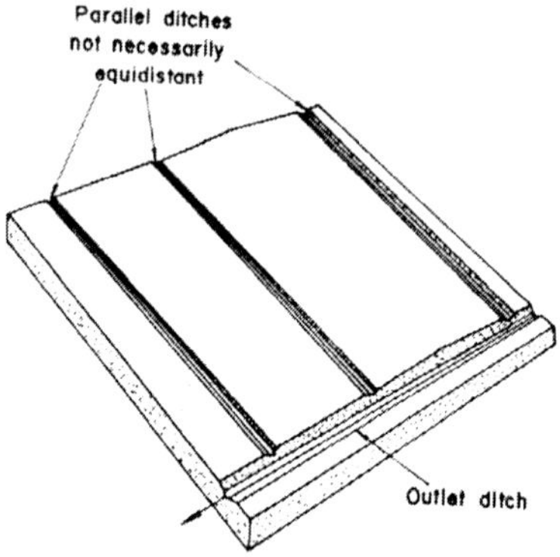

Fig. 13.4: Parallel open ditch system

13.10.2.4 Bedding System

Bedding system for land drainage is essentially a land forming process. The land is ploughed into beds, separated by shallow furrows which run in the direction of prevailing slope of the land (Fig. 13.5). Ploughing is done parallel to the furrow and all other farming operations can

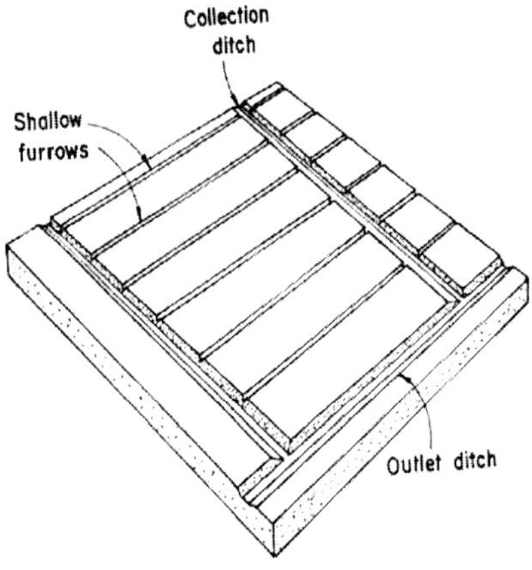

Fig. 13.5: Bedding system

be done done either parallel to the furrow or across the beds. Bedding is suitable for lands having slope up to 1.5 percent. The bed width depends on the land use, slope of the field, permeability of the soil and farming operations. The recommended bed width for various values of permeability of the soils is given in Table 13.3 (Murty, 2004).

Table 13.3: Recommended bed width of bedding systems

Permeability type	Value of permeability, cm/day	Bed width, m
Very low	0.5	8 to 12
Low	5 to 10	15 to 17
Good	10 to 20	20 to 30

Length of the bed depends on field conditions and may vary from 100 to 300 m. Maximum bed height is 20 to 40 cm. Water from the shallow furrow discharge into the field drains which are constructed at the lower end of the field and normal to the shallow furrows. The field drain discharge into field laterals which in turn discharges into main drains.

There are some demerits of this system. They are (i) furrows require regular maintenance, (ii) due to movement of top soil, there is some reduction in yields nearer the furrows could occur and (iii) the slope of the furrows may not be enough for drainage.

13.11 Open Ditches

Open ditches are easy to construct and so are mostly used for surface drainage. They can perform simultaneously the work of sub-surface drains also. The advantages of open ditches are that (i) they are easy to construct either with machines or manually, (ii) can carry large amount of water and (iii) are cheap compared to sub-surface drains. The disadvantages are (i) loss of more land for cultivation, (ii) obstruction of farming operations, (iii) require more maintenance cost and (iv) under unstable soil conditions, difficult to construct.

13.11.1 Design Criteria of Open Ditches

Some of the design considerations of open ditches have already been discussed earlier in the subsection "Points to be considered for planning a drainage system". They are (i) intensity, frequency and duration of rainfall which affects the design discharge of the ditch, (ii) depth and duration of submergence which varies from crops to crops, (iii) economic considerations and (iv) environmental aspects. Other design criteria are discharge computation which is to be handled by the ditch, alignment, capacity of the drain to carry the design discharge, safe velocity of flow to be carried by the ditch and stable side slope of the ditch. These design criteria are discussed in details below.

13.11.2 Design Discharge for Open Drains

The rate at which the open ditch has to remove the water from an area depends on rainfall, size and shape of the catchment/drainage area, characteristics of the drainage area, and nature of the crops to be grown including degree of protection needed for them from flooding. Intensity, duration as well as frequency of the rain affect the design discharge. The runoff computation procedure generated from the catchment which will be carried by an open ditch has been discussed in Chapter 3. Different methods like "Rational method", SCS Curve Number Method used to estimate the runoff rate has been discussed in Chapter 3. In flat areas (slope less than 1 per cent), consideration has to be given to the time interval for removing a certain volume of excess surface water occurring with a certain probability. The runoff for open ditch design may be expressed as drainage coefficient.

The concept of drainage coefficient is used for design of drainage system for agricultural lands. It is defined as the depth of water in cm or inch to be removed from a drainage area within a day i.e. 24 hours time period. It can also be expressed as the flow rate per unit area i.e. cumec per sq. km. Drainage coefficient is decided such that there is no appreciable loss caused to the crop grown in the area due to water. For open ditches for small agricultural areas, the value of drainage coefficient ranges from 0.6 to 2.5 cm and in extreme case the value can go up to 10 cm.

Some of the methods except Rational formula and SCS Curve Number method used to estimate the design discharge of open ditches are mentioned below.

Cypress Creek Formula: This formula in *F.P.S.* system is:

$$Q = C\ A^{p} \tag{13.1}$$

where,

Q = design discharge, ft^3/sec; A = drainage area, sq. mile; C = coefficient depending on the degree of protection and p = coefficient having a value 5.6.

This formula is empirical in nature and has been derived from data of a number of observations in U.S.A. The value of C is to be determined with local data for computation of design discharge by Eq. (13.1). The value of C is related to rainfall excess (*Re*) by the relationship as given below (Raadsma and Schulze, 1974):

$$C = 16.39 + 14.75\ \mathrm{Re} \tag{13.2}$$

Eq. (13.2) is used for south eastern part of U.S.A. The value of *Re* in Eq. (13.2) is considered in inches and is determined from the maximum 24 hours storm.

Boston Society Formula: In this method, the design discharge is given as:

$$Q = C\ A^{1/2} \tag{13.3}$$

where,

Q = design discharge, ft^3/sec; A = drainage area, sq. mile and C = coefficient.

Uppal and Sehgal (1965) gave values of C (Table 13.4) for use in Eq. (13.3) to compute the design discharge. These values are applicable for the drainage areas in Punjab and Haryana.

Table 13.4: Values of *C* for use in Boston Society Formula

Drainage area, sq. mile	Values of *C* for different values of rainfall		
	Rainfall < 20 inches	Rainfall 20–30 inches	Rainfall 30-40 inches
0 – 25	20	50	100
25 – 100	400	100	250
100 – 250	100	200	450
> 250	200	400	600

For rainfall above 40 inches, a value of *C* equal to 2500 irrespective of the drainage area is suggested.

Simplified Hydrologic Procedures: The rainfall data is analysed for frequency-duration. As drainage is to be planned taking crop into considerations, the rainfall duration-frequency for the particular crop season is required. The rainfall excess is calculated and an allowance is made for channel storage. Knowing the drainage coefficient, the capacity of the drainage system needed is calculated. Pai and Hukkeri (1979) recommend that the drains should be designed for average of 3 day maximum rainfall for 5 years return period with a runoff percentage, calculated after taking into infiltration rate, evaporation and permissible storage of water at the fields or by rainfall-runoff observations. If no dependable data are available, runoff percentage of different types of soils and intensity of vegetations as given in Table 13.5 are assumed (Pai and Hukkeri, 1979).

Table 13.5: Recommended values of runoff percentage for different types of soils

Types of soil	Values of runoff percentage
Sandy soil	20
Loamy soil (largely cultivated)	30
Loamy soil (lightly covered by crops)	40
Gardens and cultivated areas	5- 25

The period of removal of excess water from the crop field depends on the types of the crops as well as stages of their growth. Table 13.6 gives the values of recommended drainage periods for different crops (Pai and Hukkeri, 1979).

Table 13.6: Recommended drainage periods for some crops

Types of crop	Period of disposal, days	Remarks
Paddy and sugarcane	7	Panicles in case of paddy not to get submerged and depth of ponding for temporary period not to be more than 0.5 m.
Pearl millet and cotton	3	3 days rainfall to be removed in 3 days
Sorghum and maize	2	2 days rainfall to be removed in 2 days
Vegetables	1	1 day rainfall to be removed in 1day

13.11.3 Capacity of the Drain

The open ditches should be so designed that its area of cross section is adequate to carry the design discharge. For drainage ditches, the most convenient cross section for excavation and maintenance is the trapezoidal section. A typical cross section of the open ditch is shown in Fig. 13.6.

The design of the ditch section is done by knowing the design discharge it has to carry and using Manning's formula to estimate the velocity of flow in it. The design of cross section of the ditch is almost same as that of design of irrigation channels and grassed waterways. The differences in design of irrigation channel and ditch are that most efficient hydraulic section design of channel section cannot be used in ditches and the depth of computations of the ditch cannot also be fixed as deeper depths are required to get the desired grade. Moreover, if the ditch serves as outlet for sub-surface drains, then its depth will be still more.

Side slopes of the ditch should be stable so that it does not slough in while water flow in it. The side slope depends on the type of the soil. Sandy soils are more unstable than the heavy soils like clay and hence require larger side slopes. Suggested values of side slopes for different soils are given in Table 13.7 which can be used as a general guide line for design of cross section of the ditch. For loose rock and hard soil the side slope required is 0.5:1.

Table 13.7: Suggested values of side slope for open ditches

Types of soil	Values of side slope (Horizontal: Vertical)
Heavy clay and alluvial soil	1:1
Clay and silt loam	1.5 :1
Sandy soil, sandy loam and deep black cotton soil	2:1
Loose sand	3:1

The grade given to the ditches depend on the general slope of the ground and outfall conditions. The grade should be such that the velocity of flow in the channel is non-scouring and non-silting. The grade should also permit the outflow of the drainage water into a natural outlet where such facility is available. The grade should also take into consideration the velocity of water in the channel which should be more so that it is not obstructed by the weeds in the channel. This may be a bit more from the point of view of scouring action. Table 13.8 presents values of safe velocities of water in ditches in different soils (Murty, 2004).

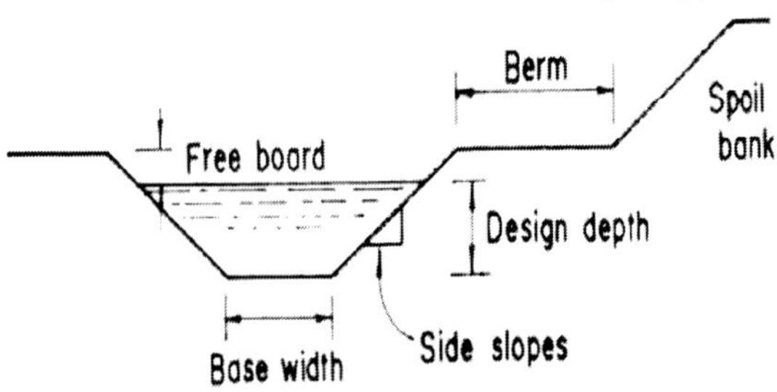

Fig. 13.6 Typical cross section of an open ditch

Alignment: Factors governing the alignment of a surface drain are as under:

(i) It is aligned along the drainage line i.e. lowest contours in the areas.

(ii) It does not traverse through ponds, marshy areas *etc.* but is connected to them through subsidiary drains.

(iii) The alignment is laid at the deepest point and ordinarily passes through as many local depressions.

(iv) It should not preferably cross the irrigation channel otherwise cross drainage system has to be adopted.

(v) Length of the drain should be the lowest.

(vi) It is down the general slope of the country for higher efficiency. Drains aligned down the slope are usually much more effective than those excavated normal to the direction of the slope of the ground.

(vii) It should not pass through habited areas as far as possible.

Table 13.8: Permissible velocities in non-vegetated channels

Soil types of the channel	Velocity, m/sec		
	Clear water	Water transporting colloidal silts	Water transporting sand, gravel *etc.*
Fine sand, colloidal	0.45	0.75	0.45
Sandy loam, non-colloidal	0.53	0.75	0.80
Silt loam, non-colloidal	0.60	0.90	0.60
Alluvial silts, non-colloidal	0.60	1.05	0.60
Ordinary farm clay	0.75	1.05	0.68
Stiff clay, very colloidal	1.13	1.50	0.90
Fine gravel	0.75	1.50	1.13
Coarse gravel, non-colloidal	1.20	1.80	1.95
Cobbles and shingles	1.80	1.80	1.50

Drainage ditches serving individual fields may be straight but drainage ditches conveying water over long distances will have some bends. At these bends, there should gradual curvature along the direction of drains. The radius of curvature (R) is expressed by the degree of the curve (D) (Eq. 13.4), which is defined as the angle subtended at the centre of a circle by a 30 cm chord

$$R = \frac{15}{Sin\ D/2} \tag{13.4}$$

The recommended range for drainage ditches is from 4 degrees from large capacity ditches and ditches with steep side slope to 20 degrees for small capacity ditches and ditches with flat side slope.

For aligning the drains, the reduced distances, *RDs*, are marked along the drain; the zero *RD* being at the outfall end and then the values of *RD* starts increasing upstream.

Minimum bank width of 1.5 to 2.0 m is adopted in case of drains with the exception that earth required for providing the required bank width, if not available from excavation, is borrowed from inside the drain section. The bank width is generally provided in one side and the side contributing the flow may be left unbanked. Where heavy spilling is expected to occur at both the sides, continuous embankments are provided on both sides of the drain. Regulated inlets are provided, wherever necessary, to allow the outside water to enter into the drain. A free board of maxium 20 percent of the depth of flow in the drain may be allowed.

Sample Calculation 1:

A drainage basin of 1000 ha is discharging through a drain at an average rate of 3 m^3/sec. Calculate the drainage coefficient. If the drainage coefficient is 3 cm, what will be the discharge to be carried by the same drain form the same area as given above?

Solution:

Volume of discharge carried by the drains in 24 hours time period = (3 x 24 x 60 x 60) m^3

Area of the drainage basin = 1000 ha = 10^7 m^2.

Drainage coefficient =Volume of discharge/ Area = $\frac{3\,x\,24\,x\,60\,x\,60}{10^7} = 0.0259\,m = 2.59\,cm$

If the drainage coefficient is 3 cm, then rate of flow through the drain is calculated as:

$$= \frac{3}{100}\,x\,\frac{10^7}{24\,x\,60\,x\,60} = 3.472\,m^3\,/\,\text{sec}$$

Sample Calculation 2:

A drainage ditch draining 10 ha of land flows at a design capacity for three days following a storm. If the system is designed using a drainage coefficient of 2 cm, how much water will be removed during this period?

Solution:

Drainage coefficient *i.e.* depth of flow to be carried by ditch in 24 hour = 2 cm = 0.02 m

Area of the drainage = 10 ha = 10^5 m^2.

Volume of drainage water to be carried by the ditch in 24 hour/ 1 day = 0.02 x 10^5 m^3

Time required to drain = 3 days

Hence, volume of water to be drained in 3 days = 3 x 0.02 x 10^5 m^3 = 6,000 m^3.

13.11.4 Procedure for Design of Ditch Section

The following steps may be used to design stable trapezoidal ditch section (Murty, 2004).

Step I: Decide what magnitude of the peak discharge/design discharge is to be carried by the channel section.

Step II: Assume certain value of permissible limiting velocity to be handled by the channel section (Table 13.8) and Manning's n for particular bed materials

Step III: With assumed velocity, Manning's n and bed/longitudinal slope, hydraulic mean radius (R) is computed by using Manning's formula.

Step IV:. Compute the value of R in terms bed width b and depth of flow d by using the relation that $R = A/P$ for the given side slope of the channel.

Step V: With the assumed velocity, work out the cross sectional area A of the channel section i.e. $A = Q/V$

Step VI: Also compute cross sectional area A in terms of bed width b and depth d for the given side slope of the trapezoidal channel section.

Step VII: The two equations in terms of b and d are solved simultaneously to find out the values of bed width b and depth of flow d.

Step VIII: Determine Lacey's f ($f = 2.5\ V^2/R$) to check the stability of the section. The Lacey's f calculated as above should not be less than 1.2.

Sample Calculation 3:

Design a trapezoidal ditch section having fine gravel bed material with side slope 1.25:1 to carry discharge of 257 cumecs. Bed slope of the ditch section is 1:6360 and Manning's n is 0.018. Water transports colloidal silts.

Solution:

Ditch has fine gravel bed material and it transports colloidal silt.

Hence, the value of the permissible velocity is1.80 m/sec (Table 13.8).

Calculate A by the formula $A = Q/V = 257/1.80 = 142.78\ m^2$.

Using Manning's formula and taking $n = 0.018$, $V = 1.80$, and $S = 1/6360$, R computed is 4.15 m.

Calculate P by formula $P = A/R = 142.85 /4.15 = 34.42$ m.

Now, $$142.85 = b\ d + 1.925\ d^2 \quad \text{(i)}$$

Also $$34.42 = b + 3.85\ d \quad \text{(ii)}$$

Solving Eqns. (i) and (ii), $d = 6.55$ m and $b = 9.25$ m.

Lacey's $f = 2.5\ V^2/R = 2.5 \times (1.80 \times 1.80)/4.15 = 1.95$ which is > 1.2 and hence channel is stable.

So, the bed width of the ditch is 9.25 m and depth of flow is 6.55 m. assume free board = 20 percent of d = 1.30 m.

13.11.5 Construction of Ditch Drain

Methods adopted for construction of open ditch depend mainly on the size of the drains and availability of the construction equipments. Small field ditches can be constructed manually. If the size of the ditch is large, then mechanical equipments are required. Some of the mechanical equipments used to excavate soil for ditch are (i) drain plough, (ii) crane shovels and (iii) scrapers.

The drain ploughs are V-shaped tools drawn using a crawler tractor. This is used to construct small ditches. The crane shovel consists of a lower travel unit which can be either wheel type or crawler type. Mounted on the lower travel assembly, it has an upper revolving frame. Various forms of end attachments are mounted on the top of the upper revolving frame. While there are several types of front end attachments, the three common types used to excavate the ditches are (i) clam shell, (ii) dragline and (iii) backhoe (Fig. 13.7).

13.11.6 Maintenance of Open Drains

Open drains work satisfactorily for many years if maintained properly. Open drains run over a larger distance like field channels used for irrigation. It passes over the fields of a number of farmers. Therefore, all the farmers should give due care for the maintenance of the drains

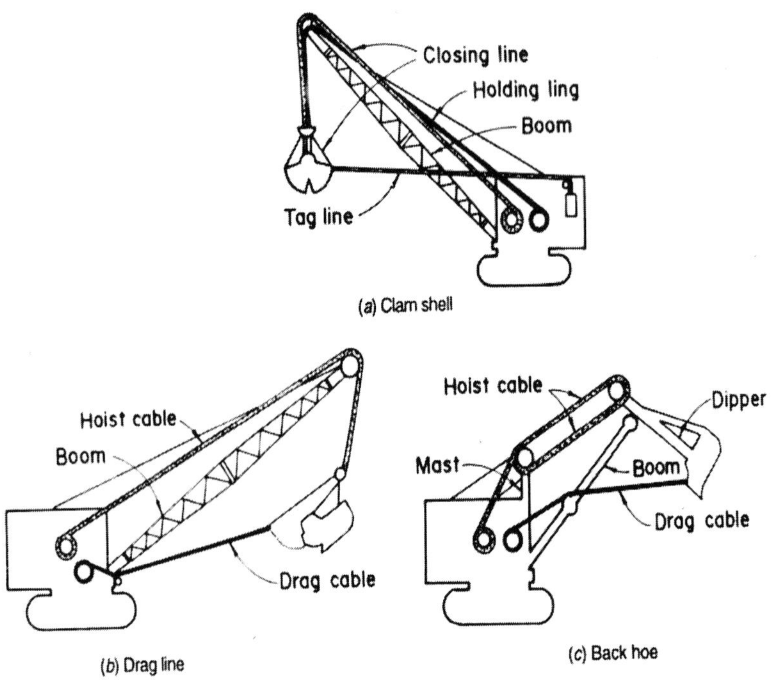

Fig. 13.7: Tools of front end attachments

passing over their own fields. That is instead of individual attention, attentions on maintenance should be on community basis. After initial construction, if the drains are not maintained properly, then there may be problems of (i) sedimentation, (ii) weed growth in the drains, (iii) bank erosion and (iv) damage to side slopes. The drains should be cleaned periodically to remove the silts which are deposited in the ditches because of silt laden water flowing in them. If not cleaned, capacity of the ditch decreases gradually and the ditches cannot carry the drainage water properly. Minimizing the watershed erosion can decrease the sediment inflow in the ditches. Construction of simple drop structures at the entry point of runoff water into the drains can also be done. These structures retain sediment to some extent on upstream side of the structure. Due care scouring and silting should be given in the ditch sections where there is abrupt change in grades of the channel and bends exist at the curves. Construction of settling basins and low check dams are recommended for such situations.

There is a general problem of weed growth in the drainage channels. The weeds decrease the flow velocity and hence carrying capacity of the drains. Weeds can be removed by (i) manually, (ii) mechanically or (iii) using chemicals. Chemical weed controls consists of spraying weedicides regularly. Removal of vegetations can be achieved using flame burners if the vegetation is green or by just ordinary burning if the vegetation is dry.

13.12 Sub-surface Drainage

Closed drains or underground drains are the sub-surface drains laid deep in the ground and then covered. They are use mainly to lower the capillary surface and water table below the ground. Agricultural lands affected by high water table generally require sub-surface drainage. Surface drains remove the excess rainwater before it reaches the root zone. Sub-surface drains lower the water table and provide a better environment including aeration in the root zone. Drainage by gravity with open drains is the simplest way of disposal of surface water. In high cost lands, open drains are expensive to construct. Besides they create no hindrance to movement of animals and farming operation as is encountered in surface drains. Closed drains are suitable for removing the water from under the soil into open drains and led away by normal methods of gravity drainage or pumped to disposal depending on the layout of the topography. Closed drains are located at a suitable depth and grade below the ground surface, usually placed in a medium of high permeability for lowering the water table to a greater depth. In low permeable stratum, suitable filter of coarse sand and shingles is provided around the drain. These drains are more expensive to construct than the surface drains.

13.12.1 Advantages of Sub-surface Drainage

Sub-surface drainage has the following advantages.

- It lowers down the water table and thereby provides proper aeration in the root zone for maximum development of plant roots.
- Increases soil structure and maintains higher temperature in the root zones and thus helps the soil micro-organisms to work better.
- Lowering of the water table decreases the soil moisture and helps in availability of the

land for early cultivation.

- It also facilitates easy movement of farm machineries in the fields because of low soil moisture.
- Removes undesirable salt from the root zones of crops, decreases waterlogging and salinity problems.
- Increases more water to enter into the soil and thereby decreasing the chances of flood hazards.
- Occupy no surface land and so makes more area available for cultivation. It is especially beneficial in high value land.
- Compared to large capacity drains as required in surface drainage method, small capacity drains are required.
- In surface drains, scouring of bed and sloughing of the sides have to be taken care of. But in sub-surface drains, no such conditions arise as the drains are permanent section.
- Compared to surface drains, it needs less maintenance cost.

13.12.2 Disadvantages of Sub-surface Drainage

Followings are the various disadvantages:

- It requires high initial investment.
- It requires steeper gradient for lay out in the under ground than the surface drains laid on ground surface.
- Repair works are costly and inconvenient.
- Being laid under ground, it cannot drain out the excess water standing on the soil surface. Only seepage water is removed and thus takes more time to remove the surface water.
- Construction is difficult and mostly is constructed by mechanical means. Requires greater care while laying the drains under ground.

13.12.3 Sub-surface Drainage Methods

In sub-surface drainage, water moves under the influence of gravity to suitable outlets. This is achieved by the following ways.

- Tile drains including perforated pipes
- Mole drains
- Drainage wells
- Deep open drains and
- Combination of tile and open drains.

Tile drains are the most efficient and permanent type of sub-surface drainage for areas where water table has permanently risen very near to the ground surface. Tile drains consists of short length pipes called tiles (30 to 90 cm long) installed at a particular depth generally ranging from 1 to 1.5 m depth below the ground surface and at suitable spacing and grade depending on the soil, climate and topography of the area. The pipes are made of concrete or burnt clay. The installed depth is generally below the effective root zone of crops. After digging the trench to the desired depth, the pipes are held end to end without any jointing. Water enters the pipe through the ends of the pipes kept close to each other and flows along the grade of the pipe. A network of tile drains laid with a grade will remove the sub-surface water and lower down the under ground water. The pipes, after they are laid inside the trench, are covered with an envelope material in certain cases and the soil is then backfilled.

Perforated pipes are just like tile drains except that they are continuous and water enters the pipes through the openings or perforations made on the body of the pipe. The pipes used in this case are *PVC*. The procedure of laying the system in the trench under ground is same as described above for tile drains.

Mole drains are cylindrical channels formed at a desired depth below the ground surface. There is no lining material and the inherent stability of the soil at the depth gives stability to the mole drains. Water enters the moles and moves along the grade to an outlet point. Special farm machinery equipments like mole plough and mole plug are used to make mole drains. It is similar to tile drainage in layout and operation which is subsequently discussed. Mole drainage is suitable for particular type of soil as the stability of the channel is most important in this method of drainage. Clay soils are generally suitable. The life span of mole drainage is 10 to 15 years depending on the structural stability of the sub-soil. Depth of mole drains vary from 45 to 120 cm and diameter varies from 7.5 to 15 cm. Smaller size moles are more durable than the bigger size moles.

Drainage wells serve the purpose of vertical drainage. Deep wells are constructed depending on the stratum and water level. These wells collect water from the surrounding aquifers and as a result the lowering of water table occurs. The wells may be pumped at regular interval so that water level in the well is pushed down. Drainage wells also collect water from the main and lateral drains of surface drainage system including open ditches which are pumped out by a pump. This way, it helps as an outlet structure for the open ditches. Drainage by wells depends on factors like aquifer conditions, water quality and soil condition. Hydraulic conductivity of the aquifer material where well are dug should be high so that more water will move towards the draining well. There should not be any intervening layer of low hydraulic conductivity impeding the movement of water. Drainage wells are most suitable in irrigated areas where the pumped water could be used as a source of irrigation.

Deep open drains can be used for sub-surface drainage. They, however, need more land area and constant maintenance. Open drains can be used both as interceptor and relief drains. The use of deep open drains as interceptor drain is more common as their length required is generally less than the relief drains and they can be used for intercepting both surface and sub-surface flows. The capacity of the open drain, when it is intercepting only the ground water, can be estimated by Darcy's law. Deep open wells are constructed at greater depth. These well are open at the soil surface like surface drains. They collect both surface and sub-surface water and helps in drainage.

At certain places combination of both tile and open drains are made to quickly remove the water from surface and sub-surface soils and thereby lowering the water table at a quicker time. This is used in the places where there is standing water on the soil surface and at the same time there ground water nearer to soil surface, thus necessitating the use of both open drains and tile drains.

13.12.4 Investigation for Sub-surface Drainage

Following information is required for design and installation of sub-surface drainage.

- Topographic map of the area where drainage works will be carried on.
- Data of soil physiochemical properties including texture, hydraulic conductivity, drainable porosity, soil salinity and alkalinity *etc.*
- Position of ground water table and its fluctuation.
- Quality of groundwater.
- Logs of soil and sub-soil materials.
- Crops to be grown and their drainage requirement.
- Irrigation practices and their requirements.

Topographic map helps to find out the land slope, undulations on the land surface, position of possible outlets, existing drainage patterns in the area, etc. and serves as a base map for preparation of bathymetric maps. Data on soil salinity and alkalinity is required to ascertain if surface drainage system is needed along with reclamation of such soils. Position and fluctuation of groundwater table will help to decide what should be the depth of tile drains to remove the sub-surface water. Logs of soil and sub soil materials are used to decide what type of drainage system is needed to be set up. It also helps to find out the hydraulic conductivity of the soil obtained from the log chart that is required for calculating the drain spacing of the tile drains.

13.12.5 Groundwater Studies for Planning Sub-surface Drainage

Three parameters of groundwater studies are required for planning of installation of sub-surface drainage. They are (i) depth and fluctuation of groundwater table, (ii) quality of groundwater and (iii) hydraulic conductivity of the sub soil material. Data on groundwater table can be measured by installing piezometers and observation wells. Piezometers and observation well are 2.5 to 5.0 cm size pipes and are inserted into the ground through auger holes. In piezometrs water enters into the pipe through bottoms only whereas in observation wells water enters into the wells both through bottoms and sides (Fig. 13.8).

Observations of water table levels in pezometers located at different depths indicate the existence of hydraulic pressure causing either upward or downward movement of water. Fig. 13.9 shows various movement of groundwater depending on the position of hydraulic pressure in the piezometers. From the data of these movements, one can say whether there is natural drainage (Fig. 13.9, a), capillary rise of water (Fig. 13.9, b) or water is moving both in upward and downward direction at a given location (Fig. 13.9 c and d).

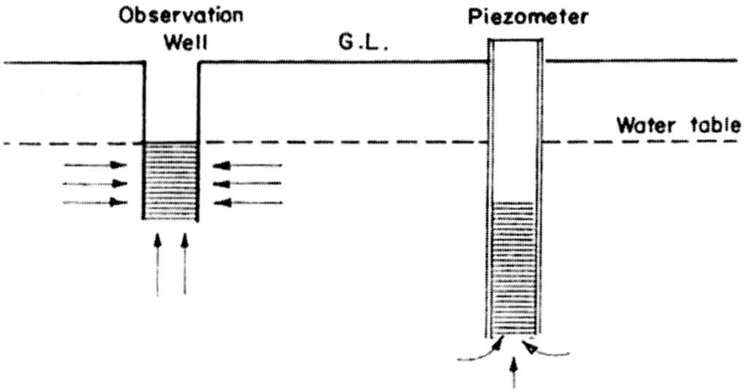

Fig. 13.8: Observation well and piezometer

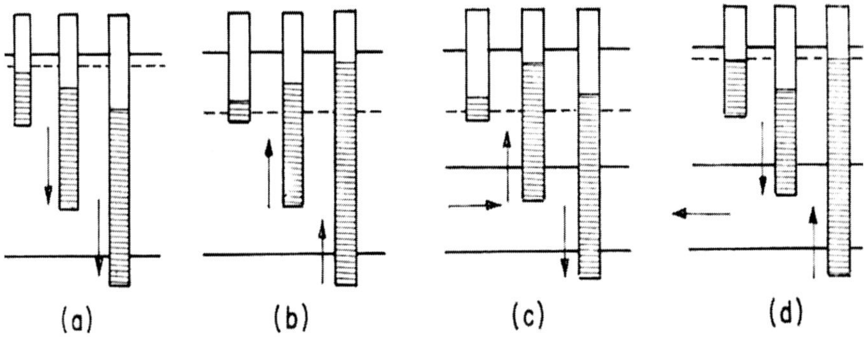

Fig. 13.9: Groundwater movement studied by piezometers

Data of depth of water table obtained by the observation wells in a grid pattern are used to prepare groundwater contour map (Fig. 13.10, a) and water table isobath map (Fig. 13.10, b) of the area.

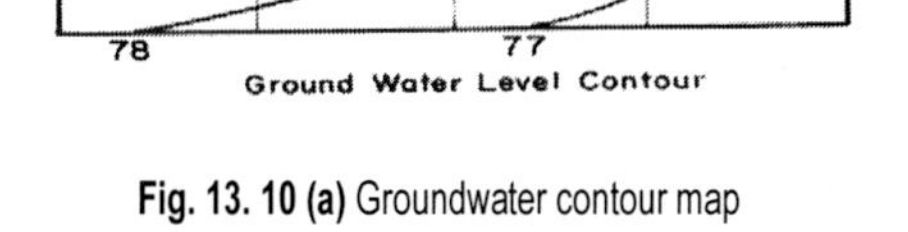

Fig. 13. 10 (a) Groundwater contour map

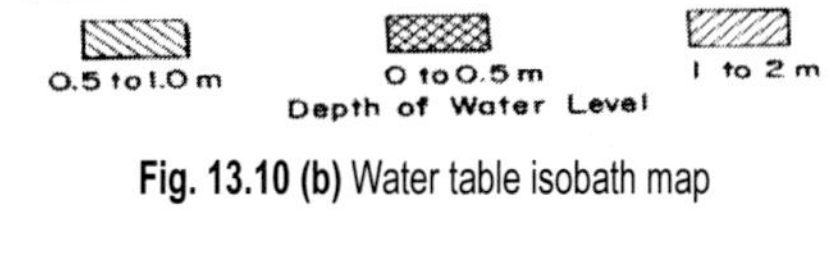

Fig. 13.10 (b) Water table isobath map

The water table contour maps are useful in ascertaining the direction of groundwater flow and location and extent of high water table areas. This map will help in lay out of the sub-surface drainage system in an area. Groundwater isobath maps indicate at a glance the areas affected by high water table problems. Prepared at regular intervals, these maps indicate the seasonal variations of affected areas. In areas where sub-surface drainage systems are installed, the isobath maps show the effectiveness of drainage system in different parts of the areas.

13.12.6 Hydraulic Conductivity

One of the important parameter used in sub-surface drainage design is the hydraulic conductivity. Hydraulic conductivity can be determined in the laboratory by disturbed soil samples. But in drainage design, it is desirable to take the field measured values of hydraulic conductivity. In the field, hydraulic conductivity of the undisturbed soil sample is determined which is more correct than the laboratory methods. Constant head permeameter and falling head permeameter are the two important laboratory methods used to determine the hydraulic conductivity. In the field, hydraulic conductivity of the soil sample can be calculated by the following methods.

- Pipe cavity method
- Using Hooghoudt's equation
- Cylinder permeameter method
- Pond-infiltration test

Procedures for determination of hydraulic conductivity in laboratory as well as in fields are beyond the scope of this book and can be read from references including (Murty, 2004).

13.12.7 Design of Tile Drain Systems

The design of tile drain systems consist in determining

- Layout of the system
- Depth and spacing of the drain
- Size of the drains
- Materials of tiles
- Grade of the tile lines
- Envelope materials and accessory structures and
- Outlet

13.12.8 Layout of Tile Drains

Drainage system comprises a main drain, its branches and subsidiary drains. Tile system is so devised as to cater for all the wet areas that could eventually be drained into one main drain. Different types of layout of tile drains are as follows (Sharma and Sharma, 2002):

- Natural system
- Parallel system
- Random system
- Herringbone system
- Grid iron system

- Double system
- Grouping system
- Intercepting system
- Composite system
- Sink hole drainage system
- Zigzag system

Natural system: It consists of a system of drains, similar to trunk system, but covering a much larger area (Fig. 13.11). The main drains are located along the depressions or low spots to conform to topography. Natural system is generally installed in areas of rolling or broken topography where drainage of isolated tract is required.

Parallel system: Parallel line system consists of a system of drains with long parallel laterals emptying into a single main drain (Fig. 13.12). It is used in poorly drained soils having uniform texture and little slope.

Herringbone system: Herringbone system is used in areas where the main or sub main lies lie in a narrow depression and laterals enter from both the sides and join at staggered manner (Fig. 13.13). On account of double drainage occurring where the laterals and mains join, it is less economical.

Grid iron system: In this system, the field drains are constructed in parallel lines along the direction of the slope which join the sub main and main at the bottom (Fig. 13.14). It is used in flat land with a uniform slope towards main and sub main.

Random system: It is used in rolling areas that have scattered wet areas slightly isolated from each other. The tiles are laid more or less at random to drain the wet areas (Fig. 13.15). The main is located at natural drainage line and individual wet spots are connected through sub-mains and laterals. Where wet spots are large, the arrangement of the sub-main and laterals for each wet place may utilize one or more of the parallel systems to provide required drainage.

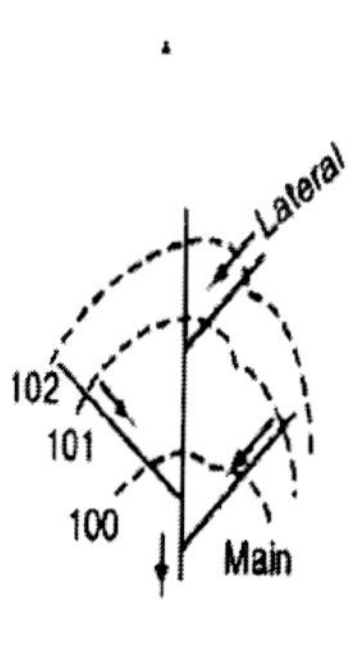

Fig. 13.11: Natural System

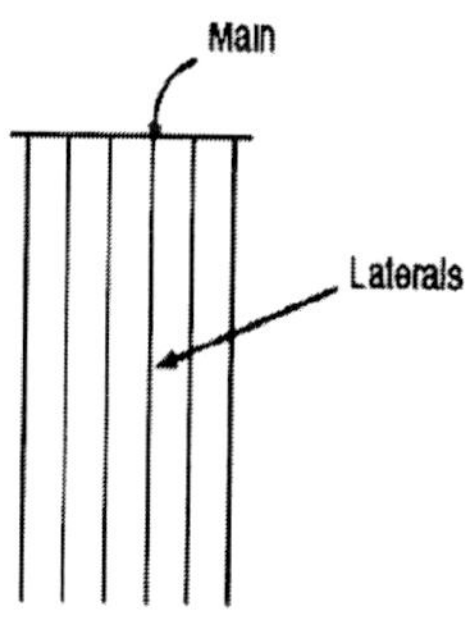

Fig. 13.12: Parallel System

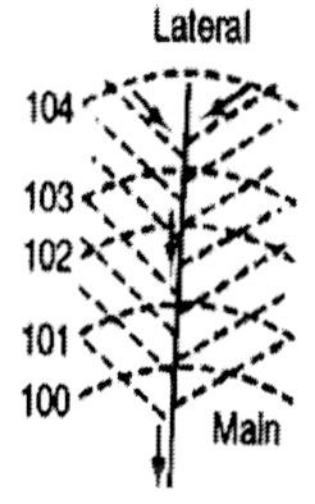

Fig. 13.13: Herringbone System

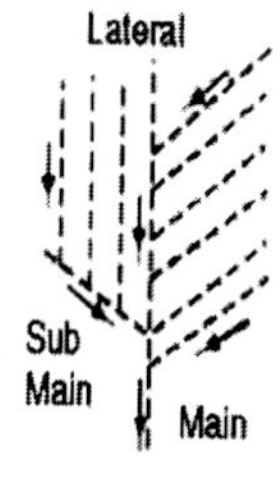

Fig. 13.14: Grid iron System

Intercepting system: In this system, tiles are placed along the hillside to intercept the seepage water that follows the upper surface of an impervious sub soil to prevent it entering into the bottom land (Fig. 13.16). The seepage plain is located and the tile line is placed approximately at the impervious layer along which the seepage water flows. It is used where the main source of drainage is from a hilly land.

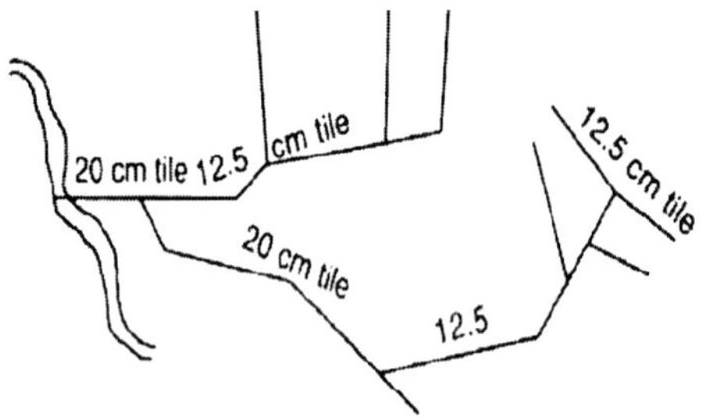

Fig. 13.15: Random system

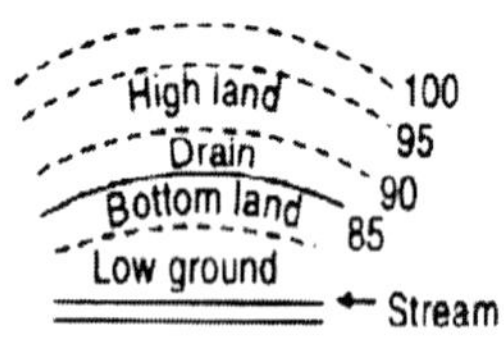

Fig. 13.16: Intercepting system

Double system: It is a system of drain similar to Herringbone system except there are two main drains in this system on each side of the depression (Fig. 13.17). It is used when the bottom of the depression is wide and sub main is a broad, flat depression, which frequently is a natural water course. It is not generally used as conditions which it needs are seldom met with in practice.

Grouping system: It is similar to the natural system except that a few laterals are provided in wet areas or ponds along the system (Fig. 13.18). The hypothesis behind working of this system is that it collects water from a number of small systems and discharges into the ditch through one outlet. It is used where the topography and wetness in the field vary.

Composite system: It is a combination of both Herringbone and grid iron system of tile drains (Fig. 13.19).

Sink hole drainage system: It is a system of drainage used to intercept seepage flow but has in addition wells dug at regular intervals to let water come up from a lower stratum and enter the drain (Fig. 13.20).

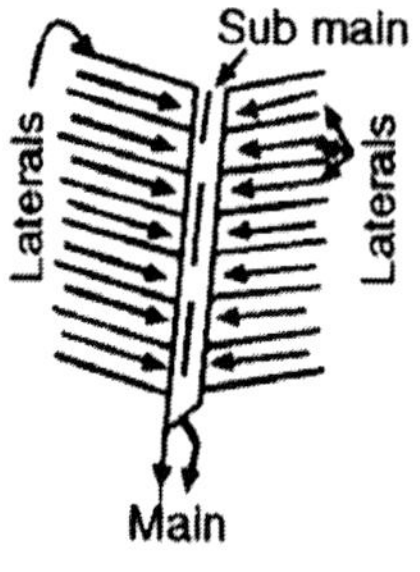

Fig. 13.17: Double system

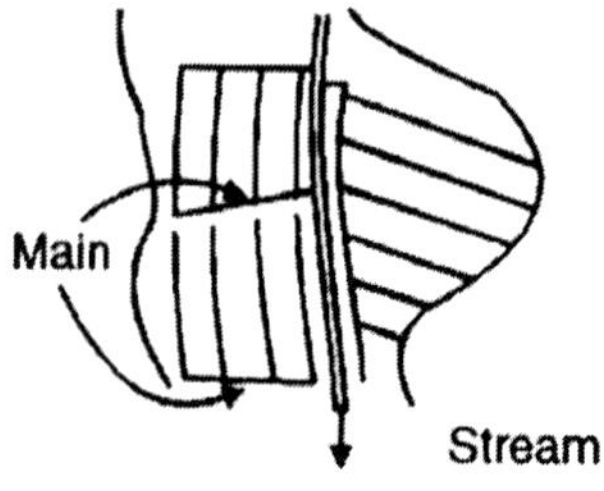

Fig. 13.18: Grouping system

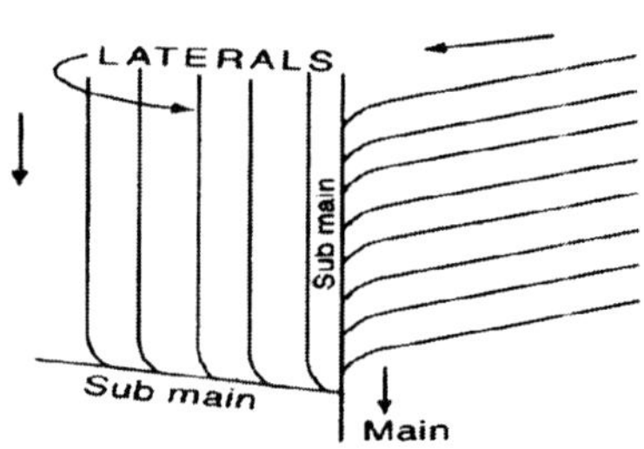

Fig. 13.19: Composite system

Zigzag system: In this system, filed drains as well as mains are constructed in zigzag pattern to reduce the high velocities of water (Fig. 13.21).

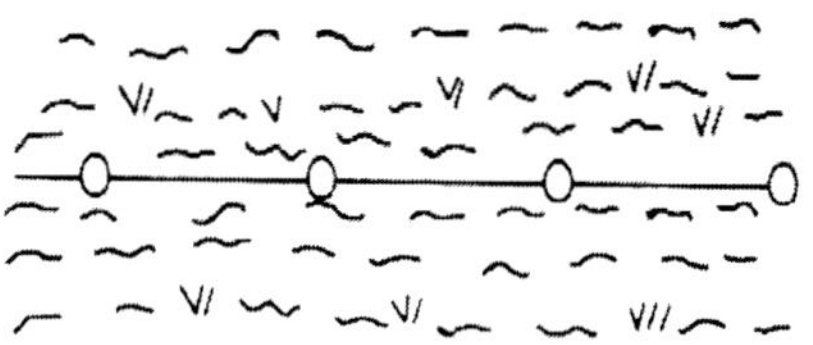

Fig. 13.20: Sink hole system

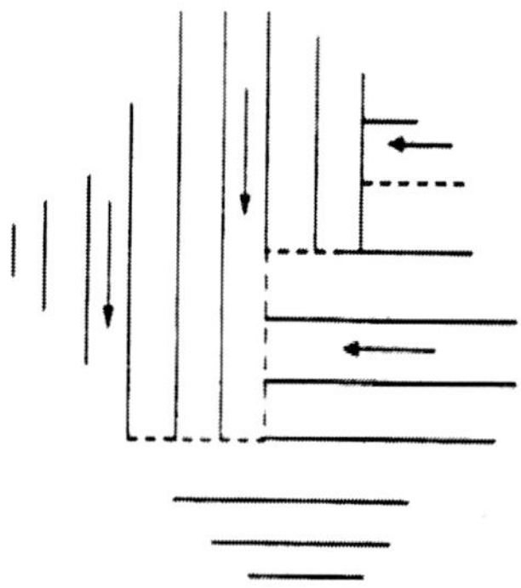

Fig. 13.21: Zigzag system

13.12.9 Depth and Spacing of Tile Drains

Depth and spacing of the tile drains are closely interrelated and depend upon the texture of the soil and its hydraulic conductivity, types of crops grown, extent of surface drainage, outlet conditions, topography of the land, salt content of the soil and the agronomic practices followed. The drain depth is reckoned from the ground surface to the bottom of the tile and is fixed from the consideration of the various factors as presented above. They are, in general, placed a t a depth 1 to 1.5 m below the ground surface. The centre of the tile drain is usually 0.3 to 0.6 m below the level up to which the water table is desired to be lowered below the root zone of the plants (Fig. 13.22).

The depth of the tile drains should be such that midway between the drains, the water table should be at a satisfactory depth. Where an impermeable bed is encountered, the tiles should be placed above such layer, except when the impermeable bed exists at a very shallow depth. If

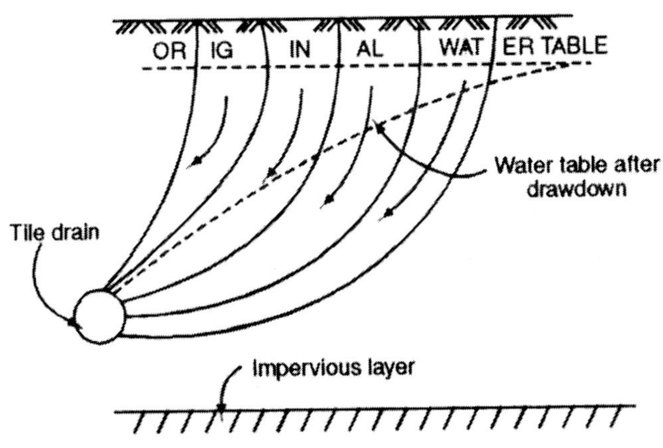

Fig. 13.22: Lowering of water table by tile drain

the tiles are placed below the impervious bed, then backfilling of the trenches should be done with permeable soil. In humid areas, water table rise quickly very near to the ground surface because of high rainfall, rate of drop of the water table is very important. However, in arid regions under irrigation, consideration should be given to maintain a minimum depth of water table for optimum crop growth. In general, deeper the drain, the wider is the spacing and lesser is the number of drains required. The following table (Table 13.9) gives the general depth and spacing of tile drains for various types of soil and their hydraulic conductivity (Schwab *et al.*, 1993).

Table 13.9: Average depth and spacing of tile drains for different types of soil

Type of soil	Hydraulic class	Hydraulic conductivity, m/day	Spacing, m	Depth, m
Clay	Very slow	0.001	9 - 15	0.9 – 1.1
Clay loam	Slow	0.001 – 0.005	12 – 21	0.9 – 1.1
Average loam	Moderately slow	0.005 – 0.02	18 – 30	1.1 – 1.2
Fine sandy loam	Moderate	0.02 – 0.06	30 – 37	1.2 – 1.4
Sandy loam	Moderately rapid	0.06 – 0.13	30 – 60	1.2 – 1.5
Peat and muck	Rapid	0.13 – 0.25	30 – 90	1.2 – 1.5
Irrigated soil	Variable	0.25 – 25.0	45 - 180	1.5 – 3.0

13.12.9.1 Hooghoudt's Equation for Spacing of Tiles

Hooghoudt's equation gives the relation between the spacing of drains and height of the water table built up between the drains due to a constant rate of irrigation or rainfall. Fig. 13. 23 indicates the physical situations of installation of tile drains. Before deriving the equation for deciding the spacing of tile drain, the following assumptions are made (Garg, 2012).

(i) The soil is homogeneous.

(ii) Darcy's law is valid for the flow.

(iii) The hydraulic gradient at any point is equal to the slope of water table above that point and

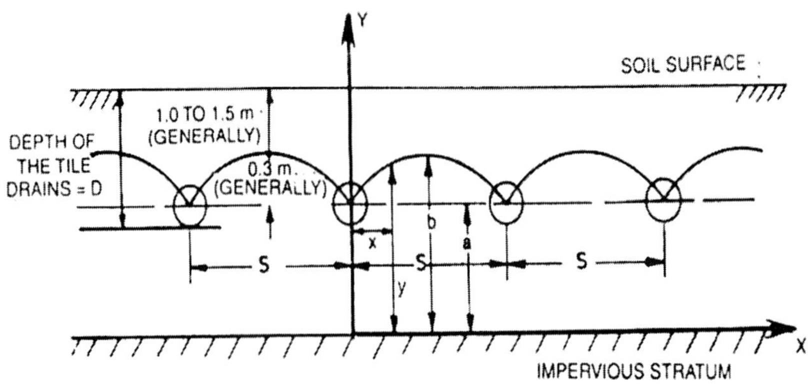

Fig. 13.23: Depth and spacing of tile drain

(iv) Flow is laminar and horizontal towards the drain.

Let spacing between the drains be S and a be depth of the impervious stratum from centre of the drain. Let the maximum height of the drained water table above the impervious stratum be b. At any distance x from the centre of the drain, let the height of the water table above the impervious stratum be y.

According to Darcy's law,

$$Q = K.I.A. \tag{13.5}$$

where,

Q = discharge of the drain in m^3/sec; K = hydraulic conductivity of the soil, m/sec; I = hydraulic gradient, m/m and A = area from which water flows towards the drain, m^2.

Discharge per unit length of the drain passing the section at y (q_y), is given as:

$$q_y = K.\frac{dy}{ds}.y \tag{13.6}$$

Assuming the inclination of the water surface to be small, such that the tangent (*i.e.* $\frac{dy}{dx}$) can be used in place of sine (*i.e* $\frac{dy}{ds}$) for the hydraulic gradient, we get

$$q_y = K.\frac{dy}{dx}.y \tag{13.7}$$

$q_y = 0$ when distance $x = S/2$ and $q_y = q/2$ when distance $x = 0$.

where

q is the total discharge per unit length carried by the drain, so that $q/2$ enters the drain from either side.

Also assuming that q is inversely proportional to the distance from the drain, we can write

$$q_y = \frac{q}{2} - \frac{q}{2}.\frac{x}{S/2} = \frac{1}{2}q.\left(1 - \frac{x}{S/2}\right) \tag{13.8}$$

Simplifying Eq. (13.8) we get,

$$q_y = \frac{q}{2S}(S - 2x) \tag{13.9}$$

Equating Eqns. (13.7) and (13.9), we get

$$K.\frac{dy}{dx}.y = \frac{q}{2S}(S-2x)$$

Rearranging and integrating we get,

$$\int \frac{q}{2SK}(S-2x)\,dx = \int y\,dy \tag{13.10}$$

Since the soil is assumed to be homogeneous, the soil hydraulic conductivity is constant. Hence integration of Eq. (13.10) can be written as

$$\frac{q}{2KS}\left(Sx - x^2\right) = \frac{y^2}{2} + C \tag{13.11}$$

When $x = 0, y = a$

Hence, putting the above limiting values in Eq. (13.11), we get $C = -a^2/2$ and putting this value of C in Eq. (13.11), we get

$$\frac{q}{2KS}\left(Sx - x^2\right) = \frac{y^2}{2} - \frac{a^2}{2}$$

On simplification we get

$$q = \frac{KS\left(y^2 - a^2\right)}{(S-x)\,x} \tag{13.12}$$

Also, when $x = S/2, y = b$

Putting the above limiting values in Eq. (13.12) and simplifying we get

$$q = \frac{4K}{S}\left(b^2 - a^2\right) \tag{13.13}$$

or $$S = \frac{4K}{q}\left(b^2 - a^2\right) \tag{13.14}$$

Eq. (13.14) is used to predict the drain spacing, S if discharge per unit length of the drain, q; soil hydraulic conductivity, K; values of the impervious stratum from centre of the drain, a. and maximum height of the drained water table above the impervious stratum b are known. Q depends on the infiltration in to the ground which should be removed by the drains. Generally a value equal to 1 percent of the average annual rainfall of a place is considered to be drained by the tile drains in 24 hours. If the average annual rainfall at a place be P_A (metres), then

$$q = \left(\frac{P_A\,x\,0.01}{24\,x\,3600}\right)(S\,x\,1)\,m^3\,/\,\text{sec}/\,m\;length\;of\;the\;drain \tag{13.15}$$

Equating Eqns. (13.12) and (13.15), and simplifying we get

$$S = \sqrt{\frac{\left(8.64\,x\,10^6\right)4\,K\left(b^2 - a^2\right)}{P_A}} \tag{13.16}$$

Spacing of the tiles can be simply calculated by using Eq. (13.16).

13.12.10 Size and Grade of the Tile Drains

Tile drains are designed according to the Manning's formula to carry discharge as per the drainage coefficient and drainage area. The drains are laid on a certain longitudinal slope varying from 0.05 to 3.0 percent. A desirable minimum working grade of 0.2 percent is preferred. Where sufficient slope is not available, the grade may be reduced to 0.1 percent. Depending on the topography, available slope of the soil and depth of the outlet, suitable value of longitudinal slope is provided to the tile drains. Their sizes can be determined by determining the discharge and using Manning's formula to find out velocity of flow in the tile drain. Minimum recommended values of tiles are 10 to 15 cm. In principle, the tiles should be adequate in size to drain out the required amount of water in 24 hours.

Diameter of the tile drains can be determined by the formula (Sharma, 1987) as:

$$d = 5.3\left(D_c\right)^{0.375}\;x\;A^{0.375}\;x\;S^{-0.1875} \tag{13.17}$$

where,

d = internal diameter of the tile drain, cm; D_c = drainage coefficient, cm/day; A = drainage area, ha; and S = hydraulic gradient or tile grade, m/m.

13.12.11 Materials of Tiles

Pipes used for tile drains are generally made from materials, such as clay, PVC or concrete. Sometimes, they may be made of bituminous fibre or steel. Corrugated plastic perforated pipes are now gaining popularity because they are portable and easy to handle. Concrete pipes can be locally made using simple moulds made of mild steel sheets. Concrete with 1:2:4 proportions should be used and it should be properly cured. Concrete pipes are withstand to freezing but may deteriorate in acidic and alkaline soils. On the other hand, good quality clay pipes do not deteriorate under the action of acidity or alkalinity and are cheaper but they are not stronger than the concrete pipes. Clay pipes should be well burnt and without any defect. Plastic pipes are also gaining popularity. The common materials are the polyvinyl chloride (*PVC*) and polythene, of which *PVC* is more popular. Plastic pipes come both in smooth and corrugated form. The smooth pipes are rigid and available in standard length. Corrugated pipes are flexible and come in coils. Corrugated pipes can withstand more resistance to outside pressure than the smooth pipes. They can also be conveniently used in the trench with the tile laying machines. Materials of pipes should be so chosen that

- They are resistant to weathering and deterioration in the soil.
- They have high density and absorb less water.
- They have uniformity in shape and wall thickness.
- They are resistant to freezing, thawing and frost damage.and
- They have sufficient strength to withstand static and impact loads transmitted from the soil above.

13.12.12 Envelope Materials and Accessory Structures

Envelope Materials: Tile drains, are usually, pipe drains made of porous earthen wares and other materials as discussed above. These drains are laid below the ground level, putting each other with open joints. The trenches, in which they are laid, are backfilled with sand and excavated materials as shown in Fig. (13.24). As far as possible, the tiles drains should not be placed in and below the impervious strata so that the impervious strata do not create hindrance in smooth flow of water from the soil placed above the tiles. However, if the tiles are placed in the less pervious strata, they are generally surrounded by graded gravel filter, called envelope filter. As shown in Fig. 13.25. The envelope filters prevent inflow of soils into the drain and it increases the effective tile diameter and thus increases the inflow rate. In soils where there is chance of heavy movement of soil particles into the tile drains and consequently clogging them, the filter materials are most to be placed around the tiles. The filters consist of different gradations, such as gravel, coarse sand, bajri *etc.* The coarsest material is placed immediately over the tile and the size is gradually reduced towards the surface. The minimum thickness of the filter is 7.5 cm.

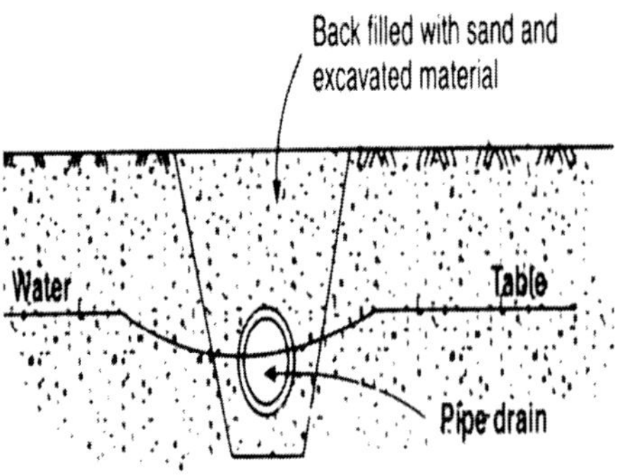

Fig. 13.24: Cross-section of tile drain without filter

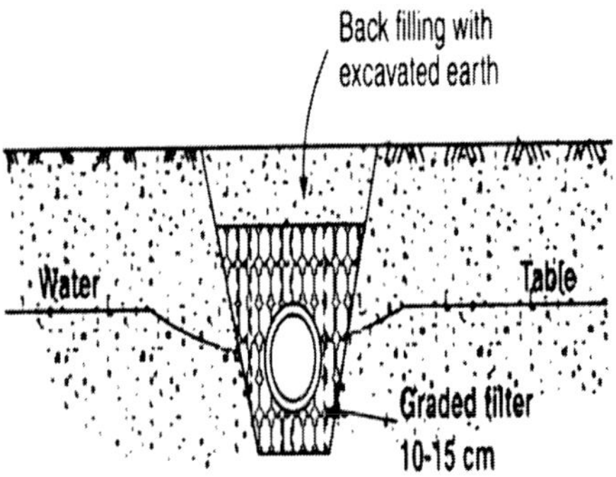

Fig. 13.25: Cross-section of tile drain with filter

The United States Bureau of Reclamation recommends the following criteria for design of gravel envelopes.

For uniform soils, $$\frac{D_{50} \text{ of envelope}}{D_{50} \text{ of soil}} = 5 \text{ to } 10 \tag{13.18}$$

For graded soils, $$\frac{D_{50} \text{ of envelope}}{D_{50} \text{ of soil}} = 12 \text{ to } 58 \tag{13.19}$$

Manholes and Sedimentation Basins: These are the vertical structures, installed at regular interval along the tile lines. They help in inspection of tile lines and cleaning the accumulated sediments in the pipe. These structures are constructed using either concrete or brick masonry. The size is sufficient so that a man can enter into them and perform inspection and cleaning works smoothly. These could be installed at the junctions of the tile lines. For causing no obstruction to farm machineries, these structures may be placed below the ground at about 60 cm depth and then covered by concrete block.

Inlet to Tile Drains: These are required to allow water to enter into the tile drains. These could also be used for flushing the tile drains. The two types of inlets used are blind inlet and surface inlet. The former is cheaper than the later one. Moreover, the blind inlet does not interfere with farming operations but there is chance of getting clogged. On the other hand, surface inlets have provisions for preventing the trash and other suspended materials entering the tile drains.

Outlets for Drain: The water drained by the tile drains is discharged into some bigger size and deep surface drains. The water from the tile drain may be discharged to these surface drains that serve as outlet drains either by gravity or by pumping and hence, the outlet may be named accordingly as gravity outlet or pump outlet (Garg, 2012).

In gravity outlet, the invert level of the tile drain is at higher elevation than the full supply level (*FSL*) of the outlet drain. In such situation, tiles discharge water to the outlet by simple gravity system. (Fig. 13.26). On the other hand, if the bed level of the outlet is higher than that of the discharging tile drain, a pump outlet is required (Fig. 13.27). It consists of an automatic controlled pump with a small sump for storage. It is costlier and requires high technical know how to operate the system.

13.12.13 Installation of Tile Drains

Installation of tile drains consists of the following steps.

(i) Excavation of trench

(ii) Laying the tiles to the prescribed grade at predetermined depth and spacing and

(iii) Putting the envelope material and backfilling the soil

Installation procedure depends on the availability of labour, material, types of tiles used, and the soil conditions. Installation of tiles should start from the outlet. Excavation of trench can be

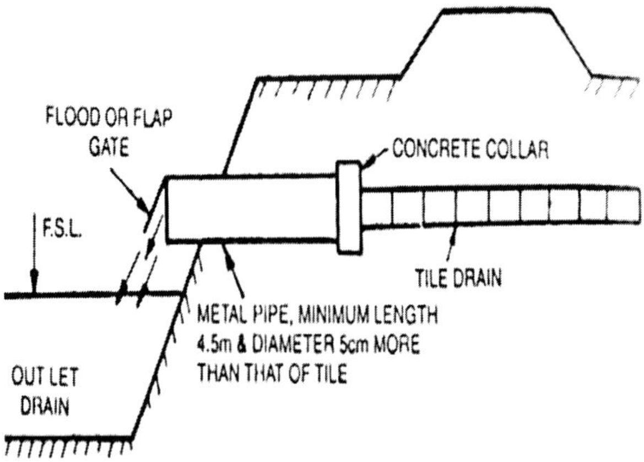

Fig. 13.26: Gravity outlet for tile drain

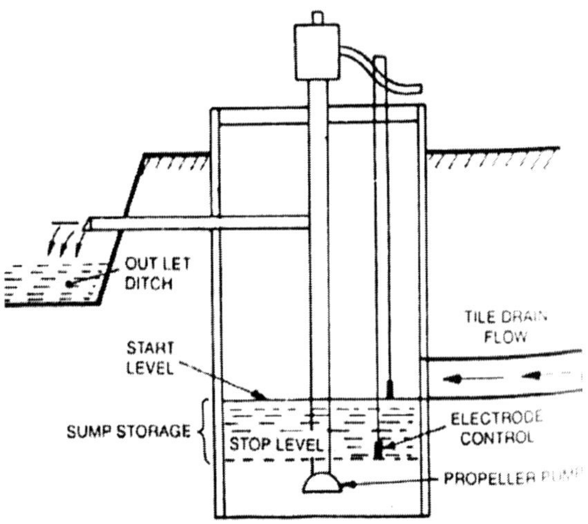

Fig. 13.27: Pump outlet for tile drain

done either manually or by using trenching machines. After excavation, tiles are laid in the trench carefully. Laying of tiles can be done manually or by machines. Some of the tools used for excavation and laying of tiles are shovel, tile hook, drain scoop and tile spades. Latest machines using laser technology are now employed for laying the tiles at correct grade at a quick time.

Sample Calculation 4:

Determine the spacing of the tile drain to handle an annual rainfall of 150 cm out of which 1 percent is the drainage coefficient. Depth of impervious layer from the ground surface is 10 m, depth of the highest position of the water table below the ground surface is 2 m and the depth of the tile drain below the ground surface is 3 m. Assume the hydraulic conductivity if the soil as 5 x 10^{-6} m/sec.

Solution:

Given annual rainfall = 150 cm = 1.5 m

Drainage coefficient is 1 percent of annual rainfall = 0.01 x 1.5 = 0.015 m/24 hours

Let the spacing of the drains be *S metres*.

Considering unit length of the drain, the drainage area = $S \, x \, 1 = S \, m^2$.

Hence, the discharge of the drain per unit length, $q = \dfrac{0.015}{24 \, x \, 60 \, x \, 60} \, x \, S = 1.73 \text{ x } 10^{-7} S \text{ m}^3/\text{sec}$.

But we know $q = \dfrac{4\,K}{S}\left(b^2 - a^2\right)$

Hence, $q = \dfrac{4\,K}{S}\left(b^2 - a^2\right) = 1.73 \text{ x } 10^{-7} S$

or, $S^2 = \dfrac{4\,K\left(b^2 - a^2\right)}{1.73} x10^7$

Depth of impervious stratum centre of the drain = a = 10-3 = 7 m

Maximum height of water table above impervious bed = b = 10 – 2= 8 m

Hydraulic conductivity = $K = 5 \text{ x } 10^{-6}$ m/sec

Hence, $S^2 = \dfrac{4\,K\left(b^2 - a^2\right)}{1.73} x10^7 = \dfrac{4\,x\,5\,x10^{-6}\left(8^2 - 7^2\right)}{1.73} x10^7 = 1734.10\,m^2$

So, drain spacing, S = $\int 1734.10 = 41.6\,m$

Sample Calculation 5:

Determine the size at the outlet of a 5 ha drainage system if the drainage coefficient is 2 cm and tile is laid at a grade of 0.5 percent. Assume the rugosity coefficient of the tile material as 0.011.

Solution:

Given drainage coefficient = 1 cm i.e. 1 cm of water is passing through the drainage area of 5 ha in 24 hours

Hence, discharge passing in the drain = $\dfrac{0.01\,x\,5\,x\,10^4}{24\,x\,60\,x\,60} = 0.00578\,m^3 / \sec$

Again we know, $Q = A\ x\ V = A\ \text{x}\ \frac{1}{n}\ x\ R^{2/3}\ S^{1/2}$

For a circular drain of size, D, area, $A =$ and wetted perimeter, $P = \pi D$

So, hydraulic mean radius, $R = A/P = D/4$

Discharge, $Q = \frac{\pi}{4}\ x\ D^2\ x \frac{1}{0.011}\ x \left(\frac{D}{4}\right)^{2/3}\ x\left(0.005\right)^{1/2}$

So, $0.00578 = \frac{\pi}{4}\ x\ D^2\ x \frac{1}{0.011}\ x \left(\frac{D}{4}\right)^{2/3}\ x\left(0.005\right)^{1/2}$

or, $D^{8/3} = 0.00229$

or, $D = 10.23$ cm ≈ 10 cm

Size of the outlet pipe may be chosen as 10 cm.

Sample Calculation 6:

In a sub-surface drainage system, the laterals were laid out 60 m apart and 200 m long and has grade of 0.2 percent. If the drainage coefficient of the area is 5 cm, what size tiles are to be used? Assume value of rugosity coefficient, $n = 0.015$.

Solution:

Drainage coefficient is given as = 5 cm.

Quantity of water to be drained in 24 hours $= \frac{5\ x\ 60\ x\ 200}{100} = 600\ m^3$

Discharge to be carried by the tile drain $= Q = \frac{600}{24\ x\ 60\ x\ 60} = 0.00694\ m^3/\text{sec}$

Grade of tile $= S = 0.2$ percent $= 0.002$

we know, $Q = A\ x\ V = A\ \text{x}\ \frac{1}{n}\ x\ R^{2/3}\ S^{1/2}$

For a circular drain of size, D, area, $A =$ and wetted perimeter, $P = \pi D$

So, hydraulic mean radius, $R = A/P = D/4$

$$\text{Discharge, } Q = \frac{\pi}{4} x\, D^2\, x \frac{1}{0.015}\, x \left(\frac{D}{4}\right)^{2/3} x\, (0.002)^{1/2}$$

$$\text{So, } 0.006944 = \frac{\pi}{4} x\, D^2\, x \frac{1}{0.015}\, x \left(\frac{D}{4}\right)^{2/3} x\, (0.002)^{1/2}$$

or, $D^{8/3} = 0.007471$

or, $D = 0.159$ m ≈ 16 cm

Size of the outlet pipe may be chosen as 15 cm since generally 10 and 15 cm pipes are available.

Question Banks

Q1. What do you mean by waterlogging?

Q2. Define the term capillary fringe. What is the height of the capillary fringe?

Q3. Irrigation projects have now caused curse in most of the irrigated commands. Do you agree with the statement? Justify the statement how it is accurse.

Q4. Describe the extent of waterlogging in some irrigation projects in India.

Q5. Waterlogging at head reach of the canal system is more serious than that at the tail reach. Explain how far it is true.

Q6. What are the different causes of waterlogging? Explain each cause.

Q7. How does the faulty irrigation management practice influence waterlogging in an irrigated area?

Q8. Briefly narrate the various ill effects of waterlogging. How does it finally affect production and productivity of crops?

Q9. What do you mean by denitrifixation? How it occurs and what is its affect on crop?

Q10. On the basis of water table and its fluctuation, arable lands can be classified into the different types. What are these types of arable lands and how are they classified?

Q11. What are the different indicators used to know whether a place is having waterlogging and drainage problem?

Q12. Describe the various preventive and curative measures required to control waterlogging.

Q13. How do the intercepting seepage drains help in reducing the water table?

Q14. What do you mean by salinisation? How is it related with waterlogging?

Q15. What are the different ways to arrest sub-soil flow of groundwater reservoir which helps in lowering the water table?

Q16. Explain how does the bio-drainage system help in lowering groundwater table?

Q17. What is land drainage? What are its objectives?

Q18. Write notes on essential requirements to be satisfied by a drain.

Q19. Describe the various points to be considered for planning a drainage system.

Q20. What are the different types of drainage?

Q21. Differentiate between the followings.

(i) Natural drainage and artificial drainage (ii) Open drain and closed drain

(iii) Surface drainage and sub-surface drainage (iv) Surface drain and seepage drain

(v) Tile drain and mole drain

Q22. How does the topography of a land affect surface drainage? How are the land shaping and grading useful in surface drainage?

Q23. Write down the various types of drainage systems used in flat areas.

Q24. Write notes with diagram about the following types of surface drainage systems used in flat areas.

(i) Random drain system (ii) Parallel filed drain system (iii) Parallel open ditch system and (iv) Bedding system

Q25. What is an open ditch and how does it helps in drainage? What are the advantages and disadvantages of open ditches?

Q26. Describe the different design criteria of open ditch drainage system?

Q27. How the design discharge of open drains is is calculated?

Q28. How the design discharge of an open drain is computed by simplified hydrologic procedures.

Q29. Write the recommended drainage periods of the following crops.
(i) Paddy (ii) Sugarcane (iii) Pearl millet (iv) Cotton (v) Sorghum (vi) Maize and (vii) Vegetables

Q30. How is the capacity of surface drains decided?

Q31. Discuss the different factors that govern the alignment of a surface drain.

Q32. A drainage basin of 1200 ha is discharging through a drain at an average rate of 3.2 m^3/sec. Calculate the drainage coefficient. If the drainage coefficient is 2.5 cm, what will be the discharge to be carried by the same drain from the same area of 1200 ha?

Q33. A drainage ditch draining 15 ha of land flows at a design capacity for two days following a storm. If the system is designed using a drainage coefficient of 2.5 cm, how much water will be removed during this period?

Q34. Mention the various steps used to design a stable trapezoidal ditch section.

Q35. Design a trapezoidal ditch section having fine gravel bed material with side slope 1.5:1 to carry discharge of 250 cumecs. Bed slope of the ditch section is 1:6000 and Manning's n is 0.018. Water flowing in the ditch transports colloidal silts.

Q36. How are the open ditches constructed in fields?

Q37. How are the maintenance works of open drains done in surface drainage system?

Q38. What is sub-surface drainage? Mention the various advantages and disadvantages of sub-surface drainage.

Q39. Mention the different methods employed in sub-surface drainage system. Explain how perforated pipes are used as sub-surface drainage system.

Q40. What is a vertical drain? How does the vertical drain work in lowering the groundwater table?

Q41. Describe the various information that is required for design and installation of sub-surface drainage system.

Q42. Narrate the parameters of groundwater studies as required for planning of installation of sub-surface drainage system in a drainage basin.

Q43. Differentiate between piezometer and observation well. What for they are used? How are the readings of piezometers helpful in study of groundwater movement?

Q44. How can you prepare a groundwater contour map and water table isobath map of an area?

Q45. Define hydraulic conductivity? What are the different methods used to measure hydraulic conductivity of soil?

Q46. What are the various parameters required for design of a tile drainage system?

Q47. With labeled diagram write down the different types of layout of tile drains in sub-surface drainage studies.

Q48. What are the factors that affect the depth and spacing of tile drain? Discuss how the depth and spacing of tile drains vary with soil texture and its hydraulic conductivity.

Q49. Derive Hooghoudt's equation for computing the spacing of tile drains.

Q50. How can you determine the size and grade of tile drain?

Q51. Describe the various materials used in construction of pipes for tile drains.

Q52. Describe the role of the envelope filters and ancillary structures in tile drainage system.

Q53. Write down the roles of gravity outlet and pumping outlet in sub-surface drainage system. Describe the situations where gravity outlet and pumping outlet are employed.

Q54. Write the procedures used to install the tile drains in fields.

Q55. Determine the spacing of the tile drain to handle an annual rainfall of 100 cm out of which 1 percent is the drainage coefficient. Depth of impervious layer from the ground surface is 10 m, depth of the highest position of the water table below the ground surface is 1.5 m and the depth of the tile drain below the ground surface is 2 m. Assume the hydraulic conductivity if the soil as 6×10^{-6} m/sec.

Q 56. Determine the size at the outlet of a 10 ha drainage system if the drainage coefficient is 3 cm and tile is laid at a grade of 0.3 percent. Rugosity coefficient of the tile material is 0.013.

References

Anonymous. 2005. Annual Report of AICRP on Water Management, OUAT, Chiplima.

Garg, S.K. 2012. Irrigation Engineering and Hydraulic Structures. Khanna Publishers, New Delhi.

Lenka, D. 2001. Irrigation and Drainage. Kalyani Publishers, New Delhi.

Michael, A.M. 1978. Irrigation Theory and Practices. Vikash Publishing House Pvt. Ltd., New Delhi.

Murty, V.V.N. 2004. Land and Water Management Engineering. Kalyani Publishers, Cuttack.

Pai, A.A. and Hukkeri, S.B. 1979. Manual on Irrigation Water Management. Department of Agriculture, Krishi Bhawan, New Delhi.

Raadsma, S and Schulze, F.E. 1974. Surface Field Drainage System. (In): Drainage Principles and Applications, Vol. IV, International Institute for Land Reclamation and Improvement. The Netherland.

Schwab, G.O., Fangmeier, D.D., Elliot, W.J. and Frevert, R.K. 1993. Soil and Water Conservation Engineering (4th Ed.), John Wiley and Sons, Inc. New York.

Sharma, R.K. 1987. A Text Book of Hydrology and Water Resources. Dhanpat Rai and Sons, Delhi.

Sharma, R.K. and Sharma, T.K. 2002. Irrigation Engineering including Hydrology, S. Chand and Company Ltd. New Delhi.

Uppal, H.L and Sehgal, S.R. 1965. Drainage Requirement of the Punjab State, International Commission of Irrigation and Drainage Bulletin, New Delhi, pp. 23-31.

14

Groundwater and Wells

14.1 Introduction

Groundwater is an important source of water for irrigation. It is the most abundant fresh water resource of the earth. Underground water reservoirs contain large stock of fresh water resources. Total quantity of groundwater within 800 m distance below the ground surface is more than 30 times the total water in all fresh water lakes, more than 60 times the total water in soil and other unsaturated rock materials, more than 300 times the water vapour in the atmosphere and 3000 times the average volume of all rivers and rivulets in the world (Lenka, 2001). But the distribution of groundwater is not uniform. At some places, it lies below a few metres whereas at other places, it is a thousand metres or more below the ground surface. In some places, there is abundant groundwater potential whereas at other places there is meager potential.

Groundwater is the water reaching the groundwater reservoir from the infiltrated water of precipitation which constitutes an important component of the hydrologic cycle. It is only the part of the subterranean water that occurs in the saturated pores of the containing rock materials underlain by water held in a zone of aeration. It should not be confused that all the water present in soil below the ground surface is groundwater.

14.2 Sub-surface Water Distribution

The soil profile lying below the soil surface up to the impervious strata/bed rock (called as sub-surface) can be divided in to (i) zone of aeration (also called as unsaturated zone) and (ii) zone of saturation. Zone of aeration is also called as vadose zone where soil water is held in suspension. Zone of aeration contains three sub zones i.e. soil water zone, intermediate zone and capillary zone. Soil water zone contains soil water, intermediate zone holds pellicular and gravitational water and the capillary zone holds capillary water. Water stored in the zone of saturation is called as groundwater. Saturated zone has the upper boundary as groundwater table and lower boundary as the impervious/bed rock. The distribution of soil profile containing different distributions of sub-surface water is depicted in Fig. 14.1.

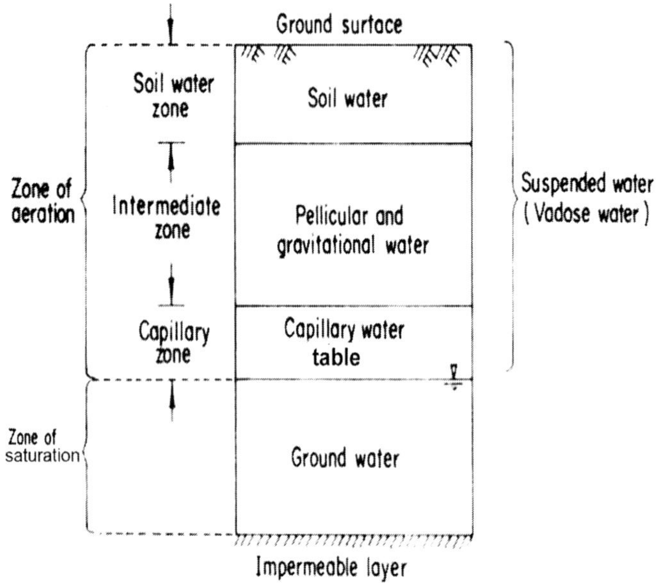

Fig. 14.1: Distribution of sub-surface water

Water table, in general, lies approximately parallel to the topographical surface above it. Near the source of recharge, the groundwater level is raised upward forming groundwater mound whereas the level gets decreased gradually as one goes away from the recharge source. This form a cone of elevation which is just opposite to the cone of depression in case of a pumping/discharge well. The action of gravity tends to make the watertable a level surface after some period. However, frequent addition and withdrawal of ground water do not permit the water table to assume a level surface. It may coincide approximately with the top of zone of saturation in coarse gravel but a few centimeters to several meters below it in fine grained soils.

14.3 Geologic Formations for Groundwater Supply

Rock is the most important geologic formation that holds and supply groundwater for various uses. These rocks are of two types i.e. (i) unconsolidated type and (ii) consolidated type. In consolidated type, the particles are held tightly by compaction and cementation whereas in unconsolidated type, materials are loose. Examples of consolidated rocks are granite, sandstone, limestone *etc.* Examples of unconsolidated rocks are gravel, sand and clay. The geologic formation, depending on the quantity of water they hold and yield to a well when discharged, may be divided into four types. They are (i) aquifer, (ii) aquifuse, (iii) aquiclude and (iv) aquitard. Of these four formations, aquifer is very important. An aquifer is defined as the geologic formation that contains sufficient amount of groundwater and also transmits and yields appreciable amount to a well constructed in it. Among all kinds of geologic formations, gravel is the best and next to gravel, the best aquifer materials are sand, sandstone and limestone. Clay and fine silts are the most unproductive materials. Porous and permeable materials form good aquifers.

14.4 Types of Aquifers

Groundwater aquifers are classified into the following three types.

- Unconfined aquifer
- Confined aquifer and
- Semi confined aquifer

Unconfined Aquifer

The top most water bearing stratum, having no confined impermeable over burden lying over it, is called as unconfined aquifer or non-artesian aquifer or watertable aquifer. Water in these aquifers is at atmospheric pressure. When wells are constructed in these aquifers, the level of the water in these wells represents the watertable in the aquifer. Wells in these aquifers are called as gravity well or watertable wells or simply wells.

Confined Aquifer

It is an aquifer confined at both its upper and lower surface by impermeable stratum. It is inclined so as to expose the aquifer somewhere to the catchment area at a higher level for the creation of sufficient hydraulic head. Confined aquifer is also called as artesian aquifer. Water in these aquifers is at pressure than the atmospheric pressure. Therefore, when a well is dug in these aquifers, water will rise to a level above the bottom of the upper confining stratum because of pressure under which water is held. The imaginary level to which water will rise in a well located in an artesian aquifer is known as piezometric level. Water is fed to the confined aquifer through recharge area. The well constructed in these aquifers are called as artesian wells. Flowing well is a special type of artesian well and is found in the areas where the piezometric surface of the confined aquifer is at lower elevation than the natural ground surface. In these well, water automatically flows out and even gush out of surface for a reasonable height.

Semi Confined Aquifer

A semi confined aquifer, also called as leaky aquifer, is a completely saturated aquifer that is bounded above by a semi pervious layer. Lowering of the piezometric head in leaky aquifer will generate a vertical flow of water from the semi pervious layer into the pumped aquifer.

A special type of unconfined aquifer is perched aquifer. Perched aquifer occurs within the unconfined aquifer. It occurs when the groundwater body is separated from the main groundwater body by a relatively impermeable stratum of small areal extent. Different types of aquifers are illustrated in Fig. 14.2. The top surface of the water held in the perched aquifer is known as the perched watertable.

14.5 Groundwater Yield

During the process of rainfall or irrigation, some amount of water infiltrates into the soil. This water moves downward and fills up the pore space present in the sub-soil in the formation. When all the pore spaces are filled up with water, then it is called as saturated soil. The water contained in these pores is drained by digging wells under the action of gravity drainage. However, it is found out that not all the water hold by the pores are drained out or yielded by the wells. Some amount of water is held by them due to their molecular attraction and this water is called as pellicular water.

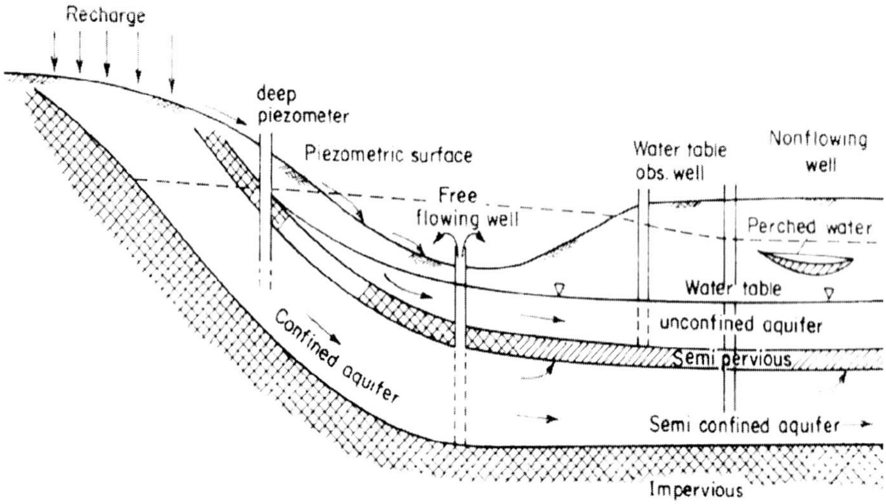

Fig. 14.2: Different types of aquifers and wells

14.6 Specific Yield

Volume of groundwater extracted by the gravity drainage from a saturated water bearing material is known as yield. Specific yield is defined as the ratio of yield to total volume of the material drained or dewatered and is given by the formula as:

$$Specific\ yield = \frac{Volume\ of\ water\ obtained\ by\ gravity\ drainage}{Total\ volume\ of\ material\ drained\ or\ dewatered} x\,100$$

14.7 Specific Retention

As discussed above, not all the water contained by the saturated material is drained out by the action of gravity drainage and some amount is retained by the material which is called as pellicular water. This quantity of water retained by the material against the action of gravity is termed as specific retention or field capacity. Like specific yield, this is expressed as the percentage of total volume of material drained. Specific retention is given by the formula:

$$Specific\ retention = \frac{Volume\ of\ water\ held\ against\ gravity\ drainage}{Total\ volume\ of\ material\ drained\ or\ dewatered} x\,100$$

Thus, the sum of specific yield and specific retention is equal to porosity.

14.8 Specific Capacity

Specific capacity of a well is the rate of flow of water from a well per unit of drawdown.

14.9 Storage Coefficient

The storage coefficient of an artesian aquifer is equal to the volume of water released from the aquifer of unit cross-sectional area and of the full height of the aquifer when the piezometric surface declines by unity. In a more common way it can be defined as the volume of water that an aquifer releases or stores per unit surface area of the aquifer per unit change in the component of head normal to that surface.

14.10 Well Irrigation

In India, ground water withdrawal from wells for irrigation has been practiced since long and it is still now an important source of irrigation. Most of the wells draw supplies from the first water bearing structure and discharge about 5 litres/sec (lps) depending on the static water table position and geological formations.

14.10.1 Advantages of Well Irrigation

Followings are the advantages of well irrigation (Panigrahi, 2011).

- It is owned by the farmers and as such has full control on sources of irrigation. It is also a handy source of irrigation.
- It is located nearer to the place where the irrigation water has to be served to crops and so conveyance loss is minimum.
- Particularly suitable for intensive irrigation to raise cash crops and is suitable to grow 2 to 3 crops from the same piece of land in a year.
- Low cost of construction.
- Reliable source of irrigation.
- It can be used as a conjunctive source with canal irrigation
- Supply is more uniform in quality and chemical composition and as such more helpful to crop growth and yield.
- Supply is free from weeds, turbidity and bacterial pollution which is associated with canal water
- Economical water use having higher delta and low duty.
- Benefit of rainfall can be availed. If there is chance of rainfall then well irrigation can be withheld.

14.10.2 Disadvantages of Well Irrigation

The disadvantages of well irrigation are (Panigrahi, 2011):

- Associated higher repair with maintenance and hence higher operating cost.

- It has low discharge and consequently small area is irrigated by it.
- Water is debris / silt free. So more fertilizer is required to be added to crop field to raise crop yields.
- If the velocity of water from well exceeds critical velocity of water in the soil that convey water to well, then there is chance of well subsidence.

14.10.3 Location of Well

- Site of open well should be in the central position of the farm so that water can be served to all parts of farms easily without more conveyance loss.
- It should be located at a higher elevation so that water can flow by gravity to all parts of farms.
- Site should have good water bearing aquifer to supply sufficient yield to well.
- If there exists any well, then the near well should be at least 75 -175 m away from it to avoid any interference of wells and avoid low yield
- It should be very much away from any source of pollution
- There should not have any adjoining boundary lines within the circle of reference of the well.

14.10.4 Open Wells

Open wells are bigger in size than that of tubewell having less depth compared to tubewell and hence is associated with fewer yields. The yield of open well varies from 1 to 5 litres/sec. The diameter of the open well generally varies from 2 to 9 m. It draws water from any one pervious stratum. However, it is recommended that the well be constructed to a depth below ground water table such that the well does not dry during summer. From financial point of view, the depth is limited to 30 m. Contribution of water to well is from both bottom and sides of wells in unlined case. If well is lined, contribution is only from bottom.

14.10.5 Types of Open Wells

Open wells may be dug, bored, driven, jetted or drilled type.

Dug Wells

Dug wells are constructed manually or by semi mechanical means. They are dug down to the water bearing strata of the aquifer. They derive water from the formation close to the ground surface. Water percolates into the wells from all sides and bottom. The earthen sides are protected by circular cement rings. The depth of wells may vary from a few metres (shallow dug well) to several meters (deep dug well) and the size varies from 0.75 to 1.2 m in diameter. In large dug wells, the size may be 6 to 10 m (Lenka, 2001). Fig. 14.3 presents the view of a shallow dug well.

Fig. 14.3 Picture of a shallow open well

Bored Wells

Bored wells are constructed by hand or by power augers. At times, it means same as the drilled wells. In order to increase the yield of an open well, a bored well is set up at the centre of dug well. Such an arrangement is called as dug-cum-bored well. For such wells to be successful, a confined aquifer should be available below the dug well.

Driven Wells

Driven wells are wells that are constructed by driving a pointed screened cylindrical section also called as strainer and the pipes are connected to it directly into the water bearing formation. Water enters the well through the strainer. Diameter of the driven well varies from 3 to 10 cm. Depth is limited to 30 m. These wells are suited to domestic use since the yield is less (80 to 200 lit/min).

Jetted Wells

Jetted wells are constructed by the cutting action of a high velocity water jet. Small size wells of diameter 3 to 8 cm is constructed with this method. Construction of wells can be done either by hand operated or power operated auger.

Drilled Wells

Drilled wells are made constructed either by percussion or rotary drills and the excavated materials are brought to surface by bailer, sand pump, suction bucket or by hydrostatic pressure. They are constructed by means of drilling a deep bore of small diameter passing through many strata. Drilled wells are also called as tubewells.

Open well can be further classified as unlined or lined well. It may also be shallow or deep well.

Unlined Wells

It is also called as kacha well having no lining either at bottom or at sides. These wells have depth 3 - 3.5 m. They are constructed in stable pervious soil so that side subsidence will not occur and groundwater can be easily tapped. These wells are cheap and easy to construct. But their longevity is less since there is chance of side collapse. The flow enters the well both from sides and bottom. Fig. 14.4 represents the view of an unlined well.

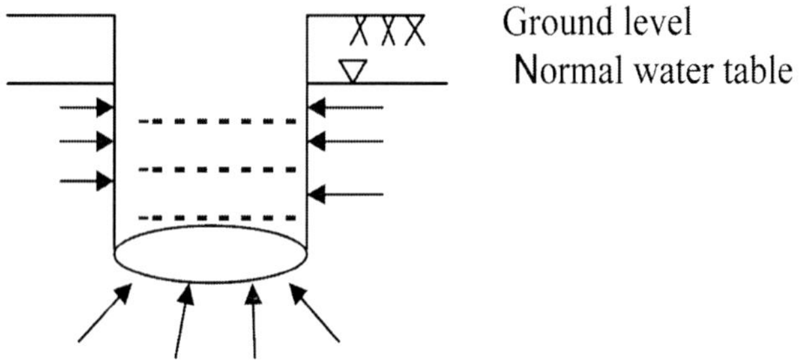

Fig. 14.4 Unlined well

Lined Wells

Depending on the types of lining materials used, lined wells may be pervious lined type or impervious lined type.

Pervious Lined Wells

Such wells have dry bricks or stone lining with open joints through which water enters the well. In sandy formation in order to avoid flow of sand surrounding the lining into the well, the space behind the lining is packed with brick ballast of size 20 -25 mm. These wells are comparatively small depth and are suitable for shallow water table and for gravel and sand formation. The flow of water into the well is both from bottom and sides. Flow is of both radial and spherical pattern. Wherever pervious wells are plugged at bottom, flow pattern is radial type only. Fig. 14.5 gives the view of pervious lined wells.

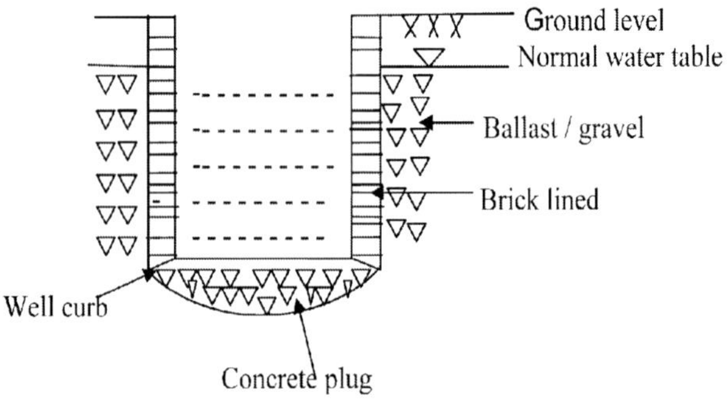

Fig .14.5: Pervious lined well

Impervious Lined Wells

In non alluvial soil, if the soil is sandy or soft and pervious, then there is chance of side collapse. So lining of sides of wells is necessary. Such wells need an impervious lining with brick masonry or stone masonry in cement or lime mortar or with reinforced concrete. Steining of 30 - 60 cm thick is supported on well curb which is also constructed of reinforced concrete.. Impervious lined well have depth ranging from 5 to 30 m depending on water bearing strata. The flow towards well is through bottom and hence spherical. A better type of impervious well draws water from deeper sandy formation through a pipe sunk down from centre of the well (Fig. 14.6).

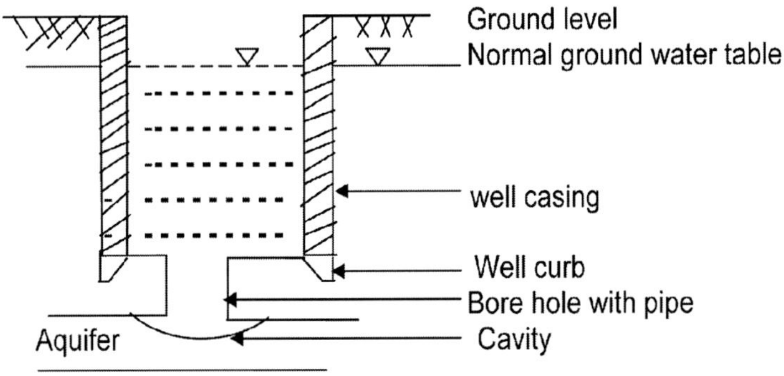

Fig .14.6: Impervious lined well

Shallow Well

A shallow well penetrates into the uppermost pervious stratum only from which it draws water. It is of less depth; yield is less and dries up quickly when discharged for a long period. It covers less area and is suitable for non rice crops.

Deep Well

It draws water from deeper layers and has high yield. It covers more area. It supplies water sufficient for paddy and other crops for irrigation. Like impervious lined well, a hole may be made at the centre of the well which draws up water from the deeper aquifer. Table 14.1 gives difference between the shallow and deep wells.

Table 14.1: Difference between shallow wells and deep well

Sl. No.	Shallow well	Deep well
(i)	Initial cost low	Initial cost high
(ii)	Cross sectional area at bottom of well less which limits high supply to well	Cross sectional area high due to cavity formation and so supply more water to well
(iii)	Suitable where ground water table is very near to the ground surface	Suitable even if ground water table is deep below ground surface.
(iv)	Yield is less	Yield is high
(v)	Quality of water is not good	Better quality water

14.10.6 Yield of an Open Well

Safe yield of a well is the yield taken out as discharge from a well without causing any mining. It is expressed as m^3/ hr or lit / min. Under stable condition, water level in the well coincides with that in the ground water table in the adjoining aquifer. But as the pumping in the well continues, water level in the pumping well falls immediately where- as the levels in adjoining aquifer remains at higher elevation. This forms a cone of depression around the well. Difference between the water level in the well and that in the aquifer is known as draw down or depression head (Slichter, 1989). The more the rate of withdrawal, the more is the depression head. The depression head is maximum at the centre of the well and gradually decreases as the radial distance from well increases. At radius of influence, it is zero. There are two tests for determining the yield of an open well. They are pumping test and recuperation test.

14.10.6.1 Pumping Test

It is an accurate method to determine the safe yield of an open well. In this method, water level in the well is decreased through heavy withdrawal till maximum safe depression head is reached. At this stage, the pump speed is regulated so that the rates of withdrawal equal the recharge. Thus constant discharge water level is maintained in the well which creates equilibrium condition. The ratio of pump discharge per hour under this equilibrium condition gives the yield of the wells at a particular draw- down.

Applying Darcy's law

$$V = K\,i = K.\ h/r \tag{14.1}$$

where V = velocity of flow of water into the well at equilibrium condition, i = hydraulic gradient equal to ratio of drawdown (h) to radial distance (r) from wells and K= constant called as hydraulic conductivity of the aquifer.

Discharge (Q) is given as:

$$Q = A.V = A.\ K.\ h/r \tag{14.2}$$

Hence, considering unit width aquifer we get $A = r$

$$Q = r.\ K.\ h / r$$

$$Q = K.\ h \tag{14.3}$$

The yield of a well can also be expressed in another way.

Equation (14.3) can be written as:

$$Q = (K/A)\ A.\ h \tag{14.4}$$

where K/A is expressed as specific yield of the well (m^3 /hr) per m^2 of the aquifer through which water enters into the well from aquifer through unit depression head. The values of K/A for different soils are given in Table 14.2 (Sharma, 1987).

Table 14.2: Values of *K/A* for different types of soils

Soil type	Value of *K/A*, (m^3 / hr per m^2 / unit depression)
Clay	0.25
Fine sand	0.5
Coarse sand	1.0

Sample Calculation 1:

Design an open well in fine sandy soil with safe yield 0.5 m^3/hr /m^2/ unit depression head so that an yield of 0.004 m^3/sec is discharged from well with depression head h = 4 m

Solution:

Using Eq (14.4) we have $Q = (K/A).\ A.\ h$

For fine sandy soil the value of (K/A) = 0.5 m^3/h_r per m^2/unit head (Table 14.2)

Q = 0.5 (m^3 /hr /m^2 / m depression) X *A* X 4 m

= 2 A m3 /hr/m^2

Given Q = 0.004 m^3 /sec = 14.4 m^3 /hr

So 14.4 m^3 /hr = 2A m^3 /hr /m^2

Hence, A = 7.2 m^2

Area $A = (\pi/4)\ .\ d^2$ where d = diameter of well

$(\pi/4)\ d^2 = 7.2\ m^2$

$$d = \frac{\sqrt{4 \times 7.2}}{\sqrt{3.142}}$$

Hence, diameter, d = 3 m.

14.10.6.2 Recuperation Test

It is used where it is difficult to maintain constant water level in the well. In this method the water level is decreased to any level below the normal level by pumping and then pumping is stopped. When pumping is stopped then water level in the well starts rising (Fig. 14.7). Time taken by water in well to attain any draw down below ground water table normal position is recorded. Then the well discharge formula (Eq 14.5) as given below is used.

$$Q = \tfrac{2.3}{t} \log_{10} \frac{h1}{h2} A.\ h \tag{14.5}$$

where t = time taken by water to attain drawdown below normal ground water table position, h_1 and h_2 are initial and final ground water table position, h = average depression head and A = area

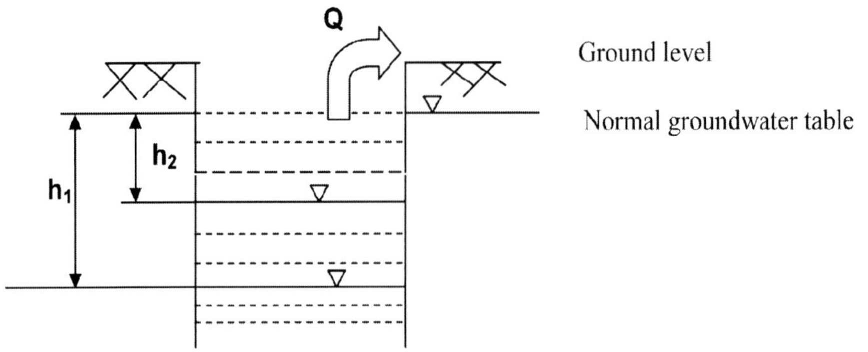

Fig. 14.7: Yield of an open well by recuperation test

of cavity of well bottom through which water enters the well given as:

$$A = \frac{4}{3}\left(\frac{\pi}{4}\right)d^2 \qquad (14.6)$$

where d = diameter of well

Sample Calculation 2:

Calculate the average yield of a well of 3 m diameter by recuperation test. The water level in well is decreased to an extent of 2.5 m. The rate of rise of water is 1.0 m/hr. Assume average depression head, h =3 m

Solution :

Given $h_1 = 2.5$ m

After one hour t= 1 hr, $h_2 = 1.5$ m

$A = (4/3) . (\pi/4) . 3^2$ and h =3 m

$$Q = \frac{2.3}{t} \times \log_{10}\left(\frac{2.5}{1.5}\right) \times \frac{4}{3} \times \frac{\pi}{3} \times 9 \times 3$$

= 9 x 3.142 x 2.3 log10 (1.67)

=26 m^3 /hr ≈ 7 lit / sec

14.11 Tubewell Irrigation

14.11.1 Advantage of Tubewell Irrigation

Advantages of tubewell irrigation are:

(i) It is an independent source of irrigation which a farmer can at his convenience dig and maintain the well

(ii) Some patches of lands which cannot be easily irrigated by other sources can be irrigated by it.

(iii) Volumetric rate of irrigation water is possible resulting in economical and optimum utilization of water.

(iv) Water loss in conveyance system is less since conveyance channels are generally lined.

(v) Since seepage and conveyance losses are minimum there is less water logging problems.

(vi) Irrigation capital investment is less compared to canal irrigation.

(vii) There is quick return from well irrigation. Canal irrigation takes 5-6 years to supply irrigation water where as tube well irrigation is possible in the same year where tube well is setup.

(viii) Optimum utilization of rain water is possible.

(ix) Water is suitable for domestic use as quality of water is good.

14.11.2 Disadvantages of Tubewell Irrigation

Following are the disadvantages of tubewell irrigation.

(i) Life of tubewell is limited to 10-15 years

(ii) Operating cost is more and so costlier than canal water

(iii) Cover less area compared to canal irrigation

(iv) Frequent repair and maintenance is required.

(v) Not well suited in areas where elevation is not available since water moves by gravity.

(vi) Tube well water carries less silts and hence crop yield is less. In order to increase crop yields, crops needs more fertilizers

(vii) Chances of theft of well components are more.

(viii) Not suitable for high water requiring crops.

14.11.3 Suitability of an Area for Tubewell Irrigation

Tube well irrigation is suitable where the sub soil formation is good for storing water and the aquifer is suitable to supply ample ground water to the well. Water bearing stratum may be confined or unconfined. If aquifer has impervious layer at the bottom and top is open to atmosphere, then it is unconfined. A confined aquifer is confined between two impervious layers. The best place for a tube well digging is sandy or sandy aquifer. If the subsoil strata has a 30-50 m thick coarse sand stratum, then it is ideal to dig well. The tube well has a low yield capacity generally 0.042 m^3 / sec., i.e 42 lps which can generally command irrigation to 160 ha of non paddy crop at 100% irrigation intensity after taking care of rainfall. A deep tube well can supply irrigation water 300 ha of non paddy crop. While selecting a site for tubewell, care should be taken so that it is dug in a high elevation to facilitate in gravity supply of water. It should also be located in the central position of the farm so that water can easily be conveyed to all parts of the farm with minimum conveyance loss. In addition to above mentioned points, lithological test, aquifer test and location of aquifer boundary should also be taken into account before selection of site of tube well construction (Ahrens, 1958; Benison, 1947; Dylla and Muckel, 1967).

14.11.4 Types of Tubewells

For irrigation point of view tubewells are of two types (i) Direct irrigation tube well and (ii) Augmentation tube well.

Direct Irrigation Tubewell

This is the tubewell which provides irrigation directly to crops in the farm. It may be classified as shallow or deep tube well.

Augmentation Tubewell

In rainy season, canals run almost full and there is no shortage of supply to crops. But in winter and summer the canal supply is reduced. At this time the ground water can be pumped in conjunction with canal water. Tube wells are used to feed to canal supply which can be used for irrigation. The augmented tube well can serve for augmentation of canal water supply or as a direct source for irrigation. It can also be used as a viable source for ground water recharge.

Tube wells can also be classified as shallow in deep tubewells.

Shallow Tubewells

As the name suggests, these are shallow in depth, usually they tap water from a few layered aquifers. They have diameter usually same from top to bottom. They cover less irrigation areas as their discharge capacity is less.

Deep Tubewells

These well are deeper in depth penetrating several multilayered aquifers. They draw water from deep confined aquifers. As augmentation tube well, they can be installed in adjacent to rivers. They are spaced at about 500 m from one another. For a given drawdown the deep tubewell yields more discharge than a shallow tubewell (Fig. 14.8).

Tubewells can further be classified as (i) cavity type wells, (ii) strainer type wells and (iii) gravel packed wells or shrouded well.

14.11.4.1 Cavity Type Tubewells

A cavity type tubewell draws water from the bottom of the well. Water does not enter into the well through the sides as is done by a screen or strainer well. The flow, thus, in a cavity well is spherical not radial. Cavity well draws water from only one water bearing stratum. Such a well is economical as it requires only plain well pipe not screened pipe which is very costly. The plain pipe is introduced into the bore hole up to a depth of water bearing stratum from which water is to be drawn.

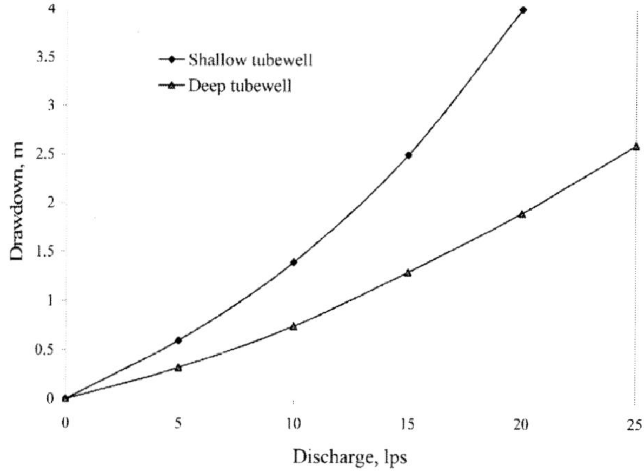

Fig. 14.8 Discharge drawdown curves of shallow and deep tubewells

A cavity type tubewell essentially consists of a pipe bored through the soil and resting on the bottom of a strong clay layer that acts as aquiclude. A cavity is formed in water bearing sand stratum lying below the clay layer from where water enters into the pipe (Fig. 14.9). The cavity is developed by drawing out apart of the sand from this layer so that such a cavity works as the

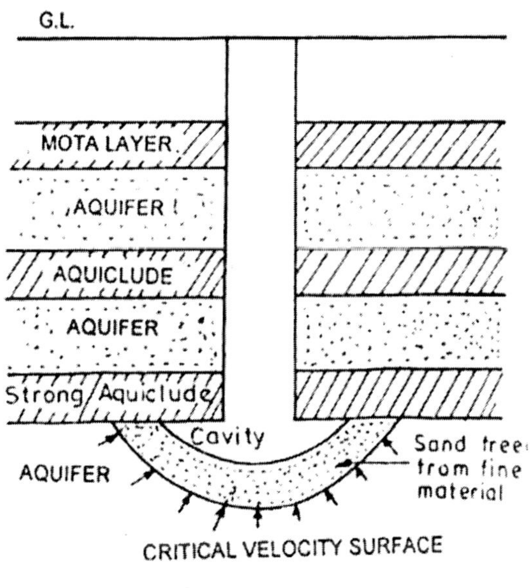

Fig. 14.9: Cavity type tubewell

storage reservoir for pumping. The cavity of such a tubewell needs to be developed carefully and slowly by using a centrifugal pump. Cavity type tubewell has low discharge and is suited for domestic purposes.

14.11.4.2 Strainer Type Tubewell

In strainer type tubewells, drilling continues through several types of layers. Drilling continues till a suitable water bearing strata is obtained. Plain pipes are introduced into the bore hole and placed in the strata having less permeability like clay and strainers are located in good water bearing strata having high permeability like sand and gravel as shown in Fig. 14.10. A bail is provided at the bottom. Water enters into the well through the strainers from the sides and the flow is radial not spherical as is in case of cavity wells. For the strainer type tubewells, drilling is to be started with pipes of larger diameter (called as casing of pipes) than the strainer and these are taken out after the strainers are installed.

A strainer essentially consists of a perforated or slotted pipe with a wire mesh wrapped around it with a small annular space between the two. The wire screen prevents the sand and other coarser particles larger than its mesh size from entering the well. This facilitates in obtaining higher flow velocities and getting clear water. A strainer type tubewell is generally unsuitable for fine sandy strata.

Types of Strainers

There are various types of strainers available for use in tubewells. Each type has its own advantages depending on its cost and efficiency to allow water to enter into the well. Some of the common types used in field are:

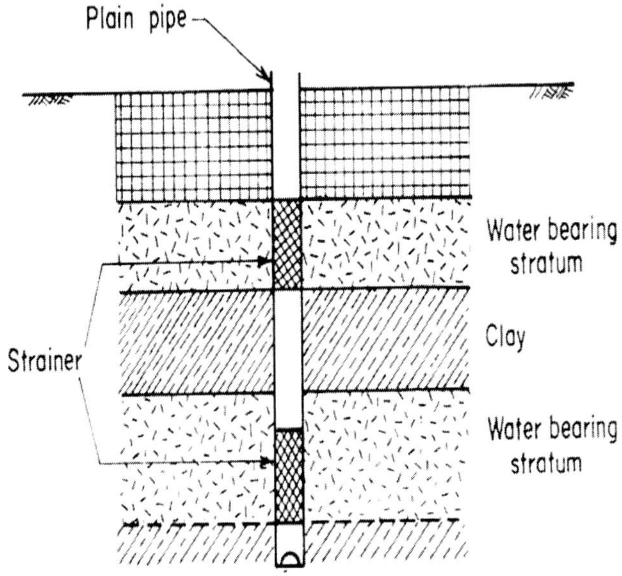

Fig. 14.10 Strainer type tubewell

- Coir rope strainers
- Agricultural strainers or Ashford strainers
- Cook-Tej brass slotted strainers
- Phoenix strainmers
- *P.V.C.* strainers
- Bamboo strainer
- Porous concrete pipe strainer
- Pre-packed filter strainer
- Coconut coir cord strainer
- Brownlie strainer
- Layne and Bowler strainer
- Legget strainer
- Polythene strainer

Coir rope strainer: Coir rope strainers are low cost strainers, generally used for shallow tubewells. Such a strainer consists of a cylindrical frame made of iron rods or bamboo strips, wound round by coir rope of 3 to 5 mm diameter. However, this strainer has low durability (3 to 5 years) and needs to be changed repeatedly. Coating the iron rods with bitumen and replacing coir rope by nylon rope can increase the life of the strainer.

Brownlie strainer: Brownlie strainer is considered to be the best type of strainer. It consists of a polygonal convoluted perforated steel pipe surrounded by a wire mesh consisting of heavy parallel copper wires woven with copper ribbons with a little space in between the two.

Layne and Bowel strainer: Layne and Bowel strainer is a strong but costly type strainer. It consists of a perforated steel or wrought iron pipe over which wedge-shaped steel wire is wound. The strainer pipe joints are made by screwed collars.

Legget strainer: In legget strainer a cleaning device, in the shape of cutters which can be turned in the slits, is provided. These cutters operate from the ground surface and clean the perforations of the strainer clogged by any solid matters.

Phoenix strainer: Phoenix strainer is made of mild steel tubes with slots of desired width. Chromium plating is done to prevent the chemical action of salts.

Agricultural strainers or Ashford strainer: It is a very delicate type of strainer and has long life. It essentially consists of a perforated pipe with a wire wound around it, to provide annular space between the perforated pipe and the wire mesh which is shouldered to the wire. The wire mesh is further protected and strengthened by a wire net over the mesh.

Cook-Tej brass slotted strainers: This type of strainer is the costliest strainer since brass is used for their construction. They are not easily corroded by water and hence have longer lives.

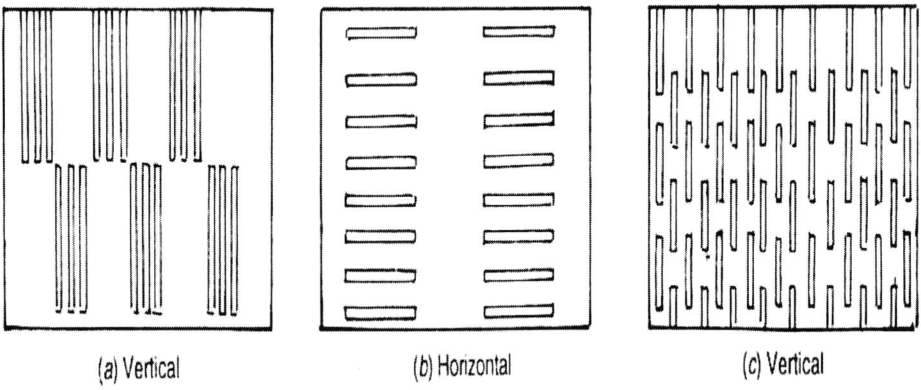

Fig. 14.11: Different patterns of slots in strainer

The slots on these strainers are about 2.5 cm in length and are about 0.5 cm apart. Width depends on the type of sand strata. The patterns of slots may be horizontal or vertical and may be in line or staggered. (Fig. 14.11).

***P.V.C.* strainer:** Strainers made with *P.V.C.* material are useful in case of aquifers with poor quality water. Slot size in such strainers should be carefully selected.

Porous concrete pipe strainer: These types of strainers are prepared from 1:6 cement gravel mixture. It is reinforced with six equally spaced longitudinal M.S. bars to the ends of which flanges are welded for making joints of the pipes. These types of strainers are not hydraulically efficient.

Pre-packed filter strainer: Its cost is very high. However, it gives satisfactory performance. It consists of a slotted pipe to the outer surface of which pea gravel is bonded by means of a special adhesive.

Coconut coir cord strainer: It is a comparatively cheaper type of strainer. Generally it is 10 cm in internal diameter. It consists of steel frame fabricated by welding the ends of eight numbers of 6 mm thick M.S. bars to the periphery of 13 cm long pieces of steel pipe, around which a coconut coir cord of 6 mm diameter is wound. .

Bamboo strainer: Bamboo strainer is a cheaper version of steel casing and brass strainer. In bamboo strainer case the brass strainer is replaced by bamboo strips made as a cylindrical frame by fixing strips of bamboo with nails around iron rings. In stead of nails, thin wires can be used

to tie the bamboo strips with the rings. The rings are placed inside and the bamboo strips are tied over it.

14.11.4.3 Gravel Packed Tubewells

These wells are improved form of strainer type tubewells. These wells are also called as shrouded wells. Shrouding refers to the filling of the surrounded strata of the strainer with a material coarser than that of the natural stratum which is usually gravel of suitable size (Fig. 14. 12). The purpose of gravel packing is to increase the effective diameter of the well as a result of which higher discharge can be obtained. In natural gravel packed well, gravels are available in the strata communicating water to the well whereas in artificial gravel packed well, gravels of suitable size are put in the space between the casing pipe and the strainer.

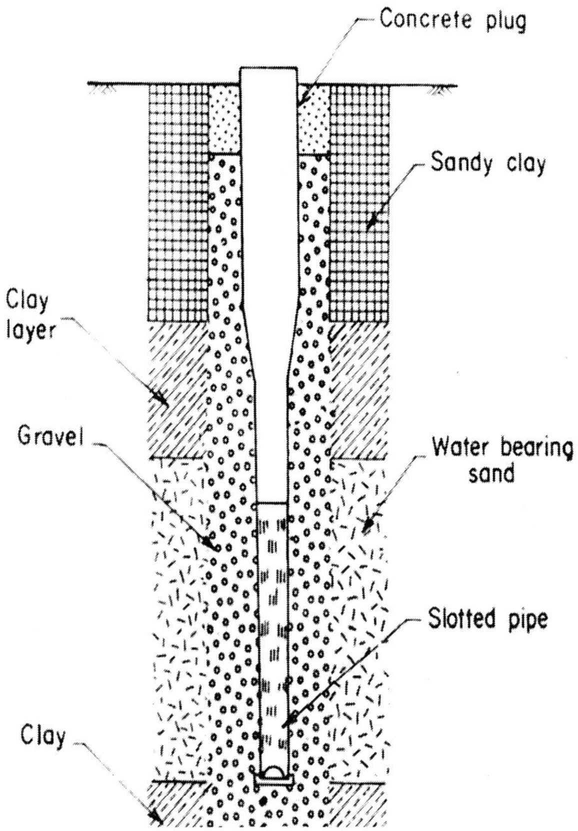

Fig. 14.12: Gravel packed well

14.11.5 Design of Tubewells

Good design of a tubewell aims to ensure an optimum consideration of performances, long service life and reasonable installation and operation cost. The tubewells should give highest yield from the aquifer and should give optimum efficiency in terms of specific capacity. Various design parameters of a tubewell/water well are (i) Well structure (ii) Well diameter (iii) Well depth (iv) Length of well screen (iv) Well screen slot opening (v) Well screen diameter (vi) Gravel pack design and (vii) Pumping arrangement. Gravel pack design is the most important part of design of tube well. Gravel pack design ensures stability of well so that side collapse as tubewell boring is avoided. At the same time, it filters water from aquifer so that good quality water is obtained from the well.

14.11.5.1 Well Structure

It consists of two main elements (i) the upper portion that houses the pumping equipments and the vertical section through which water flows upward from the aquifer to the level where it enters the pump and (ii) lower intake portion of the well. In lower intake portion, a well screen is used which allows water to enter into the well at low velocity, prevents sand and other fine materials and prevents caving-in of the loose formation materials. The intake portion consists of a bore hole drilled into the aquifer to an adequate depth.

14.11.5.2 Well Diameter

Well diameter is generally governed by the size of the pump. Diameter of the well should satisfy the following condition

(i.) Opening should be large enough to accommodate pump and pipes so that it will be easy for installation and efficient operation. The opening of the well should be such that after inserting the casing and housing pipe, 2-3 cm annular space is left for gravel packing.

(ii.) Diameter of the intake section should be large enough as to ensure good hydraulic efficiency of the well. However, the diameter of the well and well pipe is decided from the considerations of permissible flow velocities through the tube. Since, strainers are installed at different levels, the velocity of water in the well pipe of fixed size will not be constant but will be increasing towards the top. Hence, it is theoretically possible to reduce the size of the tube from the top towards the bottom such that there is a uniform velocity throughout the entire length of pipe. It is thus not necessary to have a uniform size of well diameter. In deep wells, well diameter below lowest anticipated pump setting, may be reduced to affect some economy. The lowest anticipated pump setting should be carefully estimated keeping in view future decision of ground water table. In actual practice, it is difficult to install varying size pipes, and hence, a single sized or at the most a two sized pipe may be used. Table 14.3 (Das Gupta, 1987) gives the recommended size of well casing /pipe for different well yields

Table 14.3: Recommended well casing for different wells yields

		Size of well casing	
Anticipated well yield, lpm	Nominal size of pump bowl, cm	Minimum (cm)	Optimum(cm)
Up to 400	10	12.5	15
400-600	12.5	15	20
600-1400	15	20	25
1400-2200	20	25	30
2200-3000	25	30	35
3000-4500	30	35	40
4500-6000	35	40	50
6000-10000	40	50	60

The diameter should be so chosen that it gives desired percentage of open area in the screen (15 – 18%) so that entrance velocity near the screen is within 3 to 6 cm /sec. This will reduce the well losses and hence draw down. Table 14.4 gives representative open area of well screen for various screen sizes (Das Gupta, 1987).

Table 14.4: Representative open area of well screen for various screens sizes

Nominal screen size, inches	Intake area /ft length of screen, sq inches Slot size				
	10	20	40	60	100
3	10	19	32	42	55
10	36	65	110	143	166
20	47	88	156	209	252
30	57	108	192	268	379

Screen size is designated by the size of steel pipe through which screen may be set. Slot number denotes width of openings in thousandths of an inch.

14.11.5.3 Well Depth

Expected depth of a well can be determined from well log of the place where well will be dug or from well log of a nearby well. It is desirable to have full penetration of aquifer so that specific capacity would be higher. Under such a case, the well is called as a fully penetrating well

However if the available screen length is not large enough to fully penetrate the aquifer, it should be centered between top and bottom of the aquifer. Also when the lower portion of the aquifer contains poor quality of water, well screen should be deeper to avoid pumping poor quality water.

14.11.5.4 Length of Well Screen

Optimum length of well screen depends on thickness of aquifer available, draw down and nature of aquifer materials as described below.

(a) Homogeneous Artesian/Confined Aquifer

In this case maximum draw down is the distance from initial piezometric level to top of confined aquifer. Screen length of the aquifer depends on the thickness of the aquifer as provided in Table 14.5 (Das Gupta, 1987).

Table 14.5: Length of screen as varies with thickness of aquifer

Thickness of aquifer	Length of screen
<7.5 m	70% of aquifer thickness
<7.5 – 15 m	75% of aquifer thickness
>15 m	80% of aquifer thickness

The best result is obtained by centering the screen section in the aquifer. However to obtain high yield, screen length or strainer may be provided in the entire thickness of confined aquifer

(b) Non Homogeneous Artesian Aquifer

Most permeable section of the aquifer should be screened.

(c) Homogeneous Unconfined / Water Table Aquifer

To obtain higher specific capacity, a screen as long as possible may be used. It is preferred to use screen in the bottom one third length. In some cases, bottom half may be screened.

(d) Non Homogeneous Unconfined Aquifer

Principles of design are same as case (ii). i.e. non homogeneous artesian aquifer except that screen should be located in the lower most portion of permeable section

14.11.5.5 Well Screen Slot Opening

For homogeneous aquifer slot sizes that contain 40-50% sand are preferable. For non-corrosive wtaer, select higher size opening and for corrosive water, select lower opening. For homogeneous aquifer consisting of gravels and sand, slot openings that retain 30-50% aquifer materials should be selected.

14.11.5.6 Well Screen Diameter

Diameter of the intake portion of the well can be varied without affecting specific capacity of well yield. Doubling the diameter of intake section can increase the well yield by about 10%. Screen length depends on thickness of water bearing strata. Screen opening depends on gradation of sand, where as screen diameter depends on factor that total area of screen opening must be such that the entrance velocity of water is 3 – 6 cm/sec. If velocity is 3-6 cm/sec, then friction loss in screen opening will be less, rate of incrustation will be minimum and rate of corrosion will be minimum.

14.11.5.7 Gravel Pack Design

Design of gravel pack of well consists of the following steps (Anonymous, 1966; Das Gupta, 1987; Ellithorpe, 1970):

(i) Prepare sieve analysis curves for all strata of aquifer. Determine the stratum composed of the finest sand and select the grading of gravel pack on the basis of sieve analysis.

(ii) Multiply 70% size of sand by a factor in between 4-6: 4 if formation is fine and uniform and 6 if it is coarse and non uniform. If formation sand has highly non-uniform gradation and include silt, then multiplier may be 6-9. If 70% sand is retained by a mesh of 0.013 cm, then multiply 0.013 with 5. Hence $D_{70} = 0.065$ cm.

(iii) Construct grading curve for gravel pack by trial and error. Through initial point on gravel pack curve and D_{70}, draw a smooth curve representing a material with a uniformity coefficient of 2.5 or less.

Uniformity coefficient (C_u) is expressed as:

$$C_u = \frac{D_{60}\ (40\ percent\ retained)}{D_{10}\ (90\ percent\ retained)} \tag{14.7}$$

(iv) Prepare specification of gravel pack material by selecting 4 or 5 sieve sizes that cover the spread of curve.

(v) As a final step, select a size of well screen size openings that retains 90% of gravel pack. Width of slot ranges from 1.5 – 4.0 mm and length from 5 -12.5 m. Thickness of gravel pack ≈ 10-20 cm.

14.11.6 Pumping Arrangements

To lift water from the tubewell, three types of pumps are generally used. They are (i) Centrifugal pump, (ii) Bore hole type pump and (iii) Jet pump.

A centrifugal pump lifts water from the lower level to a higher level by creating required pressure with the help of centrifugal action. The maximum suction head under which the pump practically works effectively is about 6 to 8 m. So, this pump is used in places where the

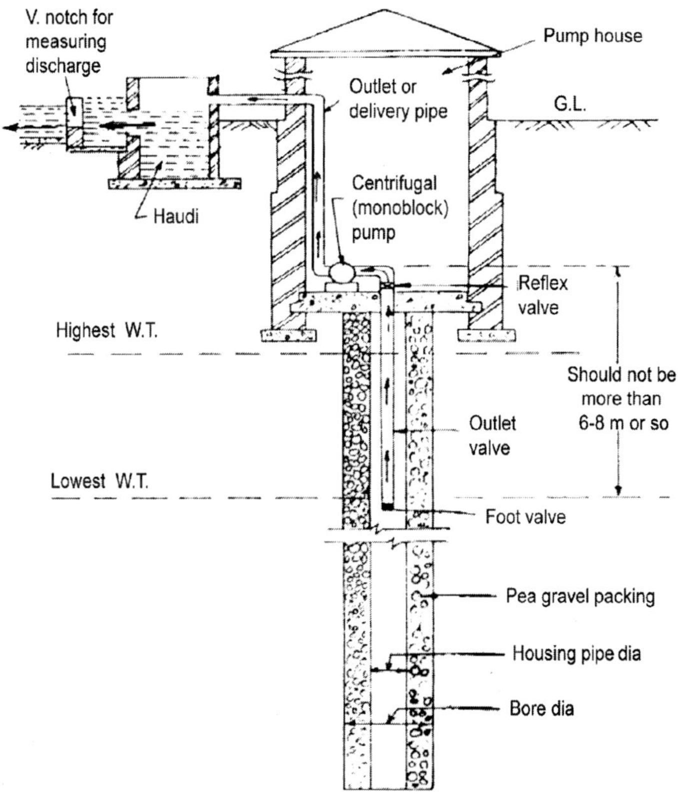

Fig. 14.13 Section of a tubewell using monoblock centrifugal pump

fluctuations of water table plus the depression head is limited to a maximum value of 8 m as shown in Fig. 14.13.

Generally monoblock centrifugal pumps are used to lift water from the well. The pump may be placed below the ground level and just above the highest water level in the well. It should not be submerged in water. Highest water level in the well is the maximum rise level of water in the well which is very near to the ground surface and it occurs in the rainy season. Centrifugal pump arrangement can be done for shallow wells not for deep wells where water is at greater depth. Such type of pumping arrangement is not now used. Rather bore hole type pump are now used to lift water from the well.

Bore hole type pump are mounted on a vertical shaft and are driven by a motor. They are small in size and can be easily lowered in the casing pipe itself. The top 20 to 30 m of the bore hole and the casing pipe is generally kept wider than the remaining normal bore so as to accommodate the pump bowl in the casing pipe.

Bore hole pumps are of two type i.e. (i) submersible pump and (ii) turbine type pump. In submersible pump, the motor and the pump are both attached together and lowered inside the bore, whereas in turbine type, the pump is driven by a direct coupled electric motor and is placed at the top of the line shaft at the ground level. A submersible type bore hole pump is very popular and relatively cheaper than the monoblock centrifugal pump

Jet pumps are also used for lifting water from the smaller tubewells installed for individual domestic supplies where water is not available within 8 m suction lift. Jet pumps can be used for suction lift of 6 to 30 m or so. A jet pump consists of centrifugal pump and a jet mechanism or ejector. In this assembly, motor and pump constitute a small unit like a centrifugal monoblock pump, and is placed at the ground level. Discharge of jet pump is about 20 lit/min against a head of 15 to 30 m above the ground level. The efficiency of jet pumps are not high and so are not generally used in irrigation tubewells. The efficiency is about 35 percent as compared to 65 to 85 percent for other types of pumps (Garg, 2012).

Horse power of the pump motor is computed as:

$$H.P_{motor}. = \frac{\omega \,.\, Q \,.\, H}{75 \ x \ Pump\ efficiency\ x\ Drive\ efficiency\ x\ Motor\ efficiency} \tag{14.8}$$

where,

ω = unit weight of water, 1000 kg/m^3; Q = discharge to be delivered, cumecs and H = total head against which the pump has to work, m.

Detail description about computation of total head and horse power of the pump has been discussed in Chapter 10 earlier.

Sample Calculation 3:

Design a tubewell to deliver 0.042 cumecs of water at a depression head of 5 m. The average water level is 10 m below the ground level in post monsoon season and 15 m in pre-monsoon season. The well log data obtained during investigation is presented below.

Depth, below ground level, m	0 - 13	13 - 18	18 - 30	30 - 50	50 - 65	65 -75	Below 75
Types of strata	Clay	Sand	Clay + kankar	Coarse sand	Clay	Medium sand	Clay + sans stone

There is a horizontal length of pipe 0f 10 m connected to the pump at ground level as deliver pipe to discharge water to the crop field.

Solution:

Design of well diameter and well pipe: Assume the velocity of flow in the well pipe to be 2.5 m/sec.

Given discharge = Q = 0.042 m^3/sec

Hence, area of cross section of the pipe, A = Discharge/Velocity = 0.042/2.5 = 0.0168 m^2

$$\text{Diameter of the pipe} = d_w = \sqrt{\frac{4\,A}{\pi}} = \sqrt{\frac{4\ x\ 0.0168}{\pi}} = 0.1462 \text{ m} = 14.62 \text{ cm}$$

Use 15 cm diameter well pipe.

For easy in operation of entering the casing, housing pipe etc and for gravel packing add another 5 cm to the well pipe diameter so that the well size becomes 20 cm. Thus a bore hole of 20 cm diameter need to be dug.

Length of the strainer: From the well log data is found that the aquifers 30 – 50 m and again 65 – 75 m are good aquifers and they are confined between the clay strata. Thus, the data reveals that it is confined aquifer.

The total thickness of the aquifer = (50 – 30) + (75 – 65) = 30 m

Assuming the homogeneous confined aquifer and since the total thickness of the aquifers is 30 m (> 15 m), 80 percent of the total aquifer thickness may be provided with well screens (Table… 14.5). Thus, the length of the strainer/well screen = 80 percent of 30 m = 24 m.

Check for entrance velocity: Assume 15 percent of the opening area in the screen as slot openings which comes to ð x 0.15 x 15% = 0.0707 m^2.

$$\text{So entrance velocity} = \frac{Disch\ arg\ e}{Length\ \ of\ \ strainer\ \ x\ slot\ \ opening\ \ area} = \frac{0.042}{24\ x\ 0.0707} = 0.024\ m/\sec$$

Entrance velocity is computed as 0.024 m/sec = 2.4 cm/sec.

Since this value lies between 2 to 3 cm/sec, the entrance velocity checked is okay.

Type of pump required: The fluctuation of water table during pre-monsson and post monsoon season is 15 – 10 = 5 m and depression head given is 5 m. Thus total head comes to 5 + 5 = 10 m which is more than 8 m. Therefore, centrifugal pump will not work. Use a bore hole type pump to lift water from the tubewell.

Horse power requirement of the pump motor: Eq. (14.8) is used to estimate the horse power requirement of the pump motor.

Total head against which motor has to work = H = Maximum depth of water table + Depression head + Velocity head + Losses

Maximum depth of water table = 15 m

Depression head = 5 m

With discharge of 0.042 m³/sec, and using 15 cm diameter well pipe, the actual velocity of flow in pipe is = $V = \dfrac{0.042\ x\ 4}{\pi\ x\ (0.15)^2} = 2.38\ m/\sec$

Velocity head = $\dfrac{V^2}{2g} = \dfrac{(2.38)^2}{2\ x\ 9.81} = 0.289\ m$

Losses of head in pipe = Loss of head due to friction in pipe (h_f) + Losses at bends in strainers etc. which may be assumed as 25 percent of h_f where h_f *is given as:*

$$h_f = \frac{f'.l.V^2}{2.g.d}$$

where $f = 0.024$ and l = length of the pipe = total vertical length of pipe including blind pipe and strainer + horizontal length of delivery pipe = 75 + 10 = 85 m

$$\text{Hence, } h_f = \frac{0.024\ x\ 85\ x\ (2.38)^2}{2.\ x\ 9.81\ x\ 0.15} = 3.93\ m$$

Losses in strainer and bends = 0.25 x 3.93 = 0.98 m

Total head losses = 3.93 + 0.98 = 4.91 m

Hence, total head, H = 15 + 5 + 0.289 + 4.91 = 25.199 = 25.2 m

Assume pump, drive and motor efficiencies as 0.65, 0.80 and 0.80, respectively.

So horse power requirement of the motor s calculated as:

$$H.P_{motor}. = \frac{\omega.Q.H}{75\ x\ Pump\ efficiency\ x\ Drive\ efficiency\ x\ Motor\ efficiency}$$

$$\frac{1000 \times 0.042 \times 25.2}{75 \times 0.65 \times 0.80 \times 0.80} = 33.9 \approx 35\text{H.P.}$$

Hence, use a 35 H.P. motor to lift water from the deep tubewell.

14.12 Investigations for Groundwater Development

For groundwater development of an area, it is essential to have knowledge on occurrence and movement of groundwater. Moreover, the hydrologic parameters of the aquifers and geological conditions of the area are required to decide the location and types of wells. Groundwater

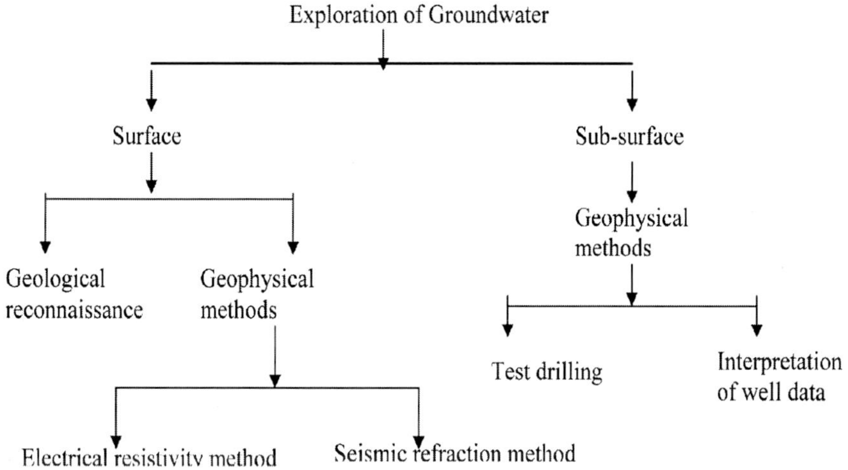

Fig. 14.14 Chart showing methods of exploration of groundwater

exploration studies will help in determining the prospective areas for setting up tubewells. This is often referred as groundwater prospecting. Methods of exploration of groundwater are given in the chart (Fig. 14.14) as below:

14.12.1 Geological Reconnaissance

It helps in understanding the geology of the area in relation to possible occurrence of groundwater. A study of structural geology in conjunction with stratigraphy to locate possible water bearing formations is required. Stratigraphy helps in locating the position and thickness of water bearing formations and confining beds. Sand, gravel etc. are the best aquifers whereas clay, shale and crystalline rocks do not yield sufficient water for explorations.

The geological studies by which the groundwater my be investigated would comprise of making a detailed geological mapping of the area , which may give accurate knowledge of the existing geological structures, including the lithological character and petrological features of various underground rocks and strata.

14.12.2 Surface Geophysical Methods

Surface geophysical methods also help to estimate where groundwater occurs. It also helps in assessing the possible areas where good quality groundwater exists. Though these methods are not always successful but they are less costly than the sub-surface investigations. These methods were first used in exploration of possible sites for petroleum and mineral industries. There are mainly two types of surface geophysical methods used for groundwater exploration. They are (i) electrical resistively method and (ii) seismic refraction method. Principles of these methods are outlined here. For detailed procedures for the tests, any standard Groundwater Hydrology Book may be referred.

14.12.2.1 Electrical Resistivity Method

Electrical resistivity makes use of the fact that water increases the conductivity of the rocks and thereby decreasing their resistivity. Hence, if it can be established geologically, that the same rock formation is existing for a certain depth and by electrical testing it is found that the resistivity is decreasing below this depth, then it can be easily concluded that the water is present below this later depth as decided by electrical testing. The resistivity of rocks at various depths can be calculated based on the following principles. .

In this method, four electrodes are inserted into the ground. Out of these four electrodes, the outer two electrodes are called current electrodes (C_1 *and* C_2) and the inner two are called as potential electrodes (E_1 *and* E_2). Current electrodes consist of metal rods driven into the ground. Potential electrodes are porous cups filled with a saturated solution of copper sulphate. Because of potential difference between the inner two potential electrodes, current will flow forming current lines. Similarly, equipotential lines are formed which are the lines joining points of equal potentials. The current lines are orthogonal to the equipotential lines (Fig. 14.15).

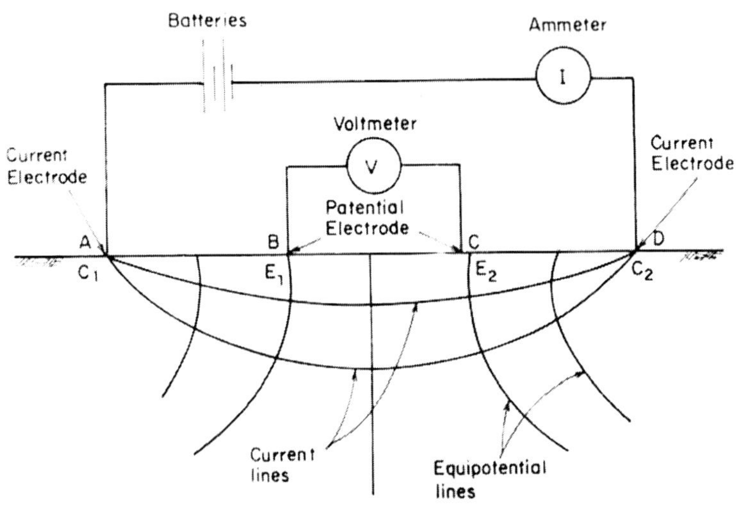

Fig. 14.15 Potential difference between two electrodes

Certain amount of current (as measured by ammeter) is allowed to flow through the outer two electrodes (C_1 *and* C_2) and the resulting potential difference between the inner two electrodes (E_1 *and* E_2) is measured by a voltmeter. When the distance AB = BC = CD = *a* (known as Wenner arrangement), the resistivity is given as:

$$\rho = 2\pi a \frac{V}{I} \tag{14. 9}$$

where,

V = potential or voltage, I = current, ρ = resistivity and a is as defined above.

When the distance BC = b and AD = L, (known as Schlumberger arrangement), the resistivity is given as:

$$\rho = \pi \frac{(L/2)^2 - (b/2)^2}{b} \frac{V}{I} \tag{14.10}$$

D.C. current is applied between the electrodes and for different spacing of the electrodes, values of *V and I* are measured. When the spacing between the electrodes is increased, a deeper penetration of the electrical field occurs and a different resistivity value id obtained. The values of the resitivity are plotted against electrode spacing for various spacing at one location and a smooth curve is plotted. Comparing the nature of this curve with the curves obtained under similar conditions, the nature of the strata and also the depth of the water table is obtained.

14.12.2.2 Seismic Refraction Method

In this method, a small wave called as seismic wave or shock wave is created at the earth's surface either by using a dynamite charge or by the impact of a heavy instrument. The time required for the sound or the shock wave to travel an unknown distance is measured. Knowing the velocity of the sound and time required for the wave to move down, strike the geologic formation and then to come back to the surface is measured. From these two known observations, one can interpret at what distance the geologic formation is present and hence gives an indication of water bearing strata. The principle underlying these seismic methods is that the velocity of the elastic waves which is caused by the shock is different in moist deposits than in the dry deposits of the same composition. Since the velocities of the shock waves depend on the type of formation and the presence of water, it may be possible to detect and estimate the presence of water table depth below ground surface from the differences in the indicated velocities recorded at several geophones.

14.12.3 Sub-surface Geophysical Methods

Geophysical well logging is an accurate and convenient way of obtaining many sub-surface data. This provides information about the nature of formations in their nearly undisturbed conditions. This information is useful in designing tubewells in an area.

Sub-surface geophysical methods consists of test drilling and obtaining well logs or bore hole logs. Logs may be defined as the records of the sub-surface investigation and provide useful interpretation regarding nature and properties of the materials occurring at various depths below the ground surface. These records may be presented in the form of tables or graphic plots with symbolic descriptions. The data making the basis of these records may be obtained by different methods and accordingly there are different types of bore hole logs. Some of them are:

- Geological logs or well logs
- Resistivity logs
- Sonic logs
- Thermal logs and
- Radiometric logs

Geological logs are those which represent the type of the strata existing at different depths and encountered during direct digging or boring of the wells (Fig. 14.16 a).

Resistivity logs are those which indicate the values of electrical resistivity of the rocks at different depths from the surface downwards in bore holes (Fig. 14.16 b).

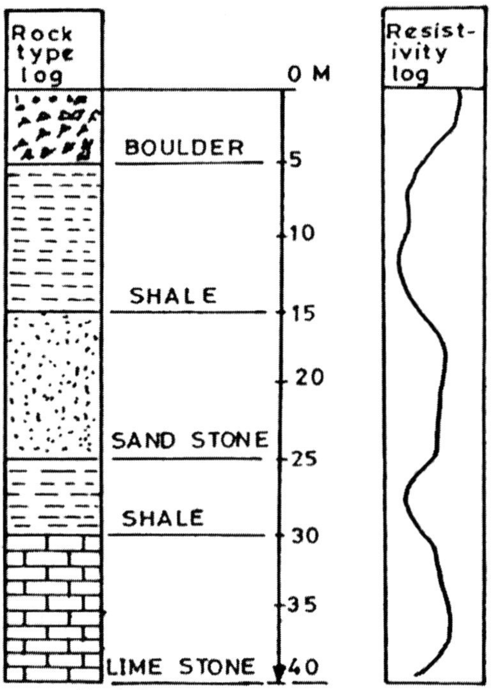

(a) Rock log or geological log. (b) Resistivity log.

Fig. 14.16: Bore hole logs

Sonic logs indicate the values of velocities of compressional waves at different depths. Thermal logs indicate the variation in temperature with depths whereas radiometric logs represent the variations of radioactivity with depths. Details of geophysical aspects of groundwater are given in Sharma and Chawla (1977) and Roscoe Moss Company (1990).

14.13 Construction of Tubewell

The construction of tubewell consists of the following operations:

- Exploratory drilling
- Drilling operations
- Shrouding of tubewell and
- Development of tubewell

14.13.1 Exploratory Drilling

It is desirable to drill a test hole first to the desired depth and collect preliminary information regarding (i) static water level, (ii) types and nature of strata met with, (iii) quality of groundwater and (iv) probable yield.

When such test drilling is done, the following procedures must be followed.

(i) The test hole drilled should be at least 10 cm in size.

(ii) A complete and accurate bore hole log of the test hole should be maintained.

(iii) Static water level in the well after some time when the rise of water level is stabilized is recorded.

(iv) The total depth of drilling shall be sufficient to determine the depth and other characteristics of the lowest aquifer as anticipated from other local considerations.

(v) During drilling, water samples from different strata are collected for chemical analysis to ascertain the quality.

(vi) In case of percussion drilling, static water level in the bore hole is measured daily in the morning before commencement of drilling operations and in the evening at close of drilling operations, to give an indication of behaviour of the aquifers met with during drilling operations.

14.13.2 Drilling Operations

The term drilling also called a boring refers to the process used in making opening in the ground for the purpose of well construction. The important methods used for drilling are as given below:

(a) Percussion drilling (also called as cable toll method) which is again of two types i.e. hand boring by rope or rod and mechanical boring with power rigs.

(b) Hydraulic rotary drilling or simply rotary drill which is again of three types i.e. direct circulation hydraulic oratory, (ii) reverse circulation hydraulic rotary and (iii) rotary air percussion drilling.

14.13.2.1 Percussion Drilling

This method of drilling is also called as cable tool method or standard method of drilling. This is called as percussion since the well hole is made by percussion i.e. by hammering and cutting. This method is best suited to well drilling in any formation from soft clay to the consolidated hard rock materials and is generally unsuitable for loose formations. The drilling can be accomplished either by hand boring by rope or by mechanical means using power rigs.

Hand boring by rope or rod: Hand boring by manual process is the simplest, cheapest and most widely used process. In this method, a pit is dug at the centre of the site which measure about 2 to 2.5 m in diameter and 2 to 2.5 m depth. The boring tube also called a casing pipe is lowered into the centre of the pit. A cutter shoe is fixed at the bottom of the casing pipe to facilitate in cutting the bed materials. After lowering the first casing pipe in the pit, it is clamped in position with wooden blocks. Casing pipe is then filled with water.

A wire rope passing over the pulley which is fixed to a tripod stand is tied to a bailer which does the actual cutting of the bed materials. For raising and lowering the bailer, the rope is tied to a winch so that the wire rope can be wound or unwound as required (Fig. 14.17). The purpose of the tripod stand is to see that the bailer moves up and down in the casing centrally. The bailer used in drilling is a steel pipe of about 6 mm thickness and is about 5 cm less in diameter than the casing pipe. The length varies from 2 m depending on the size of the bore. The bailer has a cutting edge at the bottom and a hook at the top. A flap is fitted inside the bailer just above the cutting edge. The bailer moves up and down in the casing pipe and a slightly circular motion is automatically imparted to it by the torsion in the wire rope. During

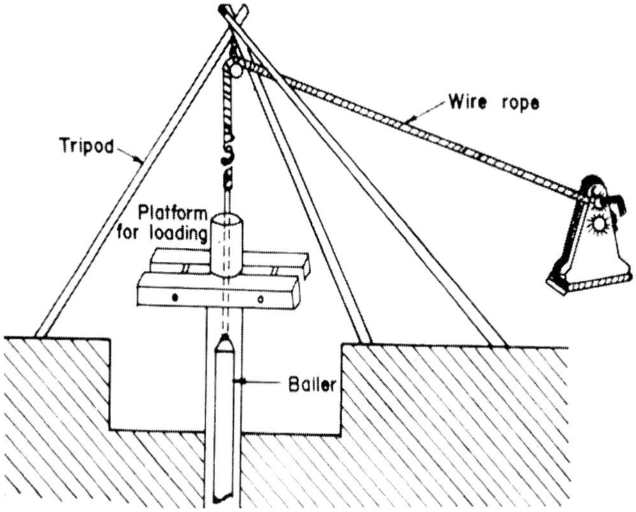

Fig. 14.17: Manual percussion boring

the downward stroke, the flap valve gets open and the cutting bed materials get pushed inside the bailer. During the upward stroke, the flap valve in the bailer is closed due to the weight of the material inside and thus the materials already entered into the bailer cannot come out. After about 30 to 40 strokes the bailer is lifted out of the casing pipe by winding the rope on the winch and the loose material retained inside the bailer is emptied out.

The casing pipe is clamped with wooden sleepers at a convenient point and is then loaded with sufficient weight so that the weight helps the casing pipe to move down automatically in the bore hole as the bailer cuts the bed materials at the bottom. The boring continues and the casing pipes are screwed one after the other till the pipes reach the bottom end of the required depth. Water is poured into the casing pipe at top so that it cleans the bore hole and comes out at the end of the pipe into the bore hole. Hand boring by rope is suitable for soft soil strata or alluvial strata. In hard rock strata, rods are used instead of wire rope to connect the drilling tools. The rod helps in rotary drilling instead of percussion drilling in hard rock strata.

Mechanical boring with power rigs: Mechanical equipments for percussion drilling consist of a portable rig mounted on a truck chasis or trailer (Fig. 14.18). The important components of this equipment are:

- A folding mast of at least 10 m high
- A two line hoist

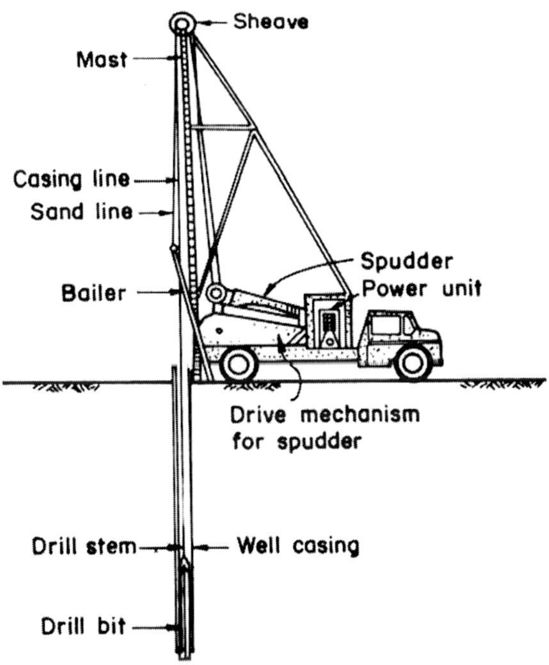

Fig. 14.18: Truck mounted drilling equipment

- A line for operating the drilling tools
- A line for extracting the bailer
- A spudder for raising and dropping the tools
- A diesel engine to provide power to continue the different operations during drilling

The drilling tools consist of (i) drilling bit, (ii) drilling stem, (iii) drilling jars and (iv) a rope socket with mandrel. The drilling bit is the actual cutting machine which breaks the hard strata and has chisel shaped straight edge. The mandrel holds the cable rigidly in the rope socket. The rope socket is joined with drilling jar. The drilling jar helps the wire rope not to get stuck up in sticky formation. The drilling stem is used to connect the drilling jars on one end and the drilling bit on the other. It also increases the weight of the drilling bit.

The process of boring is same as that in case of hand boring method. But in this case the speed of drilling is high since power rigs are used in boring. Moreover, large sized and deep boring is possible by using power rigs. The progress of percussion drilling depends on the nature of strata which is encountered. In sandy or clayey strata, progress is as much as 25 to 30 m whereas in hard rock, it may be just 1 to 2 m.

14.13.2.2 Rotary Drilling

Rotary drilling is a powerful and quick method of drilling which is suitable to almost all types of formations. In case of alluvial soil, the art of drilling is fast and in hard rock strata, the act is slow. The operation consists of rapidly rotating a column of pipes at the lower end of which a cutting bit or drill bit is attached. The bit cuts a hole slightly larger than the size of the drill pipe so that the cut materials can be removed through the annular space of the bore and drill pipes. As the drill bits rotate, simultaneously a downward thrust is also applied on the bit through the drill pipe so that the cutting points penetrate through the strata. The drilling bits have hollow shanks and have one or more centrally located openings for the flow of the drilling mud.

The rotary drill is usually stronger as it has to withstand the turning effect of the rotating drill.

The important parts of the machinery are:

- Revolving table or turn table
- Hydraulic pumps
- Line shaft and
- Engine or electric motor

The drill pipe also called a s drill rod carries the drill bit at the bottom and is screwed at the top to a pipe of square in section known as Kelley. The Kelley slides down along with the drilling rod as the bore deepens. The turn table is revolved by a bevel gear arrangements connected with the line shaft and is controlled by a clutch.

In the rotary drilling method no casing pipes are used. Instead the materials are cut, made into small pieces and are removed by maintaining a constant circulation of drilling fluid under hydraulic pressure. For removing the cut materials, two methods are used. They are (i) direct circulation hydraulic rotary method and (ii) reverse circulation hydraulic rotary method.

Direct circulation hydraulic rotary method: In this case, mud laden water is forced down under pressure through the drilling pipe and the bore hole and while it comes up, it brings along with it drill cuttings. In mud mixing, local mud or special chemical like "Aquagel" or bentonite are used. The mud used has several purposes as follows:

- It brings up the drill cuttings.
- It plasters the inside walls of the bore hole so that the walls are not collapsed
- It keeps the drilling cool at the time of drilling operation.
- It also prevents cuttings to get settle down to the bottom of the bore of the bore if circulation is interrupted.
- It speeds up the rate of drilling by softening the formation.

A slush pit is used to prepare mud and a swivel mud pump is installed for pumping the mud (Fig. 14.19). A stand by pump is used in case of rotary drilling if first pump fails to work. This is done to ensure that there will be continuous drilling till the completion of bore. During drilling if a clay stratum is encountered, than no mud is required to be added to water. Rather water used for drilling may be clear. If the formation has loose soils, than immediately casing pipes are to be entered into the bore hole so that the sides of the bore hole do not collapse. Further the pumps used in forcing the mud into the bore hole should work continuously without

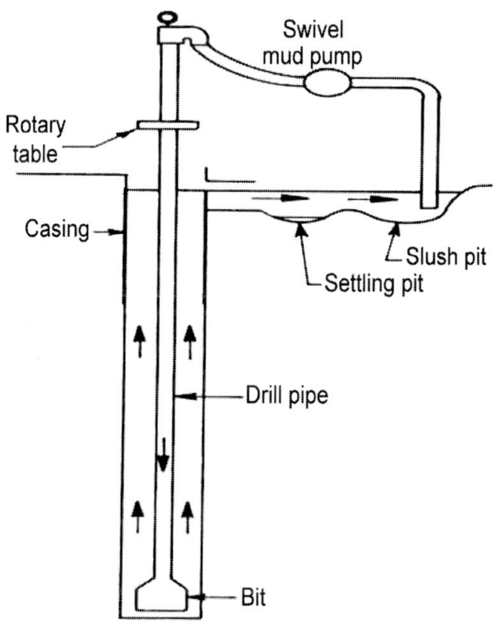

Fig. 14.19 Direct circulation hydraulic rotary method

any break. If the pump is stopped during operation than the cut materials may settle at the bottom of the bore.

Reverse circulation hydraulic rotary method: Direct circulation hydraulic rotary method is suitable for soft aquifer materials and for small size bores. In stead, a reverse circulation hydraulic rotary method is employed for drilling in heavy gravel or for drilling large sized bores. In this method, the water is pumped down through the space between the drill pipe and the bore hole. The water is drawn out through the drill pipe by means of a centrifugal pump (Fig. 14.20). The water brings along with it the drill cuttings. In case of reverse rotary method, more water is required than that required by direct rotary method.

Rotary Air percussion drilling: In conventional rotary drilling, energy is applied in two ways. They are by rotation of the bit and by static force exerted on the bit by means of weight on the bit applied by drill string. A third source of supply of energy to the bit in the form of percussion impact is added by rotary air percussion drilling. Rate of penetration of the drill bit is increased several times due to this third source of energy supply. The energy is applied by an air actuated single piston hammer. Percussion is accomplished by forcing compressed air through the hammer.

In addition to the above methods of drilling tubewells, there are some other methods like auger method, core drilling and water jet method. These methods have limited applications and are to be used in certain specific conditions. Hence, they are not discussed in this chapter.

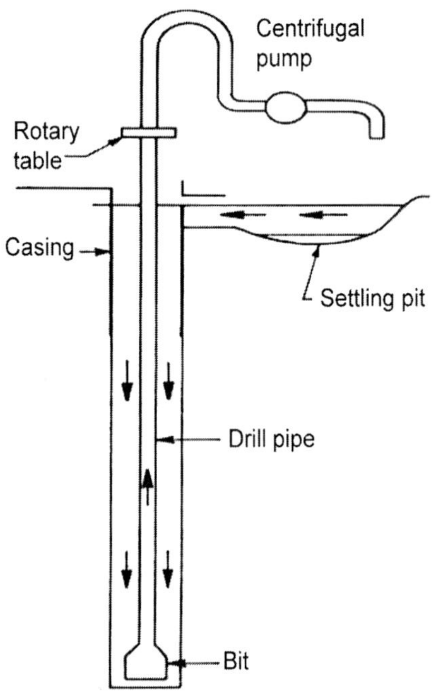

Fig. 14.20 Reverse circulation hydraulic rotary method

Choice of boring method depends on the many factors including the simplicity in construction and operation, cost of construction, convenience in handling and transporting the equipment, rate of drilling and type of formation materials. The advantages and disadvantages of each method are discussed as above. The specific advantages of each method are outlined as below.

14.13.2.3 Advantages of Cable Tool Method

The advantages are (Garg, 2012):

- A more accurate sample of formation can be obtained.
- Lesser amount of water is required during operation.
- Cable tool rig is lighter and hence easy to transport.
- Very much helpful for consolidated rocks.
- Cheaper compared to rotary method.

14.13.2.4 Advantages of Hydraulic Rotary Method

The advantages are (Garg, 2012):

- Can be used for larger size bores up to 1.5 m diameter.
- Can be best used for drilling test holes, because the hole can be abandoned with minimum cost.
- Rotary drilled wells can be gravel packed which increases its specific capacity and causes less sand trouble in getting water from the tubewell.
- Casing is to be driven only after the hole completion, and hence, can be set at any depth.
- It is a faster method of drilling and is essential suitable for unconsolidated formations.
- It can handle alternate hard and soft formations with ease at low danger of accident.

14.13.3 Shrouding of Tubewells

Shrouding of tubewells refer to gravel packing of the constructed tubewells. In this process, the annular space around the strainer is filled with gravels of size coarser than that of the natural aquifer so that the area of contact between the aquifer and the tubewell is increased which increases the specific capacity of the tubewell. The main advantages of the shrouding are (Sharma, 1987):

- It permits utilization of aquifer with poor yields characteristics through the increased effective diameter.
- It decreases the head loss of water that enters into the well.
- It prevents collapse of well due to creation of cavity around the casing.
- Drawdown is decreased due to reduced entrance friction.

- It acts as a storage basin which is of importance where most of the supply is coming from thick sandy aquifer.

Gravel packed well may be natural packed well or artificial packed well. The artificially gravel packed well differs from naturally developed well in the sense that the zone immediately surrounding the well screen is made more permeable by removing the formation materials and then packing with artificially graded coarse material. By doing so, effective diameter of well increases.

In naturally developed well, slot opening retain 40-50% of aquifer materials and allow 60-50% to enter into well during development. In gravel packed well, graded gravel is so chosen that it will retain all formation materials; a well screen opening is then chosen so that it will retain all gravels. Gravel packed well costs more than a naturally developed well. Artificial gravel pack is necessary for formation like fine uniform sand, thick artesian aquifer, loosely cemented sandstone etc. In general artificial gravel pack is needed when aquifer material is homogeneous with uniformity coefficient <3 and effective grain size <0.25 mm.

Detail procedures of design of gravel packed wells have been reported earlier in this chapter.

14.13.4 Development of Tubewells

Development of tubewell refers to operations of flushing, testing and equipping the tubewell before it is put in actual service so that good quality groundwater of sufficient quantity can be obtained from tubewell. During drilling operation, muds clog the water bearing formation which decreases the entry velocity of water into the well. Moreover, clogging increases the head loss and decreases the well efficiency. Development of tubewell clears the clogging and eradicates the above mentioned problems. The specific capacity of the well increases i.e. more discharge is obtained with less drawdown. The development causes the gravel pack and the formation around the well to settle and compact itself against the screen, thereby leading to greater stability.

The various methods of development of tubewell are based on the principle of agitating water in the formation with a view to remove the finer material near the casing and to prevent bridging of the sand particles across the screen openings and in the formation around the well.

14.13.5 Methods of Development

Depending on the formation characteristics of the aquifers, a well may be developed by the following ways:

- Over-pumping
- Back-washing
- Surging
- Compressed air
- Jetting

14.13.5.1 Over-Pumping

It is the simplest and the most commonly used method of well development. In this method, the well is pumped at a rate higher than the designed rate by large capacity and variable speed pump. Starting with slow speed, the rate of pumping is gradually increased. The rate of speed goes on increasing till sand free good quality groundwater is obtained. At this stage, sand and gravels surrounding the screen get stabilized. The discharge of water during development should be free from sand during the operating test run, with a maximum tolerance of 20 parts of sand in one million of water by volume after 20 minutes of pumping, but should at no stage, exceed 50 parts of sand per million parts of water by volume for irrigation wells.

Step draw-down test is conducted by installing a test pump in the tubewell temporarily and pumping out water at various speeds. At each rate of discharge, pumping is carried out at least for 30 minutes. If the water level and discharge are found to be fluctuating, development is carried out for some more time till discharge becomes steady and water contain tolerable limit of sands.

14.13.5.2 Back-Washing Method

Commonly used methods of back-washing development are:

(i) **Back-washing with bailer:** It is the simplest and most effective method of well development. In this method water is fed into the well as fast as possible and is removed at a faster rate with sand pump or bailer. The movement of water agitates the formation around the well. The faster the reverse movement of water through the screen, the more is the agitation in the formation. Similarly, the faster the water is bailed out, the more vigoursly the fine particles are sucked into the well from the formation.

(ii) **Back-washing under pressure:** In this method, water is fed into the well to cause outward pressure in the formation around the well. This is done either by connecting a pump or hose line from a source of supply directly to the casing pipe, or by providing top connection with a side valve outlet and suspending a pipe line to the bottom of the well for feeding water to the aquifer through the screen. Large quantity of water is fed from the hose line into the well under high pressure. After about 2 to 5 minutes, the connection is removed and the well is pumped vigorously.

(iii) **Back-washing by raw-hiding:** In this method, air lift pump is used. The pump discharges water for sometime and then it is stopped. Again it is operated and stopped. Thus pump operates intermittently to produce rapid changes in the pressure heads in the well. It is desirable to start raw-hiding at low capacity.

Procedures for Back-washing: Following procedures may be adopted for back washing.

(i) In the first procedure, the well is pumped to maximum capacity and then is suddenly stopped so that water level in the well falls down and causes reversal of flow. This agitates the formation materials around the screen. When water level in the well comes up to its original level, it is pumped again and the process continues till maximum discharge and sand free discharge is obtained.

(ii) In the second procedure, the pump is discharged at maximum rate and then is suddenly stopped. Water level in the well goes down. Before the water level in the well regains its original position, it is again pumped at maximum speed. The process causes rapid reversal of flow under a high pressure which agitates the formation materials around the screen. The alternate starting and stopping of the pump causes vigorous shaking of the aquifer materials adjacent to the screen and thereby resulting more effective well development.

(iii) In the third procedure, the pump is started and as soon as water reaches the ground level, the pump is stopped. This causes some reversal of flow in the well and agitates the formation materials around the screen. Where static water level is very deep so that water level in it falls quickly, the method works well.

14.13.5.3 Surging

Well development by surging is done by raising and lowering a plunger. The movement of the plunger is restricted in the casing pipe. Casing pipe is provided above the screened position of the well. The up and down movement of the plunger forces water alternatively into the surrounding formation and then allows it to move back into the well. The outflow portion of the surge cycle on the down stroke forces water outward through the screen to break the bridging of finer particles. The inflow portion of the surge cycle moves the fines towards the screen and into the well from where they were removed. Two types of surge plungers are used. They are (i) solid surge plunger and (ii) valve surge plunger.

Solid surge plunger is made from circular belting and plyboard blocks tightened between heavy steel washers and two faced couplings. The belting discs are cut to fit freely in the casing pipe. The valve surge plunger is convertible to solid type plunger by replacing the washers of small diameter by solid plates of diameter equal to that of wooden block. Solid type plungers do not work well in the low yielding aquifers whereas a valve type plunger works well there. The plunger should not be stopped at the bottom to avoid its becoming sand locked in the screen.

14.13.5.4 Compressed Air Method

Development of tubewell with an air compressor is commonly used method and is effective for small diameter wells. There are two general techniques. They are (i) surging and (ii) back-washing. Sometimes a combination of these two methods is used.

Back-washing techniques: It involves forcing of the well water back into the aquifer by means of compressed air introduced into the well through the top of the casing after it has been closed. When the pressure is released, the water flows back into the well through the screen to bring the fine particles from the area surrounding the well, thus ensuring their removal. The process is continued till no sand is brought in.

Surging techniques: It involves the principles of both the pumping method and the surging method as described earlier. In this method, a large quantity of air is released in the water to produce a strong surge which consequently dislodges the aquifer materials. Air compressors mounted on the trolleys and run by diesel engines are normally used for this technique. A 5 cm diameter air pipe is lowered into a 15 to 20 cm diameter discharge pipe (also called as education pipe or drop pipe). The lower portion of the air line has perforations up to a length of 2 m and the end is sealed. In the beginning the air line extends below the discharge line and is used to release a large volume of air. This causes a strong surge in the strata and dislodges the aquifer materials surrounding the well screen. The air line is raised into the discharge pipe and consequently an air jet pumping action is created causing the water to flow out. When sand free water comes out, the end of the discharge pipe is raised and the above process is repeated till the entire length of the strainer is subjected to surging and pumping action.

14.13.5.5 Jetting

Well development by jetting with water at high velocity is an effective method. The method involves operating a horizontal water jet inside the well in such a way that the high velocity water stream shoots out through the screen opening. This washes out the fine materials of the aquifer. The turbulence created by the jet brings these fine materials back into the well through the screen openings above and below the point of operation. By slowly rotating the jetting tool, and by gradually raising and lowering it, the entire surface of the screen can be covered.

14.14 Life of Tubewells and Causes of Failure

A normal tubewell lasts for about 15 to 20 years. Useful life of a tubewell depends on its (i) design, (ii) construction, (iii) types of aquifers tapped, (iv) quality of water, (v) discharge and drawdown at which it is operated and (vi) maintenance.

After a tubewell is constructed and used for some years, there may occur some problems including obtaining low yield. If the problems continue than after some years the well may completely fail to give any yield. A well which suffers from reduction in yield is termed as sick well. A sick well is inefficient and costly. The common well problems occur due to (Edward, 1966; Michael, 1978):

(i) Inflow of fine sand, silt and /or muddy water into the tube well.

(ii) Low discharge due to improper well development

(iii) Low discharge and muddy water due to improper gravel pack design and screen

(iv) High rate of pumping which result in the depletion of water in the aquifer

(v) Corrosion or incrustation problem

(vi) Mechanical failure while lowering and setting tube well assembly

(vii) Well screen not provided in good aquifer as per litholog study

(viii) Interference of well

(ix) Wells constructed nearer to barriers like mountain

The two main causes for failure of a tubewell are due to incrustation or corrosion which are discussed as below.

14.14.1 Incrustation

Incrustation of the well pipe occurs due to the deposition of alkali salts on the inside walls of the pipes. The salts that cause incrustation are calcium carbonate, calcium and magnesium sulphates and silicates. Because of incrustation, the effective diameter of the well pipe gets reduced which results in obtaining lower well yield.

Incrustation can be controlled by:

- Reducing the drawdown and hence the pumping rate
- By using screens having larger areas of openings (or larger size pipes) so as to allow some allowances for the future incrustation
- Some chemicals like acids may be used to remove the incrustation, but care should be taken such that it does not affect the strainer materials.
- By properly maintaining and periodically cleaning the well screens.

14.14.2 Corrosion

The well pipe is gradually destroyed by corrosion due to the action of acidic water on the pipe material. When chlorides and sulphates or carbon dioxide are present in water, the well pipe will get corroded. The aquifer sand, surrounding the well pipe, finds a way out into the corroded pipe through the worn out pipe walls, thus bringing sand along with water. Thus, corrosion results in excessive withdrawal of sand along with well water.

Stainless steel strainers, though costly, are good to be used as well pipes to avoid corrosion. However, cast iron or ordinary steel screens may be used but they need galvanizing to avoid corrosion. Other measures that can be used to control corrosion are:

- By reducing the drawdown and hence the pumping rate
- By reducing the flow velocity by increasing the percentage of open area or size of the well pipe
- By using thicker pipes
- By using corrosion resistant materials for the pipes and
- By using corrosion resistant coatings on the pipes such as galvanization of pipes.

14.15 Verticality of Tubewells

If vertical turbine pumps are purposed to be installed in tubewells, it should be ensured that the tubewell is vertical by means of a verticality test. This is essential to avoid the misalignment of the line shaft assembly which otherwise will result in excessive wear and tear of the shaft bearings. Installation of tubewell in vertical position will reduce the power consumption. The equipment to check the verticality of tubewell consists of (i) a tripod stand having at the top a metallic disc with a small rotating pulley at the centre, (ii) a plumb of about 40-45 cm long with diameter 0.5 cm smaller than the inside diameter and (iii) a reel of flexible steel wire. For details of verticality test of tubewells, any book on Groundwater Hydrology may be referred.

Question Banks

Q1. Differentiate between surface water and groundwater.

Q2. With a neat label sketch, describe the distribution of sub-surafce groundwater.

Q3. Differentiate between the followings:

(i) Zone of aeration and zone of saturation

(ii) Confined aquifer and unconfined aquifer

(iii) Perched aquifer and confined aquifer

(iv) Artesian well and flowing well

(v) Observation well and piezometer.

Q4. How does a flowing well originate?

Q5. Define aquifer. Give some examples of good and bad aquifer.

Q6. Why coarse sand is a good aquifer whereas clay is a bad aquifer?

Q7. With the help of hydrologic cycle, describe the occurrence of groundwater?

Q8. Define specific yield and specific retention. What is the difference between them?

Q9. Porosity of an aquifer is 50 percent. If volume of water obtained by gravity drainage is 0.3 time the total volume of aquifer drained, find out the specific yield and specific retention of the aquifer. If the total volume of the aquifer drained is 500 m^3, then calculate the volume of water held against gravity drainage.

Q10. What is storage coefficient and how is it estimated?

Q11. What is a well irrigation? What is its advantages and disadvantages?

Q12. What points need to be taken into consideration for selection of ideal site of design of well.

Q13. What is an open well? Write notes on various types of open well?

Q14. Differentiate between impervious and previous lined well.

Q15. What are the differences between shallow and deep well.

Q16. Describe the pumping test and recuperation test for determining the yield of an open well.

Q17. Design an open well in fine sandy soil with safe yield 0.45 m^3/hr /m^2/ unit depression head so that an yield of 0.0045 m^3/sec is discharged from well with depression head h = 3 m

Q18. Calculate the average yield of a well of 2,5 m diameter by recuperation test. The water level in the well is decreased to an extent of 3.0 m. The rate of rise of water is 1.0 m/hr. Assume average depression head, h =3 m.

Q19. What is tubewell irrigation? Describe the various advantages and disadvantages of tubewell irrigation.

Q20. Write short notes on various types tubewells.

Q21. Draw a typical discharge drawdown curve of shallow and deep tubewells and explain how the discharge varies with types of tubewells for a given discharge.

Q22. Describe in details about the various design parameters of tubewells.

Q23. With a label diagram describe about the structure of a tubewell.

Q24.. What are the factors on which optimum length of well screen depend on ?

Q25. What is difference between naturally developed well and artificial gravel packed well? Narrate the different design procedure of gravel packed well.

Q26. What are the differences between cavity type tubewell and strainer type tubewells?

Q27. How does a cavity well works? With a diagram, narrate the working and construction of cavity type tubewells.

Q28. Describe in details the construction and working mechanism of a strainer type tubewells.

Q29. What are the uses of strainer in tubewell. Describe various types of strainers used in tubewell irrigation.

Q30. Design a tubewell to deliver 0.045 cumecs of water at a depression head of 7 m. The average water level is 10 m below the ground level in post monsoon season and 17 m in pre-monsoon season. The well log data obtained during investigation is presented below.

Depth, below ground level, m	0-15	15-20	20-30	30-50	50-65	65-75	Below 75
Types of strata	Clay	Sand	Clay+kankar	Coarse sand	Clay	Medium sand	Clay+sans stone

There is a horizontal length of pipe of 8 m connected to the pump at ground level as deliver pipe to discharge water to the crop field.

Q31. What do you mean by groundwater prospecting? Enumerate the various methods used for investigation of groundwater development.

Q32. Write notes on the following methods of investigation of groundwater development.

(i) Electrical resistivity method (ii) Seismic refraction method (iii) Well logs (iv) Resistivity logs

Q33. Describe the various operations involved in the construction of tubewell.

Q34. How is the well construction done by percussion drilling?

Q35. What is the difference between direct circulation hydraulic rotary method and reverse circulation hydraulic rotary method?

Q36. Differentiate between the cable tool method and hydraulic rotary method of drilling of tubewell.

Q37. What do you mean by shrouding of tubewell? Enumerate the main advantages of shrouding.

Q38. What does well development refer to? What are the advantages of well development?

Q39. Describe the different methods of development of tubewell?

Q40. How is the well development by surging done? Describe the surging technique of well development.

Q41. What is the average life of a tubewell? How can the life span be increased?

Q42. What is a sick well? How can the failure of a tubewell be controlled?

Q43. What for the verticality test of tubewell is done? How is it accomplished?

References

Ahrens, T. P. 1958. Water Well Engineering. Water Well JI. 12, December.

Anonymous. 1966. Procedure used in gravel packing wells. Johnsons Drillers JI. 38:2.

Benison, E. W. 1947. Ground Water. E. E. Johnson St. Paul, Minn. 500p.

Das Gupta, A. 1987. Handouts on Ground Water Hydrology, Asian Institute of Technology, Bangkok, pp. 282.

Dylla, A. S. and Muckel, D. C. 1967. Experimental development of shallow ground water wells. Pub. University Nevada and U. S. Dept. of Agr. 23 p.

Edward, E. 1966. Ground Water and Wells. Pub. Edward E. Johnson Inc., Saint Paul, Minnesota, U. S. A., 440 p.

Ellithorpe, G. L. 1970. Current practices relative to the design and placement of artificial gravel pack for tubewells. Proc. Summer Institute on Tubewell Technology I.A.R.I., New Delhi.

Garg, S.K. 2012. Irrigation Engineering and Hydraulic Structures. Khanna Publishers, Delhi.

Lenka, D. 2001. Irrigation and Drainage. Kalyani Publishers, New Delhi.

Michael, A.M. 1978. Irrigation Theory and Practices. Vikash Publishing House Pvt. Ltd., New Delhi.

Panigrahi, B. 2011. Irrigation Systems Engineering. New India Publishing Agency, New Delhi.

Roscoe Moss Company. 1990. Handbook of Groundwater Development. John Wiley & Sons. Singapore.

Sharma, H.D. and Chawla, A.S. 1977. Mannual on Ground Water and Tubewells. Central Board of Irrigation and Power, New Delhi.

Sharma, R.K. 1987. A Text Book of Hydrology and Water Resources. Dhanpat Rai and Sons, Delhi, pp. 818.

Slichter, C. S. 1989. Theoretical investigation of the motion of ground-waters. 19th Annual Report, U.S.G.S., 295-384.

15

Well Hydraulics

15.1 Introduction

The laws governing the flow of water in wells are referred to as the hydraulics of wells. Just like surface water flows from a higher elevation to a lower elevation by gravity, groundwater also flows from a region of higher water table to a region of lower water table. This difference of water table measured by any two piezometers divided by the longitudinal distance between them is called as hydraulic gradient. Hydraulic gradient plays a dominant role in deciding the flow of groundwater. This hydraulic gradient together with the hydraulic conductivity of the aquifer material forms the main law of well hydraulics called as Darcy's law. An understanding of Darcy's law together with some other laws as discussed subsequently in this chapter is essential in order to know the discharge of a well in a particular formation .Darcy's law is valid for studying groundwater flow in alluvial formations. However, it is not valid for aquifers having hard rock areas.

15.2 Properties of Aquifer Materials

The important aquifer properties that influence the water holding and water transmitting characteristics of aquifers are discussed below.

Porosity

Porosity is expressed as the ratio of the volume of interstices (V_v) to the total volume of aquifer materials (V) is given as:

$$\eta = \frac{V_v}{V} \tag{15.1}$$

Values of porosity depend on the arrangement of the particles, shape of the grains and degree of assortment.

Specific Yield

Specific yield is defined as the ratio of volume of water drained by gravity to total volume of the material drained and is given by the formula as:

$$Specific\ yield = \frac{Volume\ of\ water\ obtained\ by\ gravity\ drainage}{Total\ volume\ of\ material\ drained\ or\ dewatered} x\,100 \qquad (15.2)$$

Specific Retention

It is expressed as ratio of volume of water held against gravity drainage to the total volume of material drained and is given as:

$$Specific\ retention = \frac{Volume\ of\ water\ held\ against\ gravity\ drainage}{Total\ volume\ of\ material\ drained\ or\ dewatered} x\,100 \qquad (15.3)$$

Thus, the sum of specific yield and specific retention is equal to porosity.

Values of porosity and specific yield of some of the aquifer materials are presented in Table 15.1.

Table 15.1: Values of porosity and specific yield of some selected aquifer materials

Aquifer material	Porosity, percent	Specific yield, percent
Clay	45-55	1-10
Sand	30-40	10-30
Gravel	30-40	15-30
Sandstone	10-20	5 - 15
Shale	1-10	0.5 - 5
Limestone	1-10	0.5 - 5

Permeability

Permeability of a material is a measure of its ability to transmit fluid like water under a given hydraulic gradient. A medium has coefficient of permeability (K) of unit length per unit time if it transmits in a unit time a unit volume of groundwater, at the prevailing viscosity through a unit cross sectional area measured at right angle to the direction of flow under a unit hydraulic

Table 15.2: Values of coefficient of permeability of selected aquifer materials

Sl. No.	Type of aquifer materials	Approximate values of K, cm/day
1.	Granite, Quartzite	0.6×10^{-5}
2.	Slate, Shale	4×10^{-5}
3.	Limestone	4×10^{-5}
4.	Sandstone	0.004
5.	Sand and gravel	0.40
6.	Only gravel	4.0
7.	Only sand	0.04
8.	Clay	0.04×10^{-5}

gradient. It is generally expressed as m/day, m/sec and cm/sec. Coefficient of permeability is also termed as hydraulic conductivity. Approximate values of K for different aquifer materials are mentioned in Table 15.2 (Garg, 2012)

Intrinsic Permeability

Intrinsic permeability *(k)* of a rock or soil is a property of the medium only. It is independent of properties of the fluids. It is related to the hydraulic conductivity, K as:

$$k = \frac{K\,\mu}{\rho\,g}$$

where,

μ = dynamic viscosity, ρ = density of fluid, and g = acceleration due to gravity and other terms are as defined earlier.

Intrinsic permeability, k has units of m^2 or darcy, equal to 0.987 $(\mu m)^2$.

Transmissivity/Transmissibility

Transmissivity/transmissibility is the rate at which water is transmitted through a unit width of aquifer under a unit hydraulic gradient. The term transmissibility introduced by Theiss is measured by the coefficient of transmissibility (T) and is defined as the rate of flow of water through a vertical strip of the water bearing material (aquifer) of unit width (b) and full depth (d) under a unit hydraulic gradient.

The relationship between hydraulic conductivity (K) and transmissibility (T) is:

$$T = K\,.\,d \tag{15.4}$$

Storage Coefficient/Storativity

It is the volume of water that an aquifer releases from or takes into storage per unit surface area of the aquifer per unit change in head. It is dimensionless. In case of unconfined aquifer, storage coefficient is equivalent to specific yield. The storage coefficient of most of the confined aquifers ranges from 10^{-5} to 10^{-3}. On the other hand, the specific yield of unconfined aquifer ranges from 0.1 to 0.3. Sometimes, the term storativity is also used to represent the term storage coefficient.

15.3 Groundwater Movement

Darcy's Law: The movement of groundwater is based on the principle of flow through the porous media. The governing equation of flow is described by hydraulic principles reported in 1856 by Henri Darcy, who investigated the flow through porous sand beds. He observed that the flow rate through the porous beds is proportional to the head loss and inversely proportional to the length of the flow path. Darcy's law forms the basis for the governing groundwater flow equations.

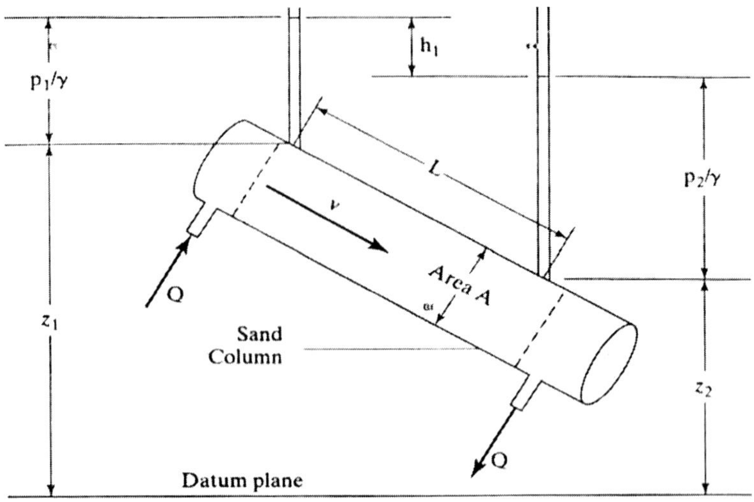

Fig. 15.1: Flow of water in inclined sand column

The experimental set up of flow of water in a sand column which is inclined to the ground surface is shown in Fig. 15.1. Two piezometers are located in the column at a longitudinal distance of L and the elevation heads of these two piezometers with reference to a datum are z_l and z_2, respectively. Let the pressure head in these two piezometers be $\frac{p_1}{\gamma}$ and $\frac{p_2}{\gamma}$, respectively. If the velocity of flow at the section 1 (inlet) is V_l and that at the section 2 (outlet) is V_2, then the velocity head at sections 1 and 2 will be $\frac{V_1^2}{2g}$ and $\frac{V_2^2}{2g}$, respectively. If the head loss due to friction in the pipe is h_l, then total energy for this system can be expressed by Bernoulli equation as:

$$\frac{p_1}{\gamma} + \frac{V_1^2}{2g} + z_1 = \frac{p_2}{\gamma} + \frac{V_2^2}{2g} + z_2 + h_l \tag{15.5}$$

where,

p = pressure, γ = specific weight of water, V = velocity of water, z = elevation head and h_l = head loss due to friction of flow in pipe.

Since the velocity of flow in porous beds are very small and the cylindrical sand column is of regular size (area of cross section of the pipe at both inlet and outlet sections are same), the velocity head at both the sections may be neglected. Under this assumption, the head loss is expressed as:

$$h_l = \left(\frac{p_1}{\gamma} + z_1\right) - \left(\frac{p_2}{\gamma} + z_2\right) \tag{15.6}$$

From Eqn. (15.6), it is revealed that the head loss is independent of the inclination of the pipe/sand column. Darcy related the flow rate to the head loss and length of column through a proportionality constant referred to as K, hydraulic conductivity which is a measure of the ability of the porous media to transmit water. Darcy's law can be stated as:

$$V = \frac{Q}{A} = -K\frac{dh}{dL} \tag{15.7}$$

The negative sign indicates that the flow of water is in the direction of decreasing head. For the above mentioned sand column, the head loss $dh = h_l$ and is given by Eqn. (15.6) and the length of the sand column, $dL = L$.

Hence, Eqn. (15.7) can be expressed as:

$$V = \frac{Q}{A} = -K\frac{\left(\frac{p_1}{\gamma} + z_1\right) - \left(\frac{p_2}{\gamma} + z_2\right)}{L} \tag{15.8}$$

Eqn. (15.8) represents the average velocity of flow in the entire cross section of sand column of area, A. The actual velocity of flow is limited to the pore channel only, so that the seepage velocity V_s is equal to the Darcy velocity divided by the porosity (η).

Thus, the actual seepage velocity is:

$$V_s = \frac{Q}{\eta \,.\, A} = -K\frac{\left(\frac{p_1}{\gamma} + z_1\right) - \left(\frac{p_2}{\gamma} + z_2\right)}{L \,.\, \eta} \tag{15.9}$$

It is to be noted that the seepage velocity is much higher than the actual Darcy velocity.

15.4 Hydraulics of Tubewell

In order to have proper design and construction of a tubewell, knowledge of well hydraulics is essential. Well hydraulics is governed by Darcy's law. When water is pumped by a well, water level in the pumping well is lowered down. This lowering of water level at any point in the aquifer is termed as drawdown or depression head. The drawdown varies with the distance from the well. It is maximum at centre of tube well and decreases as the distance from tube well increases forming a cone of depression. Draw down is zero at certain distance at which draw down of tube well is zero. This distance at which draw down of tube well is zero is called as radius of influence; R. Values of R depend on the well discharge, aquifer materials, time of pumping etc. In general value of R is maximum for a confined aquifer and minimum for an unconfined aquifer (Campbell and Lehr, 1973).

The lowering of water level in tube well causes head difference towards the well which induces water flow into the well from the surrounding water bearing formations of aquifers. The velocity of flow gradually increases towards the well. With increasing velocity, the hydraulic gradient increases as the flow converges towards the well, as a result the lowered water surface has a steeper slope towards the wells. The form of the surface resembles cone shaped depression. The main equation that governs the flow path in a tubewell is generally given by Darcy's law which states that velocity of flow through a porous media (V) is directly proportional to hydraulic gradient (i) i.e $V \alpha i$, or $V=K.\ i$, where K is hydraulic conductivity of aquifer. Since $Q = A\ V$, we can write $Q = A.\ K\ I$ (Panigrahi, 2011).

Velocity of flow through a soil depends on soil type, arrangement of grains and viscosity of water. Darcy's law is valid for laminar flow for which Reynolds number is less than 1 and perhaps as high as 10. It should be remembers that at the vicinity of well, the flow has high velocity and high hydraulic gradient that causes flow not to be laminar but turbulent. Under such case Darcy's law cannot be used.

15.5 Groundwater Flow Directions

Flow Nets:

Just like surface water flows from higher elevation to the lower elevation, in a much similar way groundwater flows from higher head to the lower head i.e. in the direction of decreasing head. Groundwater flows along the flow lines also called as stream lines. For a given set of boundary conditions, flow lines/stream lines (ψ) and equipotent lines (φ) can be mapped in two dimensions to form a flow net (Fig. 15.2).. The two sets of lines form an orthogonal pattern of small squares for isotropic system. In a few simplified cases, the differential equation governing flow can be solved to obtain the flow net (Michael, 1978).

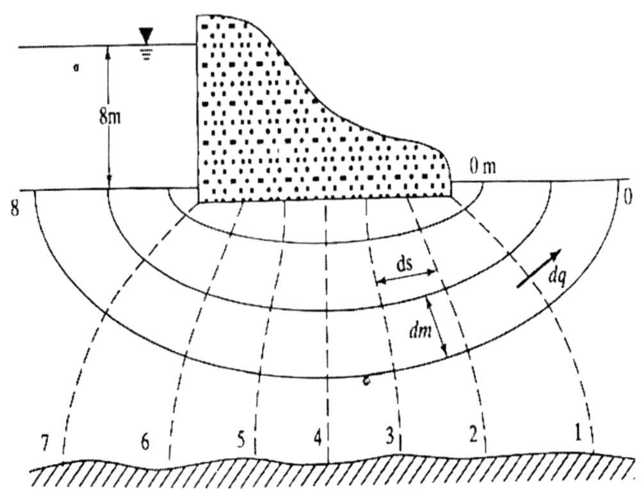

Fig. 15.2: Schematic diagram of a typical flow net

For an isotropic aquifer, equipotential lines are prepared based on the water levels in the observation wells penetrating the aquifer. Flow lines are then drawn at right angles (orthogonal) to them. For the flow net of Fig. 15.2, hydraulic gradient, i is given by

$$i = \frac{dh}{ds} \tag{15.10}$$

where,

ds = distance between two consecutive equipotential lines and dh = head loss between these two consecutive equipotential lines.

If dm is the distance between two consecutive equipotential lines, then per unit thickness of the aquifer, the constant flow rate, q between two adjacent flow lines is:

$$q = K.\frac{dh}{ds}\, dm \tag{15.11}$$

If we assume, $ds = dm$ for a square net, then for n squares between two flow lines over which total head, H is divided then we get, $h = H/n$.

If there are m numbers of flow channels, then total discharge, Q per unit thickness of the aquifer is:

$$Q = m\, q = \frac{K\, m\, H}{n} \tag{15.12}$$

The different notations of Eqn. (15.12) are already mentioned earlier.

It is to be noted that stream lines are parallel to the impermeable bed of the aquifer. Stream lines are also usually horizontal through high value of hydraulic conductivity and vertical through low values of hydraulic conductivity, because of refraction of lines across a boundary between different values of hydraulic conductivity media.

From field measurements of static water levels in wells within a basin, a water level contour map can be prepared. Flow lines sketched perpendicular to contours, show direction of movement. Contour maps of groundwater levels, together with flow lines, are useful data for locating new wells. Convex contours represent areas of groundwater recharge whereas concave contours are associated with areas of groundwater discharge. It is further to be noted that areas with wider contour spacing (flat gradients) possess higher hydraulic conductivities than those with narrow spacing (steep gradient). This means, the prospects for a productive well is better near the section having flat gradient i.e. wider contour spacing.

Sample Calculation 1:

Estimate the total flow underneath a dam of thickness 20 m as shown in Fig. 15.2. Hydraulic conductivity of the aquifer beneath dam, K is 10^{-4} cm/sec.

Solution:

Given $K = 10^{-4}$ cm/sec, m = number of flow channels = 4, n = number of squares over the direction of flow = 8, and H = total head = 8 m.

Using Eqn. (15. 12), we get $Q = \frac{K\,m\,H}{n}$

$$= \frac{10^{-4}\ x\ 4\ x\ 8}{8} = 0.0004$$

For 20 m thickness of the dam, total flow = 20 x 0.0004 = 0.008 m^3/sec = 691.2 m^3/day.

15.6 General Flow Equations

Partial differential equation for flow in porous media is obtained from Darcy's law combined with equation of continuity (equation of mass balance). Both steady state and transient flow equations can be derived.

15.6.1 Transient Saturated Flow

Consider a unit volume of porous media (Fig. 15.3) which is called as an elemental control volume. Mass balance equation which is based on law of conservation of mass involves inflow, outflow and change in groundwater storage. As per law of conservation of mass,

Mass in – Mass out = Change in storage per time.

Let Darcy velocity in x-direction be V_x and that in y and z-directions are V_y and V_z, respectively. Density of water is noted as ρ and porosity of media is η.

Mass in – mass out is referred as net flow.

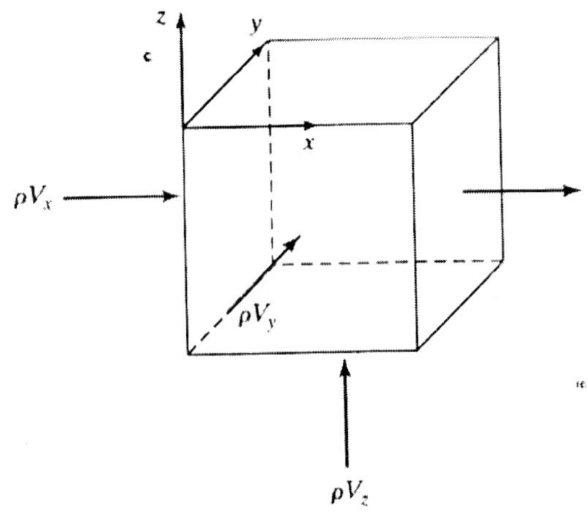

Fig. 15.3 Elemental control volume

Net flow in x-direction per unit volume of porous media = - $\frac{\partial}{\partial x}(\rho V_x)$

Thus, the net flow in y-direction per unit volume of porous media = - $\frac{\partial}{\partial y}(\rho V_y)$ and

that in z-direction per unit volume of porous media = - $\frac{\partial}{\partial z}(\rho V_z)$

Change in fluid mass storage in the control volume in unit time = $\frac{\partial}{\partial t}(\rho \eta)$

As per mass balance equation,

$$-\frac{\partial}{\partial x}(\rho V_x) - \frac{\partial}{\partial y}(\rho V_y) - \frac{\partial}{\partial z}(\rho V_z) = \frac{\partial}{\partial t}(\rho \eta) \tag{15.13}$$

Right hand side of Eqn. (!5.13) can be expanded as:

$$\frac{\partial}{\partial t}(\rho \eta) = \rho\left(\frac{\partial \eta}{\partial t}\right) + \eta\left(\frac{\partial \rho}{\partial t}\right) \tag{15.14}$$

Using Eqns. (15.13) and (15.14), we get

$$-\frac{\partial}{\partial x}(\rho V_x) - \frac{\partial}{\partial y}(\rho V_y) - \frac{\partial}{\partial z}(\rho V_z) = \rho\left(\frac{\partial \eta}{\partial t}\right) + \eta\left(\frac{\partial \rho}{\partial t}\right) \tag{15.15}$$

Eqn. (15.15) represents the transient saturated flow equation for a confined aquifer.

The first term on the right hand side of Eqn. (15.15) represents mass rate of water produced by compaction of the porous media (change in η) where as the second term on the right hand side of Eqn. (15.15) represents mass rate of water produced by an expansion of water under a change in ρ. The first term refers to the aquifer compressibility (α) and the second term relates to the compressibility of the fluid (β).

A change in ρ and η produces a change in hydraulic head, *h* and that the volume of water produced by the two mechanisms for a unit decline in head, *h* is S_s where S_s is the specific storage.

Specific storage is given by equation:

$$S_s = \rho g(\alpha + \eta \beta) \tag{15.16}$$

The mass rate of water produced (time rate of change of fluid mass storage) is given as:

Mass rate of water produced = ($\rho\, S_s\, \partial h / \partial t$). (15.17)

From Eqns. (15.15) and (15.17) we get,

$$-\frac{\partial}{\partial x}(\rho V_x) - \frac{\partial}{\partial y}(\rho V_y) - \frac{\partial}{\partial z}(\rho V_z) = \rho\, S_s\, \partial h / \partial t \tag{15.18}$$

Applying Darcy law and considering an incompressible fluid, we can write Eqn. (15.18) as:

$$\frac{\partial}{\partial x}\left(K_x \frac{\partial h}{\partial x}\right) + \frac{\partial}{\partial y}\left(K_y \frac{\partial h}{\partial y}\right) + \frac{\partial}{\partial z}\left(K_z \frac{\partial h}{\partial z}\right) = S_s \frac{\partial h}{\partial t} \tag{15.19}$$

For a homogeneous and isotropic media, $K_x = K_y = K_z = K$ and Eqn. (15.19) can be written as:

$$\frac{\partial^2 h}{\partial x^2} + \frac{\partial^2 h}{\partial y^2} + \frac{\partial^2 h}{\partial z^2} = \frac{S_s}{K} \cdot \frac{\partial h}{\partial t} \tag{15.20}$$

Eqn. (15.20) can also be written as:

$$\nabla^2 h = \frac{S_s}{K} \cdot \frac{\partial h}{\partial t} \tag{15.21}$$

For case of a horizontal confined aquifer of thickness, b, the storativity or storage coefficient, S is equal to $S_s.b$ and transmissivity, T is equal to $K.b$.

Hence, for horizontal isotropic homogeneous confined aquifer, Eqn. (15.21) can be written as:

$$\nabla^2 h = \frac{S}{T} \cdot \frac{\partial h}{\partial t} \text{ in two dimensions.} \tag{15.22}$$

15.6.2 Steady State Saturated Flow

For a steady state condition, change in storage in the elemental control volume is zero and hence right hand side of Eqn. (15.22) is zero.

Eqn. (15.22) then can be written as

$$\nabla^2 h = 0 \;\; i.e. \frac{\partial^2 h}{\partial x^2} + \frac{\partial^2 h}{\partial y^2} + \frac{\partial^2 h}{\partial z^2} = 0 \tag{15.23}$$

Eqn. (15.23) is called as Laplace equation used for steady, incomprehensible and isotopic flow case and is one of the best understood partial differential equation. The solution of Eqn. (15.23) in two dimensions can be done by graphical flow nets.

Dupuit Assumptions

For the case of unconfined groundwater flow, Dupuit developed a theory that allows for a simple solution based on several important assumptions. These assumptions are:

(i) The water table or free surface is only slightly inclined.

(ii) Streamlines may be considered horizontal and equipotential lines are vertical.

(iii) Slopes of the free surface and hydraulic gradient are equal.

15.7 Steady State Well Hydraulics

In case of steady state flow, the variation in head occurs only in space and not in time. The general flow equations as discussed above can be solved for pumping wells in both unconfined and confined aquifers under steady as well as unsteady conditions. However, for easy in computation, the boundary conditions must be kept relatively simple and aquifers must be assumed as homogeneous and isotropic.

Steady One-Dimensional Flow

Steady state groundwater flow in one dimension (x-direction) for isotropic and homogeneous aquifer can be written from Laplace equation (Eqn. 15.23) as:

$$\frac{\partial^2 h}{\partial x^2} = 0 \tag{15.24}$$

Solution of Eqn. (15.24) leads to:

$$h = \frac{-v\,x}{K} \tag{15.25}$$

where, $h = 0$ and $x = 0$ and $dh/dx = -v/K$ according to Darcy's law. This states that head varies linearly with flow in x-direction. The steady one-dimensional flow in an unconfined aquifer is derived with Dupuit's assumptions. The resulting variation in head with x is called the Dupuit Parabola which represents the approximate shape of the water table for relatively flat slopes. If steep slopes near the well exits, Dupuit's assumptions do not hold good and may lead erroneous result if used as such. However, under such case complex mathematical models are available for solution (Hansen *et al.*, 1979).

15.8 Steady Radial Flow to a Well

Assumptions:

Following assumptions are required before deriving the flow equations in well whether it is in confined or unconfined aquifer.

(i) The well is pumped at constant rate (steady state).

(ii) The well is fully penetrating the aquifer.

(iii) Aquifer is homogeneous, isotropic, horizontal and of infinite horizontal extent.

(iv) Water is released from the storage in the aquifer or other underground materials in immediate response to drop in water level or piezometric surface.

(v) Well storage is neglected.

(vi) Piezometric surface is also horizontal.

15.8.1 Confined Aquifer

Well discharge, Q at any radial distance, r from the centre of the well for steady state case with thickness of aquifer being b is given by Darcy's law as:

$$Q = 2\pi\, r\, b\, K \frac{dh}{dr} \tag{15.26}$$

Let the radius of the well be r_w.

At the periphery of the well when $r = r_{w,}$ the head, $h = h_w$ (Fig. 15. 4)

Integrating Eqn. (15.26) with boundary conditions of $r = r_{w,}$ when head, $h = h_w$ and then simplifying we get:

$$Q = 2\pi\; b\, K \frac{h - h_w}{\ln\left(r / r_w\right)} \tag{15.27}$$

Since transmissivity, $T = K.b$

Eqn. (15.27) can be written as:

$$Q = 2\pi\; T \frac{h - h_w}{\ln\left(r / r_w\right)} \tag{15.28}$$

Eqn. (15.28) represents that as radial distance, r increases, the head, h_w also increases, yet the maximum head is h_o when the radial distance, $r = r_o$ (r_o is called as radius of influence where the draw down because of well pumping is zero).

Now Eqn. (15.28) can be written as:

$$Q = 2\pi\; T \frac{h_0 - h_w}{\ln\left(r_0 / r_w\right)} \tag{15.29}$$

Eqn. (15.29) is called as Dupuit's equation.

It is to be remembered that as the head increases, drawdown decreases. The drawdown at any radial distance from the well is the difference of the original piezometric surface and the water level of an observation well at that radial distance in the aquifer.

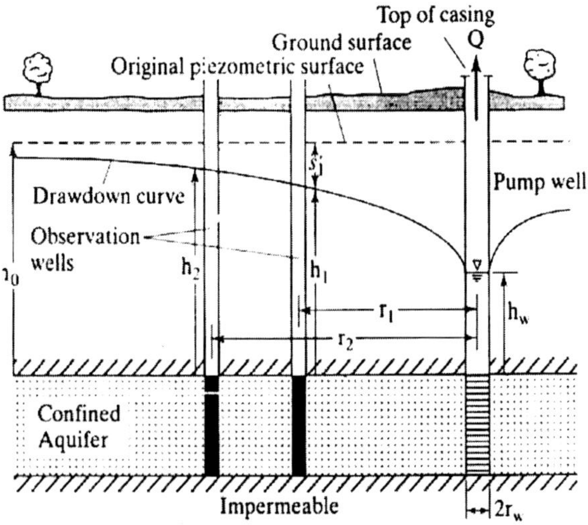

Fig. 15.4: Radial flow to a well in confined aquifer

If we consider two observation wells in the aquifer in the vicinity of the main pumping well at radial distances r_1 and r_2 where the heads are h_1 and h_2, respectively we can write Eqn. (15.28) as:

$$Q = 2\pi\ T \frac{h_2 - h_1}{\ln(r_2/r_1)} \tag{15.30}$$

Let the drawdown at observation wells at radial distances r_1 and r_2 be s_1 and s_2, respectively.

Now we can write, $s_2 = h_o - h_2$ and $s_1 = h_0 - h_1$ and using this hypothesis, we can write Eqn. (15.30) in terms of drawdown as:

$$Q = 2\pi\ T \frac{s_1 - s_2}{\ln(r_2/r_1)} \tag{15.31}$$

Eqns. Eqn. (15.30) and (15.31) are also called as Thiem's equation. The difference of Thiem's and Dupuit'equations are that in Thiem's equation we consider two observation wells to get well discharge whereas in Dupuit's equation, no observation wells are taken and the well discharge is calculated in terms of radial distances and heads at the boundary limiting conditions of near the well and at the radial distance of radius of influence.

15.8.2 Unconfined Aquifer

Applying Darcy's law for radial flow to a fully penetrating well in unconfined, isotropic, horizontal and homogeneous aquifer and using Dupuit's assumptions, we get well discharge,

Q at any radial distance, r where head is h (Fig. 15.5) as:

$$Q = 2\pi\, r\, h\, K \frac{dh}{dr} \tag{15.32}$$

Let us consider two observation wells at radial distances r_1 and r_2 where heads are h_1 and h_2, respectively (respective drawdown at r_1 and r_2 are s_1 and s_2, respectively).

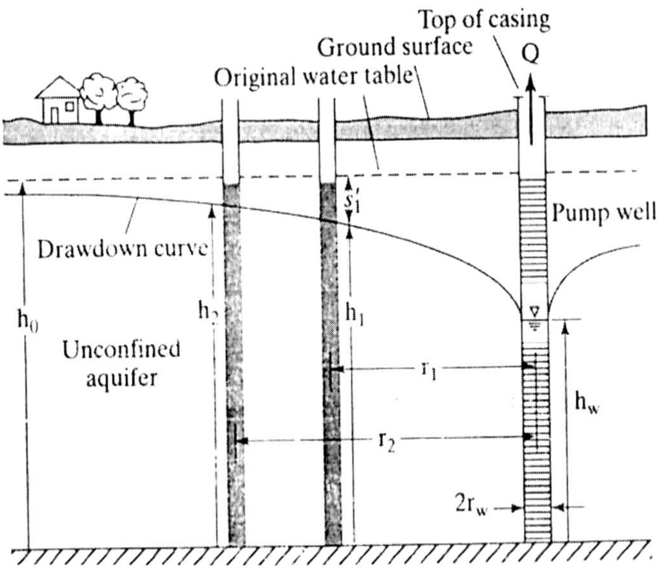

Fig. 15.5: Radial flow to a well in unconfined aquifer

Eqn. (15.32) can be integrated within the boundary limits of r_1 and r_2 where heads are h_1 and h_2, and then simplifying we get:

$$Q = \pi\, K \frac{h_2^{\,2} - h_1^{\,2}}{\ln\left(r_2 / r_1\right)} \tag{15.33}$$

In terms of drawdown, Eqn. (15.33) can be written as follows:

$$h_2^{\,2} - h_1^{\,2} = \left(h_2 - h_1\right)\left(h_2 + h_1\right)$$

$$= \left(s_1 - s_2\right)\left(h_2 + h_1\right)$$

If the amount of drawdown is small compared to the saturated thickness of the aquifer, then h_2 *and* h_1 are nearly equal and each approximately equals to the saturated thickness of aquifer h_o.

Hence, $h_2^2 - h_1^2 = (s_1 - s_2)(2h_0)$

Putting this value in Eqn. (15.33) we get,

$$Q = 2\pi K h_0 \frac{s_1 - s_2}{\ln(r_2 / r_1)}$$

$$= 2\pi T \frac{s_1 - s_2}{\ln(r_2 / r_1)} \tag{15.34}$$

Eqns. (15.33) and (15.34) are called as Thiem's formula for unconfined aquifer.

If we consider radial distance as radius of influence, r_o where the head is h_0 (this head is also equal to the saturated thickness of the aquifer) and a radial distance r_w where the head is h_w, Eqn. (15.33) becomes

$$Q = \pi K \frac{h_0^2 - h_w^2}{\ln(r_0 / r_w)} \tag{15.35}$$

Eqn. (15.35) is called as Dupuit's formula.for unconfined aquifer.

15.8.3 Unconfined Aquifer with Uniform Recharge

When there is some recharge due to rainfall or infiltration of excess irrigation water that causes deep percolation, the flow towards the well will increase.

Let us consider a cylinder of thickness *dr* and uniform recharge rate as *W.*

The increase of discharge *dQ,* through this cylinder situated at a radial distance of *r* from the centre of the main pumping well is given as:

$$dQ = -2\pi r\, dr\, W \tag{15.36}$$

Integrating above equation we get,

$$Q = -\pi r^2 W + C \tag{15.37}$$

At the well, $r = 0$, $Q = Q_w$. so that Eqn. (15.37) becomes

$$Q = -\pi r^2 W + Q_w \tag{15.38}$$

Substituting this flow in the equation for flow to the well gives

$$2\pi r K h \frac{dh}{dr} = -\pi r^2 W + Q_w \tag{15.39}$$

Integrating Eqn. (15.39) and considering the limit that at $r = r_0$, $h = h_0$ we get

$$h_0^2 - h^2 = \frac{W}{2K}\left(r^2 - r_0^2\right) + \frac{Q_w}{\pi K}\ln\frac{r_0}{r} \tag{15.40}$$

It follows that as $r = r_0$, $Q = 0$ and then Eqn. (15.38) gives

$$Q_w = \pi r_0^2 W$$

Thus, the total flow of the well equals the recharge within the circle defined by the radius of influence, r_0.

15.9 Interference Among Wells

If two or more wells are constructed in such a way that they are near to each other and their cones of depressions interact, they are said to interfere. Such interference of wells decreases the discharges of such interfering wells. Principle of linear superposition can be applied in finding the total drawdown of a number of wells (multiple well systems) such that drawdown of each well is interfering with that of the other.

Thus,

$s_t = s_1 + s_2 + s_3 + \ldots\ldots s_n$

where,

s_t is the total drawdown at a given point, $s_1, s_2 \ldots\ldots\ldots s_n$ are the drawdown at the point caused by each well individually.

Considering a well field in confined aquifer having n number of wells, the drawdown, s at a given point in the area of influence is given by the equation as (Murty, 2004):

$$s = \sum_1^n \frac{Q_i}{2\pi K b}\ln\frac{R_i}{r_i} \tag{15.41}$$

where,

R_i = distance from the ith well to a point at which the drawdown becomes negligible and r_i = distance from the ith well to the given point.

Muskat has proposed the following formulae for computation of discharge from such interfering wells. These formulae are found to give reliable results.

(a) For Confined Aquifer (Artesian Well):

For two artesian identical wells (same diameters) at a distance B apart,

$$Q_1 = Q_2 = 2\pi\ T \frac{h_0 - h_w}{\ln\left(\dfrac{r_0^{\ 2}}{r_w\ B}\right)} \tag{15.42}$$

For three artesian identical wells (same diameters) at a distance B apart in a pattern of equilateral triangle:

$$Q_1 = Q_2 = Q_3 = 2\pi\ T \frac{h_0 - h_w}{\ln\left(\dfrac{r_0^{\ 3}}{r_w\ B^2}\right)} \tag{15.43}$$

For three artesian identical wells (same diameters) at a distance B apart in a line (Fig. 15.6), the discharge in the first and the third well is (Murty, 2004):

$$Q_1 = Q_3 = \frac{2\,\pi\,T\,(h_0 - h_w)\ln\left(\dfrac{B}{r_w}\right)}{2\ln\left(\dfrac{R}{B}\right)\ln\left(\dfrac{B}{r_w}\right) + \ln\left(\dfrac{B}{2\,r_w}\right)\ln\left(\dfrac{R}{r_w}\right)} \tag{15.44}$$

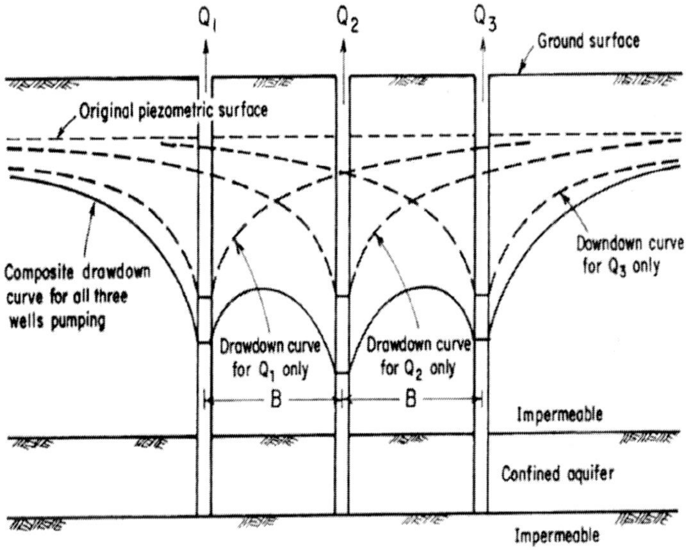

Fig. 15.6: Discharge of well when three wells are in a line

whereas the middle well gives discharge, Q_2 as (Murty, 2004):

$$Q_2 = \frac{2\pi T (h_0 - h_w) \ln\left(\frac{B}{2 r_w}\right)}{2 \ln\left(\frac{R}{B}\right) \ln\left(\frac{B}{r_w}\right) + \ln\left(\frac{B}{2 r_w}\right) \ln\left(\frac{R}{r_w}\right)} \tag{15.45}$$

(b) For Unconfined Aquifer (Gravity Well):

For two identical gravity wells (same diameters) at a distance B apart, discharge is (Sharma and Sharma, 2002):

$$\text{or}\quad Q_1 = Q_2 = \frac{2\pi K \left(\frac{h_0^{\,2}}{2} - \frac{h_w^{\,2}}{2}\right)}{\ln\left(\frac{r_0^{\,2}}{r_w B}\right)} \tag{15.46}$$

For three identical gravity wells (same diameters) at a distance B apart in a pattern of equilateral triangle, discharge is (Sharma and Sharma, 2002):

$$Q_1 = Q_2 = Q_3 = \frac{\pi K \left(h_0^{\,2} - h_w^{\,2}\right)}{\ln\left(\frac{r_0^{\,3}}{r_w B^2}\right)} \tag{15.47}$$

The various terms of the Eqns. (15.42) to (15.47) have already been defined earlier.

15.10 Well Loss and Specific Capacity

The drawdown in a pumped well represents the hydraulic head loss consisting of

(i) Aquifer head loss (also called as formation loss) and

(ii) Well loss

Well loss is caused due to flow through the well screen and flow inside the well to the pump intake. Well loss, because of turbulence in flow near the well can be considered proportional to.Q^n where n is a constant greater than 1. Value of n will be 1 for sufficiently low pumping rate so that the flow in the vicinity of the well is laminar and for discharge condition with turbulence case, n is greater than 1 (Das Gupta, 1987; Sharma, 1987).

The drawdown because of well loss can, hence, be written as:

$s_1 = C\,Q^n$

where C is well loss coefficient.

For a confined aquifer, the drawdown in the aquifer (s_2) because of a pumping well of radius, r_w and radius of influence, r_o having a discharge of Q is given as:

$$s_2 = \frac{Q}{2\pi T} \ln\left(\frac{r_0}{r_w}\right) = B\,Q \tag{15.48}$$

where, $B = \dfrac{1}{2\pi T} \ln\left(\dfrac{r_0}{r_w}\right)$

B is called as formation loss coefficient.

Total head loss (s_w) considering both aquifer loss and well loss is:

$$s_w = s_1 + s_2 = B\,Q + C\,Q^n \tag{15.49}$$

The total drawdown, s_w as shown in Fig. 15.7 is summation of drawdown due to formation loss, $B\,Q$ and well loss $C\,Q^n$.

The size of the well has little significance on well discharge. However, it has significant influence on well loss. For simplicity, in turbulence case, sometimes the value of n is taken as 2. Considering $n = 2$ Eqn. (15.49) is written as:

$$s_w = B\,Q + C\,Q^2 \tag{15.50}$$

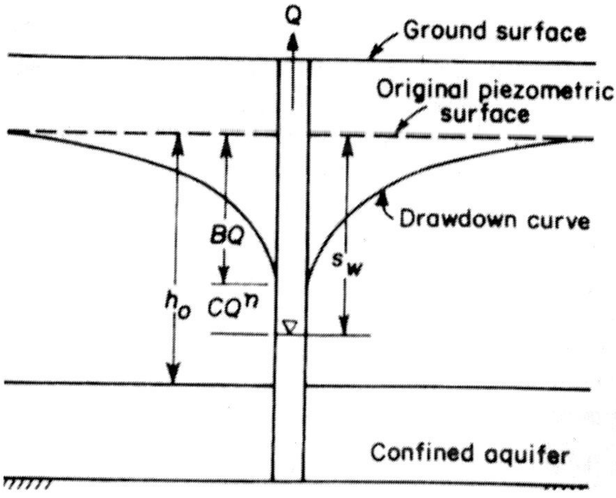

Fig. 15.7 Relation of well loss to drawdown

Eqn. (15.50) can be written as:

$$\frac{s_w}{Q} = B + C\,Q \tag{15.51}$$

Plotting s_w/Q versus Q, a linear relationship is obtained. The y-intercept of the linear line gives value of B whereas the slope gives value of C.

Specific capacity of a well is a measure of productivity of well. Larger the specific capacity better is the performance of the well. It is given by ratio of discharge to drawdown i.e. Q/s_w. It can be written as:

$$\textit{Specific capacity} = \frac{Q}{B\,Q + C\,Q^n}$$

Considering n = 2 we get,

$$\textit{Specific capacity} = \frac{Q}{B\,Q + C\,Q^2}$$

$$\text{or } \textit{Specific capacity} = \frac{1}{B + C\,Q} \tag{15.52}$$

Eq. (15.52) clearly indicates that the specific capacity of a well is not constant but decreases as the discharge increases. An increase in well loss due to clogging or deterioration of the well screen lowers the specific capacity which indicates low performance of the well.

15.11 Efficiency of Well

From the discussions as above, we may conclude that if the well loss is neglected, the discharge of a well will vary with drawdown. The discharge of the well per unit drawdown is called as specific capacity which is a measure of performance of the well. This specific capacity of a well will be different for different well designs. For determining the well efficiency, the well is pumped for varying drawdown conditions and then a graph is plotted with percent discharge (percent yield) as ordinate and percent drawdown as abscissa as shown in Fig. 15.8. The curve is called as yield-drawdown curve. The curve will be straight line up to certain value of drawdown and then it increases disproportionately to the yield. This places an efficient and optimum limit to the drawdown which may be allowed to be created in a well. This is found to be 70 percent of the maximum drawdown which can be created in a well.

The well efficiency (E_w) is a ratio of formation loss to total drawdown and is:

$$E_w = \frac{B\,Q}{B\,Q + C\,Q^n}\,x\,100 \tag{15.53}$$

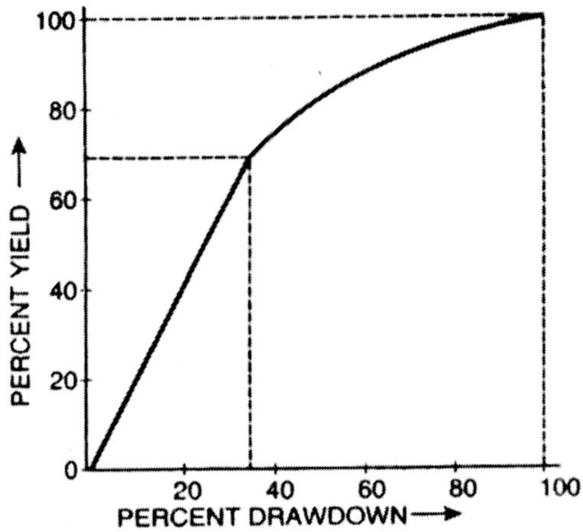

Fig. 15.8 Yield-drawdown curve

For a small discharging with laminar flow case, n is taken as 1 and then Eqn. (15.53) is written as:

$$E_w = \frac{1}{1+\frac{C}{B}} x\,100 \tag{15.54}$$

For turbulence case, the well efficiency is:

$$E_w = \left(\frac{1}{1+\frac{C\,Q^{n-1}}{B}}\right) x\,100 \tag{15.55}$$

Eqn. (15.54) reveals that the well efficiency for a laminar flow case with artesian well is independent of pumping whereas it decreases with increasing discharge for turbulence case in artesian well (Eqn. 15.55).

Efficiency of a well is generally less than 100 percent. For an adequately developed well with its screen not affected by clogging or incrustation, the well efficiency is about 80 percent.

Sample Calculation 2:

In a pumping test in medium sand and gravel, impermeable bed was found to lie at a distance of 14 m below the ground level. The normal groundwater level was at the surface. Two observation wells were located at a distance of 7.5 and 3m, respectively from the main pumping well where the drawdown were measured to be 0.3 and 1.5 m, respectively. If the discharge of the main pumping well was 5 lit/sec, calculate the value of coefficient of permeability of the soil.

Solution:

The well is present in unconfined aquifer. The saturated thickness of the aquifer, h_0 is 14 m.

At radial distance, $r_1 = 3\ m$, drawdown, $s_1 = 1.5\ m$ and hence, head, $h_1 = 14 - 1.5 = 12.5$ m.

Similarly at radial distance, $r_2 = 7.5\ m$, the drawdown, $s_2 = 0.3\ m$ and hence, head, $h_2 = 14 - 0.3 = 13.7$ m.

The discharge is, $Q = 5$ lit/sec = 0.005 m^3/sec.

Using Thiem's formla (Eq. 15.33), coefficient of permeability can be computed as

$$Q = \pi K \frac{h_2^{\,2} - h_1^{\,2}}{\ln\left(r_2 / r_1\right)}$$

$$0.005 = (\pi K)\left[\frac{(13.7)^2 - (12.5^2)}{\ln(7.5/3)}\right]$$

which gives $K = 0.0000463$ m/sec. = 0.00463 cm/sec.

Sample Calculation 3:

A well penetrates fully to an unconfined aquifer of thickness 100 m. The drawdown in the main pumping well at a discharge of 200 lit/min is 15 m. Estimate the discharge of the same main pumping well at a drawdown of 20 m assuming steady state condition. Assume homogeneous and isotropic aquifer. The radius of influence is same for both the cases.

Solution:

Given, discharge of the main pumping well, $Q = 200$ lit/min when the drawdown, $s_w = 15$ m.

Saturated thickness of the unconfined aquifer, $h_0 = 100$ m.

Drawdown in the pumping well, $s_w = 15$ m and hence head $h_w = 100 - 15 = 85$ m.

Drawdown in the second case in the same well, $s_1 = 20$ m and hence head $h_1 = 100 - 20 = 80$ m. We have to find out the new discharge at this drawdown i.e. at $s_1 = 20$ m.

Let radius of influence be r_0 and radius of well be r_w.

Now using Dupuit's formula (Eqn. 15.35) we have

$$Q = \pi K \frac{h_0^{\,2} - h_w^{\,2}}{\ln\left(r_0 / r_w\right)}$$

In the first case: $h_w = 85$ m

$$\text{Hence, } \frac{\pi K}{\ln\left(r_0/r_w\right)} = \frac{Q}{\left(h_0^{\,2} - h_w^{\,2}\right)}$$

$$= \frac{200}{(100)^2 - (85)^2} = 0.0721 \qquad \text{(i)}$$

In the second case: $h_1 = 80$ m and h_0 is same i.e. 100 m.

Now we can write the Dupuit's formula for the second case to find discharge as:

$$Q = \pi K \frac{h_0^2 - h_1^2}{\ln(r_0 / r_w)} \qquad \text{(ii)}$$

Putting the value of from Eqn. (i) in Eqn. (ii) (since for both the cases, *K*, r_0 *and* r_w are same)

$Q = 0.0721 \times [(100)^2 - (80)^2]$

= 259.60 = 260 lit/min.

Sample Calculation 4:

A 30 cm size well penetrates 20 m below the ground surface. Static water level in the aquifer is 2 m below the ground surface. After 24 hours of pumping at a discharge of 5400 lit/min, the water level in the test well at 100 m is lowered by 0.50 m and in a second well at distance of 30 m away, water level is lowered by 1 m. Estimate the drawdown in the main pumping well and transmissivity of the aquifer.

Solution:

The well penetrates 20 m below the ground surface and the piezometric level i.e. static water level is 2 m. Hence, the saturated thickness of the unconfined aquifer, $h_0 = 20 - 2 = 18$ m.

The Thiem's equilibrium formula for unconfined aquifer (Eq. 15.33) is:

$$Q = \pi K \frac{h_2^2 - h_1^2}{\ln(r_2 / r_1)} \qquad \text{(i)}$$

Here, at distance $r_2 = 100$ m, head, $h_2 = 18 - 0.5 = 17.5$ m and

At $r_1 = 30$ m, head, $h_1 = 18 - 1 = 17$ m

Given discharge, $Q = 5400$ lit/min $= 0.09$ m^3/sec.

Substituting the above values in Eqn. (i), we have:

$$0.09 = (\pi K)\left[\frac{(17.5)^2 - (17^2)}{\ln(100/30)}\right]$$

$$= 54.20\, K$$

$K = 0.001661$ m/sec = 0.1661 cm/sec.

Transmissivity, $T = K.\ h_o = 0.001661 \times 18 = 0.03$ m²/sec.

To determine the drawdown in the main well, use the above Eqn. (i) replacing $h_2 = h_1$, $h_1 = h_w$, $\pi_2 = \pi_1$ and $\pi = \pi_w$ (where $\pi_w = 0.15$m)

$$\text{or} \quad 0.09 = \pi\ x\ 0.001661\ x \left(\frac{17^2 - h_w^{\ 2}}{\ln\left(\frac{30}{0.15}\right)} \right)$$

On solution, it gives $h_w = 14.07$ m.

Sample Calculation 5:

A fully penetrating tubewell in confined aquifer has thickness 24 m. The aquifer is met at 25 m below the ground level. The static water table is 15 m below the ground level. The discharge of the main well is measured to be 3000 m³/day, when the depression head (drawdown) in it is 6 m. Permeability of the aquifer is 24.5 m/day. Assuming radius of influence of 300 m, find out the size of the tubewell.

Solution:

Given the case of a fully penetrating well in confined aquifer.

Thickness of the aquifer, $b = 24$ m.

The aquifer is met at 25 m below the ground level and the static water table is 15 m below the ground level. Hence, head of water at radius of influence, $r_0 = 300$ m is:

$h_o = 25 - 15 = 10$ m.

The drawdown at the main well, $s_w = 6$ m and so head $h_w = h_0 - s_w = 10 - 6 = 4$ m.

Using Dupuit's formula (Eqn. 15.29) we have:

$$Q = 2\pi\ T \frac{h_0 - h_w}{\ln\left(r_0 / r_w\right)} \qquad \text{(i)}$$

Given $Q = 3000$m³/day and $K = 24.5$ m/day.

Transmissivity, $T = K.\ b = 24.5 \times 24 = 588$ m²/day.

Radius of influence, $r_0 = 300$ m

Putting these values in Eqn. (i) above we have,

$$3000 = 2\pi\ x\ 588 \frac{(10-4)}{\ln\left(300 / r_w\right)}$$

Solving the above equation, we get, radius of the well, $r_w = 0.185$ m = 18.5 cm.

Size of the well = diameter of the well = 37 cm.

Sample Calculation 6:

A well of radius 30 cm is pumped at a rate of 81.6 m³/hr. Two observation wells were dug at a distance of 6 and 15 m, respectively where the drawdown were measured to be 6 and 1.5 m, respectively. The bottom of the impermeable bed up to which the well is being pumped is 90 m below the ground surface. Piezometric surface lies on the ground surface too. (a) Compute the coefficient of permeability. (b) If all the observed points were on the Dupuit's curve, what was the drawdown in the well being pumped? (c) Find out the specific capacity of the well. (d) Find out also the maximum rate at which the well can be pumped?.

Solution:

Given distance $r_1 = 6$ m where drawdown, $s_1 = 6$ m and at $r_2 = 15$ m, $s_2 = 1.5$ m.

Clearly it is the case of an unconfined aquifer and saturated thickness, h_o of the aquifer is 90 m.

Hence, head $h_1 = 90 - 6 = 84$ m and $h_2 = 90 - 1.5 = 88.5$ m

Discharge, $Q = 81.6\ \text{m}^3/\text{hr} = 1.36\ \text{m}^3/\text{sec}$

(a) Using Thiem's formula (Eqn. 15.33) we have:

$$Q = \pi K \frac{h_2^2 - h_1^2}{\ln(r_2/r_1)}$$

$$1.36 = \pi K \frac{(88.5)^2 - (84)^2}{\ln(15/6)}$$

Solution of the above gives K = 0.00051 m/min = 0.051 cm/min.

(b) Radius of the well, $r_w = 0.3$ m., $r_2 = 15$ m, $h_2 = 88.5$ m and we have to find out h_w.

Thiem's formula can now be written as :

$$Q = \pi K \frac{h_2^2 - h_w^2}{\ln(r_2/r_w)} \qquad \text{(i)}$$

Putting the above values in Eqn. (i), we have

$$1.36 = \pi \, x\, 0.00051 \frac{(88.5)^2 - h_w^2}{\ln(15/0.3)}$$

Solution of the above equation gives $h_w = 67.4$ m.

Hence, drawdown in the well = 90 – 67.4 = 22.6 m.

(c) Specific capacity of the well is the ratio of discharge per unit drawdown in the pumped well.

.

In this case we have to first find out the radius of influence (r_0) of the well.

Using Dupuit's equation (15.35) we have:

$$Q = \pi K \frac{h_0^{\,2} - h_w^{\,2}}{\ln\left(r_0 / r_w\right)}$$

$$1.36 = \pi \, x \, 0.00051 \frac{(90)^2 - (67.4)^2}{\ln\left(r_0 / 0.3\right)}$$

Solution of the above equation gives $r_0 = 20.01$ m = 20 m.

Now, specific capacity is discharge per unit drawdown. Hence, we have to find out the discharge when the drawdown in the well, $s_w = 1$ m i.e. head, $h_w = 90 - 1 = 89$ m.

Thiem's formula is used to find out the specific capacity as:

$$\textit{Specific capacity} = \pi \, x \, 0.00051 \frac{(90)^2 - (89)^2}{\ln\left(20 / 0.3\right)}$$

$$= 0.0683 \text{ m}^3/\text{min} = 68.3 \text{ lit/min.}$$

(d) Maximum discharge will occur when head in the well, h_w is zero

Now using Thiem's formula:

$$Q_{max.} = \pi \, x \, 0.00051 \frac{(90)^2 - (0)^2}{\ln\left(20 / 0.3\right)} = 3.09 \text{ m}^3/\text{min} = 3090 \text{ lit/min.}$$

Sample Calculation 7:

Two tubewells penetrating fully a 10 m thick confined aquifer are located 200 m apart. Tubewell has diameter equals to 30 cm, depression head = 5 m and radius of influence is 300 m. Coefficient of permeability is 0.1 cm/sec. (a) Find out if the wells are interfering? (b) Compute the discharge of the tubewell when one well is pumping. (c) Also find out the percentage decrease in discharge of the well if both the wells are working under a depression head of 5 m.

(d) If a third well is also located 200 m away from both the wells so that these three wells form a pattern of equilateral triangle, what will happen to the discharge of each well.

Solution:

(a) Given the radius of influence of the wells is 300 m but the two wells are located at 200 m apart. Hence, the wells are interfering.

(b) When one tubewell is working, the discharge is, Q

$$Q = 2\pi\ T \frac{h_0 - h_w}{\ln(r_0 / r_w)}$$

$$= 2\pi\ x\ 10\ x\ 0.001 \frac{5}{\ln(300/0.15)} \text{ (given depression head, } h_o - h_w = 5 \text{ m)}$$

$= 0.041$ m^3/sec

(c) When both the wells are working, the discharge of each well is (Eqn. 15.42):

$$Q_1 = Q_2 = 2\pi\ T \frac{h_0 - h_w}{\ln\left(\frac{r_0^2}{r_w B}\right)} \qquad \text{(i)}$$

Given, depression head, $(h_0 - h_w) = 5$ m. and distance between two tubewells, $B = 200$ m

Using these values and Eqn (i) as above we get,

$$Q_1 = Q_2 = 2\pi\ x\ 0.001\ x\ 10\ x \frac{5}{\ln\left(\frac{300^2}{0.15\ x\ 200}\right)} = 0.039 \text{ m}^3\text{/sec}$$

Percentage decrease in discharge is given as:

$$= \frac{Q_1 - Q_2}{Q_1} x\ 100$$

$$= \frac{0.041 - 0.039}{0.041} x\ 100 = 4.87 \text{ say 5 percent.}$$

(d) If a third well is also located 200 m away from both the wells so that these three wells form a pattern of equilateral triangle, then the discharge of each well is given by Eqn. (15.43) as:

$$Q_1 = Q_2 = Q_3 = 2\pi\ T \frac{h_0 - h_w}{\ln\left(\frac{r_0^{\ 3}}{r_w\ B^2}\right)}$$

$$= Q_1 = Q_2 = Q_3 = 2\pi\ T \frac{h_0 - h_w}{\ln\left(\frac{r_0^{\ 3}}{r_w\ B^2}\right)}$$

$= 0.037\ m^3/sec$

Hence discharge of each well decreases.

Percentage decrease in discharge is given as:

$$= \frac{Q_1 - Q_2}{Q_1} x\ 100$$

$$= \frac{0.041 - 0.037}{0.041} x\ 100 = 9.8 \text{ say 10 percent.}$$

Sample Calculation 8:

In a large basin, one tubewell having 5 cm radius is working. The permeability of the soil is 300 cm/day. Radius of influence is 1.5 km. Watertable lies 50 m above the bottom of the impermeable bed. Maximum drawdown in the well is 15 m. After some time, two more wells were dig such that all the three wells form an equilateral triangle with 250 m distance apart. Assuming size of all the three wells is same and they discharge equally, calculate percentage loss of discharge in the first well due to interference of other two wells.

Solution:

Given $r_w = 5$ cm = 0.05 m

Radius of influence, $r_0 = 1.5$ km = 1500 m.

Head, $h_o = 50$ m and $h_w = 35$ m.

Permeability of the soil = 300 cm/day = 3 m/day.

Considering unconfined aquifer, discharge of the well is given (Eqn. 15.35) as:

$$Q = \pi\ K \frac{h_0^{\ 2} - h_w^{\ 2}}{\ln\left(r_0 / r_w\right)}$$

$$= \pi \, x \, 3 \, x \, \frac{50^2 - 35^2}{\ln\left(1500/0.05\right)} = 1165.6 \text{ m}^3/\text{day}$$

When three wells are working in equilateral fashion, the discharge of each well is given by Eqn. (15.47) as:

$$Q_1 = Q_2 = Q_3 = \frac{\pi K\left(h_0^{\,2} - h_w^{\,2}\right)}{\ln\left(\dfrac{r_0^{\,3}}{r_w \, B^2}\right)}$$

$$= \frac{\pi \, x \, 3 \, x \, \left(50^2 - 35^2\right)}{\ln\left(\dfrac{1500^3}{0.05 \, x \, 250^2}\right)}$$

$= 852.8$ m^3/day

Percentage decrease in discharge is given as:

$$= \frac{Q_1 - Q_2}{Q_1} \, x \, 100$$

$$= \frac{1165.6 - 852.8}{1165.6} \, x \, 100 = 26.8 \text{ percent}$$

Question Banks

Q1. What do you mean by hydraulics of wells? State its important in the field of groundwater hydrology?

Q2. What is hydraulic gradient? How is it determined?

Q3. Define the following terms:
(i) Specific yield (ii) Specific retention (iii) Coefficient of permeability (iv) Storage coefficient (v) Intrinsic permeability (vi) Transmissivity (vii) Specific capacity (viii) Well loss (ix) Well efficiency

Q4. Though clay has more porosity, yet it is a bad aquifer. Justify why?

Q5. If a given soil sample has porosity 40 percent and specific yield is 10 percent, then what will be its specific retention?

Q6. Write down the units of coefficient of permeability and intrinsic permeability.

Q7. What is the range of the values of storage coefficient and specific yield?

Q8. What is Darcy's law? Derive the law from fundamental principle of flow through porous media.

Q9. What do you mean by depression head? How does it vary with distance from a well?

Q10. Define radius of influence of a well. What are the factors on which it depends?

Q11. What is the shape of the drawdown curve caused by pumping a well in an aquifer? Explain the nature of the curve.

Q12. Derive groundwater flow equations for both steady state and transient saturated state of a confined aquifer.

Q13. Differentiate between the followings.
(i) Isotropic and anisotropic aquifer (ii) Homogeneous and non-homogeneous aquifer
(iii) Confined and unconfined aquifer (iv) Stream line and equipotential line

Q14. Derive Laplace equation for steady flow in incomprehensible, horizontal and isotropic confined aquifer.

Q15. Enumerate the procedure of development of flow net and describe its significance.

Q16. A dam has saturated thickness 30 m. Coefficient of permeability of the aquifer is 10^{-5} m/sec. The flow net under the dam reveals that there are 4 numbers of flow channels and 8 numbers of squares over the direction of flow. The total head causing the flow is 8 m. Compute the total seepage flow underneath the dam.

Q17. What is an equipotential line? How is it prepared?

Q18. How is the direction of groundwater flow decided?

Q19. Enumerate the different assumptions proposed by Dupuit.

Q20. How is Dupuit's parabola prepared? What does it signify?

Q21. Derive the flow to a fully penetrating well in isotropic, horizontal, homogeneous, confined aquifer. Write down the various assumptions also.

Q22. Derive the flow to a fully penetrating well in isotropic, horizontal, homogeneous, unconfined aquifer. Write down the various assumptions also.

Q23. What is the basic difference on which the discharge equations are used by Dupuit's and Thiem's formula.

Q24. Derive the equations for discharge of a well with uniform recharge in an unconfined aquifer.

Q25. What do you mean by interference of tubewells? How is it determined? How does it affect the yield of well?

Q26. State how the total drawdown in a well field having n number of wells pumping at a steady state with equal discharges is computed.

Q27. Write the formulae to find out the discharge of fully penetrating wells in the following cases:
(i)For two artesian identical wells at a distance B apart in confined aquifer, (ii) For three artesian identical wells at a distance B apart in a pattern of equilateral triangle in confined aquifer (iii) For three artesian identical wells at a distance B apart in a line in confined aquifer (iv) For two identical gravity wells at a distance B apart in unconfined aquifer (v) For three identical gravity wells at a distance B apart in a pattern of equilateral triangle in unconfined aquifer

Q28. Draw a diagram for three identical fully penetrating wells in confined aquifer spaced equal distances apart in a line and show the composite drawdown curve when all the tree wells are discharging at a time.

Q29. With a neat sketch, explain how the aquifer head losses and well losses are determined. Also describe how the total drawdown is computed.

Q30. How the efficiency of a well is is determined? Narrate how the efficiency of the well can be enhanced.

Q31. In a pumping test in medium sand, impermeable bed was found to lie at a distance of 16 m below the the normal groundwater level. Two observation wells were located at a distance of 7 and 3 m, respectively from the main pumping well where the drawdown were measured to be 0.25 and 1.0 m, respectively. If the discharge of the main pumping well was 6 lit/sec, calculate the value of coefficient of permeability of the soil.

Q32. A 20 cm diameter well penetrates 19 m below the ground surface. Static water level in the aquifer is 1 m below the ground surface. After 24 hours of pumping at a discharge of 6000 lit/min, the water level in the test well at 90 m is lowered by 0.50 m and in a second well at distance of 30 m away, water level is lowered by 1 m. Estimate the drawdown in the main pumping well and transmissivity of the aquifer.

Q33. A well of radius 20 cm is pumped at a rate of 80 m^3/hr. Two observation wells were dug at a distance of 7 and 14 m, respectively where the drawdown were measured to be 5 and 1.3 m, respectively. The bottom of the impermeable bed up to which the well is being pumped is 90 m below the ground surface. Piezometric surface lies on the ground surface too. (a) Compute the coefficient of permeability. (b) Find out the specific capacity of the well. (d) Find out also the maximum rate at which the well can be pumped?.

Q34. Two tubewells penetrating fully a 12 m thick confined aquifer are located 250 m apart. Tubewell has diameter equals to 20 cm, depression head is 6 m and radius of influence is 400 m. Coefficient of permeability is 0.15 cm/sec (a) Compute the discharge of the tubewell when one well is pumping. (b) Also find out the percentage decrease in discharge of the well if both the wells are working under a depression head of 5 m. (c) If a third well is also located 200 m away from both the wells so that these three wells form a pattern of equilateral triangle, what will happen to the discharge of each well.

References

Campbell, M.D. and Lehr, J.H. 1973. Water Well Technology, Mc.-Graw-Hill Book Company, New York, pp. 681.

Das Gupta, A. 1987. Handouts on Ground Water Hydrology, Asian Institute of Technology, Bangkok, pp. 282.

Garg, S.K. 2012. Irrigation Engineering and Hydraulic Structures. Khanna Publishers, New Delhi.

Hansen, V.E., Israelsen, O.W. and Stringham, G.E. 1979. Irrigation Principles and Practices. John Wiley and Sons Inc., New York.

Michael, A.M. 1978. Irrigation Theory and Practices. Vikash Publishing House Pvt. Ltd., New Delhi.

Murty, V.V.N. 2004. Land and Water Management Engineering. Kalyani Publishers, Cuttack.

Panigrahi, B. 2011. Irrigation Systems Enginering. New India Publishing Agency, New Delhi.

Sharma, R.K. 1987. A Text Book of Hydrology and Water Resources. Dhanpat Rai and Sons, Delhi, pp. 818.

Sharma, R.K. and Sharma, T.K. 2002. Irrigation Engineering including Hydrology, S. Chand and Company Ltd. New Delhi.

16

Water Productivity

16.1 Introduction

Producing more crops, livestock, fish and forest products per unit of water use is an essence of water productivity and is highly essential for both food and environmental security. A variety of options exist for improving the productivity of water in agriculture through breeding, better management practices and supporting policies and institutions. It is imperative to identify a range of means to enhance water productivity in agriculture, assess their potential and the consequences of their adoption. The production systems to be considered can vary from purely rainfed to irrigated, and from dry zones to semi-humid monsoonal areas.

The productivity of water used in agriculture has almost doubled during 1961 to 2001, It is mainly due to increases in crop yields. Irrigated rice yields doubled and rain-fed wheat yields rose by 160% in that period, with little variation in water consumption per kilo of output. Globally, FAO estimates that water needs for food per capita halved from about 6 m^3/day to less than 3 m^3/day during 1961 to 2001 which is a significant saving and an equally significant gain for other water users.

According to an estimate, a 1% increase in water productivity in food production enables to save about 24 lit/day of water per head of population, while a 10% increase would equal current domestic water consumption. Importance of saving water in agriculture can be realised from the above observation.

16.2 What is Water Productivity?

Productivity is a ratio between a unit of output and a unit of input. Here, the term water productivity is used exclusively to denote the amount or value of product over volume or value of water depleted or diverted. The value of the product might be expressed in different terms like biomass, grain, money etc. For example, the so-called 'crop per drop' approach focuses on the amount of product per unit of water. Another approach considers differences in the nutritional values of different crops, or that the same quantity of one crop feeds more people than the same quantity of another crop. When speaking of food security, it is important to account for such criteria. There is no unique definition of productivity and the value considered for the numerator might depend on the focus as well as the availability of data. However, water productivity defined as kilogram per drop is a useful concept when comparing the productivity of water in different parts of the same system or river basin and also when comparing the productivity of water in agriculture with other possible uses of water.

At field and system level, crop water productivity is measured in terms of evapotranspiration. In irrigated agriculture in saline areas, the leaching requirement, i.e. the amount of water that needs to percolate to maintain rootzone salinity at a satisfactory level, should also be included together with evapotranspiration in the amount of water that is necessarily depleted during plant growth.

16.3 Improving Water Productivity

Improving water productivity - whether under rain-fed or irrigated conditions - requires, first, an increase in crop yields or values, i.e. the marketable yield of the crop for each unit of water transpired. It is also necessary that there should have some reduction of all outflows or losses like drainage, seepage and percolation and more effective use of rainfall, stored water, and water of marginal quality. At the plant level, higher productivity will depend mainly on germplasm improvements, such as stronger seedling, increased rooting depth and enhanced photosynthetic efficiency. Modern rice varieties are three times more productive, in terms of water use, than traditional varieties. Traditional breeding has already made progress in extending these achievements to other crops, and genetic engineering is expected to overcome long-standing obstacles to development of higher-yielding, drought-tolerant crop varieties.

In farmers' fields, higher water productivity requires changes in crop, soil and water management. Strategies include selection of appropriate crops and cultivars, use of improved planting methods (e.g. on raised beds), minimum tillage, synchronization of water applications with the most sensitive growing periods, and improved drainage for water table control. All cultural and agronomic practices that reduce water evaporation - such as variable row spacing and the application of mulches - improve water productivity. Improved nutrient management raises the yield proportionally more than it increases evapotranspiration.

It is essential to consider the various uses of water in agriculture before measures are introduced that would increase water productivity at the expense of other benefits from the same source of water, especially those benefits that accrue to the local poor and landless people.

16.3.1 Irrigation Management Practice under Deficit Supply

Deficit irrigation, in which less water is applied than that required to meet the full crop water demand, results in a small yield reduction that is less than the concomitant reduction in transpiration. In rainfed areas, availability of irrigation water at right time with required amount to meet the crop water demand is not assured. In this context, one has to judiciously use the limited supply to the crop at the correct time by appropriate method of supply of irrigation so that maximum possible yield and water productivity are obtained. Values of water productivity under field conditions increase, especially in grain crops, with the decrease in soil water supply until it is above a minimum critical level because plants actively try to economies water loss in the range from the minimum critical to the optimum moisture level (Hukkeri and Pandey, 1977). However, total production per unit area decreases as the available soil water comes down below optimum level.

Shmuely (1973) has also suggested the application of less than the desired amount of water to growing crops to increase water productivity. In areas where water supply is inadequate, limited supplemental irrigation is often practiced with a view to obtain less than maximum yield per unit of available water. The major advantage of the limited supplemental irrigation is the increase of total farm yield and water productivity by reducing the area allocated for complete rainfed farming (Stegmen *et. al.,* 1980).

Correct timing for supplemental irrigation requires knowledge of crop response to applied water at different growth stages. During some particular stages, the plants are more sensitive to moisture stress than at others and this moisture – sensitive (critical) periods differ with crops and varieties. In some cases these critical periods may also differ under different climatic and soil conditions for the same crops. These critical moisture-sensitive periods, sometimes called as critical growth stage, for various crops has been described in detail by Hukkeri and Pandey (1977) and by Hiler and Howell (1983). Application of water as supplemental irrigation at these critical growth stages is more productive than when applied at other stages because moisture stress during these periods will considerably reduce crop yield which can not be recovered by subsequent application of water of later stages.

In rain-fed agriculture, bridging crop water deficits during dry spells through supplementary irrigation stabilizes production and increases both production and water productivity dramatically. While investments in water harvesting -. e.g. construction of ditches that take the runoff to a storage reservoir - are relatively small, its effectiveness depends on many factors, including topography, soil characteristics and fertilizer availability, and-not least-the involvement of the beneficiaries in design and operation

16.3.2 Supplemental Irrigation on Increasing Water Productivity

Numerous experimental data are available on application of supplemental irrigation to different crops at different critical growth stages and their consequential affect on increase in yield, water productivity and economic returns. The results of multi-location trials carried out at different locations of India on one supplemental irrigation to rainfed crops are presented in Table 16.1 (Singh, 1983).

It is seen from Table 16.1 that in the northen part of India, wheat has maximum response to supplemental irrigation which varies from place to place.

Table 16.1: Effect of one supplemental irrigation (5 cm) on yield of rainfed cops.

Crop	Station	No of trials	Yield, q/ha		*WUE*, kg/ha/cm
			Without irrigation	With irrigation	
Wheat	Dehradun	4	21.4	35.5	282
Wheat	Jodhpur	4	19.	36.0	324
Wheat	Ludhina	4	19.2	41.1	431
Wheat	Agra	2	21.9	27.4	110
Wheat	Rawa	4	5.7	18.8	262
Barley	Varanasi	2	26.0	33.6	148
Barley	Ranchi	-	5.5	11.7	124
Sorghum	Bijapur	5	16.5	23.6	148
Sorghum	Bellary	4	4.3	13.7	182
Sorghum	Solapur	5	9.8	18.2	168
Sorghum	Hyderabad	2	13.9	19.2	106
Sorghum	Kovilpatti	3	27.8	28.9	22
Upland rice	Rawa	4	16.2	27.8	232
Bidi tobacco	Anand	4	12.1	18.1	120
Finger millet	Bangalore	-	16.2	23.2	140
Pearlmillet	Jodhpur	4	16.6	19.0	48
Pigeon pes	Jhansi	-	0.5	3.3	56
Sunflower	Bellary	-	1.3	2.9	32
Rapseed	Randhi	-	2.5	5.4	58
Chilles	Bangalore	-	2.1	4.5	48
Bidi tobacco	Vasad	2	14.6	17.5	58
Safflower	Vasad	1	15.6	20.2	92
Mstard	Vasad	1	16.5	20.4	74
Gram	Vasad	1	10.	16.3	110
Cowpea	Vasad	1	20.4	20.8	7.6

Intercropping and sequential cropping has positive affect on enhancing water productivity. Wani *et al.* (2003) conducted field experiments on grain-yield response of intercropping and sequential cropping systems to varying doses of supplemental irrigation at ICRISAT, Hyderabad. Good yield response to supplemental irrigation was obtained on alfisols in both rainy and post rainy seasons. The average irrigation water productivity which is also called as water-use efficiency or water application efficiency (*WAE* is the ratio of increase in yield due to the supplemental irrigation to the depth of water applied as supplemental irrigation) was observed to be maximum for sorghum (14.9 kg/ha/mm) and for pearl milet, it ranged from 8.8 to 10.2 kg/ha/mm. Tomato gave highest water application efficiency (*WAE*) with a value of 186.3 kg/ha/mm (Table 16.2). In sorghum + pigeon pea, intercropping system, two supplemental irrigations (*SI*), each of 40 mm gave additional gross returns of Rs. 3950/ha. The largest additional gross returns from *SI* was obtained from tomato with Rs. 13,870/ha.

Table 16.2: Grain-yield response (t/ha) of intercropping and sequential cropping systems to supplemental irrigation *(SI)* at ICRISAT.

Due to one SI of 40 mm			Due to two *SI* of each 40 mm			
Grain-yield	Increased yield due to *SI*	*WAE** (kg/ha/mm)	Grain-yield	Increased yield due to *SI*	*WAE* (kg/ha/mm)	Combined *WAE* (kg/ha/mm)
Inter cropping						
	Pearl milet			Pigeon pea		
2.353	0.403	10.0	1.197	0.423	5.3	6.8
	Sorghum			Pigeon pea		
3.155	0.595	14.9	1.22	0.535	6.7	9.4
Sequential cropping						
	Pearl milet			Cow pea		
2.577	0.407	10.2	0.735	0.425	5.3	6.9
	Pearl milet			Tomato		
2.215	0.35	8.8	26.25	14.9	186.3	127.1

Under inadequate irrigation supply condition, it is crucial to recognize the most responsive stage to supplemental irrigation that will increase the water productivity. A number of experiments have been done to assess the affect of supplemental irrigation on water use efficiency of different crops applied at critical growth stages.Venkateswaralu and Singh (1982) reported that a 5 cm supplemental irrigation given at critical stages of wheat, barley, sorghum and rice at 10 different stations of the country increased yield of these crops from 15.3 to 23.8 q/ha with efficiency of water use 20.9 kg/ha/mm.

The results of an experiment at ICRISAT, Hyderabad show that 5 cm supplemental irrigation in alfisol applied to sorghum (at grain-filling stage) and maize (at pre-tasseling stage) increased yields of both the crops by more than 100 percent. However, pear millet showed an increase of 17-59 percent and sunflower 27-32 percent (Anonymous, 1976).

Verma *et. al.* (1984) have reviewed the available literatures on supplemental irrigation to wheat and maize. Majority of research workers has suggested 5 to 6 cm supplemental irrigation to wheat at early stress especially at pre sowing stage (in case of deficient seeding zone moisture) or at *CSRI* stage of wheat. In case of maize, tesseling – silking stage has been found most critical for a supplemental irrigation. In northern India an irrigation of 5 cm at pre-sowing or at CRI stage increased rainfed wheat grain yield by 7.7 g/ha (Verma, 1981).

According to Gajri *et. al.* (1983) a supplemental irrigation to rainfed wheat at pre sowing stage increased grain yield by 30 to 81 percent. Time of application, quantity of water for irrigation, and selection of crop types are extensively important factors which need to be considered to derive the maximum benefits. Life saving irrigation at the most crop responsive stage i.e. critical growth stage shall not only protect the crops but help to stabilize the production. The results of supplemental irrigation in medium deep black soils at Bijapur over a period of 5 years indicates that crop yields with one life saving irrigation of 5 cm can enhance the yield by 40-90 percent (Table 16.3, Belgaumi, *et. al.*, 1997).

Table 16.3: Crop response to life saving irrigation in medium deep black soil at Bijapur

Crop	Yield, kg/ha		
	Control	Irrigated	Yield increase (% over control)
Sorghum (Rabi)	1845	2574	39.5
Safflower	1196	1702	42.4
Bengalgram	669	887	32.5
Sunflower	873	1429	63.7
Redgram	198	381	92.4
Hybrid cotton	1094	1524	39.3

16.3.3 Farming System Approach to Increase Water Productivity

In several regions, pond irrigation system, especially in rainfed ecosystem, plays crucial role in increasing the productivity and income generation capacity of farmers by adopting farming system approach. This system gave higher returns as compared to sole crop of paddy served by pond irrigation. The data obtained at Bijapur on different farming systems revealed that horticultural crops with ber plantations (cv. Umran and Kadaka) recorded gross income ranging between Rs. 27,962 to Rs. 37,260 per hectare with two to three supplemental irrigations as against Rs. 23,657 to Rs. 29,505 per hectare in control from the ber plantation. The next highest income was obtained from the cultivation of curry leaf (Rs. 26, 664 per hectare) with two supplemental irrigations followed by hybrid cotton (Rs. 18,840 per hectare). Intercropping of bengalgram + safflower in row proportion of 4:2 provided only Rs. 4937 per hectare with two supplemental irrigations where as the income from the arable cropping system was Rs. 3488 under control. Cultivation of mulberry, pastures (grass + legumes) under protective irrigations accounted Rs. 16,870 and Rs. 5107 per hectare, respectively against corresponding income of Rs. 14,385 and Rs. 4121 per hectare under control (Table 16.4, Anonymous, 1991).

Table 16.4: Yield and gross income of different farming systems as influenced by pond irrigation

Farming system		Average yield (kg/ha)	Gross income (Rs./ha)
Ber (Umran)	Irrigated	7452	37260
	Control	5901	29505
Ber (Kadaka)	Irrigated	5592	27962
	Control	4731	23657
Curry leaf	Irrigated	8888	26664
Mulberry	Irrigated	7216	16870
	Control	6245	14385
Pastures	Irrigated	29915	5107
Hybrid cotton	Irrigated	1884	18840
	Control	1501	15010
Bengalgram+Safflower (4:2) intercropping	Irrigated	17260	4937

16.3.4 Methods of Irrigation to Increase Water Productivity

Appropriate method of irrigation is a crucial issue especially in water scarce areas that can save a lot of precious irrigation water and enhance the water productivity. It is advisable to use micro-irrigation techniques instead of flooding irrigation as used in rice crop for irrigating the dry land crops. The micro-irrigation like drip and sprinkler has very high application efficiency to the tune of 85-90%. At Bijapur, it has been observed that drip irrigation could save 50% of water compared to the surface method of irrigation (Belgaumi, *et. al.*, 1997). The following table (Table 16.5) gives the response of a given amount of supplemental irrigation in different fruit crops by different methods of irrigation (Belgaumi, *et. al.*, 1997). From this table it is evident that the drip method of irrigation gives highest fruit yield than other methods.

Table 16.5: Response of methods of irrigation on fruit yield of different plants for a irrigation supply of 200 lit/plant

Methods of irrigation	Fruit yield, kg/plant			Percent increase over control		
	Ber	Fig	Pomegranate	Ber	Fig	Pomegranate
Drip	6.17	2.73	3.36	252	228	133
Jet	5.07	1.90	2.20	189	128	52
Surface	4.37	1.51	2.31	149	82	60
Control	1.75	0.83	1.44	-	-	-

16.3.5 Farm Pond to Enhance Water Productivity

In rainfed areas, small water harvesting structures like farm pond, jhullas, ditches etc. play a dominant role in enhancing irrigation potential and water productivity. In high rainfall areas, they harvest the adequate rainfall and runoff from the catchment and help in recycling the stored water as supplemental irrigation. However, proper design of these structures need to be undertaken so that they will harvest maximum water and reuse them for irrigation to cultivate at least two crops in a year.

The selection of catchment and farm pond and the location of the pond in the catchment is very important to maximize the returns from the harvested water. Pond is better to be selected such that there is some catchment area in the upper end of the pond and there is some service/ command area below the pond (downstream end of the pond). The crops in the catchment areas may be taken as non paddy crop in both *kharif and rabi* and the crops in the downstream end f the pond may be paddy in *kharif* and a non paddy crop in *rabi.* Selection of the crops and the cropping systems must be location specific. The surface runoff from the upper end of the catchment will be easily harvested in the pond and may not require any supplemental irrigation during kharif and one or two irrigation in rabi. However, the harvested water in the pond can be used as supplemental irrigation to crops grown in the areas at the downstream end of the pond. The irrigation frequency and amount to the crops then will depend on the then available supply and the appropriate water management practice followed for the specific crop. However, proper simulation study need to be undertaken foe selecting the catchment-command area ratio and for deciding the size of the pond. Fig 16.1 gives a representative view of the farm pond constructed in the farm having some catchment area grown with maize and command area grown with rice and other crops in the experimental farm of department of Agricultural and Food Engineering at Indian Institute of Technology, Kharagpur.

To maximize the economic returns from the harvested water of the pond, integrated farming system is advisable. The integrated farming system will help to the increase the incomes from different sources like increased yield of crops from the farms, additional income in rearing fish in the pond, returns from yield of banana, coconut, papaya, arhar etc. grown in the embankment of the ponds. A photo of a pond with cultivation of banana, papaya and arhar in the embankment of a newly constructed pond is presented in Fig. 16.2.

Fig.16.3 shows the photo of another farm pond constructed in chronically waterlogged ecosystem in the farm of regional research and technology transfer station of Chiplima, Orissa where fish are reared in the pond and banana and guava crops are grown in the embankment of the pond. In addition, the stored water of the pond is used to irrigate the rice crops. Thus, the pond acts as a viable source for integrated farming system and helps the farmers to raise their income from multi-source.

16.3.6 Land Drainage and Water Productivity

Land drainage, either surface or sub-surface drainage, help in controlling water logging and salinity problems in irrigated commands. Benefits of drainage systems have been discussed earlier. One of the significant contributions of land drainage is to lower down the groundwater table and makes the land fit for crop diversification. It helps the farmers to choose the most remunerative crop to grow in the drainage treated area and thereby enhancing water productivity. The drained water, otherwise, can also be treated and if found suitable for irrigation, can be reused as a viable source for supplemental irrigation.

16.3.7 On Farm Development to Increase Water Productivity

Achieving high values of water productivity at the farm level depends on the attention given to on-farm development and water management. When the large irrigation systems were planned and constructed, it was thought that once the water is delivered to the farmers, the objectives will be achieved. But experiences show that the farmers need considerable assistance in bringing his fields to suit irrigated agriculture (Murty, 2004). In this context, on-farm development with the following steps needs to be undertaken to enhance water productivity at field level.

- Volumetric measurement of irrigation water
- Pricing irrigation water based on volume basis
- Irrigation to crops through field channels
- Adoption of rotational method of irrigation
- Adopting conservative irrigation techniques like minimum and zero tillage and mulching
- Adopting proper irrigation scheduling
- Proper water management
- Lining of field water courses along with ancillary structures
- Adequate training and exposure to use water economically in the fields and

Fig. 16.1: Farm pond having maize in catchment area and rice in command area

Fig. 16.2: Banana + arhar + papaya grown in embankment of a farm pond

Fig. 16.3: A farm pond for multiuse at Chiplima, Orissa

In addition to the above discussed points, focus should be given to inter-basin linking system to divert water from areas with adequate availability to the areas with scarcity availability. Irrigation system management, watershed management and formation of water users association will facilitate in expanding water productivity. Adopting low cost water harvesting technology to harness the abundant rainfall resource and reuse them for irrigation in rainfed ecosystem should also be due importance.

16.4 Application Efficiency and Irrigation Uniformity

All irrigation systems have some non-uniformity. This non-uniformity influences yields and irrigation efficiency especially application efficiency. If the net amount of water applied during irrigation is the same as that required to fill the soil water deficit, then because the distribution of water is never perfect, half of the field will receive too much water while the other half will receive too less water. In such a context, a farmer will try to apply more irrigation water to his field so that a larger fraction of the field has an adequate amount. This will tend to have higher percolation and runoff and thereby declining the application efficiency. An alternative to adding extra water is to improve the uniformity of water application. A major advantage when the uniformity is improved is that the amount of deficit in the area under irrigation is reduced and the potential for water logging and salinization is also reduced.

16.4.1 Land Levelling Operation for Efficient Use of Water

Land leveling and grading help in easy, quick and uniform distribution of irrigation water in the entire field. Before irrigation is practiced in the field, land grading and leveling are very much required. Land grading literally means reshaping the field surface to a planned grade. It is necessary in making suitable field surface to con-trol the flow of water, to check soil erosion and provide for surface drainage. Land levelling operation may be grouped into three phases: (i) rough grading (ii) land levelling and (iii) land grading.

16.4.1.1 Rough Grading

It constitutes the removal of abrupt irregularities such as mounds, dunes, and ridges and filling of pits, depressions and gullies.

16.4.1.2 Land Levelling

It is well known that, the agricultural lands have non-uniform slopes and many undulations, which make the condition of soil surface difficult to supply the irrigation water uniformly and efficiently for irrigating the crops. In this situation, the necessity of land leveling or grading becomes very essential. In land grading/leveling operations, the land surfaces are prepared or modified to a plane grade, to provide a suitable top soil surface for efficient application of irrigation water and also to facilitate quick disposal of excess water from the area. The proper application of irrigation water refers to repellents the moisture content up to the root zone depth. The water applied more than the required, is not only a loss but also results the development of water logging condition, which damages the crop, and virtually reduces the crop yield. In this way, none of the irrigation method except micro-irrigation is able to supply the irrigation water uniformly all over the field. The land leveling/grading or land shaping operations remove this kind of irregularities to a large extent.

16.4.1.3 Land Grading

The operation in which land is converted to a desired grade in longitudinal as well as across-slope directions for efficient water application is called land grading.

Objectives of Land Grading

The main objectives of land grading are as under (Panigrahi, 2011):

1. To apply irrigation water very uniformly throughout the field.
2. To economize the utilization of water by minimizing the losses.
3. To augment the surface drainage.
4. To apply the fertilizers, uniformly throughout the field.
5. To increase the crop yield by uniform application of water.
6. To have uniform plant population
7. To save the labour and water

Factors Affecting Land Grading

The land grading operations are performed in the field on segment basis. Therefore, before starting the grading work, it is very essential to prepare the farm layout in respect of location of field channels, drains farm roads etc. The main factor affecting land grading is soil profile limitation. The nature of top soil, depth of sand, gravel or hard pan layer, water table depth etc. are taken into consideration under soil profile limitations, for planning of land grading work. The depth of gravel or hardpan affects the depth of cut,. However, the depth of cut should be such that it can not expose the inert materials which are not suitable for good crop. For good cultivation, it is essential to have at least 1.0 m soil depth. Hardpan or sandy layer should not be exposed. In large areas, the land may be divided into several strips across the main slope and then individual strips should be handled for grading work. The objective is achieved by removing the topsoil and placing at some other places and when land grading work has been finished, than collected topsoil should be uniformity spread throughout the area. The soil depth should always be in such extent in which roots of the crops could be easily accommodated. If the sub soil or sub-stratum is coarse textured, which have high infiltration rate and hydraulic conductivity, a higher grade is recommended so that water could run rapidly to lower reaches of field.

The land slope and soil texture are other soil profile factors that affect the land grading. The steep slope with undulating terrain and shallow soil depth make difficult to conduct the lelvelling work. A soil composed of coarser particles has greater infiltration rate resulting into rapid absorption of water. In this condition, a greater land slope is recommended. To the contrary, fine textured soils have lesser infiltration rate; as a result smaller grade is provided in such soils.

The land leveling and grading are the expensive operations due to involvement of heavy earthworks. However, the cost could be minimized by reducing the volume of earthwork and haulage distance by proper planning of land leveling/grading operations technique. The very purpose of land leveling, grading and smoothing is to help the irrigation water to be uniformly distributed over the whole area with less application losses. Increase of application efficiency depends on management factors including provision of proper irrigation and cross slope, deciding optimum plot size, soil and crop management factors and adopting suitable irrigation scheduling. In land grading work, two types of slopes are prepared, in which one is irrigation slope and other is cross slope.

16.4.1.4 Irrigation Slope

It is the slope given to the field for running the irrigation water in forward direction. Sometimes, this slope is also referred as longitudinal slope. Its range depends on the soil type, stream flow rate etc. In coarse textured soil, greater irrigation slope is recommended than the fine textured soils because greater slope minimizes the time of absorbing of water by the soil,. In addition, the irrigation slopes should also be in such a limit that the running water in the border or furrow could not cause soil erosion problem. However, based on soil types, recommended limits of longitudinal slope are given in Table 16.6.

Table 16.6: Recommended longitudinal slopes based on soil type

Sl. No	Soil type	Longitudinal slope, %
1.	Heavy soil (clay)	0.05-0.25
2.	Medium soil (Loam)	0.20-0.40
3.	Light soils(Sandy)	0.25-0.65

16.4.1.5 Cross Slope

Cross slope is perpendicular to the longitudinal slope. The limit of permissible cross slope depends on the method of irrigation used. For example, the borders and checks do not allow having the cross slope within the boundary while furrows and corrugations require a certain degree of cross slope. Generally, the cross-slope is being equal to or little less than the longitudinal slope. The maximum slope should be up to 0.5% or less in alluvial soils. Generally for eliminating the cross-slope an excessive soil cut is required, which could be reduced by dividing the entire field into different compartments and each compartment is leveled at different elevations, separated by low ridges. The volume of earthwork evolved in land grading work depends on the magnitude of diagonal slope at right angles to the direction of irrigation.

In land grading works for surface drainage, the longitudinal slope can be varied, as per requirement to control the soil erosion, reduce yardage or to avoid the exposing of extensive area of the suib soil. Planned grading should be such that each row would carry its share of surface run-off from upstream to down stream part without overtopping into the next row. The graded surface should be sloped to a specified design grade. The grade in the direction of drainage may be uniform or variable with a reverse grade. A minimum grade of 0.1% should preferably be used, if possible. However, the limit of maximum grade is decided on the basis of erosion hazard.

The magnitude of irrigation and cross slopes is decided on the basis of method of irrigation to be used. The characteristic features for the basin irrigation are:

- Soil: Medium fine textured and moderately coarse textured soil.
- Crops: Vegetable crops and paddy.
- Irrigation slope: 0.6%
- Cross slope: 0.2%

The characteristic features for the boarder irrigation are:

- Soil: Medium to fine textured soils
- Crops: Cereal and row crops.
- Irrigation slope: 0.6%
- Cross slope: 0.2%

The range of land slopes, boarder length and stream flow rate vary from soil to soil depending on the texture of soil (Table 16.7).

Table 16.7: Recommended border length, stream flow rate and slopes for different soil

Soil Texture	Border length, m	Stream flow rate, m^3/h	Slope, %
Coarse texture	60-90	20-25	0.40-0.6
Moderately coarse texture	76-150	15-20	0.4-0.6
Medium texture	90-180	13-18	0.4-0.6
Moderately fine texture	90-180	10-14	0.4-0.6
Fine texture	100-200	6-8	0.2-0.4

16.5 Water Productivity in Rice

Rice is a more water consuming crop. Producing more rice with less water is essential for food security and environmental protection in Asia. Scientists in IRRI have explored various field-level water-saving technologies, such as system rice intensification, aerobic rice and ground-cover. Each of these techniques reduces outflows - such as seepage and evaporation - and thus contributes to water productivity. In one of the water saving irrigation techniques, rice is cultivated at intermittent practice of irrigation (alternate wetting and drying) rather than following continuous ponding. They have identified the growth stages of rice in which the soil is not flooded or not even saturated. This practice of irrigation, though, may result in some yield reduction, yet they result in comparable water saving by reducing the unproductive water outflows drastically and thus increase the water productivity. These techniques generally lead to some yield decline in high-yielding lowland rice varieties, although substantial increases for some local varieties - with average yields over 8 tonnes per ha. - have been obtained using a technique called "system rice intensification", from Madagascar, in which the soil is only kept wet during the reproductive and grain producing stages.Recent results from northern China and Philippines indicate that aerobic rice under the intermittent irrigation system yields about 40% lower than those of flooded lowland systems. However, water requirements are about 60% less.

16.6 Performance of Farm Irrigation Systems

The performance of farm irrigation systems is determined by the efficiency with which water is delivered to the field and utilized to the fullest amount possible with minimum losses due to seepage and percolation, runoff, evaporation and other operational losses. The efficiency of the irrigation system will increase if the conveyance losses, application losses and other losses are reduced to a great extent.

16.6.1 Overall Efficiency

The overall efficiency of a farm irrigation system is defined as the percent of water applied to the farm that is used for irrigation on the farm. Irrigation efficiency otherwise called as overall system efficiency is defined (James, 1988) as:

$$E_i = \left(\frac{S - DP - SR - O}{S}\right) \cdot 100 \tag{16.1}$$

where, E_i = irrigation efficiency, percent; S = amount of water supplied to the farm; DP = deep percolation loss; SR = surface runoff of the farm and O = operational losses due to conveyance and storage system.

Sample Calculation 1:

A measuring structure is installed at the inlet of a canal system. The structure measures 1.5 cumecs of water and water is derived to the canal system for 10 hours continuously. If during 10 hours of operation, deep percolation loss = 2000 m^3; surface runoff = 3500 m^3 and operational losses = 1000 m^3 then determine the irrigation efficiency of the system.

Solution:

Amount of water delivered to the farm during 10 hours = 1.5 x 10 x 3600 = 54000 m^3.

Operational losses = 1000 m^3

Surface runoff = 3500 m^3

Deep percolation loss = 2000 m^3

$$\text{Hence, irrigation efficiency} = \left(\frac{54000 - 1000 - 3500 - 2000}{54000}\right) \cdot 100$$

$$= 87.96 \text{ percent.}$$

Thus, 87.96 percent of water applied to the farm is used by the crops and rest is lost due to surface runoff, deep percolation loss and operational losses.

When evaluating the performance of a farm irrigation system, it is useful to examine the efficiency of each system components. The main system efficiency components are conveyance efficiency, storage efficiency and application efficiency. The overall system efficiency is the product of above mentioned three system efficiencies and is given as:

$$E_i = \left(\frac{E_s}{100}\right) \cdot \left(\frac{E_c}{100}\right) \cdot \left(\frac{E_a}{100}\right) \cdot 100 \tag{16.2}$$

where, E_i = irrigation efficiency; E_s = water storage efficiency; E_c = conveyance efficiency and E_a = application efficiency. All the efficiency terms are expressed in percent.

16.6.2 Water Storage Efficiency

The water storage efficiency in a storage system gets reduced due to the losses like evaporation and seepage. The following formula can be used to evaluate the storage efficiency.

$$E_s = \left(1 - \frac{V_s + V_e}{V_i}\right) \cdot 100 \tag{16.3}$$

where, E_s = storage efficiency; V_i = inflow volume entering a storage system V_s = seepage volume from the system and V_e = evaporation volume of the system.

16.6.3 Water Conveyance Efficiency

Total amount of water that is delivered from a storage system never reaches a field since on the way there are losses in the channels. These losses are called as conveyance loss. The conveyance losses occur due to the earthen channels or ill maintained channels. However, by lining the channels, the conveyance losses can be reduced and so the conveyance efficiency can be increased. The conveyance efficiency is expressed as:

$$E_c = \frac{V_{df}}{V_{dp}} \cdot 100 \tag{16.4}$$

where, E_c = conveyance efficiency; V_{df} = volume of water delivered to the field and V_{dp} = volume of water diverted from the storage system.

16.6.4 Water Application Efficiency

Water application efficiency for an irrigated area is defined as the ratio of the volume of water actually used by the plants to the volume of water delivered to the field. It is to be noted that the amount of water supplied to each field is not entirely used by the crops. Some amount of the supplied water is lost due to seepage and percolation; some is lost due to runoff and other unavoidable losses. By economic application of irrigation water and scientific water management, the application efficiency can be increased. Water application efficiency is expressed as:

$$E_a = \frac{V_{ac}}{V_{df}} \cdot 100 \qquad (16.5)$$

where, E_a = water application efficiency, V_{ac} = volume of water actually used by the crops and V_{df} = volume of water delivered to the field.

16.6.5 Water Distribution Efficiency

Some times another efficiency term is used in irrigation system. That is called as water distribution efficiency and is defined as:

$$E_{de} = (1 - \frac{\bar{y}}{\bar{d}}) \cdot 100 \qquad (16.6)$$

where, E_{de} = distribution efficiency; $\bar{d}$ = average depth of water stored along the run during irrigation and $\bar{y}$ = average numerical deviation from $\bar{d}$.

16.6.6 Water Use Efficiency

In addition to the above mentioned efficiencies, often another efficiency called as water use efficiency (sometimes also called as field water use efficiency) is used. Water use efficiency is expressed as:

$$WUE = \frac{Y}{WR} \cdot 100 \qquad (16.7)$$

where, WUE = water use efficiency; Y = yield of the crop and WR = water requirement of the crop. Instead of WR if actual evapotranspiration term (*ETcrop*) is used in the denominator of water use efficiency, then it is designated as crop water use efficiency *(CWUE)*.

Sample Calculation 2:

On a certain day, 1000 m^3 of water is diverted from a farm pond. Out of this, 200 m^3 is lost as seepage loss in the conveyance system. Find out the conveyance efficiency of the system.

Solution:

Given the volume of water diverted from a farm pond = 1000 m^3.

Conveyance loss = 200 m^3

So, water actually reaching the field = 1000 – 200 = 800 m^3

Therefore, conveyance efficiency = 800 / 1000 = 0.8 = 80 percent.

Sample Calculation 3:

A stream of 135 lit/sec was diverted from a canal and 100 lit/sec were delivered to the field. An area of 1.6 ha was irrigated in eight hours. The effective root zone depth of the crop was 1.8 m. Runoff loss of the field was 432 m^3. Depth of water penetrated in the soil linearly varied from 1.8 m at up end to 1.2 m at down end of the field. Available moisture holding capacity of the soil is 20 cm/m depth of the soil. Determine the water conveyance efficiency, water storage efficiency, water application efficiency and water distribution efficiency. .

Solution:

Amount of water diverted from the canal = 135 lit/sec

Amount of water delivered to the field = 100 lit/sec

So, the conveyance efficiency = 100/135 = 0.74 = 74 percent.

Water delivered to the field = (100 x 60 x 60 x 8) /1000 = 2880 m^3

Water lost as runoff = 432 m^3

So, water stored in the root zone of the crop = 2880 – 432 = 2448 m^3

Hence, water application efficiency = 2448/2880 = 0.85 = 85 percent.

Water holding capacity of the root zone = 20 x 1.8 = 36 cm.

Moisture required in the crop root zone = $36 - \left(\frac{36 \times 50}{100}\right) = 18$ cm

Volumetrically the above amount equals = (18 /100) x 1.6 x 10000 = 2880 m^3

Hence, water storage efficiency = 2448/2880 = 0.85 = 85 percent.

Average depth of water stored along the run during irrigation = (1.8 + 1.2)/2 = 1.5 cm.

Numerical deviation from depth of penetration = {(1.8 – 1.5) + (1.5 – 1.2)}/2 = 0.3 cm.

Water distribution efficiency = (1 – 0.3/1.5) x 100 = 80 percent.

Question Banks

Q1. What is water productivity? How is it expressed?

Q2. How is the crop water productivity measured at field level?

Q3. How do the cultural and agronomic practices help in augmenting water productivity?

Q4. Describe how the irrigation is managed under deficit supply?

Q5. Define supplemental irrigation. How does the supplemental irrigation help in increasing water productivity?

Q6. State the role of intercropping and sequential cropping on water productivity.

Q7. Define water application efficiency. State how it can be increased.

Q8. Define critical growth stage of a crop. How can the irrigation be managed at the critical growth stage of crops to improve water productivity?

Q9. Methods of irrigation play a crucial role in enhancing water productivity. Discuss how?

Q10. What do you mean by farming system? How it helps in increasing water productivity.

Q11. Explain the role of farm pond in enhancing water productivity.

Q12. How can the chronically waterlogged areas be reclaimed? Discuss how this reclamation helps in increasing water productivity.

Q13. Describe the role of land drainage in augmenting water productivity.

Q14. Define the followings:

(i) Application efficiency (ii) Irrigation uniformity (iii) Land leveling (iv) Land grading (v) Rough grading.

Q15. What are the roles of land leveling and land grading in irrigation of agricultural lands?

Q16. Write the various objectives of land grading.

Q17. Enumerate the various factors that affect land grading.

Q18. How does the irrigation slope and cross slope help in distribution of irrigation water? How are they decided?

Q19. Briefly describe the various steps of on-farm development used to increase water productivity.

Q20. How can the water productivity in rice field be increased?

Q21. Briefly discuss the various factors used to evaluate the performance of farm irrigation systems.

Q22. Write notes on the followings:

(i) Over irrigation efficiency (ii) Water conveyance efficiency (iii) Water storage efficiency (iv) Water distribution efficiency (v) Water application efficiency and (vi) Water-use-efficiency.

Q23. A stream of 125 lit/sec was diverted from a canal and 100 lit/sec were delivered to the field. An area of 1.5 ha was irrigated in 10 hours. The effective root zone depth of the crop was 1.5 m. Runoff loss of the field was 430 m^3. Depth of water penetrated in the soil linearly varied from 1.6 m at up end to 1.2 m at down end of the field. Available moisture holding capacity of the soil is 20 percent. Determine the water conveyance efficiency, water storage efficiency, water application efficiency and water distribution efficiency.

Q24. A measuring structure is installed at the inlet of a canal system. The structure measures 1.2 cumecs of water and water is derived to the canal system for 10 hours continuously. If during 10 hours of operation, deep percolation loss = 1600 m^3; surface runoff = 3000 m^3 and operational losses = 800 m^3 then determine the irrigation efficiency of the system.

References

Anonymous. 1976. Annual Report of International Crop Research Institute for the Semi-Arid Tropics, Hyderabad, pp. 80.

Anonymous. 1991. Annual Progress Report of All India Co-Ordinated Research Project on Dryland Agriculture, Bijapur.

Belgaumi, M.I., Itnal, C.J. and Radder, C.D. 1997. Farm Ponds. Technical series 4, Publication Centrte, University of Agricultural Sciences, Dharwad, pp. 47.

Gajri, P.R., Verma H.N. and Parihar, S.S. 1983. Rain water harvesting and its recycling for maximization of crop production. Indian J. Soil Conserv., Vol. 9(2&3): 116-123.

Hiler, E.A. and Howell, T.A. 1983. Irrigation options to avoid critical stress. An overview in limitations to efficient water use in crop production. Madison, Wis. Soil Sc. Soc. Am.

Hukkeri, S.B and Pandey, S.L.1977. Water requirement and irrigation of crops. In: Water Requirement and Irrigation Management of Crops in India. Monograph No. 4, New Series Water Technology Centre, IARI, New Delhi: 163-273.

James, L.G. 1988. Principles of Farm Irrigation System Design. John Wiley & Sons, New York, pp. 543.

Murty, V.V.N. 2004. Land and Water Management Engineering. Kalyani Publishers, Cuttack.

Panigrahi, B. 2011. Irrigation Systems Engineering. New India Publishing Agency, New Delhi, pp. 325.

Shmuely, E. 1073. Efficient utilization of water in irrigation in arid zone irrigation. Chapmen and Hall Ltd.: 411-423.

Singh, R.P. 1983. Farm Pond. Project Bulletin No. 6, All India Co-ordinated Research Project on Dryland Agriculture, Hyderabad.

Stegman, E.C., Musick J.T. and Stewart, J.I. 1980. Irrigation Water Management. In: M.E. Jensen (Ed.), Design and Operation of Irrigation System, ASAE Monograph, St. Joseph, Michigan: 763-816.

Venkateswaralu, J and Singh, S.N. 1982. Crop management to mitigate moisture stress. In a decade of dry land agriculture research in India (1971-80). All India Co-Ordinated Research Project on Dryland Agriculture, Hyderabad: 48-50.

Verma, H.N 1981. Water harvesting for life saving irrigation of rainfed crops in the sub-mountain region of Punjab. J. of Agric. Engrg. (Indian Soc. of Agril. Engineers), Vol. 18(3&4): 64-72.

Verma, H.N., Prihar, S.S. and Singh, R. 1984. Feasibility of storage of runoff in dugout ponds and its use for supplemental irrigation in submontane Punjab. Indian J. Soil Conserv., Vol. 12(1): 31-36.

Wani, S.P; Pathak, P; Sreedevi, T.K; Singh, H.P. and Singh, P. 2003. Efficient management of rainwater for increased crop productivity and ground water recharge in Asia. Comprehensive Assessment of Water Management in Agriculture Series, Water Productivity in Agriculture: Limits and Opportunities for Improvement. Book Edited by J.W.Kijne, R. Barker and D. Molden. CABI Publishing, pp. 352.

Colour Plates

Chapter 3: Storage of Irrigation Water

Fig. 3.4: Schematic view of a silted reservoir

Chapter 8: Methods of Irrigation

Fig. 8.14: Rotating head sprinkler in operation

Fig. 8.17: Layout of drip system in farmer's field

Chapter 10: Water Lifting Device

Fig. 10.9: View of a modern wind mill

Chapter 11: Losses of Water in Canal and Control

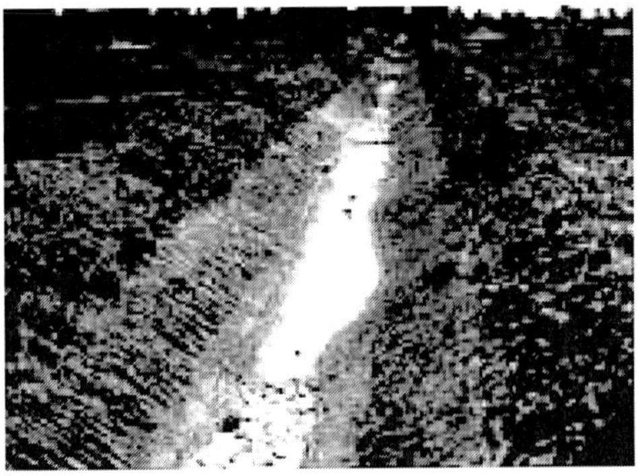

Fig. 11.3: Photographic view of an unlined channel in Hirakud canal command

Fig. 11.4: Photographic view of a concrete channel of Hirakud canal system

Chapter 16: Water Productivity

Fig. 16.1: Farm pond having maize in catchment area and rice in command area

Fig. 16.2: Banana + arhar + papaya grown in embankment of a farm pond

Fig. 16.3: A farm pond for multiuse at Chiplima, Orissa

Index

A

B

C

D

E

F

G

H

I

J

K

L

M

N

O

P

R

S

T

U

V

W

Y

Z